T0180590

Smart Innovation, Systems and Technologies

Volume 31

Series editors

Robert J. Howlett, KES International, Shoreham-by-Sea, UK
e-mail: rjhowlett@kesinternational.org

Lakhmi C. Jain, University of Canberra, Canberra, Australia, and
University of South Australia, Adelaide, Australia
e-mail: Lakhmi.jain@unisa.edu.au

About this Series

The Smart Innovation, Systems and Technologies book series encompasses the topics of knowledge, intelligence, innovation and sustainability. The aim of the series is to make available a platform for the publication of books on all aspects of single and multi-disciplinary research on these themes in order to make the latest results available in a readily-accessible form. Volumes on interdisciplinary research combining two or more of these areas is particularly sought.

The series covers systems and paradigms that employ knowledge and intelligence in a broad sense. Its scope is systems having embedded knowledge and intelligence, which may be applied to the solution of world problems in industry, the environment and the community. It also focusses on the knowledge-transfer methodologies and innovation strategies employed to make this happen effectively. The combination of intelligent systems tools and a broad range of applications introduces a need for a synergy of disciplines from science, technology, business and the humanities. The series will include conference proceedings, edited collections, monographs, handbooks, reference books, and other relevant types of book in areas of science and technology where smart systems and technologies can offer innovative solutions.

High quality content is an essential feature for all book proposals accepted for the series. It is expected that editors of all accepted volumes will ensure that contributions are subjected to an appropriate level of reviewing process and adhere to KES quality principles.

More information about this series at http://www.springer.com/series/8767

Lakhmi C. Jain · Himansu Sekhar Behera
Jyotsna Kumar Mandal
Durga Prasad Mohapatra
Editors

Computational Intelligence in Data Mining - Volume 1

Proceedings of the International Conference on CIDM, 20-21 December 2014

 Springer

Editors
Lakhmi C. Jain
University of Canberra
Canberra
Australia

and

University of South Australia
Adelaide, SA
Australia

Himansu Sekhar Behera
Department of Computer Science
 and Engineering
Veer Surendra Sai University
 of Technology
Sambalpur, Odisha
India

Jyotsna Kumar Mandal
Department of Computer Science
 and Engineering
Kalyani University
Nadia, West Bengal
India

Durga Prasad Mohapatra
Department of Computer Science
 and Engineering
National Institute of Technology Rourkela
Rourkela
India

ISSN 2190-3018   ISSN 2190-3026 (electronic)
Smart Innovation, Systems and Technologies
ISBN 978-81-322-2989-6   ISBN 978-81-322-2205-7 (eBook)
DOI 10.1007/978-81-322-2205-7

Springer New Delhi Heidelberg New York Dordrecht London
© Springer India 2015
Softcover reprint of the hardcover 1st edition 2015

Printed on acid-free paper

Springer (India) Pvt. Ltd. is part of Springer Science+Business Media (www.springer.com)

Preface

The First International Conference on "Computational Intelligence in Data Mining (ICCIDM-2014)" was hosted and organized jointly by the Department of Computer Science and Engineering, Information Technology and MCA, Veer Surendra Sai University of Technology, Burla, Sambalpur, Odisha, India between 20 and 21 December 2014. ICCIDM is an international interdisciplinary conference covering research and developments in the fields of Data Mining, Computational Intelligence, Soft Computing, Machine Learning, Fuzzy Logic, and a lot more. More than 550 prospective authors had submitted their research papers to the conference. ICCIDM selected 192 papers after a double blind peer review process by experienced subject expertise reviewers chosen from the country and abroad. The proceedings of ICCIDM is a nice collection of interdisciplinary papers concerned in various prolific research areas of Data Mining and Computational Intelligence. It has been an honor for us to have the chance to edit the proceedings. We have enjoyed considerably working in cooperation with the International Advisory, Program, and Technical Committees to call for papers, review papers, and finalize papers to be included in the proceedings.

This International Conference ICCIDM aims at encompassing a new breed of engineers, technologists making it a crest of global success. It will also educate the youth to move ahead for inventing something that will lead to great success. This year's program includes an exciting collection of contributions resulting from a successful call for papers. The selected papers have been divided into thematic areas including both review and research papers which highlight the current focus of Computational Intelligence Techniques in Data Mining. The conference aims at creating a forum for further discussion for an integrated information field incorporating a series of technical issues in the frontier analysis and design aspects of different alliances in the related field of Intelligent computing and others. Therefore the call for paper was on three major themes like Methods, Algorithms, and Models in Data mining and Machine learning, Advance Computing and Applications. Further, papers discussing the issues and applications related to the theme of the conference were also welcomed at ICCIDM.

The proceedings of ICCIDM have been released to mark this great day in ICCIDM which is a collection of ideas and perspectives on different issues and some new thoughts on various fields of Intelligent Computing. We hope the author's own research and opinions add value to it. First and foremost are the authors of papers, columns, and editorials whose works have made the conference a great success. We had a great time putting together this proceedings. The ICCIDM conference and proceedings are a credit to a large group of people and everyone should be there for the outcome. We extend our deep sense of gratitude to all for their warm encouragement, inspiration, and continuous support for making it possible.

Hope all of us will appreciate the good contributions made and justify our efforts.

Acknowledgments

The theme and relevance of ICCIDM has attracted more than 550 researchers/ academicians around the globe, which enabled us to select good quality papers and serve to demonstrate the popularity of the ICCIDM conference for sharing ideas and research findings with truly national and international communities. Thanks to all who have contributed in producing such a comprehensive conference proceedings of ICCIDM.

The organizing committee believes and trusts that we have been true to the spirit of collegiality that members of ICCIDM value, even as maintaining an elevated standard as we have reviewed papers, provided feedback, and present a strong body of published work in this collection of proceedings. Thanks to all the members of the Organizing committee for their heartfelt support and cooperation.

It has been an honor for us to edit the proceedings. We have enjoyed considerably working in cooperation with the International Advisory, Program, and Technical Committees to call for papers, review papers, and finalize papers to be included in the proceedings.

We express our sincere thanks and obligations to the benign reviewers for sparing their valuable time and effort in reviewing the papers along with suggestions and appreciation in improvising the presentation, quality, and content of this proceedings. Without this commitment it would not be possible to have the important reviewer status assigned to papers in the proceedings. The eminence of these papers is an accolade to the authors and also to the reviewers who have guided for indispensable perfection.

We would like to gratefully acknowledge the enthusiastic guidance and continuous support of Prof. (Dr.) Lakhmi Jain, as and when it was needed as well as adjudicating on those difficult decisions in the preparation of the proceedings and impetus to our efforts to publish this proceeding.

Last but not the least, the editorial members of Springer Publishing deserve a special mention and our sincere thanks to them not only for making our dream come true in the shape of this proceedings, but also for its brilliant get-up and in-time publication in Smart, Innovation, System and Technologies, Springer.

I feel honored to express my deep sense of gratitude to all members of International Advisory Committee, Technical Committee, Program Committee, Organizing Committee, and Editorial Committee members of ICCIDM for their unconditional support and cooperation.

The ICCIDM conference and proceedings are a credit to a large group of people and everyone should be proud of the outcome.

Himansu Sekhar Behera

About the Conference

The International Conference on "Computational Intelligence in Data Mining" (ICCIDM-2014) has been established itself as one of the leading and prestigious conference which will facilitate cross-cooperation across the diverse regional research communities within India as well as with other International regional research programs and partners. Such an active dialogue and discussion among International and National research communities is required to address many new trends and challenges and applications of Computational Intelligence in the field of Science, Engineering and Technology. ICCIDM 2014 is endowed with an opportune forum and a vibrant platform for researchers, academicians, scientists, and practitioners to share their original research findings and practical development experiences on the new challenges and budding confronting issues.

The conference aims to:

- Provide an insight into current strength and weaknesses of current applications as well as research findings of both Computational Intelligence and Data Mining.
- Improve the exchange of ideas and coherence between the various Computational Intelligence Methods.
- Enhance the relevance and exploitation of data mining application areas for end-user as well as novice user application.
- Bridge research with practice that will lead to a fruitful platform for the development of Computational Intelligence in Data mining for researchers and practitioners.
- Promote novel high quality research findings and innovative solutions to the challenging problems in Intelligent Computing.
- Make a tangible contribution to some innovative findings in the field of data mining.
- Provide research recommendations for future assessment reports.

So, we hope the participants will gain new perspectives and views on current research topics from leading scientists, researchers, and academicians around the world, contribute their own ideas on important research topics like Data Mining and Computational Intelligence, as well as network and collaborate with their international counterparts.

Conference Committee

Patron
Prof. E. Saibaba Reddy
Vice Chancellor, VSSUT, Burla, Odisha, India

Convenor
Dr. H.S. Behera
Department of CSE and IT, VSSUT, Burla, Odisha, India

Co-Convenor
Dr. M.R. Kabat
Department of CSE and IT, VSSUT, Burla, Odisha, India

Organizing Secretary
Mr. Janmenjoy Nayak, DST INSPIRE Fellow, Government of India

Conference Chairs

Honorary General Chair
Prof. P.K. Dash, Ph.D., D.Sc., FNAE, SMIEEE, Director
Multi Disciplinary Research Center, S 'O'A University, India
Prof. Lakhmi C. Jain, Ph.D., M.E., B.E.(Hons), Fellow (Engineers Australia),
University of Canberra, Canberra, Australia and University of South Australia,
Adelaide, SA, Australia

Honorary Advisory Chair
Prof. Shankar K. Pal, Distinguished Professor
Indian Statistical Institute, Kolkata, India

General Chair
Prof. Rajib Mall, Ph.D., Professor and Head
Department of Computer Science and Engineering, IIT Kharagpur, India

Program Chairs

Dr. Sukumar Mishra, Ph.D., Professor
Department of EE, IIT Delhi, India
Dr. R.P. Panda, Ph.D., Professor
Department of ETC, VSSUT, Burla, Odisha, India
Dr. J.K. Mandal, Ph.D., Professor
Department of CSE, University of Kalyani, Kolkata, India

Finance Chair

Dr. D. Dhupal, Coordinator, TEQIP, VSSUT, Burla

Volume Editors

Prof. Lakhmi C. Jain, Ph.D., M.E., B.E.(Hons), Fellow (Engineers Australia),
University of Canberra, Canberra, Australia and University of South Australia,
Adelaide, SA, Australia
Prof. H.S. Behera, Reader, Department of Computer Science Engineering and
Information Technology, Veer Surendra Sai University of Technology, Burla,
Odisha, India
Prof. J.K. Mandal, Professor, Department of Computer Science and Engineering,
University of Kalyani, Kolkata, India
Prof. D.P. Mohapatra, Associate Professor, Department of Computer Science and
Engineering, NIT, Rourkela, Odisa, India

International Advisory Committee

Prof. C.R. Tripathy (VC, Sambalpur University)
Prof. B.B. Pati (VSSUT, Burla)
Prof. A.N. Nayak (VSSUT, Burla)
Prof. S. Yordanova (STU, Bulgaria)
Prof. P. Mohapatra (University of California)
Prof. S. Naik (University of Waterloo, Canada)
Prof. S. Bhattacharjee (NIT, Surat)
Prof. G. Saniel (NIT, Durgapur)
Prof. K.K. Bharadwaj (JNU, New Delhi)
Prof. Richard Le (Latrob University, Australia)
Prof. K.K. Shukla (IIT, BHU)
Prof. G.K. Nayak (IIIT, BBSR)
Prof. S. Sakhya (TU, Nepal)
Prof. A.P. Mathur (SUTD, Singapore)
Prof. P. Sanyal (WBUT, Kolkata)
Prof. Yew-Soon Ong (NTU, Singapore)
Prof. S. Mahesan (Japfna University, Srilanka)
Prof. B. Satapathy (SU, SBP)

Prof. G. Chakraborty (IPU, Japan)
Prof. T.S. Teck (NU, Singapore)
Prof. P. Mitra (P.S. University, USA)
Prof. A. Konar (Jadavpur University)
Prof. S. Das (Galgotias University)
Prof. A. Ramanan (UJ, Srilanka)
Prof. Sudipta Mohapatra, (IIT, KGP)
Prof. P. Bhattacharya (NIT, Agaratala)
Prof. N. Chaki (University of Calcutta)
Dr. J.R. Mohanty, (Registrar, VSSUT, Burla)
Prof. M.N. Favorskaya (SibSAU)
Mr. D. Minz (COF, VSSUT, Burla)
Prof. L.M. Patnaik (DIAT, Pune)
Prof. G. Panda (IIT, BHU)
Prof. S.K. Jena (NIT, RKL)
Prof. V.E. Balas (University of Arad)
Prof. R. Kotagiri (University of Melbourne)
Prof. B.B. Biswal (NIT, RKL)
Prof. Amit Das (BESU, Kolkata)
Prof. P.K. Patra (CET, BBSR)
Prof. N.G.P.C Mahalik (California)
Prof. D.K. Pratihar (IIT, KGP)
Prof. A. Ghosh (ISI, Kolkata)
Prof. P. Mitra (IIT, KGP)
Prof. P.P. Das (IIT, KGP)
Prof. M.S. Rao (JNTU, HYD)
Prof. A. Damodaram (JNTU, HYD)
Prof. M. Dash (NTU, Singapore)
Prof. I.S. Dhillon (University of Texas)
Prof. S. Biswas (IIT, Bombay)
Prof. S. Pattnayak (S'O'A, BBSR)
Prof. M. Biswal (IIT, KGP)
Prof. Tony Clark (M.S.U, UK)
Prof. Sanjib ku. Panda (NUS)
Prof. G.C. Nandy (IIIT, Allahabad)
Prof. R.C. Hansdah (IISC, Bangalore)
Prof. S.K. Basu (BHU, India)
Prof. P.K. Jana (ISM, Dhanbad)
Prof. P.P. Choudhury (ISI, Kolkata)
Prof. H. Pattnayak (KIIT, BBSR)
Prof. P. Srinivasa Rao (AU, Andhra)

International Technical Committee

Dr. Istvan Erlich, Ph.D., Chair Professor, Head
Department of EE and IT, University of DUISBURG-ESSEN, Germany
Dr. Torbjørn Skramstad, Professor
Department of Computer and System Science, Norwegian University
of Science and Technology, Norway
Dr. P.N. Suganthan, Ph.D., Associate Professor
School of EEE, NTU, Singapore
Prof. Ashok Pradhan, Ph.D., Professor
Department of EE, IIT Kharagpur, India
Dr. N.P. Padhy, Ph.D., Professor
Department of EE, IIT, Roorkee, India
Dr. B. Majhi, Ph.D., Professor
Department of Computer Science and Engineering, N.I.T Rourkela
Dr. P.K. Hota, Ph.D., Professor
Department of EE, VSSUT, Burla, Odisha, India
Dr. G. Sahoo, Ph.D., Professor
Head, Department of IT, B.I.T, Meshra, India
Dr. Amit Saxena, Ph.D., Professor
Head, Department of CS and IT, CU, Bilashpur, India
Dr. Sidhartha Panda, Ph.D., Professor
Department of EEE, VSSUT, Burla, Odisha, India
Dr. Swagatam Das, Ph.D., Associate Professor
Indian Statistical Institute, Kolkata, India
Dr. Chiranjeev Kumar, Ph.D., Associate Professor and Head
Department of CSE, Indian School of Mines (ISM), Dhanbad
Dr. B.K. Panigrahi, Ph.D., Associate Professor
Department of EE, IIT Delhi, India
Dr. A.K. Turuk, Ph.D., Associate Professor
Head, Department of CSE, NIT, RKL, India
Dr. S. Samantray, Ph.D., Associate Professor
Department of EE, IIT BBSR, Odisha, India
Dr. B. Biswal, Ph.D., Professor
Department of ETC, GMRIT, A.P., India
Dr. Suresh C. Satpathy, Professor, Head
Department of Computer Science and Engineering, ANITS, AP, India
Dr. S. Dehuri, Ph.D., Associate Professor
Department of System Engineering, Ajou University, South Korea
Dr. B.B. Mishra, Ph.D., Professor,
Department of IT, S.I.T, BBSR, India
Dr. G. Jena, Ph.D., Professor
Department of CSE, RIT, Berhampu, Odisha, India
Dr. Aneesh Krishna, Assistant Professor
Department of Computing, Curtin University, Perth, Australia

Dr. Ranjan Behera, Ph.D., Assistant Professor
Department of EE, IIT, Patna, India
Dr. A.K. Barisal, Ph.D., Reader
Department of EE, VSSUT, Burla, Odisha, India
Dr. R. Mohanty, Reader
Department of CSE, VSSUT, Burla

Conference Steering Committee

Publicity Chair
Prof. A. Rath, DRIEMS, Cuttack
Prof. B. Naik, VSSUT, Burla
Mr. Sambit Bakshi, NIT, RKL

Logistic Chair
Prof. S.P. Sahoo, VSSUT, Burla
Prof. S.K. Nayak, VSSUT, Burla
Prof. D.C. Rao, VSSUT, Burla
Prof. K.K. Sahu, VSSUT, Burla

Organizing Committee
Prof. D. Mishra, VSSUT, Burla
Prof. J. Rana, VSSUT, Burla
Prof. P.K. Pradhan, VSSUT, Burla
Prof. P.C. Swain, VSSUT, Burla
Prof. P.K. Modi, VSSUT, Burla
Prof. S.K. Swain, VSSUT, Burla
Prof. P.K. Das, VSSUT, Burla
Prof. P.R. Dash, VSSUT, Burla
Prof. P.K. Kar, VSSUT, Burla
Prof. U.R. Jena, VSSUT, Burla
Prof. S.S. Das, VSSUT, Burla
Prof. Sukalyan Dash, VSSUT, Burla
Prof. D. Mishra, VSSUT, Burla
Prof. S. Aggrawal, VSSUT, Burla
Prof. R.K. Sahu, VSSUT, Burla
Prof. M. Tripathy, VSSUT, Burla
Prof. K. Sethi, VSSUT, Burla
Prof. B.B. Mangaraj, VSSUT, Burla
Prof. M.R. Pradhan, VSSUT, Burla
Prof. S.K. Sarangi, VSSUT, Burla
Prof. N. Bhoi, VSSUT, Burla
Prof. J.R. Mohanty, VSSUT, Burla

Prof. Sumanta Panda, VSSUT, Burla
Prof. A.K. Pattnaik, VSSUT, Burla
Prof. S. Panigrahi, VSSUT, Burla
Prof. S. Behera, VSSUT, Burla
Prof. M.K. Jena, VSSUT, Burla
Prof. S. Acharya, VSSUT, Burla
Prof. S. Kissan, VSSUT, Burla
Prof. S. Sathua, VSSUT, Burla
Prof. E. Oram, VSSUT, Burla
Dr. M.K. Patel, VSSUT, Burla
Mr. N.K.S. Behera, M.Tech. Scholar
Mr. T. Das, M.Tech. Scholar
Mr. S.R. Sahu, M.Tech. Scholar
Mr. M.K. Sahoo, M.Tech. Scholar
Prof. J.V.R. Murthy, JNTU, Kakinada
Prof. G.M.V. Prasad, B.V.CIT, AP
Prof. S. Pradhan, UU, BBSR
Prof. P.M. Khilar, NIT, RKL
Prof. Murthy Sharma, BVC, AP
Prof. M. Patra, BU, Berhampur
Prof. M. Srivastava, GGU, Bilaspur
Prof. P.K. Behera, UU, BBSR
Prof. B.D. Sahu, NIT, RKL
Prof. S. Baboo, Sambalpur University
Prof. Ajit K. Nayak, S'O'A, BBSR
Prof. Debahuti Mishra, ITER, BBSR
Prof. S. Sethi, IGIT, Sarang
Prof. C.S. Panda, Sambalpur University
Prof. N. Kamila, CVRCE, BBSR
Prof. H.K. Tripathy, KIIT, BBSR
Prof. S.K. Sahana, BIT, Meshra
Prof. Lambodar Jena, GEC, BBSR
Prof. R.C. Balabantaray, IIIT, BBSR
Prof. D. Gountia, CET, BBSR
Prof. Mihir Singh, WBUT, Kolkata
Prof. A. Khaskalam, GGU, Bilaspur
Prof. Sashikala Mishra, ITER, BBSR
Prof. D.K. Behera, TAT, BBSR
Prof. Shruti Mishra, ITER, BBSR
Prof. H. Das, KIIT, BBSR
Mr. Sarat C. Nayak, Ph.D. Scholar
Mr. Pradipta K. Das, Ph.D. Scholar
Mr. G.T. Chandrasekhar, Ph.D. Scholar
Mr. P. Mohanty, Ph.D. Scholar
Mr. Sibarama Panigrahi, Ph.D. Scholar

Mr. A.K. Bhoi, Ph.D. Scholar
Mr. T.K. Samal, Ph.D. Scholar
Mr. Ch. Ashutosh Swain, MCA
Mr. Nrusingh P. Achraya, MCA
Mr. Devi P. Kanungo, M.Tech. Scholar
Mr. M.K. Sahu, M.Tech. Scholar

Contents

Contents

Editors' Biography

Prof. Lakhmi C. Jain is with the Faculty of Education, Science, Technology and Mathematics at the University of Canberra, Australia and University of South Australia, Australia. He is a Fellow of the Institution of Engineers, Australia. Professor Jain founded the Knowledge-Based Intelligent Engineering System (KES) International, a professional community for providing opportunities for publication, knowledge exchange, cooperation, and teaming. Involving around 5,000 researchers drawn from universities and companies worldwide, KES facilitates international cooperation and generates synergy in teaching and research. KES regularly provides networking opportunities for the professional community through one of the largest conferences of its kind in the area of KES. His interests focus on artificial intelligence paradigms and their applications in complex systems, security, e-education, e-healthcare, unmanned air vehicles, and intelligent agents.

Prof. Himansu Sekhar Behera is working as a Reader in the Department of Computer Science Engineering and Information Technology, Veer Surendra Sai University of Technology (VSSUT) (A Unitary Technical University, Established by Government of Odisha), Burla, Odisha. He has received M.Tech. in Computer Science and Engineering from N.I.T, Rourkela (formerly R.E.C., Rourkela) and Doctor of Philosophy in Engineering (Ph.D.) from Biju Pattnaik University of Technology (BPUT), Rourkela, Government of Odisha respectively. He has published more than 80 research papers in various international journals and conferences, edited 11 books and is acting as a member of the editorial/reviewer board of various international journals. He is proficient in the field of Computer Science Engineering and served in the capacity of program chair, tutorial chair, and acted as advisory member of committees of many national and international conferences. His research interest includes Data Mining and Intelligent Computing. He is associated with various educational and research societies like OITS, ISTE, IE, ISTD, CSI, OMS, AIAER, SMIAENG, SMCSTA, etc. He is currently guiding seven Ph.D. scholars.

Prof. Jyotsna Kumar Mandal is working as Professor in Computer Science and Engineering, University of Kalyani, India. Ex-Dean Faculty of Engineering, Technology and Management (two consecutive terms since 2008). He has 26 years of teaching and research experiences. He was Life Member of Computer Society of India since 1992 and life member of Cryptology Research Society of India, member of AIRCC, associate member of IEEE and ACM. His research interests include Network Security, Steganography, Remote Sensing and GIS Application, Image Processing, Wireless and Sensor Networks. Domain Expert of Uttar Banga Krishi Viswavidyalaya, Bidhan Chandra Krishi Viswavidyalaya for planning and integration of Public domain networks. He has been associated with national and international journals and conferences. The total number of publications to his credit is more than 320, including 110 publications in various international journals. Currently, he is working as Director, IQAC, Kalyani University.

Prof. Durga Prasad Mohapatra received his Ph.D. from Indian Institute of Technology Kharagpur and is presently serving as an Associate Professor in NIT Rourkela, Odisha. His research interests include software engineering, real-time systems, discrete mathematics, and distributed computing. He has published more than 30 research papers in these fields in various international Journals and conferences. He has received several project grants from DST and UGC, Government of India. He received the Young Scientist Award for the year 2006 from Orissa Bigyan Academy. He has also received the Prof. K. Arumugam National Award and the Maharashtra State National Award for outstanding research work in Software Engineering for the years 2009 and 2010, respectively, from the Indian Society for Technical Education (ISTE), New Delhi. He is nominated to receive the Bharat Shiksha Ratan Award for significant contribution in academics awarded by the Global Society for Health and Educational Growth, Delhi.

Radiology Information System's Mechanisms: HL7-MHS and HL7/DICOM Translation

Hardeep Singh Kang and Kulwinder Singh Mann

Abstract The innovative features of information system, known as, Radiology Information System (RIS), for electronic medical records has shown a good impact in the hospital. The interoperability of RIS with the other Intra-hospital Information Systems that interacts with, dealing with the compatibility and open architecture issues, are accomplished by two novel mechanisms. The first one is the particular message handling system that is applied for the exchange of information, according to the Health Level Seven (HL7) protocol's specifications and serves the transfer of medical and administrative data among the RIS applications and data store unit. The same mechanism allows the secure and HL7-compatible interactions with the Hospital Information System (HIS) too. The second one implements the translation of information between the formats that HL7 and Digital Imaging and Communication in Medicine (DICOM) protocols specify, providing the communication between RIS and Picture and Archive Communication System (PACS).

Keywords RIS · PACS · HIS · HL7 · DICOM · Messaging service · Interoperability · Digital images

1 Introduction

The hospital information system (HIS) is a computerized management system for handling three categories of tasks in a healthcare environment [1]:

H.S. Kang (✉)
PTU, Jalandhar, Punjab, India
e-mail: hardeep_kang41@rediffmail.com

K.S. Mann
Department of IT, Gndec, Ludhiana, Punjab, India
e-mail: mannkulvinder@yahoo.com

© Springer India 2015
L.C. Jain et al. (eds.), *Computational Intelligence in Data Mining - Volume 1*,
Smart Innovation, Systems and Technologies 31, DOI 10.1007/978-81-322-2205-7_1

1. Support clinical and medical patient care activities in the hospital.
2. Administer the hospital's daily business transactions (financial, personnel, payroll, bed census, etc.).
3. Evaluate hospital performances and costs, and project the long-term forecast.

A large scale HIS consists of mainframe computers and software. Almost all HISs were developed through the integration of many information systems, starting from the days when healthcare data centers were established. Figure 1 shows the main components of a typical HIS. In the figure that HIS provides automation for such activities starts from patient registration, admission, and patient accounting. It also provides online access to patient clinical results (e.g., laboratory, pathology,

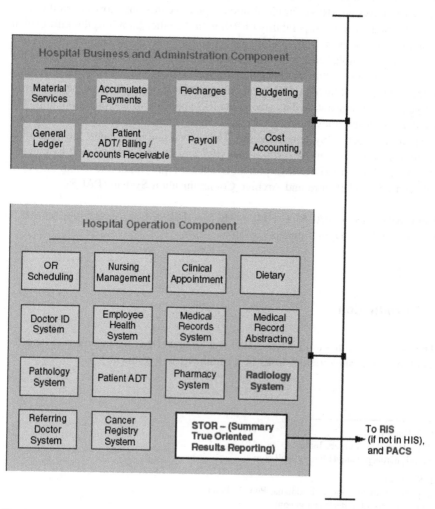

Fig. 1 Main components of a HIS [1]

microbiology, pharmacy, and radiology). The system broadcasts in real time the patient demographics, and when it encounters information with the HL7 standard or DICOM standard, it moves it to the RIS [1].

In the figure that HIS provides automation for such activities starts from patient registration, admission, and patient accounting. It also provides online access to patient clinical results (e.g., laboratory, pathology, microbiology, pharmacy, and radiology). The system broadcasts in real time the patient demographics, and when it encounters information with the HL7 standard or DICOM standard, it moves it to the RIS [1].

1.1 Radiology Information System (RIS)

Most clinical departments in a healthcare center, mainly radiology, pathology, pharmacy, and clinical laboratories, have their own specific operational requirements that differ from the general hospital operations. For this reason special information systems may be needed in these departments [1]. These information systems usually are under the umbrella of the HIS, which supports their operations. However, there are also departments that have different workflow environments that may not be covered by the HIS. So they may need their own separate information systems and must develop mechanisms to integrate data between these systems and the HIS [1]. Such is the story behind RIS, which began as a component of HIS; later an independent RIS was developed because of the limited support from HIS in handling the special data and information required by the radiology operation.

To be specific, the Radiology Information System (RIS) is designed to support both the administrative and clinical operation of the radiology department, to reduce administrative overhead, and to improve the quality of radiological examination service [1]. Therefore the RIS manages general radiology patient information, from scheduling to examination to reporting. RIS equipment consists of a computer system with peripheral devices such as RIS workstations (normally without image display capability), printers, and bar code readers. Most independent RIS are autonomous systems with limited access to HIS and some HIS offers embedded RIS as a subsystem with a higher degree of integration.

The RIS maintains many types of patient- and examination related information. Patient-related information includes medical, administrative, patient demographics, and billing information. Examination-related information includes procedural descriptions and scheduling, diagnostic reporting, patient arrival documentation, film location, film movement, and examination room scheduling [1].

RIS's services extend to an administrative and clinical level and support the interactions between the radiology department and the departments beyond this [2]. RIS is a Patient Data Management System for Radiology Exams (PDMS-RE) and manages the ordering and scheduling of patients' imaging examinations. Like all the other medical information systems that are applied in a hospital or other healthcare center, RIS is not a totally autonomous system and its interaction with other systems is necessary for the integration of the medical procedures that it serves [3, 4]. Two are the main exchange processes of RIS with other such systems [5]. On the one hand, RIS has to communicate with PACS that is responsible for the internal procedures that are performed into the radiology department [6]. These are mainly the retrieval, processing and archiving of medical imaging files. RIS needs to collect this information properly as to produce the final medical report for each examination. On the other hand, RIS interacts with HIS for the retrieval of patient's information, update of his medical record for the new exams and forwarding the data for the billing procedures accordingly.

In this paper, an advanced RIS is presented, which manages these two issues via two innovative mechanisms. The first one is HL7-Message Handling System (HL7-MHS) that provides the transfer service not only for the communication between RIS and HIS but for each kind of interaction between the different structural units of RIS too. The second one is the HL7/DICOM Translation mechanism that connects RIS and PACS according to the specifications of both protocols and is implemented as an autonomous RIS architectural component. The two mechanisms cooperate to provide RIS's services through an integrated solution.

2 Designing Issues

2.1 Functions

The designation of the provided RIS's services results from a detailed analysis of everyday-workflow in a radiology department and a thorough survey of the already proposed solutions [7, 8]. These services are supported by four functions: the patient's visits administration (**ordering**), programming of requested exams (**scheduling**), medical reports' production (**transcription**) and allocation of diagnostic files to the corresponding placers of orders (**allocation**). During the execution of each function appropriate mechanisms are triggered for the authentication and authorization of the users, the surveillance of the activities sequence and tracing of users' actions.

According to the definition of RIS two groups of users participate to RIS's functions. The first one includes the external from the radiology department users who belong to the clinics, order the imaging examinations and wait for the results (Clinics' Users—ClUs). The second group comprises the internal users that serve the requests and respond with the resulting diagnostic files (Radiology Users—RadUs). In Table 1 the RIS functions are assigned to the users.

Table 1 Users' participation to RIS functions

Ordering	Scheduling	Transcription	Allocation
ClUs	RadUs	RadUs	ClUs, RadUs

2.2 Data Classes

The information that RIS administers is structured in five data classes: Patient, Referring, Visits, Exams and Medical Reports. As shown in Fig. 2, each one is defined by a number of simple attributes or by a combination of attributes and other data classes. Their relations are implemented through the identification elements (IDs) that characterize each instance of data classes.

Patient data class comprises the demographics (identity, insurance information, and personal data) and the medical history of each patient.

Referring is the class through which ClUs request a new imaging examination with **ordering** function. Its elements are the basis for the definition of the rest data classes. Each referring is characterized by Referring_ID and this attribute is used to create integrated records that lead to valid responses for each request.

In each Referring and after the **scheduling** function a new class is assigned, a Visit.

The attributes of *Exam* class are a combination of data elements that come from both systems, RIS and PACS. *Exam* is created on the arrival of the patient at the corresponding "room" and is totally completed with the end of the *Visit*.

The Medical Reports are multimedia documents that include all the attributes of Exams, formatted to the appropriate structure. The creation of these "documents" is executed during the **transcription** function. The forwarding of medical reports to the placers, as responses to their Referrings (**allocation**), signals the completion of the whole procedure.

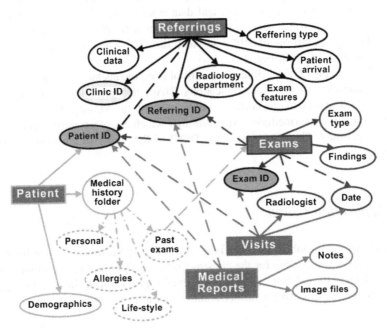

Fig. 2 RIS data classes share logically a number of basic data elements

Fig. 3 The HL7-MHS's structure for RIS messaging service

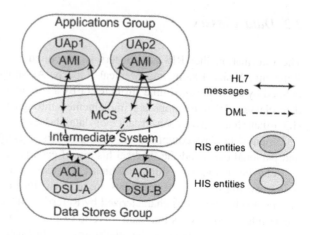

3 HL7-MHS for RIS

HL7 mechanism is based on the HL7-MHS that was designed and developed for Intra-hospital Information Systems. The detailed analysis and the specifications of this system are presented in L. Kolovou, A. Darras, D. Lymperopoulos (HL7 Message Handling for Intra-hospital Information Systems, 2nd International Conference on Computational Intelligence in Medicine and Healthcare, Costa da Caparica. Lisbon. Portugal. June 2005. Unpublished). HL7-MHS aims to serve the data exchange between applications and data stores units of all possible medical information systems that can de implemented for a hospital or other healthcare center. For the provision of RIS's services the HL7-MHS is simplified to the structure that is shown in Fig. 3.

As shown in Fig. 3, three different groups of components have been defined. The first one includes all the user applications (UAp) that are allowed to exchange information. The second one comprises only one unit, the Message Control System (MCS) and is the intermediate system through which every communicational session is performed. The data store units (DSU) of RIS and HIS compose the Data Stores Group.

3.1 Application Group

In each UAp a new embedded entity is specified, the Application Message Interface (AMI). AMI is responsible for the construction of HL7 messages, submission and receipt of them to and from the MCS and finally provision of them to the applications programs of UAp for further administration, in the proper format.

3.2 Data Stores Group

The interactions between the MCS and the DSUs are performed through the Access Queries Libraries (AQL). These libraries are embedded to each DSU and differ for each of them.

3.3 Intermediate System

MCS has the role that each open intermediate system has, according to OSI definition [9]. MCS is responsible for the proper routing of messages. In the case of a UAp-DSU session, it performs the translation of the HL7 messages to a Data Manipulation Language (DML) (e.g. SQL) format for serving the communication between the two end-system entities.

4 HL7/DICOM Translation

HL7/DICOM Translation mechanism meats the logical structure of HL7-MHS. The philosophy of 'intermediate system' is used in this case to provide the translation service, according to the specifications of HL7 and DICOM protocols [10]. In Fig. 4 the model of Translation mechanism is depicted. Following the specification that the HL7-MHS serves all the traffic of HL7 messages, the first level of the model is

Fig. 4 HL7/DICOM translation's model

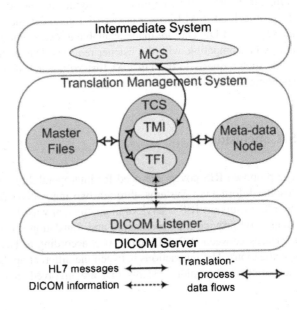

the MCS. This interacts with the Translation Management System (TMS), which is the sequent level, exchanging the HL7 messages that implement the communication between RIS and PACS. TMS consists of three different entities and serves the translation processes between the HL7 and DICOM formats properly. The last level of the model is DICOM Server and provides the 'services' that DICOM specifies in order to satisfy the processes that the upper level activates.

4.1 Translation Management System

The interconnection between the two terminal levels of the model in Fig. 4 is preformed by the Translation Control System (TCS). This includes two extra units that implement the HL7 and DICOM interfaces of the level. Translator Message Interface (TMI) has the same role and structure as AMIs in HL7-MHS [11]. It is initiated properly for managing the messages of RIS-PACS session and exchanges HL7 messages with MCS in both directions. Translator File Interface (TFI) interacts with the corresponding unit of DICOM Server, addresses syntactically compatible with DICOM Server's specifications commands and accepts the responding DICOM files.

4.2 DICOM Server

In reality, DICOM Server is not part of RIS, but is implemented in every DICOM compatible PACS. It is the unit that provides all the services that are specified by the protocol. DICOM Listener is a special unit of it and is the 'port' through which the services of DICOM are made available to every external application or program that sends compatible with the listener requests. During the translation process TFI sends such requests to the DICOM Listener, which then responds with the corresponding DICOM files.

5 Conclusion

The proposed RIS covers the need for interoperability between the different Intra-hospital Information Systems, through two innovative mechanisms that cooperate in order to support RIS's services in an efficient way. HL7-MHS and HL7/DICOM Translation provide their messaging and translation services for implementing the interfaces between RIS, HIS and PACS according to HL7 and DICOM protocols' specifications. The interactions between the different applications and data stores of the various information systems that are involved in this supra system will be performed via standardized messages and sessions. These should cover a minimum number of specifications for their cooperation through processes that are compatible with the available standards.

References

1. Huang, H.K.: PACS, and Imaging Informatics: Basic Principles and Applications, 2nd edn. Wiley-Blackwell, Hoboken (2010)
2. Broecka, R., Verhellea, F., Veldeb, R., Osteauxa, M.: Integrated Use of the DICOM Fields Information Within HIS-RIS: An Added Value by PACS in Hospital Wide Patient Management, International Congress Series, pp. 791–794 (2001)
3. Lenz, R., Kuhn, K.A.: Towards a continuous evolution and adaptation of information systems in healthcare. Int. J. Med. Informatics **73**, 75–89 (2004)
4. Short Strategic Study: Health Informatics Infrastructure, CEN/TC 251/N00-051, Health Informatics, Version 1.7 (2000)
5. Smedema, K.: Integrating the healthcare, enterprise (IHE): the radiological perspective. Medicamundi **44**, 39–47 (2000)
6. Pavlopoulos, S.A., Delopoulos, A.N.: Designing and implementing the transition to a fully digital hospital. IEEE Trans. Inf. Technol. Biomed. **3**(1), 6–19 (1999)
7. Kolovou, L., Galanopoulos, P., Lymberopoulos, D.: MRIMS: A new middleware connective level for RIS and PACS. In: 8th WSEAS CSCC (CSCC 2004), Vouliagmeni, Greece, 12–15 June 2004
8. Wong, S.T.C., Tjarndra, D., Wang, H., Shen, W.: Workflow-enabled distributed component based information architecture for digital medical imaging enterprises. IEEE Trans Inf Technol Biomed. **7**(3), 171–183 (2003)
9. Recommendations X200, Blue Book, vol. VII, ITU-CCITT (1988)
10. Kolovou, L., Galanopoulos, P., Lymberopoulos, D.: MRIMS: A new middleware connective level for RIS and PACS. In: 8th WSEAS CSCC (CSCC 2004), Vouliagmeni, Greece, 12–15 June 2004
11. Kolovou, L., Vatousi, M., Lymperopoulos, D., Koukias, M.: Advanced radiology information system. In: Proceedings of the 2005 IEEE Engineering in Medicine and Biology 27th Annual Conference Shanghai, China, 1–4 Sept 2005

Optimal Control of Twin Rotor MIMO System Using LQR Technique

Sumit Kumar Pandey and Vijaya Laxmi

Abstract In this paper, twin rotor multi input multi output system (TRMS) is considered as a prototype laboratory set-up of helicopter. The aim of studying the model of TRMS and designing the controller for it is to provide a platform for controlling the flight of helicopter. An optimal state feedback controller based on linear quadratic regulator (LQR) technique has been designed for twin rotor multi input multi output system. TRMS is a nonlinear system with two degrees of freedom and cross couplings. The mathematical modeling of TRMS has been done using MATLAB/SIMULINK. The linearised model of TRMS is obtained from the nonlinear model. The simulation results of optimal controller are compared with the results of conventional PID controller. The appropriateness of proposed controller has been shown both in terms of transient and steady state response.

Keywords Twin rotor MIMO system · Linear quadratic regulator (LQR) · Unmanned air vehicle (UAV)

1 Introduction

Recent times the development of several approaches for controlling the flight of air vehicle such as helicopter and unmanned air vehicle (UAV) has been studied frequently. The modeling of the air vehicle dynamics is a highly challenging task due to the complicated nonlinear interactions among the various variables and also there are certain states which are not accessible for the measurement. The twin rotor multi input multi output system (TRMS) is an experimental set-up that resembles with the helicopter model. The TRMS consist of two rotors at each ends of the

S.K. Pandey (✉) · V. Laxmi
Birla Institute of Technology, Mesra, Ranchi 835215, India
e-mail: skpdmk@gmail.com

V. Laxmi
e-mail: vlaxmi@bitmesra.ac.in

© Springer India 2015
L.C. Jain et al. (eds.), *Computational Intelligence in Data Mining - Volume 1*,
Smart Innovation, Systems and Technologies 31, DOI 10.1007/978-81-322-2205-7_2

horizontal beam known as main rotor and tail rotor which is driven by a DC motor and it is counter balanced by a pivoted beam [1]. The TRMS can rotate in both horizontal and vertical direction. The main rotor generates a lift force due to this the TRMS moves in upward direction around the pitch axis. While, due to the tail rotor TRMS moves around the yaw axis. However TRMS resembles with the helicopter but there is some significant differences between helicopter and TRMS. In helicopter, by changing the angle of attack controlling has been done, while in TRMS it has been done by changing the speed of rotors. Several techniques have been implemented for the modeling and control purpose of TRMS.

In [2] authors provide the detail description of dynamic modeling of twin rotor MIMO system and investigate the open loop control along longitudinal axis. In [3] the model decouples method and implementation of optimal controller has been proposed for two independent SISO systems for TRMS. The controller has been designed to tolerate some changes in system parameter. In [4] the time optimal control method has been proposed for twin rotor MIMO system. In [5] the author discuss about the sliding mode state observer controller for TRMS system. Here the Lyapunov method is used to derive the asymptotic stability conditions for robust and global sliding mode control. In [6] dynamic model is proposed to a one degree of freedom (DOF) twin rotor MIMO system (TRMS) based on a black box system identification technique. This extracted model is connected with a feedback LQG regulator. The authors describe how the system performance has been improved by using artificial non-aerodynamic forces.

In this work, dynamic and linear model for TRMS have been developed. A PID controller and an optimal state feedback controller based on LQR technique has been designed separately. The transient and steady state performance of the system has been analyzed for step input.

The paper is organized as follows. Next section deals with the modeling of the system, followed by the control technique. Section 4 deals with the results obtained and last section consists of conclusion.

2 Mathematical Modeling

According to the diagram presented in Fig. 1, the non linear equation has been derived [7, 8] and the parameters of TRMS are shown in Table 1.

$$I_1 \cdot \ddot{\psi} = M_1 - M_{FG} - M_{B\psi} - M_G \tag{1}$$

where, M_1 is the nonlinearity caused by the rotor and can be estimated as second order polynomial and due to this the torque is induced to the TRMS as given below.

Fig. 1 Twin rotor MIMO
system

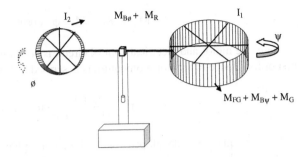

Table 1 Physical parameters of TRMS

Symbol	Parameter	Value	Unit
I_1	Vertical rotor moment of inertia	6.8×10^{-2}	kg m^2
I_2	Horizontal rotor moment of inertia	2×10^{-2}	kg m^2
a_1	Parameter of static characteristic	0.0135	N/A
a_2	Parameter of static characteristic	0.0924	N/A
b_1	Parameter of static characteristic	0.02	N/A
b_2	Parameter of static characteristic	0.09	N/A
m_g	Gravity momentum	0.32	N m
$B_{1\psi}$	Parameter of friction momentum	6×10^{-2}	N m s/rad
$B_{2\psi}$	Parameter of friction momentum	1×10^{-3}	N m s/rad
$B_{1\phi}$	Parameter of friction momentum	1×10^{-1}	N m s/rad
$B_{2\phi}$	Parameter of friction momentum	1×10^{-2}	N m s/rad
K_{gy}	Parameter of gyroscopic momentum	0.05	s/rad

$$M_1 = a_1 \cdot \tau_1^2 + b_1 \cdot \tau_1 \tag{2}$$

Considering the Fig. 1, the weight of the helicopter produces the gravitational torque about the pivot point, which is described by the following Eq. 3.

$$M_{FG} = M_g \cdot \sin \psi \tag{3}$$

The frictional torque can be estimated as following equation.

$$M_{B\psi} = B_{1\psi} \cdot \dot{\psi} + B_{2\psi} \cdot \text{sign}(\dot{\psi}) \tag{4}$$

The gyroscopic torque occurs due to coriolis force. This torque is resulted when moving main rotor changes its position in azimuth direction, and describes as the Eq. 5 given below.

$$M_G = K_{gy} \cdot M_1 \cdot \dot{\phi} \cdot \cos\psi \tag{5}$$

Here, the motor and electrical control circuit is considered as transfer function of first order. Hence the motor momentum is described in Laplace domain is as below.

$$\tau_1 = \frac{K_1}{T_{11}s + T_{10}} \cdot u_1 \tag{6}$$

Similar equation is developed for the horizontal plane motion. The net torques produced in horizontal plane motion is described by the following Eq. 7

$$I_2 \cdot \ddot{\phi} = M_2 - M_{B\phi} - M_R \tag{7}$$

where, M_2 is nonlinear static characteristic similar as main rotor.

$$M_2 = a_2 \cdot \tau_2^2 + b_2 \cdot \tau_2 \tag{8}$$

Frictional torque is calculated same as the main rotor dynamics.

$$M_{B\psi} = B_{1\phi} \cdot \dot{\psi} + B_{2\phi} \cdot \text{sign}\,(\dot{\Phi}) \tag{9}$$

M_R is the cross reaction momentum estimated by first order transfer function described by the following equation.

$$M_R = \frac{K_c \cdot (T_{0}s + 1)}{(T_{ps} + 1)} \cdot \tau_1 \tag{10}$$

Again, the D.C. motor with electrical circuit is estimated as the first order transfer function and given by the following equation.

$$\tau_2 = \frac{k_2}{T_{21}S + T_{20}} \cdot u_2 \tag{11}$$

The above mathematical model given by Eqs. 1–11, is linearized [6, 7] across equilibrium point X0 given as

$$X0 = [0 \quad 0 \quad 0 \quad 0 \quad 0 \quad 0 \quad 0]$$

The state and output vector here is given by

$$X = [\psi, \dot{\psi}, \Phi, \dot{\Phi}, \tau_1, \tau_2, M_R]^T$$
$$Y = [\psi \quad \Phi]^T$$

Here the TRMS plant is represented as below.

$$\left.\begin{array}{c} \dot{x} = Ax + Bu \\ y = C\,x \end{array}\right\} \tag{12}$$

The system considered here consists of 7 states, there are two control input and two output state namely pitch and yaw. The system matrix can be obtained by linearizing is as below.

$$A = \begin{bmatrix} 0 & 0 & 1 & 0 & 0 & 0 & 0 \\ 0 & 0 & 0 & 1 & 0 & 0 & 0 \\ -4.34 & 0 & -0.0882 & 0 & 1.24 & 0 & 0 \\ 0 & 0 & 0 & -5 & 1.4823 & 3.6 & 18.75 \\ 0 & 0 & 0 & 0 & -0.8333 & 0 & 0 \\ 0 & 0 & 0 & 0 & 0 & -1 & 0 \\ 0 & 0 & 0 & 0 & -0.0169 & 0 & -0.5 \end{bmatrix}$$

$$B = \begin{bmatrix} 0 & 0 \\ 0 & 0 \\ 0 & 0 \\ 0 & 0 \\ 1 & 0 \\ 0 & 1 \\ 0 & 0 \end{bmatrix}$$

$$C = \begin{bmatrix} 1 & 0 & 0 & 0 & 0 & 0 & 0 \\ 0 & 1 & 0 & 0 & 0 & 0 & 0 \end{bmatrix}$$

3 Proposed Control Techniques of TRMS

This section presents the control techniques using PID controller and optimal state feedback controller using LQR technique.

3.1 PID Control Technique

The conventional PID controller is used to control the horizontal and vertical movements separately. The outputs are compared with desired output value and then the error is processed to conventional PID controller as shown in Fig. 2 [8].

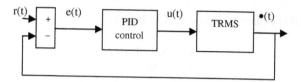

Fig. 2 PID control scheme for TRMS

Here, r(t) is the reference input, e(t) is error signal, u(t) is the control force and α(t) is the output of plant.

The controller output of conventional PID controller [9] is given as in Eq. 13

$$u(t) = K_p e(t) + K_i \int e(t)\, dt + K_d \frac{de(t)}{dt} \tag{13}$$

where,

K_p Proportional gain,
K_i Integral gain,
K_d Derivative gain.

3.2 Optimal Control Technique

Here the plant taken is time varying because the optimal control problem is formulated for the time varying system. Control input of the plant is described as shown in Fig. 3.

$$u(t) = -K(t)\, x(t) \tag{14}$$

The control input here is linear and the control energy is given by $u^T(t)R(t)u(t)$, where R(t) is the square matrix known as control cost matrix. Control energy expression is in quadratic form because the equation contains $u^T(t)R(t)u(t)$ quadratic function of u(t). The transient energy can be expressed as $x^T(t)\,Q(t)\,x(t)$,

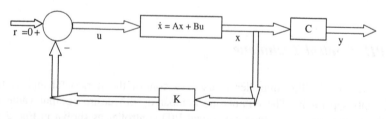

Fig. 3 Optimal control scheme of TRMS

where $Q(t)$ is square symmetric matrix called state weighing matrix [10]. Hence the objective function is

$$J(t, t_f) = \int\limits_{t}^{t_f} (x^T(t)Q(t)\,x(t) + u^T(t)\,R(t)\,u(t))dt \qquad (15)$$

where t and t_f are initial and final time respectively. The main objective here is to minimize the objective function as described by Eq. 15 by choosing an optimal value of gain matrix $K(t)$. By considering Eqs. 12 and 14.

$$\dot{x}(t) = (A - BK(t))x(t) \qquad (16)$$

$$\dot{x}(t) = A_K\,x(t) \qquad (17)$$

where $A_K = (A - BK(t))$ is close loop state dynamics matrix. The solution of Eq. 16 is

$$x(t) = \theta_K(t,\ t_0)x(t_0) \qquad (18)$$

where $\theta_K(t,\ t_0)$ is the state transition matrix of closed loop system, on substituting Eq. 18 in 15 the given objective function is as

$$J(t,\ t_f) = \int\limits_{t}^{t_f} (x^T(t)\theta_K^T(\tau, t)(Q(\tau) + K^T(\tau)R(\tau)K(\tau))\theta_K(\tau, t)x(t)d\tau \qquad (19)$$

This can be written as

$$J(t, t_f) = x^T(t)M(t, t_f)x(t) \qquad (20)$$

where

$$M(t, t_f) = \int\limits_{t}^{t_f} \theta_K(\tau, t)(Q(\tau) + K^T(\tau)R(\tau)K(\tau))\theta_K(\tau, t)d\tau \qquad (21)$$

By Eqs. 18 and 19

$$(t, t_f) = \int\limits_{t}^{t_f} (x^T(\tau)(Q(\tau) + K^T(\tau)\,R(\tau)\,K(\tau))\,x(\tau)\,d\tau \qquad (22)$$

Now differentiating Eq. 22 with respect to time 't'

$$\frac{\partial J(t, t_f)}{\partial t} = -x^T(t)(Q(t) + K^T(t)R(t)K(t))x(t) \tag{23}$$

Also partially differentiating Eq. 19 with respect to time 't'

$$\frac{\partial J(t, t_f)}{\partial t} = \dot{x}(t)^T M(t, t_f) x(t) + x^T(t)(\frac{\partial M(t, t_f)}{\partial t}) x(t) + x^T(t) M(t, t_f)\dot{x}(t) \tag{24}$$

By combining Eqs. 17 and 24

$$\frac{\partial J(t, t_f)}{\partial t} = x^T(t)(A_K(t)M(t, t_f) + (\frac{\partial M(t, t_f)}{\partial t})x(t) + M(t, t_f)A_K(t))x(t) \tag{25}$$

Now considering Eqs. 23 and 25

$$-\frac{\partial M(t, t_f)}{\partial t} = A_K(t) M(t, t_f) + A_K^T(t) M(t, t_f) + (Q(t) + K^T(t)R(t)K(t)) \tag{26}$$

Above equation describe the matrix Riccati equation for finite time duration. Optimal control gain matrix K(t) is obtained by solving Eq. 26.

$$K(t) = R^{-1}(t) B^T(t)M \tag{27}$$

By considering the closed loop system as asymptotically stable, M is a optimal matrix, Q(t) and R(t) are positive definite matrix and positive semi definite matrix are time independent. The value of Q and R are randomly chosen and varied until the output of system does not get the desired value.

The control gain matrix K has been calculated here is as below.

$$K = \begin{bmatrix} -0.1510 & 0.0044 & 0.0352 & 0.0011 & -0.54 & 0.0024 & 0.0226 \\ -0.0053 & 1 & -0.0046 & 1 & 0.0421 & -0.0056 & -2.0037 \end{bmatrix}$$

4 Simulation Results

To implement the above control techniques, the TRMS is designed using Simulink. The responses of reference inputs of the LQR controller are presented in this section which is compared with the results of the conventional PID controller. In simulation the transient and steady state response of the system is investigated such as overshoot, settling time, steady state error. The reference value for step input is taken here as 0.5 for horizontal plane and 0.2 for vertical plane. Figure 4a, b shows the

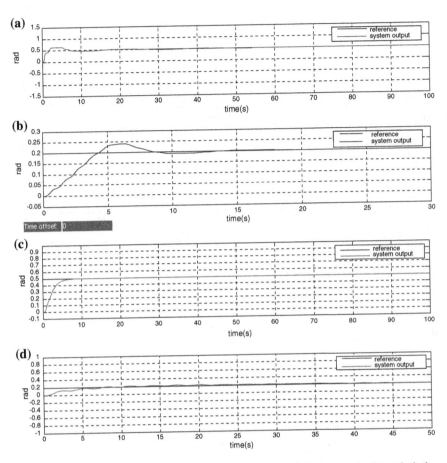

Fig. 4 **a** Step response in horizontal plane using PID control. **b** Step response in vertical plane using PID control. **c** Step response in horizontal plane using optimal control. **d** Step response in vertical plane using optimal control

response of the TRMS in horizontal and vertical plane using PID control technique. Figure 4c, d shows the response of the TRMS in horizontal and vertical plane using optimal control technique. Figure 5a, b shows the control effort of the TRMS in horizontal and vertical plane using PID control technique. Figure 5c, d shows the control effort of the TRMS in horizontal and vertical plane using optimal control technique.

Table 1 shows the physical parameters of TRMS and Table 2 depicts the characteristics of step response of TRMS using PID controller and optimal controller.

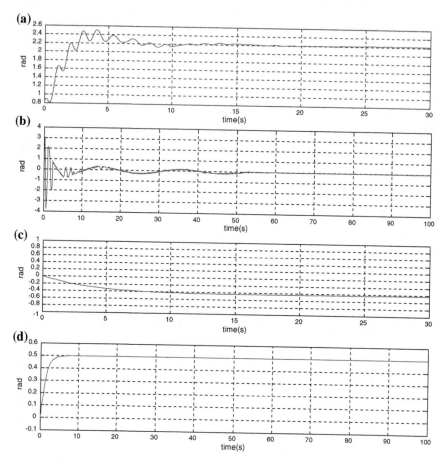

Fig. 5 **a** Control effort in vertical plane using PID control. **b** Control effort in horizontal plane using PID control. **c** Control effort in vertical plane using optimal control. **d** Control effort in horizontal plane using optimal control

Table 2 Characteristics of step response

	Plane	Reference value	Rise time (s)	Settling time (s)	Max. over shoot (%)	Steady state error
PID controller [8]	Horizontal	0.5	2.0	18	25.0	0.0
	Vertical	0.2	6.0	14	20.0	0.0
Optimal controller	Horizontal	0.5	4.4	5.7	0.0	0.0
	Vertical	0.2	5.0	8.0	8.0	0.0

5 Conclusion

In this paper, the TRMS with two degrees of freedom was considered. The mathematical modeling of TRMS has been done in MATLAB/SIMULINK. Here PID and optimal controllers has been designed to control the horizontal and vertical movements of the system. The performance of the designed controllers has been evaluated with step input. The results show that the optimal controller gives better performance in terms of both transient and steady state response as compared to the PID controller. The control effort in case of optimal controller is minimum then the PID controller.

References

1. TRMS 33-949S User Manual, Feedback instruments Ltd., East Sussex, U.K. (1998)
2. Ahmad, S.M., Chipperfield, A.J., Tokhi, M.O.: Dynamic modeling and open loop control of twin rotor multi input multi output system. J. Syst. Control Eng. (2002)
3. Wen, P., Li, Y.: Twin rotor system modeling, de-coupling and optimal control. Proceedings of the IEEE International Conference on Mechatronics and Automation, Beijing, China (2011)
4. Lu, T.W., Wen, P.: Time optimal and robust control of twin rotor system. In: IEEE International Conference on Control and Automation Guangzhou, China (2007)
5. Pratap, B., Purwar, S.: Sliding mode state observer for 2-DOF twin rotor MIMO system. In: International Conference on Power, Control and Embedded Systems, India (2010)
6. Ahmad, S.M., Chipperfield, A.J., Tokhi, M.O.: Dynamic modelling and optimal control of a twin rotor MIMO system. In: Proceedings of IEEE national aerospace and electronics conference (NAECON'2000), pp. 391–398, Dayton, Ohio, USA (2000)
7. Bennoune, A., Kaddouri, A., Ghribi, M.: Application of the dynamic linearization technique to the reduction of the energy consumption of induction motors. Appl. Math. Sci. **1**, 1685–1694 (2007)
8. Pandey, S.K., Laxmi, V.: Control of twin rotor MIMO system using PID controller with derivative filter coefficient. In: Proceedings of IEEE Student's Conference on Electrical, Electronics and Computer Science, MANIT Bhopal, India (2014)
9. Ramalakshmi, A.P.S., Manoharan, P.S.: Nonlinear modeling and control of twin rotor MIMO system. In: Proceedings of IEEE International Conference on Advanced Communication Control and Computing Technologies (ICACCCT), pp. 366–369, Ramanathapuram, India (2012)
10. Saini, S.C., Sharma, Y., Bhandari, M., Satija, U.: Comparison of pole placement and LQR applied to single link flexible manipulator. In: International Conference on Communication Systems and Network Technologies (2012)

Hybrid Synchronous Discrete Distance Time Model for Traffic Signal Optimization

Sudip Kumar Sahana and Kundan Kumar

Abstract This paper proposes a novel solution to the traffic signal optimization problem by reducing the wait time of individual vehicle users at intersections within the urban transportation system. Optimized signal timings, not only reduce the wait time of vehicle users but also improve the mobility within the system. In effect, it also reduces the ever increasing emissions and fuel consumption. A novel synchronous discrete distance-time model is proposed to frame the problem on the basis of 2-layer Stackelberg game. Thereafter, the upper layer optimization is solved using evolutionary computation techniques (ACO, GA and a Hybrid of ACO and GA). A comparative analysis done over the aforementioned techniques indicates that the hybrid algorithm exhibits better performance for the proposed model.

Keywords ACO · GA · Ant colony · Genetic algorithm · Soft computing · Traffic signal · Optimization techniques

1 Introduction

In the modern urban transportation scenario with ever increasing number of vehicles, it is widely agreed that the most cost-effective way to deal with it is by optimizing the traffic signal timings. Most of the solutions are based on a 2-layer problem model called bi-level Stackelberg game (also called Leader-Follower Model). The job of the two layers in present perspective can be described as:

S.K. Sahana · K. Kumar (✉)
Department of Computer Science and Engineering, BIT Mesra, Ranchi, India
e-mail: kundan777@gmail.com

S.K. Sahana
e-mail: sudipsahana@gmail.com

© Springer India 2015
L.C. Jain et al. (eds.), *Computational Intelligence in Data Mining - Volume 1*,
Smart Innovation, Systems and Technologies 31, DOI 10.1007/978-81-322-2205-7_3

23

- Layer 1—Signal setting problem.
- Layer 2—Vehicle routing cum load assignment problem (Stochastic User Equilibrium).

Wide range of solution methods to the signal setting problem have been discussed in the literature. Allsop [2] found mutually consistent (MC) traffic signal settings and traffic assignment for a medium size road network. Abdulaal [1] reported the formulation and solution by means of the Hooke-Jeeves' method for an equilibrium network design problem with continuous variables. Heydecker and Khoo [8] proposed a linear constraint approximation to the equilibrium flows with respect to signal setting variables and solved the bi-level problem as a constraint optimization problem.

Also there have been various forays into this field using evolutionary computation, especially GA and ACO. Lee and Machemehl [9] applied Genetic Algorithm (GA) to individual signalized intersection. Ceylan and Bell [5] proposed GA approach to solve traffic signal control and traffic assignment problem to tackle the optimization of signal timings with SUE link flows. ACO implementation of the problem has been tried by Baskan et al. [3] using a variant of ACO known as ACORSES where they tried to use heuristics to reduce the search space of the potential solution space. Putha et al. [10] discussed the advantages of using ACO over GA for traffic optimization.

All the solutions are based on a simulation model of traffic networks. The model can be discrete or continuous; macroscopic or microscopic; synchronous or asynchronous; etc. The discrete model is computationally less intensive than a continuous model evident from the success of VISSIM, TRANSYT and other such microscopic models. Cantarella et al. [4] and many others have previously proposed a discrete model but most of them were macroscopic simulation models. The proposed model works on the level of section in a link and hence can take care of various microscopic tribulations. The next section describes the proposed microscopic and discrete model.

2 Problem Formulation

2.1 Bi-level Stackelberg Model

The bi-level Stackelberg model can be represented as in Fig. 1.

The whole problem is divided into two layers; Layer 1 and Layer 2, representing the "traffic signals" and "stochastic user equilibrium" respectively. The output of Layer 1 is dependent on the output of Layer 2 whereas Layer 2 is dependent on both the output of Layer 1 and the external input. This problem is known to be one of the most attractive problems in the optimization field because of the non-convexity of the feasible region as it has multiple local optima.

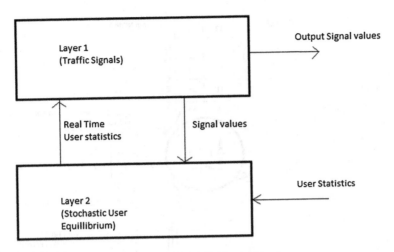

Fig. 1 Stackelberg layers (for traffic signal optimization problem)

The objective is to find the solution to the optimization of Layer 1 for which a novel discrete distance-time model is introduced.

2.2 Synchronous Discrete Distance-Time Model

An instance of the synchronous discrete distance-time model is as shown in Fig. 2.

There are two links between each node denoting the two-way traffic thus turning the road-network into a symmetric digraph. Each link is divided into a number of "time-sections", which is calculated by dividing the length of each link by the average speed of vehicles on that link. Every time-section has a number attached to it which denotes the number of vehicles (or load) in that section of the link. This value should ideally input from the Layer 2 statistics but in this case the output values of the layer 2 are simulated by random number generators for the sake of simplicity.

The unit of time (or time-quantum) is the signal duration which is constant for all intersections and all signals. The total wait-time is the sum of all the time durations that every vehicle user has to wait at respective intersections due to respective signals across all intersections. The calculation of the Total Wait Time is done in phases where one 'phase' consists of the current signal value of all the intersections. Thus, the signal duration and hence the unit time, is the duration between two phases. The signal changes synchronously over all intersections.

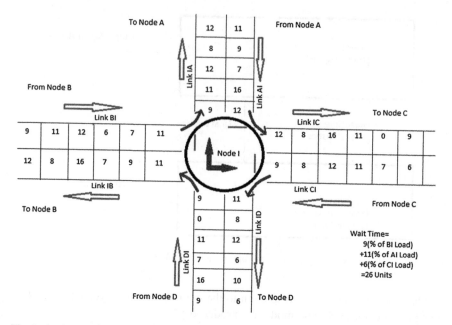

Fig. 2 An intersection in discrete distance-time model

2.3 Mathematical Formulation (for Layer 1)

The aim of layer 1 is to minimize the total wait-time of all vehicles on every intersection. The job of distributing the load based on initial Source-Destination requests so as to minimize the "total travel time" of vehicle users is relegated to layer 2.

Layer 1 can be mathematically modeled as:

$$\text{Minimize TW } (\Psi, q^*(\Psi)) = \sum_{a \in I} \sum_{b \in L(a)} q_{ab}(\Psi, q^*(\Psi)) \times t_{ab}(\Psi, q^*(\Psi))$$

Subject to:

$$\Psi_{min} \leq \Psi \leq \Psi_{max}$$

where,

TW is total wait time.
Ψ is the signal variable.
q^* is the optimized load distribution function (From layer 2).
I is the set of all Intersections.
L(a) is the set of links attached to intersection 'a'.

q_{ab} is the load waiting on link 'b' at intersection 'a'.

t_{ab} is the time for which q_{ab} waited.

3 Solution Techniques

In the following sections, three evolutionary computation solutions are provided; including the proposed hybrid algorithm.

3.1 Ant Colony Optimization (ACO)

ACO [6, 7] is inspired by the foraging behavior of ant colonies, and is simply a metaheuristic in which colonies of artificial ants cooperate in finding good solutions to difficult discrete optimization problems.

Its underlying algorithm can be generically described as:

Algorithm 1. The ACO Metaheuristic
Procedure ACOMetaheuristic
        ScheduleActivities
            ConstructAntsSolutions
            UpdatePheromones
            DaemonActions   % optional
          end-ScheduleActivities
End-procedure

The main objective of the ants is to converge on the shortest path from the source to destination on the problem graph (also called Ant Graph). ACO has been proven as a good optimizer [11, 12] for solving complex problems.

The ant-graph is presented in Fig. 3.

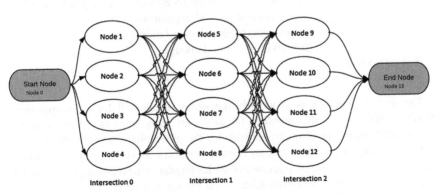

Fig. 3 Ant graph for ACO implementation

The figure shows the directed ant graph for a road network consisting of three intersections. All the intersections are four-way intersections; hence there are four nodes for each intersection describing their state corresponding to four different possible signal values. The cost of each link is determined by the wait-time of all vehicles at that intersection for one phase for the signal value described by the respective end node.

Algorithmic 'ant' deposits pheromones on its path when it travels from source node to the destination node. The shortest path gets the maximum pheromones and hence converges on the optimal solution i.e. the shortest path.

The time complexity for 'n' intersections comes out to be O(n).

3.2 Genetic Algorithm (GA)

A genetic algorithm (GA) is a search heuristic that mimics the process of natural selection. It simulates the survival of the fittest among individuals over consecutive generations for solving a problem.

After an initial population is randomly generated, the algorithm evolves through three operators:

Algorithm 2. Genetic Algorithm pseudo-code
Procedure GA
    1. **selection** which equates to survival of the fittest;
    2. **crossover** which represents mating between individuals;
    3. **mutation** which introduces random modifications.

        randomly initialize population(t)
        determine fitness of population(t)
        repeat
            select parents from population(t)
            perform crossover on parents creating population(t+1)
            perform mutation of population(t+1)
            determine fitness of population(t+1)
        until best individual is good enough
End-procedure

The first step of GA is to identify the genes which will form chromosomes (feasible solutions). In this problem, the genes are considered as the set of signal values of all intersections in the network for the given phase (also see Fig. 4).

Figure 4 shows two examples of chromosomes for a network with three intersections. For a four-way intersection the alleles (signal values) belong to the set {0, 1, 2, 3} whereas for a three-way intersection the set will be {0, 1, 2}.

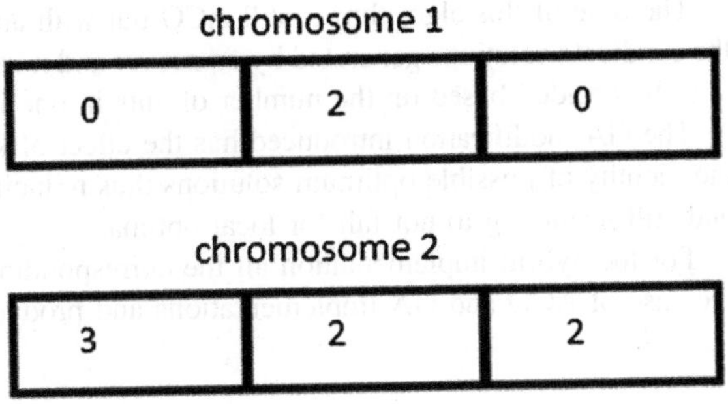

Fig. 4 Example
chromosomes with alleles for
GA implementation

chromosome 1

0	2	0

chromosome 2

3	2	2

The GA will start with an initial population of such chromosomes, called 1st generation and find the best solution so far. There after the next generation's population will be derived from the crossover and mutation of the better solutions so far and some extra chromosomes. The whole process will repeat for a designated number of generations and the best solution so far at the end will be the required optimal solution.

The over-all time complexity for this algorithm also comes out to be O(n) where 'n' is the number of intersections in the network.

3.3 Hybrid Algorithm

The underlying notion of this algorithm is to have the features of both ACO and GA to improve over their individual performances.

Algorithm 3. Hybrid algorithm
Procedure Hybrid
       Initialize GA parameters
       Initialize ACO parameters
       Repeat
              Start new generation of ants
                    ConstructAntsSolutions
                    UpdatePheromones
            End
            Apply crossover and mutation over top n solutions to generate p solutions
            Update pheromones on all p solution paths
       Until termination condition
End-procedure

The base of this algorithm is still ACO but with an extra pheromone update on the candidate solutions generated by crossover and mutation of top 'n' solutions. 'n' is to be decided based on the number of ants in one iteration.

The GA modification introduced has the effect of keeping the search of ants in the vicinity of possible optimum solutions thus reducing the effective search space and still managing to not fall for local optima.

For the hybrid implementation all the corresponding parameters are same as in the case of ACO and GA implementations and produce similar time complexity.

4 Experimental Setup

Five test cases were taken corresponding to five different networks with following specifications: 12-node network—4 intersections; 16-node network—6 intersections; 20-node network—8 intersections; 24-node network—10 intersections; 28-node network—12 intersections.

All three algorithms were run for 1,000 equivalent iterations. All the algorithms are run for limited iterations instead of going for their natural termination as per some heuristics because of the real-time nature of the problem at hand. The need is to implement the solutions in constant time. So, a reasonable number was decided based on the criterion that all three algorithms should give the optimum solution for at least the fewer nodes test-cases in reasonable number of successive runs.

Intersections were implemented as a special case of nodes (Origin-Destination nodes) with extra features like signal values, positional information of the links attached, wait time, etc.

Also there were some assumptions made to simplify the implementation maintaining the integrity of the problem. Like, all intersections were assumed to be four-way intersections, meaning the network chosen had only four-way intersections. Also, every Origin-Destination pair in the chosen networks is connected through at least one intersection. Layer 2 statistics were generated randomly to simulate its function as an input to layer 1.

The resulting programs were run on a system with Intel Pentium B980, 2.4 GHz processor; 2 GB RAM and Windows 7, in Eclipse Indigo IDE using Java JDK 1.6.

5 Results and Discussions

Results have been compared and it has been found that the proposed hybrid implementation performs better than the individual ACO or GA for the synchronous discrete distance time model.

In Fig. 5, comparison of best solutions of all three implementations over the five test case networks is shown. The ACO and hybrid implementations give almost

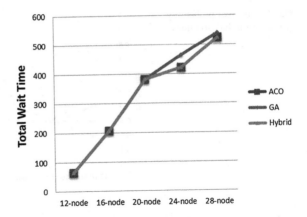

Fig. 5 Best solution comparison

exact results while the GA implementation is slightly worse off for the test case networks with higher number of nodes and intersections.

In Fig. 6, the comparison for worst solutions is shown. Here the hybrid algorithm does slightly better than both ACO and GA. Also, both the figure show near-linear relation between Total-Wait-Time and network size, validating the viability of the model for bigger networks.

Figure 7 shows the comparison of empirical probability for all the algorithms of falling in the trap of bad solutions over multiple runs, thus giving a sub-optimal solution. Here the difference between the Hybrid algorithm and others is clearly noticeable.

As the practical implementation of the problem is supposed to be in a real-time environment, it is necessary to achieve even near optimal solution in constant time; hence the limitation of iterations. So, hybrid algorithm gives better solutions than others without falling into the traps of sub-optimal solutions as often as others.

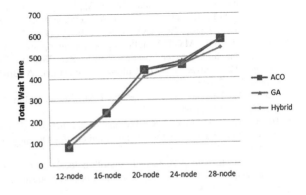

Fig. 6 Worst solution comparison

Fig. 7 Empirical probability of trapping in bad solutions

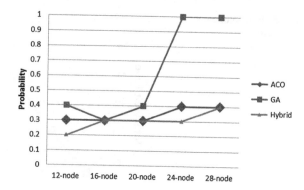

6 Conclusion and Future Work

As per the results in the previous section, it can be concluded that the Discrete Distance-Time model, when applied synchronously to the traffic signal timing optimization problem, is capable of producing good results in reasonable time for networks of various sizes. Also, the hybrid algorithm produces better results than stand-alone ACO and GA implementations.

The successful implementation of layer 2 and its integration with the Layer 1 for practical purpose can be future scope of this work.

References

1. Abdulaal, M., LeBlanc, L.J.: Continuous equilibrium network design models. Transporting Res. **13B**(1), 19–32 (1979)
2. Allsop, R.E., Charlesworth, J.A.: Traffic in a signal-controlled road network: an example of different signal timings including different routings. Traffic Eng. Control **18**(5), 262–264 (1977)
3. Baskan, O., Haldenbilen, S., Ceylan, H., Ceylan, H.: A new solution algorithm for improving performance of ant colony optimization. Appl. Math. Comput. **211**(1), 75–84 (2009)
4. Cantarella, G.E., Improta, G., Sforza, A.: Iterative procedure for equilibrium network traffic signal setting. Transp. Res. **25A**(5), 241–249 (1991)
5. Ceylan, H., Bell, M.G.H.: Traffic signal timing optimization based on genetic algorithm approach, including drivers' routing. Transp. Res. **38B**(4), 329–342 (2004)
6. Dorigo, M., Stutzle, T.: Ant Colony Optimization. A Bradford Book, The MIT Press, Cambridge (2004)
7. Dorigo, M., Di Caro, G., Gambardella, L.M.: Ant algorithms for discrete optimization. Artif. Life **5**, 137–172 (1999)
8. Heydecker, B.G., Khoo, T.K.: The equilibrium network design problem. In: Proceedings of AIRO'90 Conference on Models and Methods for Decision Support, pp 587–602, Sorrento (1990)
9. Lee, C., Machemehl, R.B.: Genetic algorithm, local and iterative searches for combining traffic assignment and signal control, traffic and transportation studies: In: Proceedings of ICTTS 98, pp 489–497 (1998)

10. Putha, R., Quadrifoglio, L., Zechman, E.: Comparing ant colony optimization and genetic algorithm approaches for solving traffic signal coordination under oversaturation conditions. Comput. Aided Civ. Infrastruct. Eng. **27**, 14–28 (2012)
11. Sahana, S.K., Jain, A.: An improved modular hybrid ant colony approach for solving traveling salesman problem. Int. J. Comput. (JoC) **1**(2), 123–127. doi:10.5176-2010-2283_1.249, ISSN: 2010-2283 (2011)
12. Sahana, S.K., Jain, A, Mahanti, P.K.: Ant colony optimization for train scheduling: an analysis. Int. J. Intell. Syst. Appl. **6**(2), 29–36. doi:10.5815/ijisa.2014.02.04, ISSN-2074-904X (print), 2074-9058 (online) (2014)

Hybrid Gravitational Search and Particle Swarm Based Fuzzy MLP for Medical Data Classification

Tirtharaj Dash, Sanjib Kumar Nayak and H.S. Behera

Abstract In this work, a hybrid training algorithm for fuzzy MLP, called Fuzzy MLP-GSPSO, has been proposed by combining two meta-heuristics: gravitational search (GS) and particle swarm optimization (PSO). The result model has been applied for classification of medical data. Five medical datasets from UCI machine learning repository are used as benchmark datasets for evaluating the performance of the proposed 'Fuzzy MLP-GSPSO' model. The experimental results show that Fuzzy MLP-GSPSO model outperforms Fuzzy MLP-GS and Fuzzy MLP-PSO for all the five datasets in terms of classification accuracy, and therefore can reduce overheads in medical diagnosis.

Keywords Fuzzy multilayer perceptron · Gravitational search · Particle swarm optimization · Breast cancer · Heart disease · Hepatitis · Liver disorder · Lung cancer · Classification · Medical data

1 Introduction

In recent years, the incorporation of soft computing approaches in medical diagnosis has achieved a new tendency to be employed successfully in a large number of medical applications. Many of the medical diagnosis procedures can be grouped

T. Dash (✉)
School of Computer Science, National Institute of Science and Technology,
Berhampur 761008, Odisha, India
e-mail: trd@nist.edu

S.K. Nayak · H.S. Behera
Department of Computer Science Engineering and Information Technology,
Veer Surendra Sai University of Technology, Burla 768018, Odisha, India
e-mail: vssut.sanjib@gmail.com

H.S. Behera
e-mail: hsbehera_india@yahoo.com

© Springer India 2015
L.C. Jain et al. (eds.), *Computational Intelligence in Data Mining - Volume 1*,
Smart Innovation, Systems and Technologies 31, DOI 10.1007/978-81-322-2205-7_4

into intelligent classification tasks [1]. These classification procedures can be (i) binary classification, where data is separated between only two classes, and (ii) multi-class classification, where data is separated among more than two classes. For example, classifying a diabetic patient is a binary classification task i.e. patient may be suffering from diabetes mellitus or diabetes insipidus. Similarly, detection of lung cancer is a type of multi-class classification based problem.

In medical data classification, methods with better classification accuracy will provide more sufficient information to identify the potential patients and to improve the diagnosis accuracy [1]. Medical database classification is a kind of complex optimization problem whose goal is not only to find an optimal solution but also to provide accurate diagnosis for diseases. And therefore, meta-heuristic algorithms such as genetic algorithm, particle swarm optimization etc., soft computing and machine learning tools such as neural networks, decision tree, and fuzzy set theory have been successfully applied in this area and have achieved significant results [2]. Artificial neural network (ANN) based approach aims to provide a filter that distinguishes the cases which do not have disease, therefore reducing the cost of medication and overheads of doctors. Back propagation neural network (BPNN) was used by Floyd et al. and Wu et al. for classification of medical data and this work achieved an overall accuracy of 50 % [3, 4]. In another study, rule extraction from ANN has been employed for prediction of breast cancer from Wisconsin dataset [5, 6]. All the above methods used back propagation learning for training the ANN, where solution got trapped in the local minima. Fogel et al. [7] attempted to solve the medical database classification problem using evolutionary computation and could achieve higher prediction accuracy than the above techniques. However, this work suffered from higher computational cost in application. Therefore, researchers tried to use an integrated fuzzy rule based approach to solve the above problem to extract fuzzy rules directly from database [8–10]. Gadaras and Mikhailov [11] presented a novel fuzzy classification framework for medical data classification. Data mining techniques such as ontology based intelligent systems, discriminant analysis, and least square SVM have been applied in medical database classification [13]. The above techniques failed for unlabelled or mislabeled databases and the accuracies in current methods are still low and insignificant enough to be adopted in medical practice.

In this research, our contribution is to develop a hybrid model combining evolutionary computation, fuzzy logic and neural network to maximize the classification accuracy and decision taking speed. In the proposed hybrid model, we combined Gravitational Search (GS) technique with Particle Swarm Optimization (PSO) to train Fuzzy Multilayer perceptron (Fuzzy MLP) for medical data classification [14–16]. The proposed algorithm has been tested for classification of five medical datasets obtained from UCI repository [17] and the results so obtained have been compared with GS based Fuzzy MLP and PSO based Fuzzy MLP.

This paper is organized as follows. Proposed methodology is given in Sect. 2. Section 3 presents experimental results and discussion. Section 4 provides the conclusion and future directions of researches.

2 Proposed Methodology

This section describes the Fuzzy MLP architecture along with the proposed hybrid training algorithm for medical data classification.

2.1 Fuzzy MLP Architecture

The Fuzzy MLP is a three layered feed forward neural network with architecture [n:m:1], where non-linear elements, called neurons, are arranged in layers. Learning is supervised for Fuzzy MLP where the computed mean squared error (MSE) at the output layer acts as the supervisor for updating knowledgebase. The proposed neural network is trained with three different evolutionary algorithms, (i) Gravitational Search (GS), (ii) Particle Swarm Optimization (PSO), and (iii) Hybrid of GS and PSO (GSPSO).

Data fuzzification: The first phase of the method is fuzzification of the input data which can be done by use of a fuzzy membership function (MF). In fuzzy set theory, there exists no perfect rule for selecting a fuzzy MF. Researches always consider different MF for different problems. This work will adopt spline based or S-shaped function as primary MF. This MF puts the input dataset in a range $[a, b]$. The equation for S-shaped MF has been given in Eq. 1.

$$f(x, a, b) = \begin{cases} 0 & x < a \\ 2\left(\frac{x-a}{b-a}\right)^2 & a \leq x \leq \frac{a+b}{2} \\ 1 - 2\left(\frac{x-b}{b-a}\right)^2 & \frac{a+b}{2} \leq x \leq b \\ 1 & x > b \end{cases} \tag{1}$$

2.2 Proposed GSPSO Algorithm

In GSPSO algorithm, following major steps are followed. All agents are initialized in two dimensional search spaces. According to law of gravity, a gravitational force acts on each particle due to mutual attraction among them. Hence, in second step, the force, the acceleration and the position of the particle will be calculated. The above steps are repeated until the termination criteria are satisfied.

The following equations are used for various calculations during the above steps. The gravitational constant G is calculated using Eq. 2. In the following equation, α is a small constant, G_o is the initial gravitational constant, '*iterator*' is the current iteration, '*maxiteration*' is maximum iteration to be performed.

$$G(t) = G_o \exp\left(-\alpha \frac{Iteration}{MaxIteration}\right) \tag{2}$$

Fitness of the candidate solution in this work is calculated by considering the mean squared error (MSE) as given in Eq. 3, where N is the total number of training instances in the medical dataset, '*Target*' is the target class for the current instance and '*Output*' is the computed output class for the current instance. Mass of a particle is calculated by using Eq. 4, where Fit_i the fitness is value of agent 'i', in the current epoch evaluation and '*worst*' is defined as given in Eq. 4 below.

$$Fitness = MSE = \frac{1}{N}\sum_{i=1}^{N}(Target(i) - output(i))^2 \tag{3}$$

$$M_i = \frac{Fit_i - worst}{\sum_{j=1}^{n}(Fit_i - worst)} \tag{4}$$

GSPSO Pseudo-code for training Fuzzy MLP:

Fuzzify *the input medical dataset*
Initialize *Fuzzy MLP parameters based on input data: n, m, weights, biases*
Initialize *GSPSO parameters: maxiteration, G, particle initial position, inertia weights, c_1, c_2, search space dimension, F, M, X, a*
Set iterator=0
While *iterator ≤ Maxiteration*
   iterator=iterator+1
   Update G using Equation-2
   For *each agent*
      Initialize *weights (W) and biases (b) of Fuzzy MLP*
   End for
   Calculate fitness
   *If(obtained fitness is better than **gbest**)*
      *set **gbest** to obtained fitness*
   End if
   Update updated F, M, X, a
End while

The force acting on object i from object j is defined as given in Eq. 5. Similarly, summing up all the individual forces acting on particle i, we will get the resultant force on the particle i as given in Eq. 6. In the following equations, R_{ij} is defined as the Euclidean distance between two objects i and j. The parameters r_1, r_2 and θ are arbitrary random values in the range $[0, 1]$. Therefore, resultant force acting on particle i and the acceleration of this particle is calculated using Eqs. 7 and 8 respectively.

$$F_{ij} = G_o \left(\frac{M_i M_j}{R_{ij} + \theta} \right) (x_j - x_i) \tag{5}$$

$$F_i = \sum_{j=1, j \neq i}^{n} r_j F_{ij} \tag{6}$$

$$F_i = G_o M_i \sum_{j=1, j \neq i}^{n} \frac{M_j}{R_{ij} + \theta} (x_j - x_i) \tag{7}$$

$$a_i = \frac{F_i}{M_i} \tag{8}$$

Velocity and position of particle i are calculated using Eqs. 9 and 10 respectively, where $V_i(t)$ is the velocity of agent i at iteration t, c_1' and c_2' are acceleration coefficients, $gbest$ is the best fitness so far, $X_i(t)$ is the position of particle in iteration t, a_i is the acceleration of particle i, θ_1 and θ_2 are random numbers in the range $[0, 1]$.

$$V_i(t+1) = w V_i(t) + c_1' \theta_1 a_i c_i(t) + c_2' \theta_2 (gbest - X_i(t)) \tag{9}$$

$$X_i(t+1) = X_i(t) + V_i(t+1) \tag{10}$$

3 Results and Discussion

All the simulations are carried out in MATLAB R2010a which is installed in a PC having Windows 7 OS and 2 GB main memory. The processor is Intel dual core and each processor has an equal computation speed of 2 GHz (approx.)

3.1 Dataset Description

The proposed model has been tested with five medical datasets obtained from UCI machine learning repository. Table 1 shows a summary of these datasets. It should be noted that number of attributes in Table 1 is the number of input attributes plus one class attribute. Further details and properties of all the mentioned five datasets can be obtained from UCI repository.

Table 1 UCI medical dataset properties

Dataset	No. of instances	No. of attributes	No. of classes
WBC	699	11	2
Heart disease	270	14	2
Hepatitis	155	20	2
ILPD	583	10	2
Lung cancer	32	57	3

3.2 Simulation Parameters

Number of hidden units in the Fuzzy MLP is set to *2n*, where n is the total number of input units. Number of population is 30 for all the cases. Inertia weight (w) is set to 2; w_{max} and w_{min} set to 0.9 and 0.5 respectively. The coefficients, c_1 and c_2 are set to 2 each; gravitational constant G_o is set to 1. All the simulations are allowed to run for 50 epochs. For all the three models, GS based Fuzzy MLP (Fuzzy MLP-GS), PSO based Fuzzy MLP (Fuzzy MLP-PSO) and proposed GSPSO based Fuzzy MLP (Fuzzy MLP-GSPSO), MSE is noted against each epoch during training. Testing results are classification accuracy, simulation time. Tables 2, 3 and 4 show the simulation results for all the five tested datasets. All the three models are compared for the above four parameters. It should be noted that each simulation is run for ten times to avoid any biasness towards results.

Figure 1a–e given in Appendix give a comparison of all the three models based on convergence of error during training. Table 2 shows mean MSE obtained by all the three models for various datasets. It can be seen that Fuzzy MLP-GSPSO outperforms the other two models for all the datasets except lung cancer dataset. However, the classification accuracy obtained by the proposed model for the lung cancer dataset is higher than that obtained by the GS and PSO models (shown in Table 3). The simulation time is also a crucial parameter for comparison which is presented in Table 4. Proposed GSPSO model achieved the best results i.e. 82 % accuracy for the WBC dataset and 81 % accuracy for ILPD datasets. However, it could achieve only 67 % accuracy when tested for hepatitis dataset.

Table 2 MSE (mean ± std. deviation) for five tested medical datasets

Dataset	Fuzzy MLP-GS	Fuzzy MLP-PSO	Fuzzy MLP-GSPSO
WBC	0.1785 ± 0.0098	0.1690 ± 0.0155	0.1461 ± 0.0123
Heart disease	0.1826 ± 0.0118	0.1819 ± 0.0121	0.1331 ± 0.0124
Hepatitis	0.2520 ± 0.0248	0.2442 ± 0.0258	0.2212 ± 0.0032
ILPD	0.2047 ± 0.0305	0.2043 ± 0.0372	0.1164 ± 0.0560
Lung cancer	0.1332 ± 0.0735	0.1150 ± 0.0386	0.1333 ± 0.0012

Table 3 Classification accuracy (mean ± std. deviation) for five tested medical datasets

Dataset	Fuzzy MLP-GS	Fuzzy MLP-PSO	Fuzzy MLP-GSPSO
WBC	72.6960 ± 5.621	75.5700 ± 4.597	81.6953 ± 2.001
Heart disease	76.2987 ± 0.004	79.2963 ± 0.020	76.9697 ± 1.350
Hepatitis	46.2885 ± 4.665	52.4186 ± 0.320	66.3700 ± 2.740
ILPD	67.6150 ± 3.721	67.5800 ± 0.001	80.6600 ± 6.330
Lung cancer	47.6525 ± 8.976	50.0000 ± 0.000	71.8750 ± 0.000

Table 4 Simulation time (training time + testing time) for five tested medical datasets

Dataset	Fuzzy MLP-GS	Fuzzy MLP-PSO	Fuzzy MLP-GSPSO
WBC	115.94 + 0.027	38.990 + 0.026	114.48 + 0.026
Heart disease	56.530 + 0.015	17.266 + 0.013	57.750 + 0.011
Hepatitis	44.700 + 0.008	12.880 + 0.008	45.300 + 0.008
ILPD	100.60 + 0.022	33.610 + 0.022	102.50 + 0.024
Lung cancer	43.790 + 0.007	10.720 + 0.006	43.750 + 0.006

It can also be seen that the Fuzzy MLP-PSO has better training speed than that of Fuzzy MLP-GS and Fuzzy MLP-GSPSO. However, as training is carried out before adopting the model physically, therefore, it is of least importance. The real decision speed is dependent on the testing time, which is the time taken by the models to classify the input instance. Table 4 also shows that the proposed GSPSO model could achieve the best result within 0.5 s of time, which is also an advantage of adopting the proposed model in medical practices.

4 Conclusion and Future Works

In this paper, a new hybrid of gravitational search and particle swarm called GSPSO has been proposed for classification medical data using Fuzzy MLP. Five medical datasets are used to evaluate the performance of the proposed model. The results are compared with GS and PSO based Fuzzy MLP models. For all the datasets, the GSPSO model shows better performance in terms of error convergence, classification accuracy and decision speed. To ensure applicability of this model, further comparison of Fuzzy MLP-GSPSO with other evolutionary models is a part of future works.

Appendix

See Fig. 1a–e.

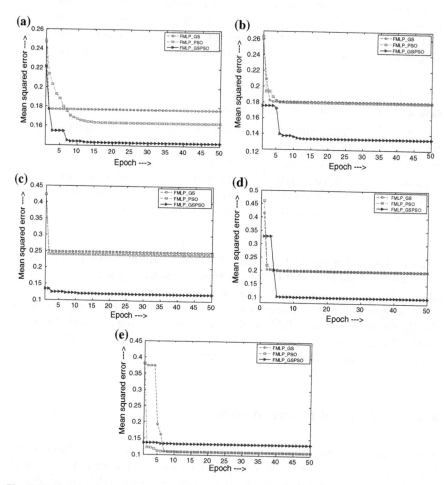

Fig. 1 Evolution curve; **a** WBC; **b** Heart disease; **c** Hepatitis; **d** ILPD; **e** Lung cancer dataset

References

1. Fan, C.-Y., Chang, P.-C., Lin, J.-J., Hsieh, J.C.: A hybrid model combining case-based reasoning and fuzzy decision tree for medical data classification. Appl. Soft Comput. **11**, 632–644 (2011)
2. Bojarczuk, C.-C., Lopes, H.-S., Freitas, A.-A.: Genetic programming for knowledge discovery in chest-pain diagnosis. IEEE Eng. Med. Biol. Mag. **19**(4), 38–44 (2000)
3. Floyd, C.E., Lo, J.Y., Yun, A.J., Sullivan, D.C., Kornguth, P.J.: Prediction of breast cancer malignancy using an artificial neural network. Cancer **74**, 2944–2998 (1994)

4. Wu, Y.-Z., Giger, M.-L., Doi, K., Vyborny, C.J., Schmidt, R.A., Metz, C.E.: Artificial neural networks in mammography: application and decision making in the diagnosis of breast cancer. Radiology **187**, 81–87 (1993)
5. Setiono, R., Huan, L.: Understanding neural networks via rule extraction. In: Proceedings of the International Joint Conference on Artificial Intelligence, pp. 480–487, Morgan Kauffman, San Mateo, CA (1995)
6. Setiono, R.: Generating concise and accurate classification rules for breast cancer diagnosis. Artif. Intell. Med. **18**, 205–219 (2000)
7. Fogel, D.B., Wasson, E.C., Boughton, E.M.: Evolving neural networks for detecting breast cancer. Cancer Lett. **96**(1), 49–53 (1995)
8. Pulkkinen, P., Koivisto, H.: Identification of interpretable and accurate fuzzy classifiers and function estimators with hybrid methods. Appl. Soft Comput. **7**, 520–533 (2007)
9. Chang, P.C., Liao, T.W.: Combining SOM and fuzzy rule base for flow time prediction in semiconductor manufacturing factory. Appl. Soft Comput. **6**(2), 198–206 (2006)
10. Song, X.-N., Zheng, Y.-J., Wud, X.-J., Yang, X.-B., Yang, J.-Y.: A complete fuzzy discriminant analysis approach for face recognition. Appl. Soft Comput. **10**, 208–214 (2010)
11. Gadaras, I., Mikhailov, L.: An interpretable fuzzy rule-based classification methodology for medical diagnosis. Artif. Intell. Med. **47**(1), 25–41 (2009)
12. Lee, C.S., Wang, M.H.: Ontology-based intelligent healthcare agent and its application to respiratory waveform recognition. Expert Syst. Appl. **33**(3), 606–619 (2007)
13. Polat, K., Gunes, S., Arslan, A.: A cascade learning system for classification of diabetes disease: generalized discriminant analysis and least square support vector machine. Expert Syst. Appl. **34**(1), 482–487 (2008)
14. Rashedi, E., Nezamabadi-pour, H., Saryazdi, S.: GSA: a gravitational search algorithm. Inf. Sci. **179**, 2232–2248 (2009)
15. Eberhart, R., Kennedym, J.: A new optimization using particle swarm theory. In: Sixth International Symposium on Micro Machine and Human Science, MHS'95, pp. 39–43, IEEE (1995)
16. Dash, T., Behera, H.S.: Fuzzy MLP approach for non-linear pattern classification. In: International Conference on Communication and Computing (ICC-2014), Bangalore, India. Computer Networks and Security, pp. 314–323, Elsevier Publications (2014)
17. Bache, K., Lichman, M.: UCI machine learning repository. University of California, School of Information and Computer Science, Irvine, CA, http://archive.ics.uci.edu/ml (2013)

HOG Based Radial Basis Function Network for Brain MR Image Classification

N.K.S. Behera, M.K. Sahoo and H.S. Behera

Abstract Fully automated computer-aided diagnosis system is very much helpful for early detection and diagnosing of brain abnormalities like cancers and tumors. This paper presents two hybrid intelligent techniques such as HOG+PCA+RBFN and HOG+PCA+k-NN, which consists of four stages namely skull stripping, feature extraction, dimension reduction and classification. For efficient feature extraction Histograms of Oriented Gradients (HOG) method is used to extract the required feature vector and then the proposed techniques are used to classify images as normal or abnormal. The results show that the proposed technique gives an accuracy of 100 %, sensitivity of 99 % and specificity 100 %.

Keywords Principal component analysis · Histograms oriented gradients · Magnetic resonance imaging · Radial basis function network · Skull stripping

1 Introduction

Magnetic Resonance Imaging (MRI) is excellent imaging technique which provides rich information about anatomical structures of different body parts like brain and breast of human [1]. Use of artificial intelligence in computer aided diagnosing system is now pervasive across a wide range of medical research area, such as brain tumors, heart diseases, cancers, gastroenterology and other clinical studies [2]. Brain MR image classification include the classification of various diseases, such as

N.K.S. Behera (✉) · M.K. Sahoo · H.S. Behera
Department of Computer Science Engineering and Information Technology,
Veer Surendra Sai University of Technology, Burla 768018, Odisha, India
e-mail: nksbehera@gmail.com

M.K. Sahoo
e-mail: mk186.sahoo@gmail.com

H.S. Behera
e-mail: mailtohsbehera@gmail.com

© Springer India 2015
L.C. Jain et al. (eds.), *Computational Intelligence in Data Mining - Volume 1*,
Smart Innovation, Systems and Technologies 31, DOI 10.1007/978-81-322-2205-7_5

glioma overlay, huntingtons chorea, meningioma, metastic adenocarcinoma and metastic bronchogenic carcinoma.

Recent works shows supervised machine learning algorithms such as k-nearest neighbors (k-NN) [2], support vector machine (SVM) [3], feed forward neural networks (FF-ANN) and unsupervised techniques such as self-organization map (SOM) [4], fuzzy c-means (FCM) [5] are used for classification of human brain MR images. In this study another supervised technique, such as radial basis function network (RBFN) to classify the T2-weighted brain MR images into normal or particular pathological cases.

Initially the input T2-weighted brain MR images are preprocessed and then segmented to remove the skull around the brain tissues. Then the skull stripped image used for feature extraction by HOG feature extraction algorithm [6] and then PCA method is applied to reduce the feature matrix. Proposed classifiers are trained using these reduced features and then classify the testing data into proper classes.

The contribution of this paper is the integration of a robust feature extraction algorithm, a proper segmentation technique, an efficient classifier and one popular feature reduction technique to perform a more accurate and robust MR image classification. Then the performance and accuracy rate is compared with the k-NN based classification methods by considering different statistical measures. Then the proposed method HOG+PCA+RBFN and HOG+PCA+k-NN has higher classification accuracy then DWT+PCA+k-NN [2].

The structure of this paper is organized as follows. A details description of our proposed method is presented in Sect. 2. An experiment in Sect. 3 briefs the structure and training approach of the radial basis function network (RBFN) for classification. Section 4 demonstrates the rapidness and effectiveness of the proposed algorithm and Sect. 5 concludes this paper.

2 Methodology

The proposed hybrid techniques are based on the following techniques: skull stripping, HOG feature extraction, dimension reduction, and RBFN classifier. This method consists of four stages: (1) skull stripping, (2) HOG feature extraction, (3) dimension reduction, and (4) supervised classification (Fig. 1).

2.1 Skull Stripping

Skull stripping (Fig. 2) is a process of brain image segmentation, where extra cerebral tissues such as skull, eyeball and skins are removed from the brain tissues [7]. A hybrid skull stripping approach based on the combination of watershed algorithm and deformable surface models is followed to remove the skulls from the brain images. The details procedure for skull stripping of the image is as follows [8].

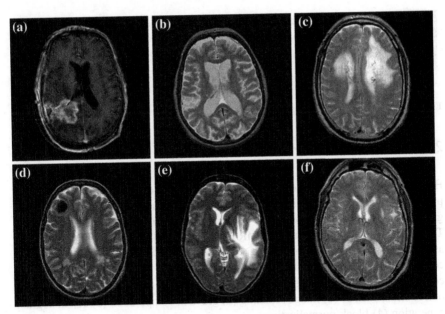

Fig. 1 **a** Glioma overlay. **b** Huntingtons Chorea. **c** Meningioma. **d** Metastic adenocarcinoma. **e** Metastic bronchogenic carcinoma. **f** Normal images

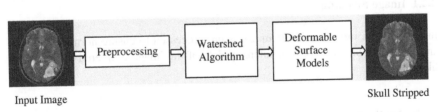

Input Image   Skull Stripped

Fig. 2 Skull stripping approach

First the input brain MR T2-weighted image I is converted form gray scale to rgb I_{rgb} and then certain morphological operations are carried out. Then watershed algorithm is applied to segment the image into different connected components by calculating the image intensity gradients [8, 9]. Deformable surface algorithm is applied to estimate local brain parameters to locate true brain boundary by taking segmented brain volumes as input. Then in the skull stripped image the skull, eye sockets are removed and are taken as input for feature extraction.

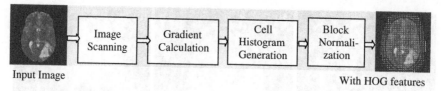

Input Image

With HOG features

Fig. 3 HOG feature extraction technique

2.2 HOG Feature Extraction

HOG feature extraction algorithm is an object detection method used in image processing and computer vision to generate features from the histograms of the gradient orientations of the localized portions of the input brain MR image [6, 10]. This method divides the input brain MR image into small connected regions known as localized cells. Histograms of the edge orientations or gradient directions are compiled for each pixel within the localized cells and then these histograms are combined to represent the descriptor [11]. The HOG descriptor (Fig. 3) is consists of five steps, such as (1) image scanning (2) gradient computation (3) cell histogram generation (4) block normalization.

2.2.1 Image Scanning

In this step the input brain MR image is pre-processed to ensure normalized color and gamma value and color or intensity data of the image is filtered using kernel filters [6].

2.2.2 Gradient Computation

The gradients of the feature vector are calculated using $n \times n$ Sobel mask. Suppose I_s is the skull stripped brain MR image. S_x and S_y are two images which contain the horizontal and vertical derivative approximation as mentioned in Eq. 1.

$$S_x = \begin{bmatrix} -1 & 0 & +1 \\ -1 & 0 & +1 \\ -1 & 0 & +1 \end{bmatrix} * I_s \ and \ S_y = \begin{bmatrix} +1 & +1 & +1 \\ 0 & 0 & 0 \\ -1 & -1 & -1 \end{bmatrix} * I_s \tag{1}$$

The resulting the gradient magnitude is the combination of gradient approximations as given in Eq. 2.

$$S = \sqrt{(S_x^2 + S_y^2)} \ and \ \theta = q\tan 2(S_x + S_y) \tag{2}$$

where S and θ are the magnitude and directions of the gradient approximations respectively.

2.2.3 Cell Histogram Generation

In this step the cell histograms are generated by selecting an orientation based histogram channel based on the magnitude S and direction θ of the gradient computation. Then these histograms are locally normalized and are combined to spatially connected blocks [11]. Geometry of each block can be rectangular (R-HOG) or circular (C-HOG). Then the components of normalized cell histograms from all region blocks are combined into a vector I_{hog} known as HOG descriptor [6, 10].

2.2.4 Block Normalization

Suppose I_{hog} is the non-normalized vector containing all the histograms in a given block and $|U|_k^n$ is k-form for k = 1, 2,..., n and ε is a small constant. Then the normalization factor L_n-factor is used to find the normalized feature vector I_{nh}.

$$L_n\text{-}factor = \frac{I_{hog}}{\sqrt{(|U|_k^n + \varepsilon)}}. \tag{3}$$

2.3 Dimension Reduction

Principal component analysis (PCA) is a popular subspace projection technique, which took a large set of feature space as input and transforms it into a lower dimensional feature space. PCA uses the largest eigenvectors of the correlation matrix to reduce the dimensionality of the large data sets [12]. This feature reduction technique takes the normalized HOG features I_{nh} as input and then uses orthogonal transformation to convert into a set of uncorrelated variables called principal components. In the transformation the first principal component has the largest possible variance than others [12, 13]. The normalized HOG features I_{nh} having size n × p, where n rows represents repetitions of experiment and p represents a particular kind of datum. $L_{(k)} = (L_1, L_2,..., L_p)_{(k)}$ is a p-dimensional loading vector that map each row vector $I_{nh}(i)$ of I_{nh} to a net principal component scores $X_{(i)} = (x_1, x_2,..., x_p)$ given in Eq. 4.

$$X_{(i)} = I_{nh}(i) \times L_{(k)} \tag{4}$$

where I_{nh} each loading vector L must be a unit vector and individual variables of T is considered over successively inherits the maximum possible variance from I_{nh}. The first loading vector $L_{(1)}$ has to satisfy [13],

$$L_{(k)} = \frac{\arg\max}{\|L\|=1}\left\{\sum(X)^2_{(i)}\right\} = \frac{\arg\max}{\|L\|=1}\left\{\sum(I_{nh}(i) \times L)^2\right\} \tag{5}$$

Equivalently.

$$L_{(k)} = \frac{\arg\max}{\|L\|=1}\left\{\sum(I_{nh}(i) \times L)^2\right\} = \frac{\arg\max}{\|L\|=1}\left\{\|I_{nh} \times L\|^2\right\}$$
$$= \frac{\arg\max}{\|L\|=1}\left\{L^T I_{nh} L I_{nh}\right\} = \frac{\arg\max}{\|L\|=1}\left\{\sum\frac{L^T I_{nh}^T L I_{nh}}{L^T L}\right\} \tag{6}$$

The successive kth component can be calculated by the subtracting the first $k-1$ principal component from I_{nh} [13].

$$I_{nh} = I_{nh} - \sum_{s=1}^{k-1} I_{nh} L_{(s)} L_{(s)}^T \tag{7}$$

After faithful transformation the $X = I_{nh} \times L$, which maps a feature matrix I_{nh} from an original space of $n \times p$ variables to a new space of $n \times r$.

2.4 RBFN Classifier

Radial basis function network (RBFN) comprises one of the most used feed forward neural network model trained using supervised learning technique and uses radial basis function as the transfer function. RBFN (Fig. 4) typically consists of three layers, an input layer, a hidden layer and an output layer [14].

The hidden layer neurons contain a non-linear radial basis activation function whose outputs are inversely proportional to the distance from the center of the neuron [15]. The final output of the RBFN is a linear combination of the neuron

Fig. 4 RBFN with input features

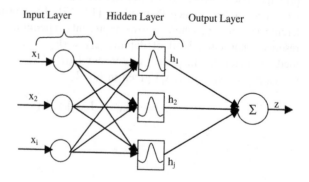

parameters and radial basis function of the inputs. The reduced HOG feature vector $X = (x_1, x_2, ..., x_i)$ is taken as input of the RBFN and fed into each i-th input node then it is forwarded to hidden layer, where it is passes through the j-th neurons having radial basis function as the activation function.

$$h_j = f_j(x) = \exp[-\|x - c_j\|^2/(2\sigma^2)] \tag{8}$$

where c_i is the centre vector for the radial basis function, h_i is the outputs from the hidden layer and $\|x - c\|^2$ is the square of the distance between the input feature vector x.

$$Z = \frac{1}{m}\sum_{n=1}^{m} u_n \times h_n \tag{9}$$

where z is the output of the RBFN, which specifies the class to which the image is belonging.

3 Experiments and Discussions

The proposed hybrid techniques have been implemented on a real human brain MRI dataset achieved from the Harvard Medical School website. The input dataset contains 8 normal and 40 abnormal brain MR images and each image is axial T2-weighted, 256 × 256 × 3 pixels used for both testing and training. Figure 1 presents some samples from the input dataset for normal and abnormal brain: a—glioma overlay, b—huntingtons chorea, c—meningioma, d—metastic adenocarcinoma, e—metastic bronchogenic carcinoma, f—normal brain.

All computations of HOG+PCA+RBFN and HOG+PCA+k-NN classification were performed on a personal computer with 2.00 GHz Intel Pentium-Dual core processor and 1 GB memory, running under Windows-8.1 operating system. Above classification techniques are implemented and trained in MATLAB-8.1 using a combination of Neural Network Toolbox, Image Processing Toolbox, Statistical Toolbox and Wavelet Toolbox. Figure 5 shows that the input human brain MR image data is segmented to remove non-brain tissues such as skull, skins and eyeball sockets from brain tissues known as skull stripping. Watershed algorithm and deformable surface models are used for the skull stripping purpose after pre-processing of the input images. Then HOG algorithm is applied to extract features from the input images and these features are reduced into necessary features using PCA method. One of the classifier from RBFN and k-NN is selected and then reduced feature is used as the input of the classifier. After that the classifier is trained and tested using the training and testing dataset respectively.

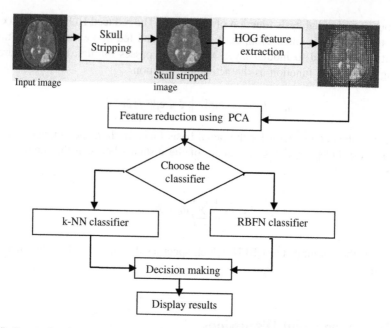

Fig. 5 Proposed techniques HOG+PCA+RBFN and HOG+PCA+k-NN

4 Results Analysis

In this section, several statistical parameters such as confusion matrix, correct classification rate (CCR), specificity, sensitivity, prevalence, positive predictive power (PPP), negative predictive power (NPP), likelihood ratio positive (LRP), likelihood ratio negative (LRN), false positive ratio (FPR), and false negative ratio (FNR) are used to evaluate the performance of the proposed techniques. These terms are defined as follows [16].

The following confusion matrix represents the details about the actual and predicted cases for HOG+PCA+RBFN and HOG+PCA+k-NN. Table 1 represents the confusion matrix for HOG+PCA+RBFN for both training and testing images.

Correct classification rate (CCR) is defined as the ratio between the numbers of cases which are correctly classified to the total number of cases of the testing data as given in Eq. 10.

$$CCR = \frac{(T_P + T_n)}{(T_p + F_p + T_n + F_n)} \tag{10}$$

Sensitivity is defined as the probability that the diagnostic test is positive, which means the person has the disease as given in Eq. 11.

Table 1 Confusion matrix for HOG+PCA+RBFN

	Training			Testing	
	Actual+	Actual−		Actual+	Actual−
Predicted+	40	0	Predicted+	39	1
Predicted−	0	8	Predicted−	0	8

$$Sensitivity = \frac{T_p}{(T_p + T_n)} \tag{11}$$

Specificity is defined as the probability that the diagnostic test is negative, which means that the person does not have the disease as given in Eq. 12.

$$Specificity = \frac{T_n}{(T_n + F_p)} \tag{12}$$

Positive predictive power (PPP) is defined as the probability of positive prediction as given in Eq. 13.

$$PPP = \frac{T_p}{(T_p + F_p)} \tag{13}$$

Negative predictive power (NPP) is defined as the probability of negative prediction as given in Eq. 14.

$$NPR = \frac{T_n}{(T_n + F_n)} \tag{14}$$

Prevalence is defined as the percentage of persons with a given diseases over a period of time as given in Eq. 15.

$$Prevalence = \frac{(T_p + F_p)}{(T_P + T_n + F_P + F_n)} \tag{15}$$

False positive ratio (FPR) is also known as the expectancy and is defined as the probability of falsely rejecting the null hypothesis for particular test cases as given in Eq. 16.

$$FPR = \frac{F_p}{(T_p + F_n)} \tag{16}$$

False negative ratio is defined as the ratio of persons affected by a disease whose results wrongly show that they are disease free given in Eq. 17.

Table 2 Comparison of results between different methods such as HOG+PCA+RBFN, HOG+PCA+k-NN and DWT+PCA+k-NN

Accuracy measures	HOG+PCA+RBFN		HOG+PCA+k-NN		DWT+PCA+k-NN	
	Training (%)	Testing (%)	Training (%)	Testing (%)	Training (%)	Testing (%)
CCR	100	100	97.91	97.91	95.83	95.83
Sensitivity	83.33	83.33	81.25	81.25	79.16	79.16
Specificity	100	100	88.88	88.88	80.00	80.00
Prevalence	83.33	83.33	81.25	81.25	83.33	83.33
LRP	84.17	84.17	82.07	82.07	100	100
LRN	82.33	82.33	87.08	87.08	97.70	97.70
PPP	100	100	97.66	97.66	95.00	95.00
NPP	100	100	100	100	100	100
FPR	0	0	2.63	2.63	5.26	5.26
FNR	0	0	0	0	0	0

$$FNR = \frac{F_n}{(T_p + F_n)} \qquad (17)$$

Likelihood ratio positive (LRP) is defined as the ratio of the probability of a person who has the disease testing positive to the probability of a person who does not have the disease testing positive as given in Eq. 18.

$$LRP = \left| \frac{Sensitivity}{(1 - Specificity)} \right| \qquad (18)$$

Likelihood ratio negative (LRN) is defined as the ratio of the probability of a person who has the disease testing negative to the probability of a person who has the disease testing positive as given in Eq. 19.

$$LRN = \left| \frac{(1 - sensitivity)}{Specificity} \right| \qquad (19)$$

where T_p, T_n, F_p, and F_n represents the true positive, true negative, false positive and false negative respectively. Table 2 shows the comparison of performance between HOG+PCA+RBFN and HOG+PCA+k-NN in terms of several statistical parameters.

Table 2 shows the comparison between the HOG+PCA+k-NN and HOG+PCA+RBFN with DWT+PCA+k-NN method [2]. For this dataset the proposed method HOG+PCA+RBFN is more accurate and efficient than other methods such as HOG+PCA+RBFN and DWT+PCA+k-NN [2].

5 Conclusion

In this study, proposed hybrid techniques are employed on T2-weighted brain MR image dataset for classification of normal/abnormal images. According to the experimental results, HOG+PCA+RBFN produced 100 % classification accuracy rate and 83.33 % sensitivity, while HOG+PCA+k-NN classifier produce 97.91 % classification accuracy rate and 81.25 % sensitivity. The stated results show that proposed HOG+PCA+RBFN is more accurate and robust classifier and applicable for real world bio-medical classification of brain MRI images. Limitations of this technique are that fresh training is required whenever new image is added into the image database. Application of this technique for classification of tumors and lesions of CT scan image is the topic of future research.

References

1. Zhang, Y., Dong, Z., Wu, L., Wang, S.: A hybrid method for brain image classification. Expert Syst. Appl. **38**(8), 10049–10053 (2011)
2. El-Dahshan, E.S.A., Hosny, T., Salem, A.B.M.: Hybrid intelligent techniques for MRI brain images classification. Digital Signal Process. **20**(2), 433–441 (2010)
3. Chaplota, S., Patnaika, L.M., Jagannathanb, N.R.: Classification of magnetic resonance brain images using wavelets as input to support vector machine and neural network, Biomed. Signal Process. Control **1**(1) 86–92 (2006)
4. Jiang, J., Trundle, P., Ren, J.: Medical image analysis with artificial neural networks. Comput. Med. Imaging Graph. **34**(8), 617–631 (2010)
5. Kannan, S.R., Ramathilagam, S., Devi, R., Hines, E.: Strong fuzzy c-means in medical image data analysis. J. Syst. Softw. **85**, 2425–2438 (2012)
6. Dalal, N., Triggs, B.: Histograms of oriented gradients f or human detection, CVPR-2005. In: IEEE Computer Society Conference on Computer Vision and Pattern Recognition, vol. 1, pp. 886–893 (2005)
7. Segonne, F., Dale, A.M., Busa, E., Glessner, M., Salat, D., Hahn, H.K., Fischl, B.: A hybrid approach to the skull stripping problem in MRI. NeuroImage **22**, 1060–1075 (2004)
8. Hahn, H.K., Peitgen, H.O.: The skull stripping problem in MRI solved by a single 3D watershed transform. In: Proceedings of MICCAI, LNCS, vol. 1935, pp. 134–143. Springer, Berlin (2000)
9. Galdames, F.J., Jaillet, F., Perez, C.A.: An accurate skull stripping method based on simplex meshes and histogram analysis for magnetic resonance images. J. Neurosci. Methods **206**, 103–119 (2012)
10. Lowe, G.D.: Distinctive image features from scale-invariant key points. Int. J. Comput. Vis. **60**(2), 91–110 (2004)
11. Kobayashi, T.: BoF meets HOG: feature extraction based on histograms of oriented pdf gradients for image classification. In: 2013 IEEE Conference on Computer Vision and Pattern Recognition, pp. 747–754 (2013)
12. Warmuth, M.K., Kuzmin, D.: Randomized online PCA algorithms with regret bounds that are logarithmic in the dimension. J. Mach. Learn. Res. **9**, 2287–2320 (2008)
13. Abdi, H., Williams, L.J.: Principal component analysis. Wiley Interdisc. Rev.Comput. Stat. **2**, 433–459 (2010)
14. Park, J., Sandberg, I.W.: Universal approximation using radial-basis-function networks. Neural Comput. **3**(2), 246–257 (1991)

15. Chen, S., Cowan, C.F.N., Grant, P.M.: Orthogonal least squares learning algorithm for radial basis function networks. IEEE Trans. Neural Netw. **2**, 2 (1991)
16. Lalkhen, A.G., McCluskey, A.: Clinical tests: sensitivity and specificity. Oxf. J. Med. CEACCP **8**(6), 221–223 (2004)
17. Sathyaa, P.D., Kayalvizhi, R.: Optimal segmentation of brain MRI based on adaptive bacterial foraging algorithm. Neurocomputing **74**(14), 2299–2313 (2011)
18. Kuruvilla, J., Gunavathi, K.: Lung cancer classification using neural networks for CT images. Comput. Methods Programs Biomed. **113**(1) 202–209 (2014)
19. Jones, R.D., Lee, Y.C., Barnes, C.W., Flake, G.W., Lee, K., Lewis, P.S., Qian, S.: Function approximation and time series prediction with neural networks. In: Proceedings of the International Joint Conference on Neural Networks, pp. 17–21 (1990)

Hybrid CRO Based FLANN for Financial Credit Risk Forecasting

S.R. Sahu, D.P. Kanungo and H.S. Behera

Abstract Modern financial market has become capable enough to provide its services to a large number of customers simultaneously. On the other hand, the exponential hike in financial crises per year has uplifted the demand for precise and potential classifier models. In this work a hybrid model of clustering and neural network based classifier has been proposed, i.e. FCM-FLANN-CRO. Three financial credit risk data sets were applied to the processing and the model is evaluated using the performance metrics such as RMSE and Accuracy. The experimental result shows the proposed model outperforms its MLP counterpart and other two non-hybrid models. The proposed model provides its best result with 97.05 % of classification accuracy.

Keywords Classification · Functional link artificial neural network (FLANN) · Multilayered perceptron (MLP) · Credit risk forecasting · Chemical reaction optimization (CRO) · Clustering · Fuzzy C-means (FCM)

1 Introduction

Financial risks are unfavourable outcomes of any financial assets which includes credit risk, business risk, operational risk, etc. [1]. During the last decade, financial risk had shown a very eye-catching behaviour. For example, the high yield bond issues have an increasing graph from 1990 to 2013. The number of bond issues filed in the years 1990 and 2013 are 19 and 618 respectively, and the Principal amount

S.R. Sahu (✉) · D.P. Kanungo · H.S. Behera
Department of Computer Science Engineering and Information Technology,
Veer Surendra Sai University of Technology, Burla 768018, Odisha, India
e-mail: mailsomuhere@gmail.com

D.P. Kanungo
e-mail: dpk.vssut@gmail.com

H.S. Behera
e-mail: hsbehera_india@gmail.com

© Springer India 2015
L.C. Jain et al. (eds.), *Computational Intelligence in Data Mining - Volume 1*,
Smart Innovation, Systems and Technologies 31, DOI 10.1007/978-81-322-2205-7_6

involved are $1,060 and $335,200 (In millions) respectively [2]. With high level data storage techniques the financial institutions are now able to provide their services to numerous customers. A great deal of decision making is involved in classifying the customer portfolios to predict the possibility of failure of potential counterparty [3]. Classifier models play a vital role in the decision making of the credit admission evaluation process. Thus, these fields have been very important and very widely studied [4]. Traditionally the credit risk assessments had to rely on personal judgments, which bring the problem of biasing from the previous experience. On the other hand, classification algorithms are used to provide quicker and less erroneous outcome which requires comparatively very low human resource with no personal biasing [5].

Various classification models have been proposed to incorporate the large demand of classifier models, such as statistical models [6], nonparametric statistical models [7], artificial intelligence methods [8], etc. In recent years, attempts were made to cross-fertilize engineering and life science methodology to innovate advanced inter-operable systems. These techniques are basically advanced technology whose core concepts are based on natural/biological/physical phenomenon's [9]. Some of the most popular ones are particle swarm optimization [10], ant colony optimization [11], and chemical reaction optimization [12], etc. In a similar attempt to address the financial classification problem, Marinaki et al. [9] implemented novel honey bee optimization technique for the feature selection of credit risk assessment issue. An important step towards obtaining classifier models with a high accuracy rate is the appropriate selection of independent variables, i.e., feature selection. Recently, Marinakis et al. [13] implemented ACO and PSO for the purpose of feature selection of the credit risk assessment problem.

In another investigation Tsai [3] hybridized the performance of supervised and unsupervised classifier to obtain better accuracy. And the result shows that the ensemble of self organizing map (SOM) and multilayered perceptron (MLP) provides the best results [3]. This paper focuses on implementing a hybrid of fuzzy c-means (FCM) clustering and FLANN classifier for the prediction of credit risk assessment problem. The rest of the paper is divided into different sections as follows: Sect. 2 describes the proposed methodology which includes a detailed description of FCM, FLANN, and CRO. Section 3 addresses the experimental setup, experimental result and discussion. Section 4 provides the conclusion and the future work.

2 Proposed Methodology

2.1 Fuzzy C-Means Clustering

FCM is one of the most widely studied fuzzy clustering techniques, which was proposed by Dunn [14] in 1973 and eventually modified by Bezdek [15] in 1981. And the salient feature of FCM is that a data point can be the member of all clusters with corresponding membership values. Let us suppose that M-dimensional N data points represented by $x_i (i = 1, 2 . . . N)$, are to be clustered. Assume the number of clusters to be made, that is, C, where $2 \leq C \leq N$. The level of fuzziness (f) should be chosen appropriately such that $f > 1$. Then the membership function matrix U is initialized having a size $N \times C \times M$ in a random manner such that $U_{ijm} \epsilon [0, 1]$ and $\sum_{j=1}^{C} U_{ijm} = 1.0$, for each i and a fixed value of m. Then the cluster centre CC_{jm} is determined for jth cluster and its mth dimension using the following equation:

$$CC_{jm} = \frac{\sum_{i=1}^{N} U_{ijm}^{f} x_{im}}{\sum_{i=1}^{N} U_{ijm}^{f}} \tag{1}$$

After that the Euclidian distance between ith data point and jth cluster centre with respect to mth dimension is calculated using the following expression:

$$D_{ijm} = \left\| x_{im} - CC_{jm} \right\| \tag{2}$$

Then the membership function matrix is updated using the equation:

$$U_{ijm} = \frac{1}{\sum_{c=1}^{C} \left(\frac{D_{ijm}}{D_{icm}} \right)^{\frac{2}{f-1}}} \tag{3}$$

If the distance becomes zero then the data point will coincide with the cluster centre and the membership function value will be 1.0. This procedure is repeated until stopping criterion is encountered. In this work a hybridization of FCM clustering and classification using FLANN model is proposed. Figure 1 provides the block diagram of the proposed model.

2.2 Functional Link Artificial Neural Network (FLANN)

FLANN was proposed by Pao et al. [12] in the year 1989. It is a single layer and flat artificial neural network. Let X be the dataset matrix and it has K number of data patterns. Let a kth pattern is given to the functional expansion block of the FLANN model as shown in Fig. 2, then for each element in the pattern, the block will

Fig. 1 Block diagram of proposed model

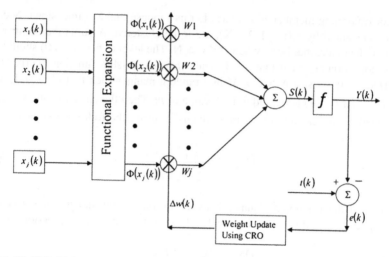

Fig. 2 FLANN-CRO model

expand the term into 2N + 1 trigonometric term. Out of which 2N terms are trigonometric terms and the last one is the element itself. Let there are J number of features present in the pattern, then for each pattern is expanded into 'm' = J* (2*N + 1) elements. Now this m expanded terms are multiplied with the corresponding weights. Then these m weighted elements are summed up to get a summation term 'S'. Now the activation function is applied on the 'S' to get the output for the ith pattern, i.e. $Y(i)$.

Once output is obtained it is compared with the corresponding target to find out the error e(i). Once after all the patterns processed by the model, then if the convergence condition is not met then we upgrade the weight and the bias using different learning algorithm such as Chemical Reaction Optimization (CRO).

2.3 Chemical Reaction Optimization

Chemical reaction optimization (CRO) is recently proposed in 2010 by Lam and Li [17]. In CRO, it is assumed that the reaction is taking place in a container. A population consisting of random numbers of unstable molecules is the reactants of the chemical reaction. The unstable molecule, i.e. vector of Weights and bias of FLANN, are bombarded with each other and with the surrounding, i.e. with the wall, thereby under-going reaction. Let W represent the weight matrix of the FLANN-CRO model, then the characteristics of the four elementary reactions are given in the Table 1.

The pseudo code of the CRO algorithm is given below, where the 'P_size' is the size of the population 'P', 'best_mole' represents the molecule having the highest P. E., 'max_gen' represents the maximum number of generations allowed. Halting Criteria: in this work the elementary reactions are executed 'max_gen' is reached.

Table 1 Characteristics features of elementary reactions of CRO

Elementary reaction	Equation for the reaction
Unimolecular reaction (involves one molecule)	
Decompose	$W \rightarrow W1' + W2'$ (4)
On wall ineffective collision	$W \rightarrow W'$ (5)
Intermolecular reaction (involves more than one molecule)	
Inter molecular ineffective collision	$W1 + W2 = W1' + W2'$ (6)
Synthesis	$W1 + W2 = W'$ (7)

Pseudo code for proposed CRO Algorithm

```
Load the dataset in D.
Initialize the 'p' randomly.
Evaluate the population, i.e. find the P.E. or fitness value of each molecule.
While Halting criteria is not satisfied do
    For I=1 to P_size do
        Generate two random number c1 and c2 in the range [0, 1].
            If c1>0.5 then          //Uni-molecular collision
                W=best_mole
                If c2>0.5 then
                    [W1', W2']=Decomposition (W)
                    Evaluate W, W1', W2' and assign the best one to P(I).
                Else                    //On-wall ineffective
                    W'=Onwall_in_col(W)
                    Evaluate W and W' and assign the best one to P(I).
                End if
            Else            // Inter- molecular collision
                If P(I) is better than P(I+1) then
                    W1=P(I)
                Else
                    W1=P(I+1)
                End if
                W2=best_mole
                If c2>0.5 then                  //Synthesis
                    W'=Synthesis(W1, W2)
                    Evaluate W1, W2 and W' and assign the best two to P(I)and P(I+1).
                Else            //Inter-molecular ineffective
                    [W1', W2']=Inter_mol_col(W1, W2)
                    Evaluate W1, W2, W1' and W2'and assign the best two to
                    P(I)and P(I+1).
                End if
            End if
            I=I+1
    End for
End While
```

3 Experimental Analysis

3.1 Experimental Setup

All the experiments were carried out using MATLAB 2013a on an Intel(R) Pentium (R) DUAL CPU system with 2.0 GHz processor. The total experiment consists of steps such as data collection, data pre-processing, clustering, classification with 10 fold cross validation. In this section the above are described in detail.

3.1.1 Data Collection and Pre-processing

In this work three well known credit risking data sets namely German Credit Dataset, Australian Credit Approval Dataset and Qualitative Bankruptcy Data. These data sets were collected from the UCI Machine Learning Repository. The German credit dataset has 1,000 instances and has 20 independent attributes and one class attribute. The Australian credit approval data has 690 instances and 15 attributes including the class attribute. The qualitative bankruptcy dataset has 175 instances and has 6 independent attribute and a class attribute. Initially the data sets are normalized in the range 0–1 so as to avoid any discrepancy due to the rawness of the data. The following min-max mapping equation is used to normalize the data. Let X is a dataset then it can be normalized as follows:

$$Normalized\ X_i = \frac{X_i - X_{min}}{X_{max} - X_{min}} \tag{8}$$

3.1.2 Processing of Proposed Model

The normalized data sets are given to the FCM module for clustering. After clustering the output clustered data is fed into the Cross validation module where the clustered data undergoes 10-fold cross validation process. In this process the data are randomly divided into 10 equal sub-sets. Then out of 10 sub-sets one subset is extracted and kept for the testing of the module. The rest 9 subsets consist the training set and fed to the model FLANN-CRO model for classification. After completion of the training the updated weights of the FLANN-CRO model are frozen and the testing data is feed to the model for testing with the frozen weight. The performance matrices are calculated and stored. This processes is repeated for 10 times so that each time one subset will act as the testing set and the remaining 9 set will be combined together to provide the training set. Moreover, the result obtained, is collected and their average gives the final result of the model. The performance metrics used in the work are described in the next section.

3.1.3 Performance Metrics

In this work four performance matrices [3] are used to for the model evaluation. Their details are described in Table 2.

Where N represents the number of data patterns present in the data set, Y_i represents the output of the ith pattern using proposed model and T_i stands for the target of the ith pattern. TP stands for true positive and it is the number of correctly classified positive instance or abnormal instance. TN is the number of correctly classified normal instances. And FP and FN are the number of abnormal and normal instances which are misclassified as normal and abnormal instances respectively.

Table 2 Performance metrics

Performance matrices	Features	Equations
Root mean square error	Performance measure during training phase	$RMSE = \sqrt{\frac{1}{N}\sum_{i=1}^{N}(Y_i - T_i)}$ (9)
Classification accuracy	The most widely used performance metrics for testing	$Acc = \dfrac{TP + TN}{TP + TN + FP + FN}$ (10)
Sensitivity	Measures the capability in recognizing abnormal record	$Sensitivity = \dfrac{TP}{TP + FN}$ (11)
Specificity	Measures the capacity of recognizing normal record	$Specificity = \dfrac{TN}{TN + FP}$ (12)

3.2 Result and Discussion

In this work a hybridized model i.e., FCM-FLANN-CRO model is used to classify three credit risk data. And to evaluation and analysis of the result three additional models are implemented. Those are MLP-CRO, FLANN-CRO and FCM-MLP-CRO. Table 3 shows the experimental results of three data sets, which are processed by FLANN-CRO and MLP-CRO model. And from the analysis of the results, it is evident that FLANN-CRO model provides better results than the MLP-CRO model. FLANN-CRO model provides the best result among the two models for Qualitative Bankruptcy data, where accuracy is 0.9657 and RMSE is 0.1215, Specificity has a value of 0.9409 and Sensitivity is 0.9571.

Table 4 shows the classification results of the proposed model i.e., FCM-FLANN-CRO and its MLP counterpart. The result analysis of the proposed hybrid model shows that the hybrid model outperforms the single classifier model in most of the cases. Moreover, FCM-FLANN-CRO model outperform all other models with respect to Overall accuracy. From the analysis, it is also evident that FLANN model outperforms the MLP model in most of the cases.

Table 3 Testing results of FLANN-CRO and MLP-CRO model

Data sets	RMSE	Classification accuracy	Sensitivity	Specificity
FLANN-CRO				
German credit data	0.2090	0.9240	0.9100	0.7900
Australian credit data	0.2295	0.9388	0.9784	0.9837
Qualitative bankruptcy	**0.1215**	**0.9657**	**0.9371**	**0.9409**
MLP-CRO				
German credit data	0.2493	0.7310	0.8271	0.7967
Australian credit data	0.2405	0.8774	0.8124	0.8549
Qualitative bankruptcy	0.1740	0.8935	0.8143	0.8827

Table 4 Testing Result of FCM-FLANN-CRO and FCM-MLP-CRO model

Data sets	RMSE	Classification accuracy	Sensitivity	Specificity
FCM-FLANN-CRO				
German credit data	0.1936	0.9505	0.9205	0.9265
Australian credit data	0.2373	0.9446	0.8667	0.9170
Qualitative bankruptcy	**0.1114**	**0.9706**	**0.9405**	**0.9556**
FCM-MLP-CRO				
German credit data	0.2166	0.8263	0.8579	0.9048
Australian credit data	0.2355	0.9314	0.7475	0.7924
Qualitative bankruptcy	0.0689	0.9228	0.9857	0.9900

4 Conclusion and Future Work

The demand of highly potential classifier has an exponentially increasing graph. To fulfil the demand numerous classifier have been designed, which are competing with each other in order to attain better accuracy in classification. In this work a hybrid model of clustering and classification has been proposed. Two financial credit risk and one bankruptcy data sets were used for the classification purpose. The performance results of FCM-FLANN-CRO model was compared with that of its MLP counterpart and two non-hybrid models, those are MLP-CRO and FLANN-CRO. The experimental result shows that the proposed model outperforms all other models in most of the cases. But as this field is a very challenging field, various optimization techniques can be applied to further enhancement of the performance of the classifier.

References

1. Doumpos, M., Zopounidis, C., Pardalos, P.M.: Multicriteria sorting methodology: application to financial decision problems. Parallel Algorithms Appl. **15**(1–2), 113–129 (2000)
2. Public Company Bankruptcy Filing Report, New Generation Research, Inc. Available from http://www.bankruptcydata.com/default.asp
3. Tsai, C.F.: Combining cluster analysis with classifier ensembles to predict financial distress. Inf. Fusion **16**, 46–58 (2014)
4. Shin, K.-S., Lee, T.S., Kim, H.J.: An application of support vector machines in bankruptcy prediction model. Expert Syst. Appl. **28**, 127–135 (2005)
5. Peng, Y., Wang, G., Kou, G., Shi, Y.: An empirical study of classification algorithm evaluation for financial risk prediction. Appl. Soft Comput. **11**, 2906–2915 (2011)
6. Altman, E.I., Avery, R.B., Eisenbeis, R.A., Sinkey Jr, J.F.: Application of classification techniques in business: banking and finance. JAI Press Inc., CT (1981)
7. Chatterjee, S., Barcun, S.: A nonparametric approach to credit screening. J. Am. Stat. Assoc. **65**, 150–154 (1970)
8. Baesens, B., Setiono, R., Mues, C., Vanthienen, J.: Using neural network rule extraction and decision tables for credit-risk evaluation. Manage. Sci. **49**, 312–329 (2003)

9. Marinaki, M., Marinakis, Y., Zopounidis, C.: Honey bees mating optimization algorithm for financial classification problems. Appl. Soft Comput. **10**, 806–812 (2010)
10. Kennedy, J., Eberhart, R.: Particle swarm optimization. In: Proceedings of 1995 IEEE International Conference on Neural Networks, vol. 4, pp. 1942–1948 (1995)
11. Dorigo, M., Stutzle, T.: Ant colony optimization. A Bradford Book, The MIT Press, Cambridge (2004)
12. Lam, Y.S., Li, V.O.K.: Chemical-reaction-inspired metaheuristic for optimization. IEEE Trans. Evol. Comput. **14**(3), 381–399 (2010)
13. Marinakis, Y., Marinaki, M., Doumpos, M., Zopounidis, C.: Ant colony and particle swarm optimization for financial classification problems. Expert Syst. Appl. **36**, 10604–10611 (2009)
14. Dunn, J.C.: A fuzzy relative of the ISODATA process and its use in detecting compact well-separated clusters. J. Cybernet. **3**, 32–57 (1973)
15. Bezdek, J.C.: Pattern recognition with fuzzy objective function algorithms. Kluwer Academic Publishers, Norwell (1981)
16. Pao, Y.H.: Adaptive pattern recognition and neural networks. Addison-Wesley, Reading (1989)
17. Hand, D.J., Till, R.J.: A simple generalization of the area under the ROC curve to multiple class classification problems. Mach. Learn. **45**(2), 171–186 (2001)

Improved Mean Variance Mapping Optimization for the Travelling Salesman Problem

Subham Sahoo and István Erlich

Abstract This paper presents an improved Mean Variance Mapping Optimization to address and solve the NP-hard combinatorial problem, the travelling salesman problem. MVMO, conceived and developed by István Erlich is a recent addition to the large set of heuristic optimization algorithms with a strategic novel feature of mapping function used for mutation on basis of the mean and variance of the population set initialized. Also, a new crossover scheme has been proposed which is a collective of two crossover techniques to produce fitter offsprings. The mutation technique adopted is only used if it converges towards more economic traversal. Also, the change in control parameters of the algorithm doesn't affect the result thus making it a fine algorithm for combinatorial as well as continuous problems as is evident from the experimental results and the comparisons with other algorithms which has been tested against the set of benchmarks from the TSPLIB library.

Keywords Mean variance mapping optimization · Travelling salesman problem · Combinatorial optimization

1 Introduction

Optimization techniques aim for a better solution, whether single objective or multi-objective. In multi-objective problems, it generally involves a compromise between two solutions of all the objective functions. It is quite unconventional that simple

S. Sahoo (✉)
Department of Electrical and Electronics Engineering,
Veer Surendra Sai University of Technology,
Burla 768018, India
e-mail: subhamsahoo50@gmail.com

I. Erlich
Department of Electrical Engineering and Information Technologies,
University of Duisburg, Essen, Germany
e-mail: istvan.erlich@uni_due.de

© Springer India 2015
L.C. Jain et al. (eds.), *Computational Intelligence in Data Mining - Volume 1*,
Smart Innovation, Systems and Technologies 31, DOI 10.1007/978-81-322-2205-7_7

problems are difficult to solve as the combinations increase heavily with the size of the concerned problem; as well as computationally broadening. The objective of traveling salesman is to find out the shortest distance for a salesman, where he intends to visit a set of cities exactly once, and then returning back to his destination. The search space can have $(n - 1)!/2$ possible sequences as per the n-city symmetric TSP. Out of the possible sequences, it also contains an optimal solution. The search gets easier for small value of n, but as the value of n increases, the search gets arduous.

TSP is one of the most studied problems in combinatorial optimization. It has got various applications such as computer wiring, scheduling hence solving TSP has made an impact for practical importance as well as active research [1]. Even complex problems can be solved with ease, and thus enabling real world optimization problems. Every algorithm has got its own convergence characteristics to find out the optimal solution in the search space which decides the effectiveness and reliability of the algorithm. To find out the shortest distance travelled between a numbers of cities or the cost covered during the tour, many optimization techniques have been applied in the past, such as the Bee colony Optimization (BCO) [2], Cuckoo search Algorithm (CSA) [3], Firefly algorithm [4], Monkey Search Algorithm [5], Ant Colony Optimization (ACO) [6] Discrete Cuckoo Search Algorithm [7]. Many factors such as the ease of implementation and producing improved solutions with increasing iterations decide the ability to solve combinatorial problems. However, the success rate as well as compatibility of algorithms differs for every problem.

Heuristic optimization algorithms have always found a knack in handling real-world complex problems. Effort has been put on various variants, which includes novel schemes for information exchange, population size management. But handling large scale problems with a multiple local optima is still an issue which requires the need of new techniques with computational efficiency and easy adaptability. MVMO is an algorithm which uses a special mapping function for mutation operation on the basis of the mean and variance of the best solutions obtained and saved in the solution archive, works best in search dynamics. The basic implementation of MVMO has been practiced for single particle approach thus making it faster and reduced risk of premature convergence. However, practicing MVMO with multi particle strategy further helped in search capability which has been thoroughly tested in the paper.

In this paper, a novel technique subjecting to the determination of minimum distance/cost for the travelling salesman problem which uses MVMO algorithm with modified crossover features which has been tested on a set of benchmarks from the TSPLIB library.

2 Improved MVMO Algorithm

Mean-Variance mapping optimization technique has been discovered in the recent past [8], and has been used for many real world optimization problems. It starts with a normal initialization of all its variables for a set of total number of particles (NP). After initialization, the variables are normalized since the range of the search space is restricted to [0, 1]. Following the normalization, mutation operation via mapping operation making sure the offspring will never violate the search boundaries. Before fitness evaluation, the variables are again de-normalized. Remaining steps can be followed from the flowchart in Fig. 1.

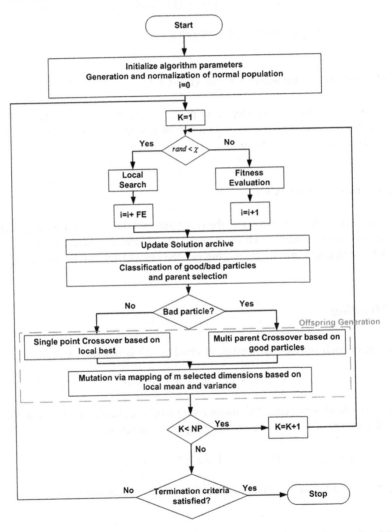

Fig. 1 Hybrid MVMO overall procedure. *K* particle counter, *i* fitness evaluation number, *NP* total number of particles, *FE* number of fitness evaluations, *rand* random number in [0, 1]

2.1 Crossover

The parent child vector set is created by combining the inherited elements from parent vector and similar dimensional set (via selection) that undergo mutation through mapping based on mean and variances calculated from the solution archive. The two crossover techniques can be used alternatively to produce fitter offsprings everytime the objective function is tested. The decision is based on these sets of continuously varying quantities: (i) tau_{scc}, tau_{cc}; (ii) scc_{rand}, cc_{rand}. Another component of the path, which can be updated as per the formula:

$$tau_x = tau_x + Q \times f_x(x) \tag{1}$$

where f_x represents the fitness of the offspring.

2.1.1 Sequential Constructive Crossover (SCC)

This crossover technique was adopted to provide increased rate of convergence. It considers better edges between the parents as the judging measure than the shortest distance between two cities to form the offsprings [9].

2.1.2 Cyclic Crossover (CC)

This crossover operator works under the constraint that the created offsprings works for each city name coming from the parent [10]. Hence, one of the fitter probable path is to be updated into tau_{cc} carrying out the circular crossover operation.

2.2 Mutation

The value of each selected dimension of child vector is determined by

$$x_t = h_x + (1 - h_l + h_o) \cdot x_r^* - h_o \tag{2}$$

where x_r^* is a randomly generated number with uniform distribution between [0, 1], where the values of h represent the mapping transformation.

$$h(x_{mean}, s_1, s_2, x) = x_{mean}(1 - e^{-x \cdot s_1}) + (1 - x_{mean})e^{-(1-x)s_2} \tag{3}$$

where h_x, h_l and h_o are the mapping function outputs accountable for

$$h_x = h(x = x_r^*), \quad h_o = h(x = 0), \quad h_l = h(x = 1) \tag{4}$$

Hence, x_r is always within [0, 1]. s_r is the shape factor with v_r as variance, f_s as the shaping factor.

$$s_r = -\ln(v_r) \cdot f_s \tag{5}$$

3 Experimental Results

The improved MVMO was tested on several benchmark functions from the electronic library TSPLIB. All the instances in the library has been previously optimized by other algorithms and hence has been used for reliable comparison for solving TSP in this paper. The programs were performed on a Dell XPS L502X personal computer equipped with Intel® Core™ i5-2430 M, 2.40 GHz and 4 GB RAM, under Windows 7 Home Premium, 64 bit OS. The implementation of all programs was done in MATLAB® Version 2009b.

3.1 Random Search

To test the reliability of the proposed method, two methods have been adopted—one for random matrix and another for benchmark functions of TSPLIB library. So a random matrix consisting of n number of cities in the TSP program has been used to show the comparison between the convergence characteristics of the improved MVMO and MVMO for a fixed number of iteration. For random search, large data set has not been taken into consideration.

3.2 TSPLIB Library Functions

The benchmark functions were tested against various algorithms and the results shown in Tables 1 and 2. The values shown in the table are a result of 50 independent runs of each algorithm, many performance metrics has been added to show the deviation from the optimal values taken from TSPLIB library.

$$\text{Percentage Deviation } (\%) = \frac{\text{Concerned length } - \text{ Optimal length}}{\text{Optimal length}} \times 100$$

Table 1 Computational results of Improved MVMO for 20 TSP benchmark functions

Instance	Optimal solution	Best	Average	Worse	PDBest (%)	PDAve (%)
kroE100	22,068	22,068	22,094.52	22,134	0.0	0.01
pr107	44,303	44,303	44,310.48	44,362	0.0	0.02
bier127	118,282	118,282	118,327.36	118,592	0.0	0.03
pr136	96,772	96,781	97,003.46	97,147	0.0	0.02
kroA150	26,524	26,524	26,547.05	27,125	0.0	0.08
kroA200	29,368	29,377	29,479.2	29,876	0.0	0.02
kroB200	29,437	29,437	29,477.24	30,042	0.0	0.02
ts225	126,643	126,643	126,658.36	126,848	0.0	0.01
a280	2,579	2,579	2,590.72	2,614	0.0	0.04
lin318	42,029	42,176	42,348.94	42,679	0.03	0.76
pr1002	259,045	264,249	272,484.38	290,087	0.02	0.05

In Fig. 2, the benchmark function "eil101" has been used to show the convergence characteristic of the comparison between both the algorithms—Improved MVMO and MVMO.

To test the efficiency of the algorithm, large data sets has also been tested and the results has been put up in Table 2. The improvement of the performance of improved MVMO over MVMO is due to the two crossover strategies put up for the TSP problem as well as tuned parameters. The significant reason improved MVMO accounts for its better convergent behavior over other metaheuristic algorithms is that it incorporates local search, and the proposed crossover scheme to explore the search diversity. A clear analysis of comparison of PDAve (%) for different benchmark functions used in Table 1 has been shown in Fig. 3.

After testing the results of some selected benchmark functions in TSPLIB library, the results of some other algorithms such as Ant Colony System (ACS) [11], S-CLPSO [12] (Set Based Particle Swarm Optimization) and GSA-ACS-PSOT [13] has been compared with improved MVMO for different instances. Table 2 summarizes the results obtained over 50 runs of all the algorithms mentioned above.

Hence, all the results speak for a good balance between exploration and exploitation in MVMO; thus making it an efficient algorithm. It is evident from Table 2 that the error percentages are lowest for improved MVMO. Also, the average values of each function are more optimized than the rest of the algorithms.

Table 2 Comparison of results of Improved MVMO with other algorithms

Instance	Optimal solution	Imp-MVMO			ACS			S-CLPSO			GSA-ACS-PSOT		
		Best	Average	SD	Best	Average	SD	Best	Average	SD	Best	Average	SD
eil76	538	538	538.08	0.0	540	543.26	0.01	538	540.68	0.005	538	541.78	0.008
st70	675	675	675	0.0	678	681.68	0.01	675	678.92	0.005	676	679.28	0.007
ch130	6,110	6,110	6,126.83	0.27	6,118	6,132.84	0.3	6,112	6,149.04	0.6	6,116	6,142.56	0.5
d198	15,870	15,872	15,892.46	0.01	15,884	15,924.84	0.03	15,896	15,945.64	0.07	15,912	15,964.22	0.1
pr299	48,386	48,403	48,433.26	0.01	48,421	48,458.04	0.02	48,405	48,437.86	0.01	48,412	48,442.34	0.016
lim318	42,029	42,176	42,348.94	0.76	42,196	42,452.78	0.84	422,184	42,410.46	0.80	42,214	42,568.92	1.24

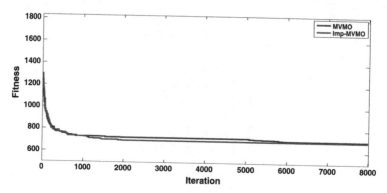

Fig. 2 Comparison of convergence characteristics of improved MVMO and MVMO-eil101

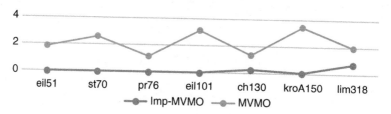

Fig. 3 PDAve (%) for 50 runs for 7 TSP functions

4 Conclusion

In this paper, the improved MVMO has been used for a combinatorial problem, the travelling salesman problem by introducing a new crossover strategy and the high convergence rate of the algorithm. It has been compared with MVMO; compared with other hybrid algorithms. The results of comparison have proved that the improved MVMO has outperformed every other algorithms used for TSP for some benchmark functions. This can be explained by the local search minima; efficient use of scaling factor in mapping function on basis of mean and variance for n particles while mutating as well as the crossover scheme used to produce fitter offsprings everytime. The general MVMO has also been applied to the particular problem statement but the results pointing out that the crossover strategy used in MVMO giving away too much deviation from the optimized value compared to the proposed optimization technique. The goal of the algorithm is to design efficient and composed ideas in the large area of emerging metaheuristic algorithms. Also, the change in the value of the control parameters doesn't alter the fitness of the functions used, which is another typical feature of MVMO.

References

1. Arora, S.: Polynomial time approximation schemes for euclidean traveling salesman and other geometric problems. J. ACM **45**(5), 753–782 (1998)
2. Teodorovic, D., Lucic, P., Markovic, G., Orco, M.D.: Bee colony optimization: principles and applications. In: 8th Seminar on neural network applications in electrical engineering, pp. 151–156. IEEE (2006)
3. Yang, X.S., Deb, S.: Cuckoo search via levy flights. In: World Congress on Nature and Biologically Inspired Computing (NaBIC 2009), pp. 210–214. IEEE (2009)
4. Yang, X.S.: Firefly algorithms for multimodal optimization. In: Stochastic Algorithms: Foundations and Application (SAGA 2009). Lecture notes in computer sciences, vol. 5792, pp. 169–178 (2009)
5. Mucherino, A., Seref, O.: Monkey search: a novel metaheuristic search for global optimization. In: Data mining, systems analysis, and optimization in biomedicine (AIP conference proceedings), vol. 953, pp. 162–173. American Institute of Physics, Melville, USA (2007)
6. Dorigo, M., Gambardella, L.M., et al.: Ant colonies for the travelling salesman problem. BioSystems **43**(2), 73–82 (1997)
7. Ouaarab, A., Ahiod, B., Yang X.S.: Discrete cuckoo search algorithm for the travelling salesman problem. Neural Comput. Appl. (2013)
8. Erlich, I, Venayagamoorthy, G.K., Nakawiro, W: A mean-variance optimization algorithm. In: Proceedings of 2010 IEEE World Congress on Computational Intelligence, pp. 1–6, Barcelona, Spain (2010)
9. Ahmed, Z.H.: Genetic algorithm for the travelling salesman problem using sequential constructive crossover. IJBB **3**(6) (2010)
10. Oliver, I., et al.: A study of permutation crossover operators on the travelling salesman problem. In: Proceedings of the Second International Conference on Genetic Algorithms, pp. 224–230 (1987)
11. Hlaing, Z., Khine, M.: Solving travelling salesman problem by using improved ant colony optimization algorithm. Int. J. Inf. Educ. Technol. **1**(5), 404–409 (2011)
12. Chen, W., et al.: A novel set-based particle swarm optimization method for discrete optimization problems. IEEE Trans. Evol. Comput. **14**(2), 278–300 (2010)
13. Chen, S.M., Chien, C.Y.: Solving the traveling salesman problem based on the genetic simulated annealing ant colony system with particle swarm optimization techniques. Expert Syst. Appl. **38**(12), 14439–14450 (2010)

Partial Segmentation and Matching Technique for Iris Recognition

Maroti Deshmukh and Munaga V.N.K. Prasad

Abstract One of the main issues in iris segmentation is that the upper and lower region of the iris is occluded by the eyelashes and eyelids. In this paper, an effort has been made to solve the above problem. The pupil boundary is detected using adaptive thresholding and circular hough transform (CHT). The iris boundary is detected by drawing arcs of different radius from the pupil center and finding maximum change in intensity. The annular region between the iris inner and outer boundary is partially segmented. The segmented iris is transformed into adaptive rectangular size strip. Features are extracted using scale-invariant feature transform (SIFT). The experimental results show that the proposed technique has achieved high accuracy and low error rate.

Keywords Iris · SIFT · EER

1 Introduction

An iris based biometric system is used to authenticate an individual with the help of iris textural patterns. A typical iris recognition system includes image acquisition, preprocessing, feature extraction and matching [1, 2]. Preprocessing consists of segmentation, normalization and enhancement. Among these fundamental modules, the iris segmentation module plays rather a critical role in the overall system

M. Deshmukh (✉) · M.V.N.K. Prasad
Institute for Development and Research in Banking Technology,
Hyderabad 500057, India
e-mail: marotideshmukh100@gmail.com

M.V.N.K. Prasad
e-mail: mvnkprasad@idrbt.ac.in

M. Deshmukh
School of Computer and Information Sciences, University of Hyderabad,
Hyderabad 500046, India

© Springer India 2015
L.C. Jain et al. (eds.), *Computational Intelligence in Data Mining - Volume 1*,
Smart Innovation, Systems and Technologies 31, DOI 10.1007/978-81-322-2205-7_8

performance because it isolates the valid part of an iris in the input eye image [3]. The iris has a very complex structure and contains many distinctive features such as ridges, rings, zigzag, collarette, crypts, furrows, corona, arching ligaments, freckles, etc. [4, 5]. Iris recognition systems have been deployed in several critical applications, such as fast border checking and recently in the UIDAI program in India. Traditional security systems rely on knowledge (e.g., passwords and the personal identification numbers) and tokens (e.g., keys and identity cards) which could be shared, lost, and/or hacked [6].

The rest of the paper is organized as follows. Section 2 introduces related work. Section 3 describes proposed method. Section 4 describes the experimental result and analysis. Finally, Sect. 5 concludes the paper.

2 Literature Survey

Nguyen et al. [7] proposed robust iris segmentation. For iris inner boundary detection a median filter is used to smooth the image as well as eliminate unexpected noise. The threshold is used for converting black and white image. A morphological scheme is applied to eliminate unwanted regions. To detect the iris outer boundary, an expanding active contour scheme is utilized. A Sobel vertical edge detector is used for the left and right side for iris edge enhancement. Ross et al. [8] proposed iris segmentation using Geodesic Active Contours. The iris segmentation scheme employing geodesic active contours to find out the iris from the surrounding structures. The active contours can assume any shape. Active contours segments multiple objects simultaneously and reduce some of the concerns associated with traditional iris segmentation models. Active contour scheme elicits the iris texture pattern in an iterative fashion and is guided by both local and global properties of the iris image. Li et al. [9] proposed to use Figueiredo and Jain's Gaussian Mixture Models (FJ-GMMs) to model the underlying probabilistic distributions of both valid and invalid regions on iris images. For detecting possible features it found that Gabor Filter Bank (GFB) provides the most unique information. Simulated Annealing (SA) technique to optimize the parameters of GFB in order to obtain the best recognition performance. Daugman [10] adopted 2D Gabor filters to demodulate the iris phase information and extract features. Boles and Boashash [11] presented the zero crossing of a one-dimensional wavelet transform to represent distinct levels of a concentric circle in an iris image, and two dissimilarity functions were used for matching the iris features obtained. Wildes et al. [12] used Laplacian pyramids to analyze the iris texture and combine the features from four different resolutions. A normalized correlation was then calculated to decide whether the input image and an enrolled image belonged to the same class.

Zhu [13] proposed an efficient algorithm of iris feature extraction based on Scale-invariant feature transform (SIFT). The direct application of the scale invariant feature transform (SIFT) method would not work well for iris recognition because it does not take advantage of the characteristics of iris patterns.

Belcher [14] proposed the region-based SIFT approach to iris recognition. This method does not require polar transformation, affine transformation or highly accurate segmentation to perform iris recognition and this new method is scale invariant. Alonso-Fernandez [15] used the Scale Invariant Feature Transform (SIFT) for recognition using iris images. To extract characteristic SIFT feature points in scale space and perform matching based on the texture information around the feature points using the SIFT operator.

3 Proposed Method

The proposed technique involves the following basic modules: pupil boundary detection, iris boundary detection, partial iris segmentation, normalization and matching. During the preprocessing phase, the input iris image is used to find the pupil and iris boundary. Most of the iris part is not visible because of upper and lower eyelids and eyelashes. Lower eyelashes are less in comparison to upper eyelashes so, we have taken more iris area in the lower side of iris image and less iris area in the upper side of iris image. The recognition area of the proposed technique is shown in Fig. 1.

3.1 Pupil Boundary Detection

An appropriate threshold helps to determine the region of interest of the pupil. A static value of the threshold may fail for different images taken under varying illumination conditions. The highest intensity value contributing to the pupil neither exceeds 0.5 times the highest grayscale value nor drops below the 0.1 times the highest grayscale value [16]. The input iris image is shown in Fig. 2a. Adaptive threshold and hole filling operation is used to detect pupil region as shown in Fig. 2b. Circular Hough transform uses radius range to detect the pupil radius and center of the pupil. The pupil boundary is detected using adaptive threshold and Circular Hough Transform (CHT) as shown in Fig. 2c.

Fig. 1 Recognition area of proposed iris recognition

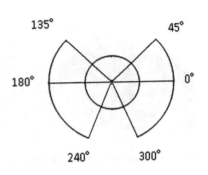

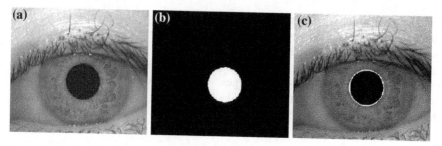

Fig. 2 Pupil boundary detection. **a** Iris image. **b** Adaptive thresholding. **c** Pupil boundary

3.2 Iris Boundary Detection

A median filter is used to remove noise from input iris image so that we can clearly
identify the iris border. The enhanced iris image after applying median filter is
shown in Fig. 3a. To find out the iris boundary we draw arcs in left and right region
of the iris image of different radii from the pupil center as shown in Fig. 3b. The
intensities lying over the arcs of the line are summed up. Among the summed arc
lines, the arcs having a maximum change in intensity with respect to the previously
drawn arc is the iris boundary as shown in Fig. 3c.

3.3 Partial Iris Segmentation

A typical iris biometric system performs better for the ideal image, which is
acquired under controlled conditions. However, its performance degrades when
upper and lower part of the iris is occluded by the eyelashes and eyelids. In order to
solve this problem only left and a right region of the iris is segmented. The pupil
and iris boundary is shown in Fig. 4a. The partial segmented iris is shown in
Fig. 4b.

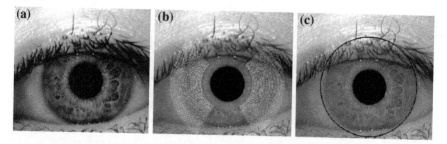

Fig. 3 Iris boundary detection. **a** Enhanced iris image. **b** Concentric arcs. **c** Iris boundary

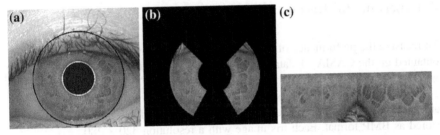

Fig. 4 Partial segmentation and normalization. **a** Pupil and iris boundary. **b** Partial segmented iris. **c** Normalized iris

3.4 Iris Normalization

Normalization is the process of converting the annular shaped iris region into rectangular shape, so that the transformation from cartesian coordinates to polar coordinate takes place [17]. Daugman [18] proposed the rubber sheet model assigns to each point in the iris, indifferently of size and pupil dilation, a pair of dimensionless real coordinates. To overcome aliasing artifacts, the proposed scale based approach normalizes the iris image by converting it from cartesian space to non-uniform polar space as shown in Fig. 4c.

3.5 Matching

Lowe [19] first proposed SIFT to get distinctive features in images. Using SIFT, key points are detected and each key point is described with a 128 bit feature vector. Based on key points extracted from each image, the same key points are selected as matching pairs and the number of matching pairs is used to measure the similarity of these two iris images. The best candidate match for each key point is found by identifying its nearest neighbor. The nearest neighbor is defined as the key point with minimum euclidean distance for the invariant descriptor vector. The suitable threshold is selected after testing the matching results of the whole iris database. Two iris images will be classified in the same class if the number of matching pairs is bigger than the threshold, otherwise these two iris images will be classified into different classes. Matching score is given by Eq. 1.

$$Matching\ Score = \frac{M}{\left(\frac{D1+D2}{2}\right)} \tag{1}$$

where M is the number of matched points, D1 is the number of detecting key points in the first image and D2 is the number of detecting key points in the second image.

4 Experimental Result

To measure the performance of the iris recognition system, experimental results are obtained on the CASIA-v1 database [20]. Iris images of CASIA-v1 were captured with a homemade iris camera. CASIA-v1 contains total 756 iris images from 108 eyes. Seven images are captured in two sessions for each eye, where three samples are collected in the first session and four in the second session. All images are stored as BMP format. Each iris image with a resolution 320 × 280.

To evaluate the performance of the iris biometric system the following measures are used.

False Acceptance Rate (FAR): FAR is the percentage of invalid inputs which are incorrectly accepted

False Rejection Rate (FRR): FRR is the percentage of valid inputs which are incorrectly rejected

Equal Error Rate (EER): EER is the point where FAR and FRR are equal.

$$EER = \frac{FAR + FRR}{2}$$

Total Error Rate (TER): The Total Error Rate (TER) consists of the sum of the False Accept Rate (FAR) and the False Reject Rate (FRR).

$$TER = FAR + FRR$$

Genuine Acceptance Rate (GAR): GAR is the fraction of genuine scores exceeding the threshold T. It is defined as

$$GAR = 1 - FRR$$

Receiver Operating Characteristic (ROC): ROC curve depicts the dependence of FAR with GAR for change in the value of threshold.

Figure 5a shows the FAR and FRR curves, from which it can be seen that when the threshold value of the proposed technique with normalization is 0.1196 then the EER is 2.21. Figure 5b shows the FAR and FRR curves, from which it can be seen that when the threshold value of the proposed technique without normalization is 0.1457 then the EER is 7.03. The ROC curve for the CASIA-v1 database with normalization is shown in Fig. 6a and without normalization is shown in Fig. 6b. It depicts the dependence of FAR with GAR for change in the value of threshold.

Fig. 5 FAR and FRR graph of proposed method. **a** FRR versus FAR graph with normalization of proposed method. **b** FRR versus FAR graph without normalization of proposed method

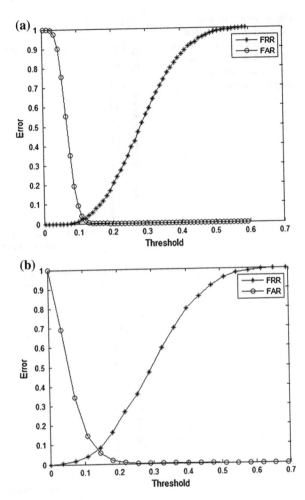

Table 1 shows the accuracy of the proposed technique is better than [15, 17]. Table 2 shows that the EER obtained with the proposed technique is less than that obtained in [8, 15, 16]. Tables 3 and 4 shows GAR, FAR, FRR, EER and TER for the proposed method with and without normalization respectively.

Fig. 6 ROC curve of proposed method. **a** ROC curve with normalization of proposed. **b** ROC curve without normalization of proposed method

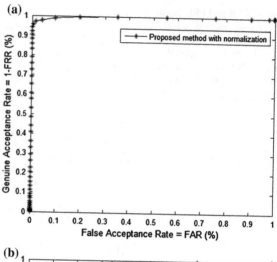

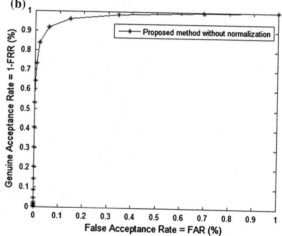

Table 1 Comparison of accuracy for iris recognition

Method	Accuracy (%)
Thumwarin et al. [21]	94.89
Hemal Patel et al. [22]	94.86
Proposed method without normalization	**91.56**
Proposed method with normalization	**97.12**

Table 2 Comparison of EER for iris recognition

Method	EER (%)
Boles et al. [11]	24.68
Patel et al. [22]	5.14
Yang et al. [23]	3.01
Proposed method without normalization	**7.03**
Proposed method with normalization	**2.21**

Table 3 Optimal threshold of GAR, FAR, FRR, EER and TER for iris recognition with normalization

Threshold	GAR (%)	FAR (%)	FRR (%)	EER (%)	TER (%)
0.0880	99.49	19.49	0.51	10.00	20.00
0.0985	98.97	9.34	1.03	5.18	10.37
0.1196	**97.12**	**1.54**	**2.88**	**2.21**	**4.42**
0.1301	95.88	0.55	4.12	2.33	4.67
0.1406	94.34	0.17	5.66	2.91	5.83

Table 4 Optimal threshold of GAR, FAR, FRR, EER and TER for iris recognition without normalization

Threshold	GAR (%)	FAR (%)	FRR (%)	EER (%)	TER (%)
0.0364	99.59	69.22	00.41	34.81	69.63
0.0729	98.35	34.27	01.65	17.96	35.92
0.1093	96.09	14.44	03.91	09.17	18.35
0.1457	**91.56**	**05.63**	**08.44**	**07.03**	**14.07**
0.1822	83.74	02.03	16.26	09.14	18.29

5 Conclusion

We have proposed partial iris segmentation and matching technique for iris recognition. The upper and lower region of the iris are not visible because of the eyelashes and eyelids, so for iris segmentation only left and right region of the iris is considered. A CASIA-v1 iris image database has been used to evaluate the performance of our proposed technique. The proposed iris recognition system performs with higher accuracy and a lower equal error rate.

References

1. Ma, L., Tan, T., Wang, Y., Zhang, D.: Personal identification based on iris texture analysis. Pattern Anal. Mach. Intell. **25**, 1519–1533 (2003)
2. Huang, J., You, X., Tang, Y.Y., Du, L., Yuan, Y.: A novel iris segmentation using radial-suppression edge detection. Sig. Process. **89**, 2630–2643 (2009)
3. Aligholizadeh, M.J., Javadi, S., Sabbaghi-Nadooshan, R., Kangarloo, K.: Eyelid and eyelash segmentation based on wavelet transform for iris recognition. Image Sig. Process **3**, 1231–1235 (2011)
4. Chen, W.-K., Lee, J.-C., Han, W.-Y., Shih, C.-K., Chang, K.-C.: Iris recognition based on bidimensional empirical mode decomposition and fractal dimension. Inf. Sci. 439–451 (2013)
5. Sahmoud, S.A., Abuhaiba, I.S.: Efficient iris segmentation method in unconstrained environments. Pattern Recogn. **46**, 3174–3185 (2013)

6. Jan, F., Usman, I., Agha, S.: Reliable iris localization using Hough transform, histogram-bisection, and eccentricity. Sig. Process. **93**, 230–241 (2013)
7. Nguyen, K., Fookes, C., Sridharan, S.: Fusing shrinking and expanding active contour models for robust iris segmentation. In: Information Sciences Signal Processing and their Applications (ISSPA), pp. 185–188 (2010)
8. Shah, S., Ross, A.: Iris segmentation using geodesic active contours. Inf. Forensics Secur. IEEE Trans. **4**, 824–836 (2009)
9. Li, Y.-H., Savvides, M.: An automatic iris occlusion estimation method based on high-dimensional density estimation. Pattern Anal. Mach. Intell. IEEE Trans. **35**, 784–796 (2013)
10. Daugman, J.G.: High confidence visual recognition of persons by a test of statistical independence. Pattern Anal. Mach. Intell. **15**, 1148–1161 (1993)
11. Boles, W.W., Boashash, B.: A human identification technique using images of the iris and wavelet transform. Sig. Process. **46**, 1185–1188 (1998)
12. Wildes, R.P., Asmuth, J.C., Green, G.L., Hsu, S.C., Kolczynski, R.J., Matey, J.R., McBride, S.E.: A machine-vision system for iris recognition. Mach. Vis. Appl. **9**, 1–8 (1996)
13. Zhu, R., Yang, J., Wu, R.: Iris recognition based on local feature point matching. In: International Symposium on Communications and Information Technologies (ISCIT), pp. 451–454 (2006)
14. Belcher, C., Du, Y.: Region-based SIFT approach to iris recognition. Opt. Lasers Eng. **47**, 139–147 (2009)
15. Alonso-Fernandez, F., Tome-Gonzalez, P., Ruiz-Albacete, V., Ortega-Garcia, J.: Iris recognition based on SIFT features. In: Biometrics, Identity and Security (BIdS), pp. 1–8 (2009)
16. Mehrotra, H., Sa, P.K., Majhi, B.: Fast segmentation and adaptive SURF descriptor for iris recognition. Math. Comput. Model. **58**, 132–146 (2013)
17. Ramkumar, R.P., Arumugam, S.: A novel iris recognition algorithm. In: International Conference on Computing, Communication and Networking Technologies (ICCCNT), pp. 1–6 (2012)
18. Daugman, J.: How iris recognition works. Circ. Syst. Video Technol. **14**, 21–30 (2004)
19. Lowe, D.G.: Distinctive image features from scale-invariant keypoints. Int. J. Comput. Vis. **60**, 91–110 (2004)
20. CASIA-v1 Iris Image Database: http://www.idealtest.org
21. Thumwarin, P., Chitanont, N., Matsuura, T.: Iris recognition based on dynamic radius matching of iris image. In: Electrical Engineering/Electronics, Computer, Telecommunications and Information Technology (ECTI-CON), pp. 1–4 (2012)
22. Patel, H., Modi, C.K., Paunwala, M.C., Patnaik, S.: Human identification by partial iris segmentation using pupil circle growing based on binary integrated edge intensity curve. In: Communication Systems and Network Technologies (CSNT), pp. 333–338 (2011)
23. Yang, G., Pang, S., Yin, Y., Li, Y., Li, X.: SIFT based iris recognition with normalization and enhancement. Int. J. Mach. Learn. Cybernet. **4**, 401–407 (2013)

Medical Data Mining for Discovering Periodically Frequent Diseases from Transactional Databases

Mohammed Abdul Khaleel, G.N. Dash, K.S. Choudhury
and Mohiuddin Ali Khan

Abstract Medical data mining has witnessed significant progress in the recent past. It unearths the latent relationships among clinical attributes for finding interesting facts which helps experts in health care in decision making. Recently, frequent patterns in transactional medical databases that occur periodically are exploited to know the temporal aspects of various diseases. In this paper we modified K-means algorithm to extract yearly and monthly periodic frequent patterns from medical datasets. The datasets contain electronic health records of 2012 and 2013. Periodical frequent patterns between these years and monthly patterns were extracted using the proposed methodology. To achieve this we used the notion of making temporal view that is instrumental in adapting K-means for this purpose. We built a prototype to test the algorithm and the empirical results reveal that the proposed methodology for knowledge discovery related periodic frequent diseases is useful. The application can be reused to have lasting implications on health care industry for improving quality of services with strategic and expert decision making.

Keywords Data mining · Medical data mining · Periodic frequent diseases · K-means

M.A. Khaleel (✉) · G.N. Dash · K.S. Choudhury
Sambalpur University, Orissa, India
e-mail: khaleel_dm@yahoo.com

G.N. Dash
e-mail: gndash@sunivac.in

K.S. Choudhury
e-mail: kschoudhury@gmail.com

M.A. Khan
Utkal University, Orissa, India
e-mail: moinkku@gmail.com

© Springer India 2015
L.C. Jain et al. (eds.), *Computational Intelligence in Data Mining - Volume 1*,
Smart Innovation, Systems and Technologies 31, DOI 10.1007/978-81-322-2205-7_9

1 Introduction

Periodic frequent patterns were first introduced by Tanbeer et al. as explored in [1]. A Frequent pattern when occurs on at regular intervals can be called as periodic frequent pattern. Periodic frequent pattern will be useful for decision making when it satisfied the support and confidence provided by domain expert. Domain expert is the human expert who has specific knowledge in the given domain with respect to the usage of periodic frequent patterns. Generally support and confidence can be used to filter out unwanted patterns. Support and confidence can be computed as follows.

$$\text{Support} = \text{number of records of A with B/total number of records}$$

$$\text{Confidence} = \text{number of records of A with B/the total number of records with A}$$

Pattern Growth approach is used in [2] for periodic frequent pattern mining on transactional databases. This approach proved to be effective. Fizzy mining algorithm is used in [3] for medical data mining using time-series datasets. In [4] association rule mining is sued for discovering adverse drug reactions. The similar line of study was made in [5] where drug event monitoring helped pharmaceutical industry to optimize their services.

In [6–8] Apriority algorithm is used for medical data mining. Especially it is used for discovering frequent disease patterns in medical data. In [9, 10] experiments are made on mining streaming medical data. The researchers here focused on different methods that could withstand the nature of data which is streaming continuously from sources. In [11] semantics based approach is followed for discovering cases pertaining to smoking as it is hazardous to health. This has helped to counsel patients based on the patterns identified. FP-Growth algorithm was used in [12] for discovering periodically frequent patterns from medical datasets. This research helped in making strategic decisions on the administration part and let the decisions taken by the leadership to be implemented by management. As leadership is more about having wisdom to make well informed decisions, the medical data mining is able to help the leaders in order to make expert decisions. These decisions lead to improving quality of service in health care domain.

In this paper the contribution is to explore possibilities of finding periodic frequent diseases between datasets pertaining to the years 2012 and 2013. We also focused on finding monthly periodic frequent diseases. Towards achieving it, we employed a K-means variant that could discover periodic frequent diseases from electronic health records. The algorithm also makes use of given support and confidence in order to increase accuracy of decision making. The research has implications on healthcare domain as it can improve the quality of service in the domain when strategies are identified and executed based on the results of the experiments in this paper. The rest of the paper is structured as follows. Section 2 presents review of literature on medical data mining especially discovering periodic frequent patterns in medical data mining. Section 3 presents proposed methodology that discovers periodic frequent patterns. Section 4 provides experimental results while Sect. 5 concludes the paper.

2 Related Works

Medical data mining has witnessed considerable research in the field of frequent pattern mining. This section focuses on the review of literature on periodically frequent diseases. Surana et al. [2] explored an efficient approach with empirical study on mining periodic frequent patterns from transactional databases. They could mine patterns based on user-interest. They also focused on "rare item problem" that arise from the results of periodic frequent pattern mining. They made experiments on such mining with single constraint and multiple constraints. With single constraint their approach does not comply with downward closure property while with multiple constraints it could comply with downward closure property. They used Pattern—Growth algorithm for mining periodic frequent patterns. Altiparmak et al. [13] performed data mining on medical database using clinical trials datasets. The data contained in the datasets is of type high-dimensional, heterogeneous time-series data. The experimental results revealed hidden groups of data that needs to be subjected to biomedical analysis. In 2006 Chen et al. [3] used time—series data pertaining to bioinformatics and medical treatment. They proposed a fuzzy mining algorithm that is domain specific. Their approach first converts data into angles and then converted them into continuous subsequences using sliding window before generating frequent patterns finally using Apriori like fuzzy mining algorithm. They also employed post processing for getting rid of redundant patterns.

Evidence based research is quite common in health care domain. Towards this, Cameron et al. [11] in 2012 focused on connecting dots in huge amount of medical data in order to discover new insights that will have impact on patient care, treatment and diagnosis. They used unstructured clinical notes as dataset and explored Smoker Semantic Types (SST) for evidence-based scientific approaches in medical data mining. Thus their research has helped to diagnose smoking cases and take steps to counsel the patients as smoking leads to cancer besides plethora of cardiovascular disorders. Noma and Ghani [12] in 2012 proposed a knowledge discovery model for medical data mining using FP-Growth algorithm. The experiments made by them on medical data could discover the symptoms of giddiness, vertigo, and tinnitus for accurate diagnosis of diseases. Bone conduction test results and tone audiometry related datasets were used for experiments. Knowledge utilization and knowledge evaluation are two important features of their framework. Ilayaraja and Meyyappan [7] in 2013 employed Apriori algorithm to medical data in order to mine frequent diseases. This could help in better decision making with respect to planning and execution of strategies to combat diseases. Thus the quality of service in healthcare domain gets improved.

Previous papers also focused on medical data mining. In [14] made a survey of data mining techniques used in medical data mining for discovering locally frequent diseases. Later in [6] we employed modified Apriori algorithm to mine locally frequent diseases. The study on discovering temporally frequent diseases from medical database is in [15]. In [16] we used modified Karmalego algorithm in order to mine temporally frequent diseases from health care data. In [17] we studied the techniques used for discovering periodically frequent patterns in medical databases. For all these experiments we used datasets obtained from a Reputed Private Hospital. In this paper the focus is on mining periodically frequent patterns using modified K-means algorithm.

3 Methodology for Mining Periodic Frequent Diseases

We studied several techniques in literature. We were realized the simplicity of K-means for discovering trends or patterns in medical data. Basically K-means is one of the top 10 data mining algorithm that has been around for more than a decade used for clustering. In this paper we adapted a modified version of K-means that could help in discovering periodically frequent diseases. K-means is an evolutionary algorithm that makes given objects into k groups after analyzing them. The general mechanism of K-means is as shown in Fig. 1.

As can be seen in Fig. 1, the K-means algorithm needs dataset and number clusters as input and have an iterative process of determining centroid, finding

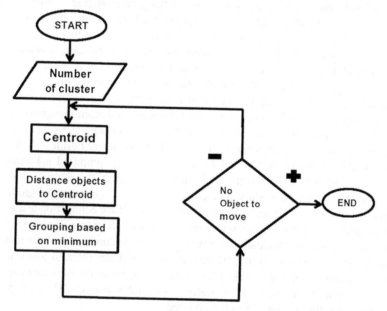

Fig. 1 Illustrates flow of K-means algorithm

distance of objects from centroid and grouping based on the distance. It converges into given number of clusters. We used this algorithm with required modifications in order to discover periodically frequent diseases from medical data sets.

3.1 Proposed Methodology

The aim of the paper is to build an algorithm or technique that produces periodically frequent diseases both at year and month levels. Towards it we implemented the methodology described here. First of all we collected medical data pertaining to years 2012 and 2013 from Reputed Private Hospital. The dataset has around 65,000 records in each year. The medical datasets were collected in the form of spreadsheets. The diseases considered are Pain and Swelling upper left tooth, Rec. Tonsillitis, snoring, mouth breathing, Decayed teeth upper right, Fever, Cough, Sore throat, Body ache, Painful lesion on toes, Rect. headache, More in occipital region, Fever, Cough, Rhino, heart-ear ache, Ref. Feeding, Trauma to right fore arm due to falling, With Cough/Fever, 52 Y/o Male PT. RT Knee pain weeks ago, Decayed tooth Upper right and upper left, RB Elbow pain after carrying heavy object, Runny/Nose/Snoring, Earache and Discharged F/Up, Odynophagia, and Unproductive cough. Then we have done some preprocessing in order to associate numeric values with discrete data. Then we employed the modified K-means algorithm that made clusters. From analyzing these clusters, periodic frequent diseases were discovered. Following is the modified K-means algorithm. As can be seen in algorithm, it is evident that the preprocessing is required in order to work with numeric values for clustering. The process of associating numeric values with discrete data is done in the preprocessing step. Then the discovery of periodically frequent patterns is done in two phases. In the first phase modified K-means is applied to obtain yearly periodic frequent patterns while the second phase focuses on extracting monthly periodic frequent patterns. In either case the basic steps of K-means algorithm are followed. However, the algorithm makes use of the temporal view of the database which provides time related details to K-means for making decisions on clustering pertaining

Algorithm: Modified K-Means
Inputs: Medical dataset **db**, number of clusters **k**
Output: Periodically frequent patterns Py and Pm
Preprocess:
Associate numeric value with discrete data **db'**
Process
Obtain temporal views **Ty** and **Tm** of the **db'**
Step1: Mining yearly periodic frequent diseases
(a) Define initial cluster means based on Ty of **db'** and **k**
(b) Use Euclidian distance measure to cluster means to associate other objects to form clusters

(c) Adjust centroid of each cluster

(d) Repeat step (b) and (c) until yearly periodic frequent patterns are produced

(e) Return **Py**

Step2: Mining monthly periodic frequent diseases

(a) Define initial cluster means based on **Tm** of **db'** and **k**

(b) Use Euclidian distance measure to cluster means to associate other objects to form clusters

(c) Adjust centroid of each cluster

(d) Repeat steps (b) and (c) until monthly periodic frequent patterns are produced

(e) Return **Pm**

4 Experimental Results

Empirical study has been made on medical data set obtained from Reputed Private Hospital. Around 65,000 records are present in each of 2012 dataset and 2013 dataset with attributes pertaining to health. We built a prototype application for medical data mining which demonstrates mining of periodically frequent diseases. The environment used for the application development is Java programming language, a PC with 4 GB RAM, Core 2 Dual process running Windows 7 operating system. Experiments are made to know periodic frequent diseases of years 2012 and 2013 followed by monthly frequent periodic patterns pertaining to diseases. Table 1 shows the periodical frequency of diseases compared between the years 2012 and 2013. Minimum support used to obtain the results is 0.3. However, a subset of dataset is used for experiments. As per the 0.3 support the periodically frequent diseases are computed using temporal view of the clusters made.

As can be seen in Table 1, it is evident that various hidden patterns of periodic frequent patterns are discovered by an algorithm. The diseases that appear in the results table include Pain and Swelling upper left tooth; Rec. Tonsillitis, snoring, mouth breathing; decayed teeth upper right; Fever, Cough, Sore throat, Body ache; Painful lesion on toes; Rect. headache, More in occipital region; Fever, Cough, Rhino heart-ear ache, Ref. Feeding; Trauma to right fore arm due to falling; With Cough/Fever; 52 Y/o Male PT. RT Knee pain weeks ago; Decayed tooth Upper right and upper left; RB Elbow pain after carrying heavy object; Runny/Nose/Snoring; Earache and Discharged F/Up; Odynophagia, Unproductive cough. The results reveal many latent facts. For instance there is frequency change among all diseases between 2012 and 2013. The disease "Rect. headache, More in occipital region" has almost equal frequency between the 2 years. It also happens to be highest frequency while the least frequency is with "Trauma to right fore arm due to falling" in year 2012 and 2013. For this disease the frequency also happens to be same.

Table 1 Periodic frequency of diseases between 2012 and 2013

Disease	Frequency in 2012	Frequency in 2013
Pain and swelling upper left tooth	5	6
Rec. tonsillitis, snoring, mouth breathing	4	7
Decayed teeth upper right	9	6
Fever, cough, sore throat, body ache	6	8
Painful lesion on toes	5	12
Rect. headache, more in occipital region	11	10
Fever, cough, rhino heart-ear ache, Ref. Feeding	4	6
Trauma to right fore arm due to falling	3	3
With cough/fever	5	6
52 Y/o Male PT. RT knee pain weeks ago	8	5
Decayed tooth upper right and upper left	6	7
RB Elbow pain after carrying heavy object	4	9
Runny/nose/snoring	9	3
Earache and discharged F/Up	8	5
Odynophagia, unproductive cough	5	8

The ailment "Painful lesion on toes" has highest frequency in 2013 while "Rect. headache, more in occipital region" has highest frequency in 2012. As can be seen in Table 2, some interesting periodically frequent patterns are revealed in the results. The frequency here is computed based on the total number of diseases occurred in given year. For instance April has least number of diseases in 2012 while the same occurs in January in 2013. The highest number of diseases is record in November in 2012 while the same occurs in February and December in 2013. From this it is understood that the periodical frequent patterns change from time to time. The experimental results are graphically presented in Fig. 2. They reflect

Table 2 Monthly frequent periodic diseases	Month	No. of diseases in 2012	No. of diseases in 2013
	January	12	03
	February	09	09
	March	06	06
	April	02	09
	May	08	08
	June	15	06
	July	08	08
	August	16	16
	September	14	05
	October	09	09
	November	19	10
	December	13	09

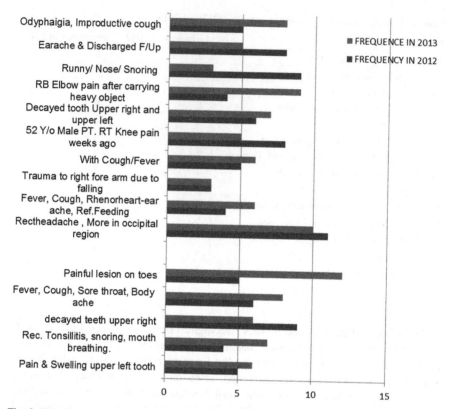

Fig. 2 Periodically frequent diseases between 2012 and 2013

various hidden periodically frequent patterns between 2012 and 2013 besides revealing monthly frequent periodic patterns of diseases.

As can be seen in Fig. 2, the results revealed that periodically frequent patterns have different frequencies in 2012 and 2013. As discussed earlier, the results reveal different trends in 2 years besides showing monthly periodic frequent patterns as well as reflected in Fig. 2.

As shown in Table 2 the number of diseases occurred in April 2012 is very less compared to any other month in that year. Highest number of diseases occurred in November 2012. August, June, September, December and January show next highest number of diseases occurred in decreasing order. May and July shows equal number of diseases occurred. Table 2 shows monthly periodic frequent diseases in 2013.

As shown in Table 2 the number of diseases occurred in January 2013 is very less compared to any other month in that year. Highest number of diseases occurred in August 2013. November has second highest number of diseases. Interestingly same number of diseases frequency is found in the months February, April, October and December. In the same fashion similar frequency of monthly periodic pattern is found in the months May and July. The second least number of diseases is found in September 2013.

5 Conclusion

In this paper we studied the methods used for mining periodically frequent patterns that can help in making strategic decisions in healthcare domain to improve quality of services. We have built the notion of temporal view both yearly and monthly. Then we modified K-means algorithm that makes use of these temporal views in order to generate yearly and monthly periodic frequent patterns. We built a prototype application to demonstrate the efficiency of the proposed algorithm. The empirical results reveal that the extracted periodic frequent patterns pertaining to diseases can help in making well informed decisions pertaining to strategic planning and execution on administration part of the health care domain. Thus the research has its impact on making expert decisions that will improve quality of services in healthcare industry. In future we work on the medical data mining using different programming paradigm such as Map Reduce which leverages parallel programming and cloud computing. Towards it the existing medical data mining techniques are to be altered to harness the power of modern computing facilities.

References

1. Tanbeer, S.K., Ahmed, C.F., Jeong, B.-S., Lee, Y.-K.: Discovering periodic-frequent patterns in transactional databases. In: PAKDD '09: Proceedings of the 13th Pacific-Asia Conference on Advances in Knowledge Discovery and Data Mining, pp. 242–253 (2009)
2. Surana, A., Kiran, R.U., Reddy, P.K.: An efficient approach to mine periodic-frequent patterns in transactional databases. In: Center for Data Engineering, International Institute of Information Technology, pp. 1–12
3. Chen, C.H., Hong, T.P., Vincent, S.M.: Tseng: a less domain-dependent fuzzy mining algorithm for frequent trends. In: IEEE, pp. 1–6. 16–21 July 2006
4. Jin, H., Chen, J., He, H., Williams, G.J., Kelman, C., O'Keefe, C.M.: Mining unexpected temporal associations: applications in detecting adverse drug reactions. IEEE 12(4), 1–13 (2008)
5. Huang, J., Huan, J., Tropsha, A.: Semantics-driven frequent data pattern mining on electronic health records for effective adverse drug event monitoring. In: IEEE, pp. 1–4 (2013)
6. Khaleel, M.A., Pradhan, S.K., Dash, G.N.: finding locally frequent diseases using modified apriori algorithm. Int. J. Adv. Res. Comput. Commun. Eng. 2(10) (2013)
7. Ilayaraja, M., Meyyappan, T.: Mining medical data to identify frequent diseases using apriori algorithm. In: IEEE, pp. 1–6 (2013)
8. Agrawal, R., Srikant, R.: Mining sequential patterns. In: Proceedings of the 11th International Conference on Data Engineering (1995)
9. Feng, W.U., Quanyuan, W.U., Yan, Z., Xin, J.: Mining frequent patterns in data stream over sliding windows. In: IEEE, pp. 1–4 (2009)

10. Lin, J., Li, Y.: Finding approximate frequent patterns in streaming medical data. In: IEEE, pp. 1–6 (2010)
11. Cameron, D., Bhagwan, V., Sheth, A.P.: Towards comprehensive longitudinal healthcare data capture. In: IEEE, pp. 1–8 (2012)
12. Noma, N.G., Abd Ghani, M.K.: Discovering pattern in medical audiology data with FP-growth algorithm. In: IEEE, pp. 1–6 (2012)
13. Altiparmak, F., Ferhatosmanoglu, H., Erdal, S., Trost, D.C.: Information mining over heterogeneous and high-dimensional time-series data in clinical trials databases. IEEE 10(2), 1–10 (2006)
14. Khaleel, M.A., Pradham, S.K., Dash, G.N.: A survey of data mining techniques on medical data for finding locally frequent diseases. Int. J. Adv. Res. Comput. Sci. Softw. Eng. 3(8) (2013)
15. Khaleel, M.A., Pradhan, S.K., Dash, G.N., Mazarbhuiya, F.A.: A survey of data mining techniques on medical data for finding temporally frequent diseases. Int. J. Adv. Res. Comput. Commun. Eng. 2(12) (2013)
16. Khaleel, M.A., Pradhan, S.K., Dash, G.N.: Finding temporally frequent diseases using modified karmalego algorithm. Int. J. Comput. Eng. Appl. 5(2) (2014)
17. Khaleel, M.A., Pradhan, S.K., Dash, G.N.: A survey on medical data mining for periodically frequent diseases. Int. J. Adv. Res. Comput. Commun. Eng 3(4) (2014)

Bandwidth Enhancement by Direct Coupled Antenna for WLAN/GPS/WiMax Applications and Feed Point Analysis Through ANN

Sakshi Lumba, Vinod Kumar Singh and Rajat Srivastava

Abstract This paper presents a direct coupled slotted microstrip antenna with enhanced antenna gain and wider bandwidth. A typical design has been implemented and the results are presented and discussed. The different antenna geometries are simulated through IE3D Zeland simulation software for the comparative analysis of bandwidth. The proposed antenna has dual frequency band having dual bandwidth 3.38 % (1.392–1.44 GHz) and 69.5 % (1.733–3.58 GHz) which is suitable for WLAN/GPS/WiMax applications.

Keywords Direct coupling · Enhance bandwidth · Compact microstrip patch · Ground plane · Gain

1 Introduction

Microstrip patch antennas have drawn the attention of researchers due to its light weight, low profile, low cost and ease of integration with microwave circuit. But the major drawback of rectangular Microstrip antenna is its narrow bandwidth and lower gain [1, 2]. The bandwidth of Microstrip antenna may be increased using several techniques such as use of a thick or foam substrate, cutting slots or notches like U slot, E shaped H shaped patch antenna, introducing the parasitic elements either in coplanar or stack configuration, and modifying the shape of the radiator

S. Lumba (✉) · V.K. Singh
S.R. Group of Institutions, Ambabai, Jhansi, Uttar Pradesh, India
e-mail: sakshilumba@gmail.com

V.K. Singh
e-mail: Singhvinod34@gmail.com

R. Srivastava
Bundelkhand Institute of Engineering Technology, Jhansi
Uttar Pradesh, India
e-mail: raj.sriv89@gmail.com

© Springer India 2015
L.C. Jain et al. (eds.), *Computational Intelligence in Data Mining - Volume 1*,
Smart Innovation, Systems and Technologies 31, DOI 10.1007/978-81-322-2205-7_10

Fig. 1 Geometry of proposed microstrip antenna

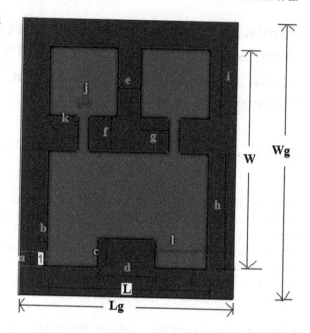

patch by introducing the slots [3–5]. In the present work the bandwidth of Microstrip antenna is increased by direct coupling and it is obtained that the bandwidth of direct coupled C slotted rectangular Microstrip antenna is 'ten times' greater than simple rectangular Microstrip antenna. Directly coupled C slotted rectangular Microstrip antenna with Microstrip line feed is shown in Fig. 1. The width of the Microstrip line was taken as 2.5 mm and the feed length as 1 mm. The patch is energized electromagnetically using 50 ohm Microstrip feed line [6–8]. There are numerous substrates that can be used for the design of microstrip antennas and their dielectric constants are usually in the range of $2.2 \leq \varepsilon_r \leq 12$. The proposed antenna has been designed on glass epoxy substrate ($\varepsilon_r = 4.4$). The design frequency of proposed antenna is 2.4 GHz. The frequency band (1.733–3.58 GHz) of proposed antenna is suitable for broad band applications (1.605–3.381 GHz) such as military, wireless communication, satellite communication, global positioning system (GPS), RF devices, WLAN/Wi-Max application [9–12].

2 Antenna Design and Specification

All the dimensions of proposed antenna have been calculated very carefully by using the Eqs. 1–4. The design frequency of presented antenna is 2.4 GHz. The proposed antenna design is compared with three different structures that are shown in Figs. 2 and 3 having same dimensions and line feed at the same position as in the proposed antenna. It is found that rectangular patch antenna with two direct coupled

Fig. 2 Geometry of direct coupled T slotted MSA

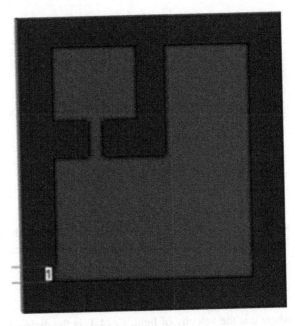

Fig. 3 Geometry of direct coupled double T slotted MSA

Table 1 Antenna design specifications

S. No.	Parameters	Value	S. No.	Parameters	Value
1.	Design frequency (f_r)	2.4	10.	b	1.0
2.	Dielectric constant (ε_r)	4.4	11.	d	10
3.	Substrate height	1.6	12.	e	4.0
4.	Patch width	38.0	13.	f	6.0
5.	Patch length	28.3	14.	g	5.0
6.	Ground plane width	47.6	15.	h	20
7.	Ground plane length	37.9	16.	i	12
8.	Feed coordinates	5, 4	17.	j	2.0
9.	a	2.5	18.	l	9.0

structures and with C slot provides largest bandwidth. For designing a rectangular Microstrip patch antenna, the length and width are calculated in Table 1.

$$W = \frac{c}{2f\sqrt{(\varepsilon_r + 1)/2}} \tag{1}$$

where c is the velocity of light, ε_r (4.4) is the dielectric constant of substrate(glass epoxy), f_r(2.4 GHz) is the antenna design frequency, W is the patch width, and the effective dielectric constant ε_{reff} is given as [2, 4]

$$\varepsilon_{eff} = \frac{(\varepsilon_r + 1)}{2} + \frac{(\varepsilon_r - 1)}{2}\left[1 + 10\frac{h}{W}\right]^{-\frac{1}{2}} \tag{2}$$

At h = 1.6 mm, the extension length ΔL is calculated as [10, 13, 14]

$$\frac{\Delta l}{h} = 0.412\frac{\left(\varepsilon_{eff} + 0.300\right)\left(\frac{W}{h} + 0.262\right)}{\left(\varepsilon_{eff} - 0.258\right)\left(\frac{W}{h} + 0.813\right)} \tag{3}$$

By using the above mentioned equation we can find the value of actual length of the patch as [7, 10]

$$L = \frac{c}{2f\sqrt{\varepsilon_{eff}}} - 2\Delta l \tag{4}$$

3 Comparative Analysis of Antenna Geometries

To obtain the optimum bandwidth of the proposed direct coupled C slotted antenna, different structures are simulated sequentially through Zeland IE3D simulation software and bandwidth are compared in Table 2. By this computation work it is

Table 2 Comparison of bandwidth of different antenna designs

S. No.	Antenna structures	Frequency band	Band width (%)
1.	Design III direct coupled with T slot MSA	1.695–2.549	40.24
2.	Design I direct coupled with double T slot MSA	1.906–3.648	62.72
3.	Design II direct coupled with double T slot having C shape MSA	1.733–3.58	69.52

clear that the direct coupled double T slot with C shape MSA has been presented maximum bandwidth which is shown in Fig. 8.

4 Neural Network Designing

4.1 Data Set Generation

This is the most important step in designing a neural network.

4.1.1 Generation of Input Data Set

Different values of design frequency or operating frequency (f_r), antenna patch length (L) and patch width (W) are taken as input to the neural network. Antenna dimensions can be calculated by using Eqs. 1–4 as describe above. The operating frequency range is from 1.2 to 2.5 GHz. Within this frequency range 90 sets of input data were generated. The input data set is given in Table 3 [13–15].

Table 3 Comparison of results obtained from IE3D and ANN model

Fr (GHz)	W (mm)	L (mm)	Y_1 from IE3D	Y_1 from ANN
2.39	38.19	28.42	6.0	5.9
2.37	38.51	28.67	6.0	5.9
1.18	77.36	59.10	10.5	10.97
2.94	31.05	22.82	7.0	7.07
3.00	30.42	22.34	7.0	7.07
1.73	52.76	39.84	10	10.8
2.6	35.11	26.01	7.0	6.96
1.82	50.15	37.79	10	10.8
1.93	47.29	35.55	5.0	5.08
1.88	48.55	36.54	5.0	5.09
1.1	82.98	63.51	10.5	10.79
2.05	44.53	33.38	5.0	5.10
1.44	63.39	48.16	11	10.78

4.1.2 Generation of Target Data Set

To generate target data set for proposed neural network the proposed direct coupled C slotted rectangular microstrip antenna is simulated through IE3d simulation software for its different dimensions. A set of 90 data's of output are generated through IE3D. During the generation of data set, 90 designs of proposed antenna for different values of length and width were constructed with operating frequency varies from 1.2 to 2.5 GHz. Here the important thing is that some dimensions of proposed antenna should be kept constant and some varies according to the variations in length and width. The proposed antenna with operating frequency 2.4 GHz, length L = 28.3 and width W = 38.03 is taken as reference antenna [16–18].

Along X axis
Let the increment or decrement in length of the proposed antenna is L1.
(It is clear that if the frequency is greater than 2.4 GHz then the proposed antenna length and width decreases from the present value and vice versa).
Then the dimensions 'q' = 'l' increases or decreases by L1/2.
And the dimensions 'k' = 'n' = 'g' = 'o' increases or decreases by L1/4.
Along Y axis
Let the increment or decrement in width of the proposed antenna is W1.
Then the dimension't' = 'h' increases or decreases by W1/2.
And the dimensions 'm' = 'i' increases or decreases by W1/2.

In this manner it can be constructed the proposed antenna with different value of length and width. Figures 2 and 3 shows the geometry of proposed antenna at operating frequency 2.5 GHz and at 1.4 GHz respectively. The dimensions that are showing in both the figures are same as in the proposed antenna with operating frequency 2.4 GHz.

Since the antenna is fed by 1 mm × 2.5 mm microstrip line feed, the proposed antenna with different dimensions (design frequency range 1.2–2.5 GHz) is analyzed for maximum bandwidth using IE3D simulation software. The patch is placed in such a manner that it is always at a distant 4.8 mm from each side of the ground plane. With this analysis it is found that proposed antenna with different dimensions provide maximum bandwidth along Y axis with feed length 1 mm and width 2.5 mm. Hence the feed coordinate along X axis is fixed that is (5, 4) and the feed coordinates along Y axis is obtained by investigation. Let the Y coordinates are Y_1 and Y_2. Here $Y_2 = Y_1 + 2.5$

Now from this it is clear that we have to find Y1. This is our target value to neural network. In above figure one example of feed coordinates is taken. From this it is clear that if all four coordinates are known then feed can be placed at proper position at which antenna gives highest bandwidth.

The four combinations of coordinates are given as:

$$(X_1, \ Y_1), (X_1, \ Y_2), (X_2, \ Y_1), (X_2, \ Y_2)$$

Here $X_1 = 5$, $X_2 = 4$ and $Y_2 = Y_1 + 2.5$

Y_1 is obtained from analysis for maximum bandwidth through IE3D simulation software. It should be noted that within 1.1–1.53 GHz design frequency range antenna is operated or analyzed in the frequency range 0–3 GHz.

4.2 Architecture of Proposed Neural Network

The architecture and training of proposed neural network is shown in Figs. 4, 5, 6 and 7

The other specifications of the proposed neural network are given as:

Network type → Feed Forward Back Propagation

Number of layers → 3

Number of neurons in hidden layer → 14

Transfer function → TANSIG.

Fig. 4 Training of neural network

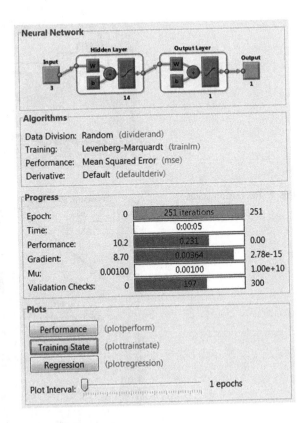

Fig. 5 Training performances showing minimum MSE

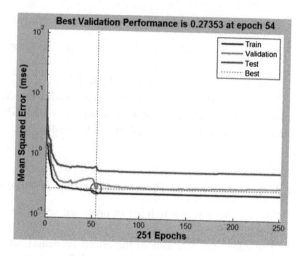

Fig. 6 Neural network training result

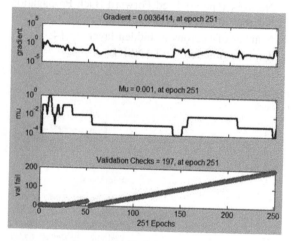

Fig. 7 Regression states

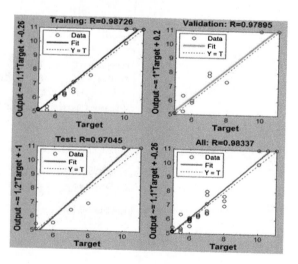

Training function → TRAINLM (Levenberg-Marquardt)
Adaption learning function → LEARNGDM
Performance → MSE (mean square error)
Number of epoch's → 251
Iterations → 251
Gradient → 2.78e − 15

5 Simulation Results Obtained from IE3D and Neural Network

5.1 Proposed Antenna Results from IE3D

The performance specifications of proposed antenna like gain, efficiency and radiation pattern is shown in the Figs. 8, 9, 10 and 11. Figure 8 shows optimum return loss versus frequency plot of proposed microstrip antenna giving the maximum bandwidth of 69.5 % (1.733–3.58 GHz). Figure 10 shows the gain versus frequency plot of optimum design having the gain of 5.11 dBi. Figure 11 shows efficiency versus frequency plot of proposed antenna having maximum efficiency of 97.21 %.

5.2 Proposed Antenna Results from Neural Network

Table 3 shows the result and accuracy of proposed neural network. The data that is used for comparison of neural network results with IE3D results is not included in training of neural network.

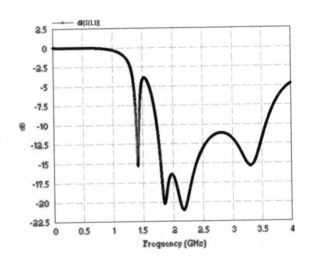

Fig. 8 Return loss versus frequency graph

S. Lumba et al.

Fig. 9 2D radiation pattern of antenna

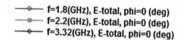

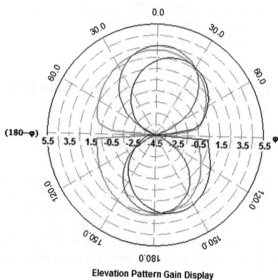

Elevation Pattern Gain Display
(dBi)

Fig. 10 Gain versus frequency plot

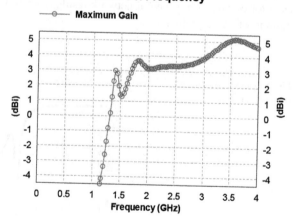

Fig. 11 Efficiency graph of
proposed antenna

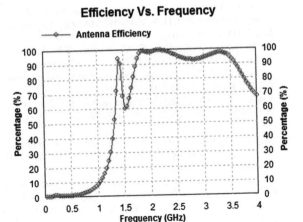

6 Conclusion

In this paper, wide band C slotted compact antenna using glass epoxy material is proposed and presented. The proposed antenna design provides wider bandwidth with compact size and shows adequate gain and efficiency. In this article the enhancements of the bandwidth of C slotted Microstrip antenna by direct coupling, together with the analysis for those enhancements for maximum bandwidth for different antenna dimensions has been presented. It is also studied that developments of artificial neural network that directly provides the feed point coordinates at which antenna gives maximum bandwidth. For this purpose we used 'Feed Forward Neural Network with Levenberg-Marquardt training algorithm. The proposed design is best suitable for WLAN/GPS/WiMAX.

References

1. Balanis, C.A.: Antenna Theory, Analysis and Design. Wiley Inc, Hoboken, New Jersey (2005)
2. Wong, K.-L.: Compact and Broadband Microstrip Antennas. ISBNs: 0-471-41717-3 (Hardback); 0-471-22111-2 (Electronic), Wiley (2002)
3. Ali, Z., Singh, V.K., Ayub, S.: A neural network approach to study the bandwidth of microstrip antenna. IJARCSSE **3**(1), 64–69. ISSN: 2277 128X (2013)
4. Jain, S., Singh, V.K., Ayub, S.: Bandwidth and gain optimization of a wide band gap coupled patch antenna. IJESRT. ISSN: 2277-9655 (2013)
5. Zhao, G., Zhang, F.-S., Song, Y., Weng, Z.B., Jiao, Y.-C.: Compact ring monopole antenna with double meander lines for 2.4/5 GHz dual-band operation. Prog. Electromagn. Res. PIER **72**, 187–194 (2007)
6. Srivastava, S., Singh, V.K., Ali, Z., Singh, A.K.: Duo triangle shaped microstrip patch antenna analysis for WiMAX lower band application. Procedia Technol **10**, 554–563 (2013). (Elsevier)
7. Pathak, R.S., Singh, V.K., Ayub, S.: Dual band microstrip antenna for GPS/ WLAN/ WiMAX applications. IJETED **2**(7). ISSN: 2249-6149 (2012)

8. Hu, C.-L., Yang, C.-F., Lin, S.-T.: A compact inverted-f antenna to be embedded in ultra-thin laptop computer for LTE/WWAN/WI-MAX/WLAN applications. IEEE Transactions AP-S/USRT,978-1-4244-9561 (2011)
9. Chakraborty, U., Chatterjee, S., Chowdhury, S.K., Sarkar, P.P.: A compact microstrip patch antenna for wireless communication. Prog. Electromagn. Res. C **18**, 211–220 (2011)
10. Alkanhal, M.A.S.: Composite compact triple-band microstrip antennas. Prog. Electromagn. Res. PIER **93**, 221–236 (2009)
11. Singh, V.K., Ali, Z., Ayub, S.:Bandwidth optimization of compact microstrip antenna for PCS/DCS/Bluetooth application. Cent. Eur. J. Eng. **4**(3), 281–286. ISSN: 1896-1541 (2014). (Springer)
12. Singh, V.K., Ali, Z., Singh, A.K., Ayub, S.: Dual band triangular slotted stacked microstrip antenna for wireless applications. Cent. Eur. J. Eng. **3**(2), 221–225. ISSN: 1896-1541 (2013). (Springer)
13. Singh, V.K., Ali, Z., Singh, A.K., Ayub, S.: A compact wide band microstrip antenna for GPS/DCS/PCS/WLAN applications. Intell. Comput. Netw. Inform. 183–204 (2013). (Springer)
14. Singh, V.K., Ali, Z., Singh, A.K., Ayub, S.: Dual band microstrip antenna design using artificial neural networks. Int. J. Adv. Res. Comput. Sci. Softw. Eng. (IJARCSSE) **3**(1), 74–79. ISSN: 2277 128X (2013)
15. Ali, Z., Singh, V.K., Ayub, S., Singh, A.K.: A neural network approach to study bandwidth microstrip antenna. Int. J. Adv. Res. Comput. Sci. Softw. Eng. (IJARCSSE) **3**(1), 64–69. ISSN: 2277 128X (2013)
16. MATLAB Simulink Help, the Math Works Inc., MATLAB 7.12.0 (R2011a)
17. Thakare, V.V., Singhal, P.K.: Bandwidth analysis by introducing slots in microstrip antenna design using ANN. Prog. Electromagn. Res. M **9**, 107–122 (2009)
18. Thakare, V.V., Singhal, P.: Neural network based CAD model for the design of rectangular patch antennas. JETR **1**(7), 129–132 (2009)

Optimal Power Flow Using PSO

Prashant Kumar and Rahul Pukale

Abstract This paper presents an efficient technique to solve optimal power flow based on PSO in which the power transmission loss function is used as the problem objective while considering both the real and reactive as a sub problem. The proposed method is used for solving the non-linear optimization problems while minimizing the objective voltage stability margin is also maintained. The proposed technique is tested on IEEE 57 bus system.

Keywords Particle swarm optimization · Power loss minimization · Reactive power · Voltage control · Voltage stability

1 Introduction

The optimal power flow is one of the nonlinear constrained and occasionally combinational optimization problems of power systems. The optimal power flow problems have been developed continually since its introduction by Carpentier [1]. Optimal power flow problem solution aims to optimize objective function via optimal adjustment of the power system control variables while satisfying both equality and inequality constraints OPF problem has been solved by a number of optimization methods such as nonlinear programming [2–7], quadratic programming [8, 9] and linear programming [10–12]. These methods have failed to handle non-convexities and non-smoothness problems. And evolutionary algorithm has more or less success in solving these problems. Recently, a new evolutionary computation technique called PSO has been proposed. PSO is a derived population based optimization technique first introduced by Eberhart and Kennedy [13] based on the sociological behaviour associated with the bird flocking and fish behaving, It takes into consideration the global and local information to determine its flying direction at next step. PSO has been successfully applied to several optimizing

P. Kumar (✉) · R. Pukale
Depatrment of Electrical Engineering, AMGOI, Kolhapur, Maharashtra 416112, India
e-mail: prashant2685@gmail.com

problems such as reactive power and voltage control, economic dispatch and determining generator contributions to transmission system. The original PSO is developed to solve continuous problems. After a little modification it can handle continuous and discrete variables easily. PSO is an unconstrained optimization method and to include the constraints the penalty factor approach is used to convert the constrained optimization to an unconstrained optimization problem.

2 Checking the PDF File Optimal Power System Optimization Sub Problem

2.1 Real Power Minimization Sub Problem

It is a traditional economic load dispatch problem which is done for minimising the generation cost while satisfying both equality and inequality constraints.

2.2 Reactive Power Minimization Sub Problem

Reactive power should be minimised to provide better voltage profile as well as to reduce the total System Transmission loss. Thus the objective of the reactive power is the minimization of real power transmission loss. The reactive power of the system is closely related to the voltage stability problem. With increasing in load, the voltage at the load bus may drop quickly which may lead to voltage instability. Voltage collapse occurs due to loss of generation and increase in demand. Voltage collapse can be accurately detected with the help of voltage stability indices. So the approached method uses a voltage stability index and checked how far the system from voltage instability is [14].

In the proposed method, first we have optimised the real power sub-problem. As the two objectives are complementary, finally we got a solution where both the objectives are optimised.

3 Problem Formulation

The conventional economic dispatch minimizes fuel cost which can be expressed mathematically as

$$F_t = \sum_{i=1}^{N_g} F_i = \sum_{i=1}^{N_g} \left(a_i P_{Gi}^2 + b_i P_{Gi} + c_i \right) \tag{1}$$

where

N_G total number of generators
P_{Gi} output power of ith generator
a_i, b_i, c_i fuel cost coefficient of ith generator.

3.1 Reactive Power Optimization Subsystem

The objective function for the optimization of reactive power sub problem can be formulated as

$$F_i = \sum_{k=1}^{NL} P_{Loss}$$

$$= \sum_{i=1}^{N_{bus}} \sum_{j=1}^{N_{bus}} g_{ij} \left(v_i^2 + v_j^2 - 2v_i v_j \cos \theta_{ij} \right) \tag{2}$$

where

N_{bus} Total number of buses
g_{ij} Conductance of line i–j
θ_{ij} Angle difference of ijth line
v_i Voltage magnitude of ith line

In order to minimize both the total fuel cost F_t and total real power loss F_i simultaneously subject to following constraints.

3.2 Equality Constraint

The equality constraints are basically the real and reactive power balance equations in each and every bus which are power flow equations. Mathematical formulation of equality constraints as
Real Power Balance

$$P_{Gi} - P_{Qi} = V_i \sum_{j=1}^{N_g} V_j \left(g_{ij} \cos \theta_{ij} + b_{ij} \sin \theta_{ij} \right) \tag{3}$$

Reactive Power Balance

$$Q_{Gi} - Q_{Di} = V_i \sum_{j=1}^{Ng} V_J \left(g_{ij} \cos \theta_{ij} - b_{ij} \cos \theta_{ij} \right) \tag{4}$$

where

P_{Gi} Real Power generation at bus i

Q_{Gi} Reactive Power generation at bus i

P_{Di} Real Power demand at bus i

Q_{Di} Reactive Power demand at bus i

N_B Total number of buses

Q_{ij} Angle of bus admittance element ij.

3.3 Inequality Constraints: Variable Limitations

1. Active and Reactive Generator Limits

$$P_{Gi}^{min} \leq P_{Gi} \leq P_{Gi}^{max}$$

$$Q_{Gi}^{min} \leq Q_{Gi} \leq Q_{Gi}^{max}$$

for i = 1, 2...N_g

2. Transformer Tap setting

$$T_i^{min} \leq T_i \leq T_i^{max}$$

Operating Limits:

$$V_i^{min} \leq V_i \leq V_i^{max}$$

$$\delta_i^{min} \leq \delta_i \leq \delta_i^{max}$$

for i = 1, 2...N_{bus}

Voltage Security Limits:

Index: $I_{fi} > 0.85$ for i = 1, 2...N_{bus}

I_{fi} is voltage stability index for ith bus. Voltage stability index occurs at $I_{fi} \leq 0.5$.

Lines overload Limits

$$S_i \leq S_i^{max}$$

for i = 1, 2...N_{line}.

4 Solution Strategy

The proposed solution strategy can be explained with the help of flow chart (Fig. 1). We need to maintain the voltage magnitude unchanged while minimizing the real power loss we can minimize the fuel cost via optimal adjustment of control variables. After cost is minimized, the reactive power sub problem minimizes the total transmission loss by keeping PV bus voltage magnitude constant. Thus the total cost also decreases after second objective minimization

Fig. 1 Flow chart of proposed method

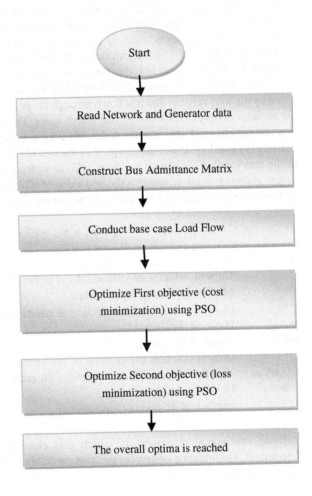

5 Overview of Particle Swarm Optimization

The salient features are as follows:

(a) It is mainly concern with biologically motivated general search technique such as fish schooling.
(b) PSO serves as simple and powerful tool for solving optimization problems.
(c) This technique is best suited for multidimensional discontinuous non-linearities.

In an evolutionary computation particle swarm optimization is an exciting new methodology. It is the population based optimization tool. It is motivated from the behavior of social systems, such as fish schooling and birds flocking. The basic assumption behind the PSO algorithm is, the birds find food by flocking and not individually. The observation is that the information is owned jointly in the flocking. The swarm has the population of random solutions. Each called a particle (agent) is given by random velocity and it is flown through the problem space. In a PSO system the particles exists in the n-dimensional search space. The position of the each particle in a two dimensional search space is described by x–y axis and correspondingly the velocity is vx and vy. In the search space all the particles have memory and keep track of the previous bets is known as pbest, with its fitness value. Another value of the swarm is known as gbest, which is the best value of all the particles best. The particles update its velocity and position using the following Eqs. 5 and 6.

$$v_i^{n+1} = w_i v_i^n + c_1 \times r_1 \times (pbest_i - s_i^n) + c_2 \times r_2 \times (gbest_i - s_i^n) \qquad (5)$$

v_i^n	Current velocity of individual i at iteration n,
v_i^{n+1}	Modified velocity of individual i,
s_i^n	Current position of individual i at iteration n,
$pbest_i$	Pbest of individual i,
$gbest$	Gbest of the group,
w_i	Weight function for velocity of individual i,
c_1 and c_2	Acceleration coefficient concern with each term,
r_1 and r_2	Uniform random values,
n	number of iterations (generations).

Using that equation, a velocity which gets close to P_{bests} and G_{best} can be determined. Modification of current position (searching point in the solution space) can be stated as

$$s_i^{n+1} = s_i^n + v_i^{n+1} \qquad (6)$$

Figure 2 shows the concept of searching points with modification. The Discrete variables which involved in Eqs. (5) and (6) was subjected to some little modification. But the state variables which used in the algorithm have no limitation. The salient aspects of search procedures are as follows:

(a) In this technique no of searching points are used and it will nearer to the optimal point using their P_{bests} and the G_{best}.
(b) Diversification which corresponds to the search procedure is denoted by RHS of (5).
(c) The expanded PSO can be easily applied to discrete problem.
(d) In these searching procedures there is no inconsistency of using state variables with its axis for XY positions and velocities.
(e) The concept of searching procedure is well explained for XY-axis (two dimension space) only. But, it is suitable for n-dimension problem.

The above concept of (b) is as follows [15]. The RHS of (5) having three terms. The first one is the previous velocity of the individual. For getting changes in the velocity of the individual second and third terms are used. The individual will be in "flying" until it reaches its boundary. The first term corresponds to the searching procedure .If it is not present, the velocity of the "flying" individual is determined by pbest and current best position of it. So the first term corresponds to the intensification. The concept of searching points with particles in a search space is shown in Fig. 2. The concept of selection is utilized for expanding the original PSO for getting high quality solutions [16].

S_i^k	searching point (current position)
s_i^{k+1}	searching point (modified)
v_{orig}	Velocity of current position
v_{mod}	Velocity of modified position
v_{pbest}	Velocity corresponds to Pbest
v_{gbest}	Velocity corresponds to G_{best}.

6 Penalty Function Approach to Handle Constraints

Penalty method can be used to convert unconstrained optimization techniques to constrained optimization techniques by adding penalty. We use this method as because the basic PSO is an unconstrained optimization problem. That's why we need penalty factor method for handling such constraints. Figure 3 shows the concept for searching points of particles in solution space [6].

Now the change of constrained problem into an unconstrained problem:

Fig. 2 Concept of searching point with modification

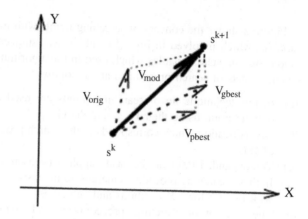

Fig. 3 Concept of searching points of particles in a solutions space

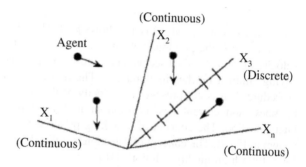

$$\text{Min } T(x) = f(x) + r_k P(x)$$

where

f(x)　objective function of the constrained problem
r_k　a scalar denoted as the penalty or controlling parameter
P(x)　function which impose penalties for infeasibility
T(x)　(pseudo) transformed objective.

Transformation equation:

$$T(x) = f(x) + r_k \left(\sum_{i=1}^{m} \max[0, g_i(x)]^2 \right) + \left(\sum_{j=1}^{n} [h_j(x)]^2 \right) \qquad (7)$$

where

$g_i(x) \le 0$'s are the inequality constraints.

7 Simulation Results

The proposed algorithm is applied to IEEE 57 bus system as shown in Fig. 4 [17]. Tables 1, 2 and 3 gives the details of the system [17]. Tables 4 and 5 gives the optimal result [6–12]. Figures 5, 6 and 7 show the characteristics plot of IEEE bus system [14].

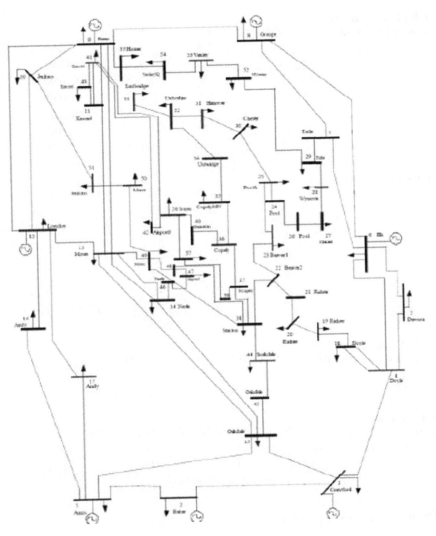

Fig. 4 Network model for IEEE 57 bus system

Table 1 Description of the IEEE 57 bus system

System	IEEE 57
Number of buses	57
Number of lines	80
Number of generators	7
Number of tap positions	17
Number of shunt positions	3

Table 2 Generator data IEEE 57 bus system

Unit	a_i	b_i	c_i
1	115	2.00	0.0055
2	40	3.50	0.0060
3	122	3.15	0.0050
4	125	3.05	0.0050
5	120	2.75	0.0070
6	70	3.45	0.0070
7	150	1.89	0.0050

Table 3 Cost coefficient of IEEE 57 bus system

Bus No.	P_G^{min}	P_G^{max}	Q_G^{min}	Q_G^{max}
1	0.2	0.5	0.0	0.0
2	0.15	0.9	0.5	−0.17
3	0.1	5	0.6	−0.1
4	0.1	0.5	0.25	−0.08
5	0.12	0.5	2	−1.4
6	0.1	3.6	0.09	−0.03
7	0.5	5.5	1.55	−0.5

Table 4 Minimum of all 25 independent runs using PSO

Fuel cost ($/h)	Ploss (p.u)
754.4	0.2

Table 5 Comparison of results with different methods for fuel cost and power loss

S. No.	Methods used	Fuel cost ($/h)	Ploss (p.u)
1	EP	752.6	0.35
2	PSO	754.4	0.20

EP Evolutionary Programing

The parameters adopted in PSO are:

Population size = 60
Initial inertia weight w = 1.2
Constriction factor = 0.792
Maximum iteration = 10.

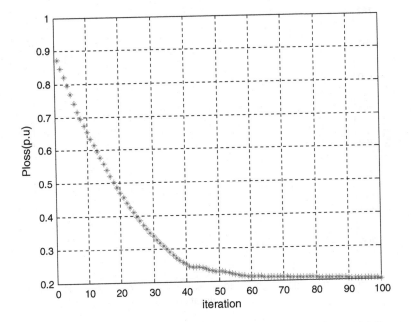

Fig. 5 Convergence characteristics by PSO for 57 bus system

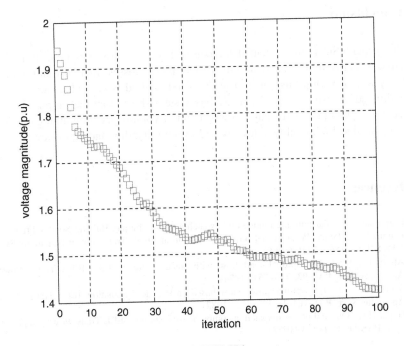

Fig. 6 Voltage magnitude versus iteration for IEEE 57 bus system

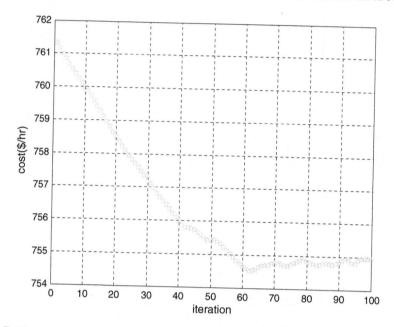

Fig. 7 Generation cost versus iteration for IEEE 57 bus system

8 Conclusion

In this paper PSO is implemented for real and reactive power optimization. The reactive power optimization with aims of real power losses, fuel cost and voltage stability is deal within this paper. The proposed method is very efficient and fast. The simulation results to the IEEE 57 bus system shows that the proposed technique give more optimized result as compared to Evolutionary Programing method. PSO could find high quality solutions with more reliability and efficiency.

References

1. Carpienter, J.: Contribution e l'etude do Dispatching Economique, Bulletin Society Francaise
2. Dommel, H., Tinny, W.: Optimal power flow solution. IEEE Trans. Power Apparatus Syst. **PAS-87**(10), 1866–1876 (1968)
3. Alsac, O., Stott, B.: Optimal load flow with steady state security. IEEE Trans. Power Apparatus Syst. **PAS-93**, 745–751 (1974)
4. Shoults, R., Sun, D.: Optimal power flow based on P-Q decomposition. IEEE Trans. Power Apparatus Syst. **PAS-101**(2), 397–405 (1982)
5. Happ, H.H.: Optimal power dispatch: a comprehensive survey. IEEE Trans. Power Apparatus Syst. **PAS-96**, 841–854 (1977)

6. Mamandur, K.R.C.: Optimal control of reactive power flow for improvements in voltage profiles and for real power loss minimization. IEEE Trans. Power Apparatus Syst. 3185–3193 (1981)
7. Habiabollahzadeh, H., Luo, G.X., Semlyen, A.: Hydrothermal optimal power flow based on a combined linear and nonlinear programming methodology. IEEE Trans. Power Syst. **PWRS-4** (2), 530–537 (1989)
8. Burchett, R.C., Happ, H.H., Vierath, D.R.: Quadratically convergent optimal power flow. IEEE Trans. Power Apparatus Syst. **PAS-103**, 3267–3276 (1984)
9. Aoki, K., Nishikori, A., Yokoyama, R.T.: Constrained load flow using recursive quadratic programming. IEEE Trans. Power Syst. **21**, 8–16 (1987)
10. Abou El-Ela, A.A., Abido, M.A.: Optimal operation strategy for reactive power control modelling. Simul. Control A **41**(3), 19–40 (1992)
11. Stadlin, W., Fletcher, D.: Voltage versus reactive current model for dispatch and control. IEEE Trans. Power Apparatus Syst. **PAS-101**(10), 3751–3758 (1982)
12. Mota-Palomino, R., Quintana, V.H.: Sparse reactive power scheduling by a penalty-function linear programming technique. IEEE Trans. Power Syst. **1**(3), 31–39 (1986)
13. Kennedy, J., Eberhart, R.: Particle swarm optimization. Proc. IEEE Int. Conf. Neural Networks **4**(27), 1942–1948 (1995)
14. Sinha, A.K., Hazarika, D.: A comparative study of voltage stability indices in a power system. Int. J. Electr. Power Energy Syst. **22**(8), 589–596 (2000)
15. Shi, Y., Eberhart, R.: Parameter selection in particle swarm optimization. In: Proceedings of 7th Annual Conference Evolutionary Program, pp. 591–600 (1998)
16. Angeline, P.: Evolutionary optimization versus particle swarm optimization: philosophy and performance differences. In: Proceedings of 7th Annual Conference Evolutionary Program, pp. 601–610 (1998)
17. Saraswat, A., Saini, A.: Multi-objective optimal reactive power dispatch considering voltage stability in power systems using HFMOEA. Eng. Appl. Artif. Intell. (EAAI) **26**(1), 390–404 (2013)
18. Sun, D.I., Ashley, B., Brewer, B., Hughes, A., Tinney, W.F.: Optimal power flow by Newton approach. IEEE Trans. Power Apparatus Syst. **PAS-103**(10), 2864–2875 (1984)
19. Santos, A., da Costa, G.R.: Optimal power flow solution by Newton's method applied to an augmented lagrangian function. IEE Proc. Gener. Transm. Distrib. **142**(1), 33–36 (1995)
20. Rahli, M., Pirotte, P.: Optimal load flow using sequential unconstrained minimization technique (SUMT) method under power transmission losses minimization. Electr. Power Syst. Res. **52**, 61–64 (1999)
21. Yan, X., Quintana, V.H.: Improving an interior point based OPF by dynamic adjustments of step sizes and tolerances. IEEE Trans. Power Syst. **14**(2), 709–717 (1999)
22. Momoh, J.A., Zhu, J.Z.: Improved interior point method for OPF problems. IEEE Trans. Power Syst. **14**(3), 1114–1120 (1999)
23. Lai, L.L., Ma, J.T.: Improved genetic algorithms for optima power flow under both normal and contingent operation states. Int. J. Electr. Power Energy Syst. **19**(5), 287–292 (1997)
24. Yuryevich, J., Wong, K.P.: Evolutionary programming based optimal power flow algorithm. IEEE Trans. Power Syst. **4**, 1245–1250 (1999)
25. Ozcan, E., Mohan, C.: Analysis of a simple particle swarm optimization system. Intell. Eng. Syst. Artif. Neural Networks **8**, 253–258 (1998)
26. Clerc, M.: The swarm and the queen: towards a deterministic and adaptive particle swarm optimization. In: Proceedings of the 1999 Congress on Evolutionary Computation, Washington D.C., pp. 1951–1957 (1999)

Video Retrieval Using Local Binary Pattern

Satishkumar Varma and Sanjay Talbar

Abstract Local binary pattern (LBP) operator is defined as gray-scale invariant texture measure. The LBP operator is a unifying approach to the traditionally divergent statistical and structural models for texture analysis. In this paper the LBP, its variants along with Gabor filters are used as a texture feature for content-based video retrieval (CBVR). The combinations of different thresholds over different pattern using Gabor filter bank are experimented to compare the retrieved video documents. The typical system architecture is presented which helps to process query, perform indexing, and retrieve videos form the given video datasets. The precision and mean average precision (MAP) are used over the recent large TRECViD 2010 and YouTube Action video datasets to evaluate the system performance. We observe that the proposed variant features used for video indexing and retrieval is comparable and useful, and also giving better retrieval efficiency for the above available standard video datasets.

Keywords Texture · Local binary pattern · Gabor filter · Query processing · Video indexing · Video retrieval

1 Introduction

The image contains various objects distributed across the various parts of the image. The image content is full of objects varying in size and shape. However in case of video frames, the content includes not only objects of deferent size and shape but

S. Varma (✉)
Department of Information Technology, PIIT, New Panvel, Navi Mumbai, India
e-mail: varmasl@yahoo.co.in

S. Talbar
Department of Electronics and Telecommunication, SGGSIE&T, Vishnupuri,
Nanded, India
e-mail: sntalbar@yahoo.com

© Springer India 2015
L.C. Jain et al. (eds.), *Computational Intelligence in Data Mining - Volume 1*,
Smart Innovation, Systems and Technologies 31, DOI 10.1007/978-81-322-2205-7_12

also the object location changes over time. The summation of local patterns together can help to represent the content of the image or frame depending on the context. In case of video frames these local patterns play important role in video indexing and retrieval. Each frame contains specific pattern which are extracted either globally or locally. The global pattern extraction helps to describe entire frame whereas the local pattern extraction describe only part of the image.

The image contains various objects distributed across the various parts of the image. The image content is full of objects varying in size and shape. However in case of video frames, the content includes not only objects of different size and shape but also the object location changes over time. The summation of local patterns together can help to represent the content of the image or frame depending on the context. In case of video frames these local patterns play important role in video indexing and retrieval. Each frame contains specific pattern which are extracted either globally or locally. The global pattern extraction helps to describe entire frame whereas the local pattern extraction describe only part of the image.

1.1 Literature Survey

The content-based image retrieval (CBIR) utilizes the visual contents of an image such as color [1], shape [1, 2] texture [3], transforms coefficients [4–6], moments [7] etc. in order to represent and index the image. In all this the computational complexity and retrieval efficiency are the key objectives in the design of CBIR system [8].

The recently proposed LBP features are designed for texture description for various applications such as recognition and CBIR. Ojala et al. proposed the LBP [9] and these LBPs are converted to rotational invariant for texture classification [10]. Pietikainen et al. proposed the rotational invariant texture classification using feature distributions [11]. The LBP operator is defined as gray-scale invariant texture measure and used for classification.

Ahonen et al. [12] and Zhao and Pietikainen [13] used the LBP operator facial expression analysis and recognition. Heikkila et al. proposed the background modeling and detection by using LBP [14]. Huang et al. [15] proposed the extended LBP for shape localization. Heikkila et al. [16] used the LBP for interest region description.

1.2 Contribution and Outline of This Paper

Existing methods that helps us to retrieve only images by using simple LBP texture feature along with filters like Gabor. Also LBP has been used for segmentation and recognition purpose. In this paper we use different thresholds over LBP applied on video frames filtered using Gabor filter banks on different color planes such as

RGB, and we design a feature set that is able to accommodate maximum content of the video key frame for CBVR from the given standard video datasets.

The organization of the paper as follows: In Sect. 1.1, a brief review of video retrieval and related work is given. Section 2, presents a system architecture and approach used for video indexing and retrieval. Section 3 presents the feature indexing using LBP. The experimental results and discussions are given in Sect. 4. Based on above work conclusion are derived in Sect. 5.

2 System Architecture

The system architecture in generalized form is presented in Fig. 1. The system architecture is divided into four units. The four units are frame extraction, Gabor filter bank selection, local binary pattern, feature vector formation, and feature indexing. The frame extraction block consists of four functions namely video selection, key frame extraction, color processing and block formation. The block formation step helps to process the local object as local features.

The different Gabor filter banks (described in Sect. 3.1) are applied on each key frame to describe the presence of edges and bars of the corresponding orientation and spatial frequency. The LBP unit helps to compute the texture features in all local regions of the key frame.

3 Local Binary Pattern

The LBP operator is a unifying approach to the traditionally divergent statistical and structural models for texture analysis. The local detail such as points, edges, curved edges, patterns, etc. are various local binary patterns used for image/frame retrieval. LBP computes texture features in all local regions in the given frame.

As LBP features correspond to all local regions of a frame/keyframe, we describe them by their statistical distribution. The example of the LBP operator introduced by Ojala et al. [9] is shown in Fig. 2.

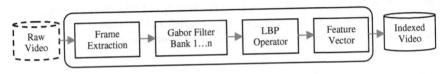

Fig. 1 The system architecture

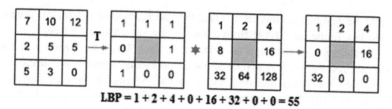

$$\text{LBP} = 1 + 2 + 4 + 0 + 16 + 32 + 0 + 0 = 55$$

Fig. 2 The example of local binary pattern

3.1 Gabor Filter Bank

The Gabor filter is basically a Gaussian (with variances Sx and Sy along x and y-axes respectively). The deferent coordinates and variables are used to present the different piece of information like {I, Sx, Sy, f, θ, γ, λ, φ} passed to filter function. A bank of Gabor filters with different preferred coordinates and variables applied to a frame are given here as an example in Table 1.

Gabor filters proves to be very useful texture analysis and is widely adopted in the literature. Texture features are found by calculating the mean and variation of the Gabor filtered image. Rotation normalization is realized by a circular shift of the feature elements so that all images have the same dominant direction.

3.2 Feature Indexing

The LBP example given in Fig. 2 is traditional method used most of the areas mentioned in above section. It can be easily observed that this is certainly not a good idea. In this system architecture, we have used the additional threshold (i.e. mean) to experiment and evaluate the performance. In this threshold an average value is taken. The threshold used is given in Eq. 1. In the modified Eq. 1, the where B_k is Kth block of the ith frame F_i of the jth video V_j in given video dataset.

$$\text{Threshold} = \text{mean}\left(B_k\left(V_j(F_i)\right)\right). \tag{1}$$

The GLBP along with traditional approach (represented as GLBPmd) and mean value (represented as GLBPmn) are incorporated in the proposed system architecture given in Fig. 1. The LBP is applied over sub-blocks of Gabor filtered frame. The row and column pixels are used for feature vector formation.

Table 1 The details of parameters used in Gabor filter bank

Parameter	Sample values
{I, Sx, Sy, f, θ, γ, λ, φ}	{I, 2, 4, 16, pi/6, 0.5, 12, pi/3}

Fig. 3 The 31st keyframe of 152nd video in TRECVid 2010 test collection

3.3 An Example

In this section, we present an example to understand the approach used for CBVR. The GLBP are used to index the given frame/keyframe of a video. To understand the feature extraction, let us consider a keyframe. This keyframe (shown in Fig. 3) is 31st keyframe of 152nd video in TREC10 test collection.

The above selected keyframe (320 × 240) is resized of size to 240 × 240. Now using this resized keyframe, in GLBPmn the mean value is used as a threshold (shown in Eq. 1). The GLBPmn yields 380 elements in feature vector. The first 24 normalized elements out off 380 feature elements of this key frame are given in Eq. 2. The feature elements in GLBP approach depends on the block size of the given frame. Also, it depends upon the local threshold of the key frame.

$$\begin{aligned} mn = {} & 0.0033;\ 0.0058;\ 0.0064;\ 0.0053;\ 0.0072;\ 0.0077;\ 0.0068;\ 0.0069; \\ & 0.007;\ 0.0075;\ 0.0079;\ 0.0083;\ 0.0088;\ 0.0091;\ 0.0095;\ 0.0108; \quad (2) \\ & 0.0131;\ 0.0132;\ 0.0132;\ 0.0141;\ 0.0153;\ 0.0208;\ 0.0197;\ 0.0183 \end{aligned}$$

4 Experimental Results

In this work, the YouTube Action Dataset 2009 (YTAD09) [17] and the recent TRECViD Video Dataset 2010 (TREC10) datasets are used to evaluate the performance of proposed system. YouTube dataset used in [17] contains 11 action categories: basketball shooting, biking/cycling, diving, golf swinging, horseback riding, soccer juggling, swinging, tennis swinging, trampoline jumping, volleyball spiking, and walking with a dog. The details of the dataset used are shown in Table 2.

The YouTube dataset is very challenging due to large variations in camera motion, object appearance and pose, object scale, viewpoint, cluttered background, illumination conditions, etc. For each category, the videos are grouped into 25 groups with more than 4 action clips in it.

Table 2 Details of standard video test collection used for performance evaluation

Dataset used	No. of video	No. of keyframe	Frame size
TRECViD 2010	250	11,000	320 × 240
YTAD 2009	275	12,100	320 × 240

4.1 Video Retrieval Results

In this section, the performance comparison of GLBPmd and GLBPmn over two datasets (mentioned in Table 2) is presented. The comparative results for the GLBP features are shown in the Table 3. The system performance is measured with the help of precision and mean average precision (MAP) [17–19]. The precision is measured at document cut-off 10, 20 and 30 (i.e. the top 10, 20 or 30 documents). The MAP is measured for entire document. This is denoted as P@10 (read as 'P at 10'), P@20 and P@30.

The first column in Table 3 gives the information about dataset used. In each of the two datasets, there are three retrieved results for three different input video queries. These video queries are selected randomly. The result of GLBPmd is shown using 2nd column through 4th column. The performance of GLBPmd and GLBPmn in term of MAP is shown using 5th column through 9th column. The performance improvement with respect to MAP is shown in the last column.

For instance, using GLBPmd feature elements, the first query from YTAD09 dataset gives the precision P@10, P@20 and P@30 as 0.87, 0.63 and 0.44 respectively. The MAP for the same query is 0.0024. Whereas, in case of GLBPmn, the first query from YTAD09 dataset gives the precision P@10, P@20 and P@30 as 0.93, 0.65 and 0.46 respectively. The MAP for the same query is 0.0026. The GLBPmn achieved a higher MAP of 0.0026 compared to 0.0024 achieved by GLBPmd. The improvement by the GLBPmn scheme is 0.08.

4.2 Performance Comparison

The performance comparison of GLBP features with respect to retrieval time is shown in Fig. 4a. The performance comparison of GLPB1 and GLBPmn using MAP is shown in Fig. 4b. The improvement by GLBPmn is relative to the MAP of the GLBPmd representation. The significant improvement in the performance is observed for both the datasets (see in Figs. 4 and 5).

Overall, the results indicate that the GLBPmn is superior to the basic GLBPmd of the keyframe. The proposed GLBPmn gives better results because it considers mean value of each block elements whereas for a given frame the GLBPmd over a keyframe blindly uses center value as a threshold to capture local details. The space required after indexing and the time required for indexing is given in Fig. 5.

Table 3 The performance comparison of GLBPmd and GLBPmn over dataset (a) YTAD09 and (b) TREC10

Method dataset	GLBPmd				GLBPmn				
	P@10	P@20	P@30	MAP	P@10	P@20	P@30	MAP	Impr.
YTAD09	0.87	0.63	0.44	0.0024	0.93	0.65	0.46	0.0026	0.08
	0.73	0.55	0.39	0.0103	0.83	0.70	0.51	0.0179	0.74
TREC10	1.00	1.00	0.73	0.0021	1.00	1.00	0.73	0.0022	0.05
	0.97	0.77	0.66	0.0019	0.93	0.82	0.63	0.0021	0.11

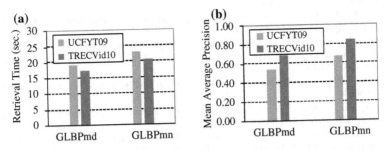

Fig. 4 Performance of GLBPmd and GLBPmn using **a** retrieval time and **b** MAP

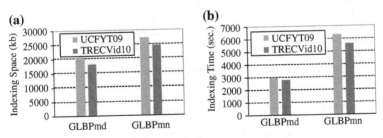

Fig. 5 Performance of GLBPmd and GLBPmn using **a** indexing space and **b** indexing time

5 Conclusion

In this paper, the mean and median thresholds over LBP are experimented to index and retrieve video from publicly available two standard video test collections. The local change in the keyframe using mean threshold rather than median threshold over GLBP was recorded. Also different Gabor filter banks are experimented to improve the results. In general, the local change within the video frames is more frequent than the global change between the two consecutive frames. Also, the two different thresholds give different details of the objects in the frame. The GLBPmn scheme (using mean threshold) is more suitable as compared to GLBPmd (traditional median threshold) for video documents.

Acknowledgments I would like to thank my teachers and the colleagues of SAKEC, DBIT, and PIIT for encouraging me for implementation and writing papers.

References

1. Swain, M.J., Ballard, D.H.: Color indexing. Int. J. Comput. Vis. **7**(1), 11–32 (1991)
2. Jain, A.K., Vailaya, A.: Image retrieval using color and shape. Pattern Recogn. **29**(8), 1233–1244 (1996)
3. Manjunath, B.S., Ma, W.Y.: Texture feature for browsing and retrieval of image data. IEEE PAMI **8**(18), 837–842 (1996)
4. Ngo, C.W., Pong, T.C., Chin, R.T.: Exploiting image indexing techniques in DCT domain. Pattern Recogn. **34**, 1841–1851 (2001)
5. Talbar, S.N., Varma, S.L.: iMATCH: image matching and retrieval for digital image libraries. In: 2nd International Conference on Emerging Trends in Engineering and Technology, pp. 196–201 (2009)
6. Mali, K., Gupta, R.D.: A wavelet based image retrieval. **3776**, 557–562 (2005). ISBN 978-3-540-30506-4
7. Varma, S.L., Talbar, S.N.: IRMOMENT: image indexing and retrieval by combining moments. IET Digest **38** (2009)
8. Smeulders, A.W.M., Worring, M., Satini, S., Gupta, A., Jain, R.: Content-based image retrieval at the end of the early years. IEEE Trans. Pattern Anal. Mach. Intell. **22**(12), 1349–1380 (2000)
9. Ojala, T., Pietikainen, M., Harwood, D.: Performance evaluation of texture measures with classification based on Kullback discrimination of distributions. In: Proceedings of the 12th IAPR International Conference on Pattern Recognition, **1**, 582–585 (1994)
10. Ojala, T., Pietikainen, M., Harwood, D.: A comparative study of texture measures with classification based on feature distributions. Pattern Recogn. **29**, 51–59 (1996)
11. Ojala, T., Piettikainen, M., Maenpaa, T.: Multiresolution gray-scale and rotation invariant texture classification with local binary patterns. IEEE Trans. Pattern Anal. Mach. Intell. **24**(7), 971–987 (2002)
12. Ahonen, T., Hadid, A., Pietikainen, M.: Face description with local binary patterns: applications to face recognition. IEEE Trans. PAMI **28**(12), 2037–2041 (2006)
13. Zhao, G., Pietikainen, M.: Dynamic texture recognition using local binary patterns with an application to facial expressions. IEEE Trans. PAMI **29**(6), 915–928 (2007)
14. Heikkila, M., Pietikainen, M.: A texture based method for modeling the background and detecting moving objects. IEEE Trans. on PAMI, vol. 28 (4), pp. 657–662 (2006)
15. Huang, X., Li, S.Z., Wang, Y.: Shape localization based on statistical method using extended local binary patterns. In: Proceedings of International Conference on Image and Graphics, pp. 184–187 (2004)
16. Heikkila, M., Pietikainen, M., Schmid, C.: Description of interest regions with local binary patterns. Elsevier J. Pattern Recogn. **42**, 425–436 (2009)
17. Liu, Jingen, Luo, Jiebo, Shah, Mubarak: Recognizing realistic actions from videos in the wild. IEEE Int. Conf. CVPR **2670**, 1996–2003 (2009)
18. Snoek, C.G.M., Worring, M., Koelma, D.C., Smeulders, A.W.M.: A learned lexicon-driven paradigm for interactive video retrieval. IEEE Trans. Multimedia **9**(2) (2007)
19. Muller, H., Muller, W., Squire, D., Marchand-Maillet, S., Pun, T.: Performance evaluation in content-based image retrieval: overview and proposals. PR Lett. **22**(5) (2001)
20. Smeaton, A.F., Kraaij, W., Over, P.: The TREC VIDeo retrieval evaluation (TRECVID): a case study and status report. In: RIAO (2004)

Mobile Agent Communication Protocols: A Comparative Study

Ajay Rawat, Rama Sushil and Lalit Sharm

Abstract Mobile Agent is considered as a striking technology for distributed applications. Its mobility feature makes it difficult to track and consequently performing communication between them becomes difficult. In multi mobile agent environment, a reliable communication is still a big challenge. In this paper, we have addressed the pros and cons of the existing different Mobile Agent communication protocols with their limitations. This paper also presents their parametric comparative study in tabular form.

Keywords Mobile agent · Communication protocol · Reliable communication

1 Introduction

In computer network, a Mobile Agent (MA) represents users and it acts on their behalf to perform some computation to fulfill their goals, therefore they are called as agents. MA is a program which can migrate autonomously from one node to another node in widely distributed heterogeneous networks [1]. Mobility is a basic characteristic of a MA. It has features like self-contained, identifiable, autonomy,

A. Rawat (✉)
Department of CIT, University of Petroleum and Energy Studies,
Dehradun 248007, India
e-mail: rawat.ajay@hotmail.com

R. Sushil
Department of Computer Science and Application, DIT University,
Dehradun 248009, India
e-mail: ramasushil@yahoo.co.in

L. Sharm
Department of Computer Application, G.B. Degree College,
Rohtak 124001, India
e-mail: lalitsharma2004@gmail.com

131

© Springer India 2015
L.C. Jain et al. (eds.), *Computational Intelligence in Data Mining - Volume 1*,
Smart Innovation, Systems and Technologies 31, DOI 10.1007/978-81-322-2205-7_13

learning, adaptive, reactive and mobility etc. [2]. They can be considered as an object in which code, data and execution state is wrapped together.

MA is free to travel between the nodes in the network, so once they created in one platform they can suspend their execution, transfer their state and code to migrate to another platform in the network. Its state depicts its attributes which facilitates what to perform, when it recommence its execution at the destination node. MA Code is a logic or function necessary for an agent execution [3].

Due to its mobility and autonomous behavior, MAs are appropriate to various computing areas like network management, distributed systems [4], information retrieval, monitoring, notification, information dissemination, parallel processing, internet, intrusion detection, telecommunication network service and secure broking etc. It provides several advantages like network load reduction, execute autonomously and asynchronously, overcome network latency, adapt dynamically, heterogeneous, robustness and fault tolerance.

Next section discusses the need of communication between the MAs. Section 3 describes the various existing MA communication protocols. Section 4 presents dimension evaluation and comparison of different MA communication protocols based on the different identified parameters. Section 5 concludes the paper and the last Sect. 6 presents the future research work.

2 Mobile Agent Communication

In Mobile Agent System (MAS), agents coordinate and collaborate with each other [5]. Communication is an essential feature of MAS, which facilitates MAs to interact with each other through exchange of knowledge and sharing of information [6]. To achieve a large objective or huge task, different agents collectively work together, so they have to communicate with each other to complete that task. In an agent oriented paradigm, there is a concept of master (parent) and slave (child) agent [7]. Master agent subdivides the huge task into smaller subtasks and delegates them to its slave agents, which helps in completing the task fast and easily. During this process parent and child agents need to communicate with each other to complete the given task in a reliable and efficient manner.

In order to communicate, agents must be able to deliver, receive messages, parse messages and understand the messages. Communication mechanism between mobile agents is not that simple as the communication method in computer network [8]. In most of the interactive applications where the cooperating MAs exchange partial results and valuable information, they have to make decisions while migrating in the network. For example in a network where MAs are used for information retrieval, it is effective to use MA communication between them to share partial results while each of the agents performs their own task.

3 Existing Communication Protocols for Mobile Agents

In MA paradigm several communication protocols have been proposed by the researchers. To the best of my knowledge, all existing protocols are discussed below.

In *Session Oriented Communication* Protocol [9], a session is needed to setup before an agent communicates with other agent. After establishing session the Remote Procedure Call or Message Passing technique is used to communicate between the agents. Once communication is done, session is terminated. In this protocol agents are identified either by unique global agent identifier or by the badge concept. 'Mole' [10] MAS uses this protocol.

Central Server Scheme (CS) [11] uses a dedicated and centralized single location based server, to maintain and track the location of all MAs. A mobile agent notifies its location to central server before leaving the current node and after reaching to the new node. For example if a sender MA 'Y' wants to send messages to another MA 'X; sender first sends the message to the central server, which forwards the message to the MA 'X' [12]. The SeMoA (Secure Mobile Agent) toolkit implements a central server solution, named Atlas [13].

Home Server Scheme [11] is an extension of CS and scales better as it is having more than one tracking servers. At Home Agency (HA where the agent was started) each MA has its own CS. HA maintains the database of all agents that uses this host as home agency. The central server is called 'Home Server' which binds the MAs name to their HA's address. This scheme is analogous to mobile IP which routes IP packets to mobile hosts. On each migration, MA must inform its new location to HA. To send a message, sender sends message to home server, home server sends message to agent's HA, and finally HA forwards the message to the actual location of the MA from the database. SPRINGS and Aglet [14] uses this protocol.

In *Forwarding Proxy Protocol* [15] location management is done through 'Forwarding Pointers' and message delivery is through forwarding approach. It creates a proxy on each node that maintains the agent location information. Whenever an agent migrates, it updates its new location to the current node's proxy. This proxy helps in tracking the agent to deliver the message. But if the path proxy is long it becomes difficult to deliver the message and the cost of the communication increases. e.g. If agent 'X' wants to migrate from node n1 to n2, it leaves a forwarding pointer at n1 proxy. If there is no agent pointer then home is contacted. Voyager [16] and Epidaure use this protocol.

The *Shadow Protocol* [17] is a combination of Forwarding and Home Proxy Pointer protocols. It uses the concept of shadow which acts as a placeholder and keeps a record of agents. MA updates its location to its linked shadow according to Time To Live (TTL). If TTL still remains, a MA produces a forwarding proxy at each and every visited node. Subsequently when the TTL finishes, the MA updates its current location [18] to its shadow, which curtails the length of search path. Associated shadow of and MA is responsible for message delivery. Shadow protocol uses path proxies to trace the agent for message delivery. If the shadow is removed from the agent then it is considered as orphan agent. 'Mole' uses this protocol.

Message Delivery Protocol (MDP) [19] facilitate in accurately tracking the agent location. It uses hierarchy tree like structure retain all agent's addresses. It consists of domains where the group of agents is placed and each domain contains a gateway where root is the top of it. On agent migration server in a domain updates the agent's address of new location. It is a simple technique but the hierarchical tree structure is complex to design in a large network like internet.

Distributed Snapshot [20] broadcasts the message with two states: FLUSHED and OPEN to each node in incoming channel. For the first time when message arrives at a node, it is stored locally and after that message is propagated to all outgoing channels with the flushing process. Message is then delivered to all MAs.

Resending-Based Protocol [21] uses the concept of sliding-window acknowledgement mechanism (used by TCP) for communication between agents. After sending the message, it buffers the unacknowledged message and computes the expected estimated round trip time for the acknowledgement. If the time exceeds, the sender retransmit the message as it retains all the messages that have not been acknowledged by the receiver. On the other hand receiver maintains the reception window for the acknowledgement. It suffers from communication overhead on frequent migration. MStream (Mobile Stream) [21] MAS uses this protocol.

Agent Communication Transfer Protocol (ACTP) [22] uses several different other protocols communication and message exchange between MAs. ACTP is an application layer internet protocol. It is a 'shield' protocol which hides low level networking details from MA. In this protocol sender agent delivers ACL (Agent Communication Language) messages to ACTP and ACTP is responsible for message delivery. It provides two levels of encapsulation, *'higher'* where communication process networking details are concealed from the agents and *'lower'* where application protocol implementation is hidden from ACTP. It uses 'Name Server', which provides agent's low level information like domain, logical name, physical address and data authentication to ACTP for interaction with the agent.

In *Blackboard Protocol* (BB) [23] every node has a pool information space (Tuple space or Blackboard) to exchange messages between MA. Whenever a MA wants to send a message to other agent, irrespective of the receiver agent location sender agent simply places the message to their blackboard. This helps in providing the asynchronous message delivery. To obtain message MA has to move to the respective nodes, which create unnecessary communication overhead.

The *Broadcast Protocol* [24] is used for group communication. In order to deliver message, it broadcasts message to all nodes in a network without knowing the location information of MA. If MA is residing in its home node, protocol delivers the message to it directly otherwise it leaves message in its mailbox. Whenever MA wants to access the message it moves to the respective node. In this protocol, as the number of nodes increases the cost of communication also increases and become impractical to use. GCS-MA [25] uses this protocol.

Search-By-Path-Chase (SPC) protocol [26] supports multi-region environment for mobile communication. In this protocol, location management is done through Forwarding Pointers and Location Servers. Message is delivered through 'Direct approach'. Region Agent Register (RAR) or a Site Agent Register (SAR) is used to store location information in a distributed fashion. RAR stores location information of all the agents residing within a region. SAR stores the information of agents which are residing at a node or those which have visited the node. The links provided at RAR and SAR facilitate in achieving location management and message delivery.

Message Efficiently Forwarding Schema (MEFS) [27] is a combination of two schemes named Central Server and Forwarding Pointer. It resolves the problem of location tracking, avoidance and detection for message delivery. Each host keeps track of created agent and currently residing agent on it. To keep this information updated, each migrated agent first unregister it, place a forwarding pointer in the current node, informing the destination address to the initial host and register with new host as well. It introduces Chasing Message Register, over Speed Blocking to attain synchronization and resolve the message chasing problem.

Adaptive and Reliable Protocol (ARP) is derived from 3-D framework proposed by Cao et al. [28]. In this protocol location management is done through Forwarding Pointer and Location server. Message is delivered through Mailbox [29, 30]. Mailbox is attached to each MA. An agent needs to remember the location of its mailbox. At the time of agent migration, its mailbox also moves with it. A message first reaches to mailbox and then MA agent can retrieve the message whenever needed. In transferring of mailbox high overhead is incurred [31].

Proxy-Based Protocol [32] is an improvement over the HSS protocol. It includes an additional agent called 'proxy agent' in the home-server protocol. In this scheme, a cluster of interconnect computers called 'Domain', shares a Central Directory Services Database (CDSD), which contains privacy and security information with user account. Every domain contains minimum one proxy agent to offer communication facilities to all other MAs in that particular domain. Before dispatching messages over the network, messages are sent through these proxies to MAs. On the basis of receiver's latest knowledge, the proxy agent selects where the message should move.

Session Initiation Protocol (SIP) [33] reduces the MA communication fault rates which enhance its reliability to locating MA and deliver messages. This protocol proposes two models Synchronous Invite Model (SIM) and Asynchronous Subscribe and Notify Model (ASNM) [34]. Normal communication between MAs is handled by SIM. For extreme cases like moving agents or when agents migrate frequently ASNM is used to transmit message.

Distributed Sendbox Scheme [35] uses SIP. It uses home server 'H' as a centralized supervisor to keep track of all the MAs and manages them in an MAS. In this protocol every agent has to register themselves at 'H' to update their location after reaching to their destination and have to unregister at 'H' before they migrate from their location. To identify agent name and its location a unique SIP URL <agent_id@hostname> is used where agent_id is unique which remain unchanged and host name changes as per the agent's current location. This Protocol offers two different modes: Direct and Forward. Direct mode is used when the recipient agent resides on a node (i.e. the agent is stationary) and Forward mode is used when the agent moves frequently. In the latter case protocol retains the message in the buffer called 'Sendbox' and the message is delivered to the agent when it arrives at the node.

Reliable Asynchronous Message Delivery Protocol (RAMD) [36] consists of Blackboard technique for asynchronous message delivery. It also deals with location management of MAs. In a particular region, it is the responsibility of the Region Server (RS) to deliver messages to all MAs. Each RS consists of blackboard where the messages to be delivered are stored. Every time when receiver agent migrates, it updates its location in RS, RS checks the message in the blackboard for the receiver. If there is a message, it is delivered to the receiver agent. RAMD tightly binds the message delivery at the time of agent migration. Oddugi MAS uses this protocol.

Reliable Communication Protocol (RCP) [37] is a communication protocol for multi-region environment [38]. It uses region server, which maintains the MA location information within the region and a Blackboard in a region server for storing the message to be delivered. It also deals with cloned MAs, parent-child mobile agent location management and message delivery in distributed environment.

Reliable User Datagram Protocol (RUDP) [39] communication Protocol is an enhancement over standard UDP [40], which is more efficient than TCP. It is based on, technical characteristics of UDP protocol, which can be used in high MA communication architecture. It implements the mechanism of packing/unpacking, ordering guarantee, recognition technology, data retransmission, and error control and data security.

A protocol proposed by Yousuf and Hammo [41] uses 'client agent' to request the service and 'server agent' to provide services. It consists of different regions; each region is having CS (central server) which comprises of Global Agent Location Table (GALT). GALT keeps track of agent's location of other region. Local Agent Location Table (LALT) keeps track of agent's location residing locally in current region. It uses global naming scheme. On agent creation it registers its unique name at name server and LALT. In case an agent transit to other region, its new location is registered to GALT with '0' status till agent reaches its destination. To establish a connection a 'client agent' fetch 'server agent' name from name server, CS uses LALT for local communication or GALT for remote communication. It uses 3 different schemes for agent mobility management; (1) New location addressing in a multicast message to all interested region, (2) Forward to all using successor and predecessor, (3) Forward to nearest one.

Ahn [42] proposed an atomic MA protocol for group communication. This protocol improves scalability by allowing MA to select only few number of nodes as Agent Location Manager (ALM) based on predefined strategies such as location updating, message delivery cost security, inter-agent communication pattern, network latency and topology. It gives an assurance of reliable agent communication even if the ALM fails [43]. To achieve maximum scalability it replicates the MA location information. It caches agent group location to shorten the message delivery time to targeted MA and decrease the message forwarding load on ALM.

4 Dimensional Evaluations and Comparison of Communication Protocol

From the above different MA communication protocols, we find some parameters to compare and evaluate these communication protocols. The identified parameters are Reliability, Asynchronous, Timeliness, Location Transparency and Scalability. These parameters are defined below.

4.1 Reliability

Reliability in MA communication protocols mean that messages should be delivered without failure even during the agent migration. The agent's mobility behavior creates "Tracking problem" [44] which results in message delivery failure. The asynchronous nature of message forwarding and agent migration may cause message loss or chasing problem. During agent's migration message cannot be delivered, so protocol must solve this problem for reliable message delivery in timely manner.

4.2 Asynchronous

In MA communication, Asynchronous means sender agent sends a message to the destination agent, during this phase destination agent may start or continues (not blocked) its own task execution [45], thereby it may migrate. A MA performs its task asynchronously and does not need a permanent connection to its home node.

4.3 Timeliness

Timeliness [46] can be described as a level of significance that is entrusted upon time, stated as expired or out of date based upon the previous interactions that have taken place between agents. It is the work of the communication protocol to ensure that messages are delivered in a timely manner to different agents.

4.4 Location Transparency

Mobile agent can send message to the other agents without having prior knowledge of the receiver agent's location. Therefore location tracker module of the MAS has to keep a record of the location of MAs continuously.

Table 1 Parametric values of some existing MA communication protocols

Protocol	Reliability		Asynchronous	Timeliness	Location transparency	Scalability
	Tracking problem	Message delivery				
HSS	×	×	Δ	Θ	Θ	×
FP	×	×	Θ	×	Θ	×
Broadcast	Δ	Δ	Δ	Θ	×	×
Shadow	×	×	Δ	Δ	Θ	×
SPC	×	×	Δ	Δ	Θ	Δ
ARP	×	Δ	Θ	×	Θ	×
MDP	Θ	Δ	Θ	Θ	Θ	Δ
MEFS	Θ	Θ	Δ	Δ	Θ	×
Session oriented	×	×	×	Θ	Θ	×
Distributed snapshot	Δ	Δ	Θ	Θ	×	×
Resending protocol	×	×	Θ	×	Θ	×
CS	×	×	Δ	Θ	Θ	×
Blackboard	×	×	Θ	×	×	Θ
SIP	Θ	×	Δ	×	×	×
Distributed Sendbox	Θ	×	Δ	×	×	×
RAMD	Θ	Θ	Θ	Δ	Θ	Θ
RCP	Θ	Θ	Θ	Δ	Θ	Θ
MailBox	×	×	Θ	×	Θ	×

Θ: Yes, ×: No, Δ: in some cases

4.5 Scalability

Scalability of a communication protocol means that when the number of MAs increases the communication protocol should response without performance degradation. Table 1 presents different existing communication protocols with their parametric values.

5 Conclusion

Only few existing communication protocols like SPC, RCP and BSPC consider multiregional computing environments. Limited protocols, like RCP and RAMD deal with location management and message delivery together. Few protocols are taking care of cloned mobile agents and parent-child mobile agents for communication. In most of the communication protocols like Central Server, Home Server protocol the 'triangle problem' results in communication failure. Most of the protocols are not giving any consideration to keep the network and memory overheads low. Communication overhead (includes both location updates and message delivery) is high in most of the existing protocols e.g. HSS, BB, ARP, Resending based protocol. Only few existing protocols guarantee message delivery to MAs during their migration e.g. RCP, RAMD and MEFS. Forwarding Proxy and Shadow protocols follow path proxies to deliver messages, ARP protocol cannot retrieve message until MA deliver message in mailbox so they do not provide timeliness. Broadcast protocol suffers with high cost as it transmits multiple messages. There is no helping tool or standard criteria for the selection of an application suitable communication protocol. To solve these above problems, a selection criteria and a novel Mobile Agent Communication Protocol is to be developed for multi-region Mobile Agent Platform.

6 Future Work

In future we intend to develop a generic framework to aid the designing of application suitable and Mobile Agent Platform independent communication protocols. For efficiency measurement and comparison purpose a performance matrix is also planned to develop, which will include the parameters like; interaction overhead, scalability, location update overhead, migration overhead, message complexity etc. Using the above proposed generic framework, it is planned to design and validate a more efficient novel communication protocol and then measuring its efficiency and performing its comparison with existing protocols.

References

1. Milojicic, D.: Mobile agent applications. IEEE Concurrency 7(3), 80–90 (1999)
2. White, J.: Mobile Agents White Paper (1996)
3. Lange, D.B., Oshima, M.: Seven good reasons for mobile agents. Commun. ACM 42(3), 88–89 (1999)
4. Kumar, K.S., Kavitha, S.: Analysis of mobile agents applications in distributed system. In: Proceedings of the International Conference on Internet Computing and Information Communications, pp. 491–501. Springer, India (2014)
5. Sreekanth, V., Ramchandram, S., Govardhan, A.: Inter-agent communication protocol for adaptive agents. Int. J. Future Comput. Commun. 2(6), 692–696 (2013)
6. Virmani, C.: A comparison of communication protocols for mobile agents. Int. J. Adv. Technol. 3(2), 114–122 (2012)
7. Ahmed, S., Aamer, N.: A survey on mobile agent communication protocols. In: Proceedings of the IEEE International Conference on Emerging Technologies, pp. 1–6 (2012)
8. Hidayat, A.: A review on the communication mechanism of mobile agent. Int. J. Video Image Process. Netw. Secur. 11(1), 6–9 (2011)

9. Baumann, J., Hohl, F., Radouniklis, N., Rothermel, K., Straber, M.: Communication Concepts for Mobile Agent Systems. Mobile Agents, pp. 123–135. Springer, Berlin (1997)
10. Baumann, J., Hohl, F., Radouniklis, N., Straber, M., Rothermel, K.: Mole-concepts of a mobile agent system. World Wide Web J. **1**(3), 123–127 (1998). (Special issue on Applications and Techniques of Web Agents)
11. Wojciechowski, P.T.: Algorithms for location-independent communication between mobile agents. In: Proceedings of the AISB Symposium on Software Mobility and Adaptive Behaviour, York, UK (2001)
12. Braun, P., Rossak, W.R.: Mobile Agents: Basic Concepts, Mobility Models, and the Tracy Toolkit. Elsevier, Amsterdam (2005)
13. Roth, V., Jalali, M., Pinsdorf, U.: Secure Mobile Agents (SeMoA) (2007)
14. Lange, D.B., Mitsuru, O.: Programming and Deploying Java Mobile Agents Aglets. Addison-Wesley Longman Publishing, Amsterdam (1998)
15. Desbiens, J., Lavoie, M., Renaud, F.: Communication and tracking infrastructure of a mobile agent system. In: Proceedings of the 31st Hawaii IEEE International Conference on System Science, vol. 7, pp. 54–63 (1998)
16. Glass, G.: ObjectSpace voyager the agent ORB for java. In: Worldwide Computing and Its Applications-WWCA. Springer, Berlin, pp. 38–55 (1998)
17. Baumann, J., Rothermel, K.: Shadow approach: an orphan detection protocol for mobile agents. In: Proceedings of the 2nd International Workshop on Mobile Agents, LNCS, vol. 1477, pp. 2–13. Springer, Berlin (1998)
18. Sushil, R., Bhargava, R., Garg, K.: Location update schemes for mobile agents. J. Comput. Sci. Int. Sci. J. Braz. **7**(2), 37–43 (2008)
19. Lazar, S., Weerakoon, I., Sidhu, D.: A scalable location tracking and message delivery scheme for mobile agents. In: Proceedings of the 7th IEEE International Workshops on Enabling Technologies: Infrastructure for Collaborative Enterprises, pp. 243–248 (1998)
20. Murphy, A.L., Picco, G.P.: Reliable communication for highly mobile agents. Auton. Agent. Multi-Agent Syst. **5**(1), 81–100 (2002)
21. Ranganathan, M., Beddnarek, M., Montgomery, D.: A reliable message delivery protocol for mobile agents. In: Agent Systems, Mobile Agents, and Applications. Springer, Berlin, pp. 206–220 (2000)
22. Artikis, A., Pitt, J., Stergiou, C.: Agent communication transfer protocol. In: Proceedings of the 4th International Conference on Autonomous Agents, ACM, pp. 491–498 (2000)
23. Cabri, G., Leonardi, L., Zambonelli, F.: Mobile-agent coordination models for internet applications. IEEE Comput. **33**(2), 82–89 (2000)
24. Deugo, D.: Mobile agent messaging models. In: Proceedings of the 5th IEEE International Symposium on Autonomous Decentralized Systems, pp. 278–286 (2001)
25. Xu, W., Cao, J., Jin, B., Li, J., Zhang, L.: GCS-MA: a group communication system for mobile agents. J. Netw. Comput. Appl. **30**(3), 1153–1172 (2007)
26. Stefano, A.D., Santoro, C.: Locating Mobile Agents in a Wide Distributed Environment. IEEE Trans. Parallel Distrib. Syst. **13**(8), 153–161 (2002)
27. Jingyang, Z., Zhiyong, J., Chen, D.X.: Designing reliable communication protocols for mobile agents. In: Proceedings of the 23rd IEEE International Conference on Distributed Computing Systems Workshops, pp. 484–487 (2003)
28. Cao, J., Xu, W., Chan, A.T.S., Li, J.: A reliable multicast protocol for mailbox-based mobile agent communications. In: Proceedings of the IEEE International Conference on Autonomous Decentralized Systems ISADS, pp. 74–81 (2005)
29. Cao, J., Feng, X., Das, S.K.: Mailbox-based scheme for mobile agent communications. IEEE Comput. **35**(9), 54–60 (2002)
30. Cao, J., Zhang, L., Yang, J., Das, S.K.: A reliable mobile agent communication protocol. In: Proceedings of the 24th IEEE International Conference of Distributed Computing Systems, pp. 468–475 (2004)

31. Cao, J., Feng, X., Lu, J., Das, S.K.: Design of adaptive and reliable mobile agent communication protocols. In: Proceedings of the 22nd IEEE International Conference on Distributed Computing Systems, pp. 471–472 (2002)
32. Zhou, X.Y., Arnason, N., Ehikioya, S.A.: A proxy-based communication protocol for mobile agents: protocols and performance. In: Proceedings of the IEEE International Conference on Cybernetics and Intelligent Systems, vol. 1, pp. 53–58 (2004)
33. Rosenberg, J., Schulzrinne, H., Camarillo, G., Johnstan, A., Peterson, J., Sparks, R., Handley, M., Schooler, E.: Session Initiation Protocol (SIP), vol. 23. Internet Engineering Task Force, RFC 3261 (2002)
34. Tsai, H.H., Leu, F.Y., Chang, W.K.: Mobile agent communication using SIP. In: Proceedings of the IEEE Symposium on Applications and the Internet, pp. 274–279 (2005)
35. Tsai, H.H., Leu, F.Y., Chang, W.K.: Distributed sendbox scheme for mobile agent communication. In: Proceedings of the IEEE International Conference on Mobile Business, pp. 545–550 (2005)
36. Choi, S., Kim, H., Byun, E., Hwang, C., Baik, M.: Reliable asynchronous message delivery for mobile agents. IEEE Internet Comput. 10(6), 16–25 (2006)
37. Choi, S., Baik, M., Kim, H., Byun, E., Choo, H.: A reliable communication protocol for multiregion mobile agent environments. IEEE Trans. Parallel Distrib. Syst. 21(1), 72–85 (2010)
38. Sushil, R., Bhargava, R., Garg, K.: Design, validation, simulation and parametric evaluation of a novel protocol for mobile agents in multiregional environment. J. Comput. Sci. 4(3), 256–271 (2008)
39. Yong-Qiang, Z., Hong-Bin, G.: Design and implementation of rudp protocol for multiple mobile agent communication. In: Proceedings of the IEEE International Conference of Computer Application and System Modeling, ICCASM, vol. 8, pp. 614–618 (2010)
40. Postel, J.: User Datagram Protocol, ISI (1980)
41. Yousuf, A.Y., Hammo, A.Y.: Developing a new mechanism for locating and managing mobile agents. J. Eng. Sci. Technol. 7(5), 614–622 (2012)
42. Ahn, J.: Atomic mobile agent group communication. In: Proceedings of the 7th IEEE Conference on Consumer Communications and Networking Conference, pp. 1–5 (2010)
43. Ahn, J.: Mobile agent group communication ensuring reliability and totally-ordered delivery. Comput. Sci. Inf. Syst. 10(3) (2013)
44. Perkins, C., Myles, A., Johnson, D.B.: IMHP: a mobile host protocol for the internet. Comput. Netw. ISDN Syst. 27(3), 479–491 (1994)
45. Tanenbaum, A.S., Steen, M.V.: Introduction to Distributed Systems: Distributed Systems Principles and Paradigms. Prentice Hall, Upper Saddle River (2002)
46. McDonald, J.T., Yasinsac, A., Thompson, W.: Trust in mobile agent systems. Florida State University, Technical Report (2005)

Automatic Vehicle Identification: LPR with Enhanced Noise Removal Technique

Charulata Palai and Pradeep Kumar Jena

Abstract Automatic Vehicle Identification using license plate image is having versatile range of applications. It includes traffic law enforcement, toll fee collection, tracking the vehicle's route. It is more challenging to automate the identification of Indian vehicles by using it's number plates. The font type and font size of the letters used randomly, the dealer's logo or other artefacts may be available on the plate. In this proposed algorithm it is emphasise on extraction of the number plate irrespective of the angular inclination of the image i.e. extraction of region of interest (ROI) and removal of the additional artefacts then the character image segmentation followed by character recognition. The character recognition is done using correlation coefficient matching. The system is adaptive to the new fonts or regional characters by extended training set.

Keywords Vehicle identification · License plate recognition · LPR noise removal · Character recognizer · Region of interest · Image segmentation · Template matching

1 Introduction

The Automatic Vehicle Identification (AVI) using License Plate Recognition (LPR) becoming more popular now days. There are few basic reasons for the same: (a) The traffics are highly populated with number of vehicles (b) Fast and improved processing of digital image and video data (c) Well defined traffic system. First the captured image of the vehicle is scanned for license plate detection then the

C. Palai (✉)
Department of CSE, NIST, Berhampur, Odisha, India
e-mail: charulatapalai@gmail.com

P.K. Jena
Department of MCA, NIST, Berhampur, Odisha, India
e-mail: pradeep1_nist@yahoo.com

© Springer India 2015
L.C. Jain et al. (eds.), *Computational Intelligence in Data Mining - Volume 1*,
Smart Innovation, Systems and Technologies 31, DOI 10.1007/978-81-322-2205-7_14

143

character recognition is performed to identify vehicle by the number available in its licence plate. Now days the automatic license plate recognition is getting popular for automatic toll fee collection, maintaining traffic activities, law enforcement [1].

The LPR process consists of three main phases: License Plate Extraction from the captured image, image segmentation to extract individual characters image and character recognition. All the above phases of License Plate Recognition is most challenging as it is highly sensitive towards weather and lighting condition, license plate placement and the other artefact like frame, symbols or logo which is placed on licence plate picture. In India the license number is written either in one row or in two rows. Hence additional analysis required to categories the number plate information into either one or two rows type.

There are many approaches derived for license plate extraction and recognition. Some of the related works are as follows; Shidore and Norote [2] proposed LPR for Indian vehicles. They used Sobel filter, Morphological operations and connected component analysis for licence plate extraction. Character segmentation was done by using connected component and vertical projection analysis. Ozbay and Ercelebi [3] proposed Automatic vehicle identification by using edge detection algorithm, smearing algorithm and some morphological algorithm. Chang et al. [4] proposed LPR by using fuzzy disciplines to extract license plate from input image. Licence number identification was done by using Neural networks. The overall rate of success for this algorithm was 93.7 %. Kasaei et al. [5] proposed new morphology based method for Iranian car plate detection and recognition. They used Morphological operations for localizing license plate. Template matching scheme was used to recognize the digits and characters from the plate. Senthilkumaran and Rajesh [6] and Kulkarni et al. [7] used feature based number plate localization to extract number plate [8], Image scissoring method used for segmentation [9] and statistical feature extraction used for character recognition [10]. Wang et al. [11] proposed vertical projection segmentation approach based on the distribution character segmentation and for recognition character feature based template matching algorithm is used.

Many people have worked on LPR using fuzzy logic and neural network approach, they focused to improve the accuracy level but the processing time is least concerned, again the performance of the algorithm with a noisy back ground is hardly discussed.

All the existing algorithms discussed works well for an ideal number plate image as claimed by the authors. In Indian vehicles many times we find some additional logo or sticker of the distributer some time the letters are combination of different size and style or the number is written in two lines.

In this paper we proposed an improved LPR for the AVI system. In addition to the speed and accuracy factors, we successfully removed the any other artefacts up to half of the size of the standard letter used in the number plate.

In this work the balance between time complexity and accuracy is maintained. The edge detection method and vertical and horizontal processing is used for number plate localization. The edge detection is done with 'Roberts' operator. The connected component analysis (CCA) with proper threshold is used for segmentation. Then after for character recognition [12, 13] we used template matching by correlation function [14]. The level of matching is improved by using an enhanced database.

2 Methodology

A simple and efficient algorithm proposed to recognise license plate of vehicles automatically. First of all an input image is captured by a high quality camera from distance of 3–5 m. Then it passed through pre-processing module, where input image is prepared for further processing. The pre-processed image is send to the extraction module where number plate area will be extracted. The extracted number plate is segmented into characters. In next module the connected component analysis with some proper thresh holding is applied for image segmentation. Then the segmented letters are pre-processed and sent to the recognition module. Here identification is done by using template matching. The proposed model is shown in Fig. 1.

Fig. 1 Proposed model

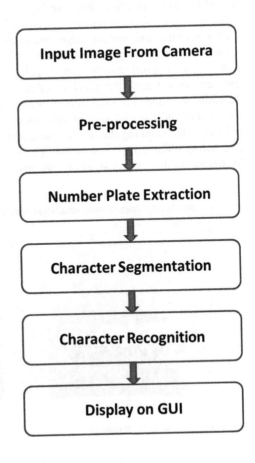

2.1 Pre-processing

The proposed technique for AVI starts with pre-processing of input image. The basic operation during pre-processing is noise removal and image enhancement in order to highlight the desired area. In case the image is captured low intensity it is under exposed or if the image is capture in bright day light it is over exposed to enhance it we have used the histogram equalization method.

2.2 Number Plate Extraction

License Plate Detection is the most challenging phase of AVI system. In order to remove unnecessary information from the image, it requires only edges of the image as the input for further processing. Initially the original image is converted into gray-scale image, shown in Fig. 2. Then it is binarized by determining a threshold for different intensities in the image, after this the 'ROBERTS' approach is used for edge detection, shown in Fig. 3.

Then to determine the region of license plate from the pre-processed image we used horizontal and vertical edge processing. First the horizontal histogram is calculated by traversing each column of an image. The algorithm starts traversing with the second pixel from the top of each column of image matrix. The difference between second and first pixel is calculated. If the difference exceeds certain threshold, it is added to total sum of differences. Then it traverses until the end of the column and the total sum of differences between neighbouring pixels are calculated. After scanning the whole image a matrix of the column-wise sum of the thresholing differences is calculated. The same process is carried out for vertical histogram. In this case, instead of columns, rows are processed.

Fig. 2 Binarization using threshold

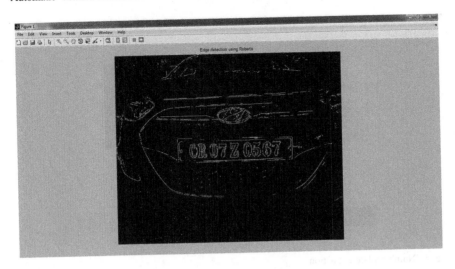

Fig. 3 Edge detection using roberts

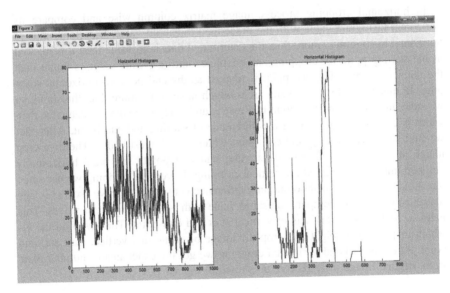

Fig. 4 Horizontal and vertical histogram

The analysis of horizontal and vertical histogram show a threshold value which is 0.434 times of maximum horizontal histogram value represents the number plate region i.e. our ROI. The next step is to extract the ROI using cropping the number plate area. The cropping is implemented in two steps first we first crop the image horizontally which gives value of x_1 and x_2 then crops it vertically by calculating values of y_1 and y_2 as shown in Fig. 4, the ROI is the area defined within (x_1, y_1) and (x_2, y_2).

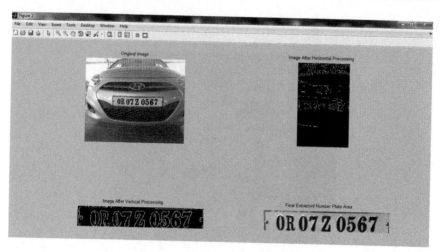

Fig. 5 Number plate extraction

In horizontal cropping process image matrix column wise and compare its horizontal histogram value with the predefined threshold value. If certain value in horizontal histogram is more than the determined threshold value then it is marked as the starting point for cropping and continues until a greater value is found than the threshold value. The last point is marked as the end point of horizontal scanning. In this process we get many areas which have value more than threshold, so we store all starting and end point in a matrix and compare width of each area, then width is calculated considering the difference of starting and end point. After that we find set of that staring and end point which map largest width. Then we crop image horizontally by using that starting and end point. This new horizontally cropped image is processed for vertical cropping. In vertical cropping we use same threshold comparison method but only difference is that this time we process image matrix row wise and compare threshold value with vertical histogram values. This helps to find the different sets of vertical start and end point. Then we find the set which maps largest height and crop the image by using that vertical start and end point. After vertical and horizontal cropping we got the exact area of number plate from original image in RGB format, shown in Fig. 5.

2.3 Character Segmentation

The number plate area is a part of original image, which is extracted from grayscale image. After this it is converted into a binary image using the same predefined threshold. To segment characters we used connected component analysis (CCA) with minor modification. Since it has been observed even for the images with taken with certain angle of slope have less or no effect on the size of the characters. Then to improve the execution time we determined the connected component and labelled them. Here to well justify the required characters only the connected components having their sum of pixel value is more than 2 % of area defined within ROI are labelled and other are removed from the list. Then after with the help of number of connected components and their labelling we cropped each letter separately.

2.4 Removal of Additional Artefacts

The pre-processed image is segmented into different characters. However many times after segmentation some extra dark pixels are found which not characters but the impression of nuts and bolts, black bar above and below the character or a small logo of the vendor. To handle such artefacts we have used a fuzzy classifier which considers the ratio of the size of the segment with respect to the average segment size, the gray level ratio which deals with the coloured artefacts and the match percentile with the most suitable character in the set. The removal different types of noise are shown in the result section.

2.5 Character Recognition

For the character recognition template matching technique is used. In this method a database of standard size characters is prepared. The template matching is carried out by using correlation coefficient which measures correlation coefficient between two images. The correlation coefficient determines how closely two images are similar. The range of correlation coefficient is −1.0 to 1.0. Then each segmented letter is resized up to the size of 24*42 which is same as that of database image. The correlation coefficient is calculated for each extracted letter with the database image and stored in a matrix. The position of every character is fixed in database. The index value of maximum correlation gives the matched character, shown in Fig. 6.

The images which are taken in the extreme lighting condition are either over or under exposed to the light. It is difficult to extract the number plate region from such image. Hence to overcome this difficulty it is required to normalize the image before processing. The normalisation is usually done in number of steps by subtracting the mean and dividing by standard deviation, as explained below. Let $f(x, y)$ is image to be searched and $t(x, y)$ is template image, and then the normalised cross correlation between the image pair is defined as

$$r = \frac{1}{n} \sum_{x,y} \frac{(f(x, y) - \bar{f})(t(x, y) - \bar{t})}{\sigma_f \, \sigma_t} \tag{1}$$

Fig. 6 Segmented characters before and after removal of unwanted character

where, r = Correlation coefficient
n = Number of pixels in $f(x, y)$ and $t(x, y)$
$\bar{f}$ = Mean value of $f(x, y)$
$\bar{t}$ = Mean value of $t(x, y)$
σ_f = Standard deviation of $f(x, y)$
σ_t = Standard deviation of $t(x, y)$

$$\sigma_f = \sqrt{1/n \sum_x \sum_y \left(f(x, y) - \bar{f} \right)^2} \qquad (2)$$

and

$$\sigma_t = \sqrt{1/n \sum_x \sum_y \left(t(x, y) - \bar{t} \right)^2} \qquad (3)$$

To improve matching efficiency total 70 different characters are taken in a database. It incorporates the multiple copies of the few special characters, which differs highly when written in different fonts.

3 Result Analysis

This AVI system is simulated in MATLAB version 7.10.0.499(R2010a). Images can be captured 3–5 m for better result. Table 1 shows the accuracy of the match in percentile for the different operations such as image extraction, segmentation and character recognition.

Table 1 Experimental results

Module	No. of correct detection	Accuracy (%)
Extraction	205/230	89.13
Segmentation	195/205	95.12
Recognition	191/195	97.94

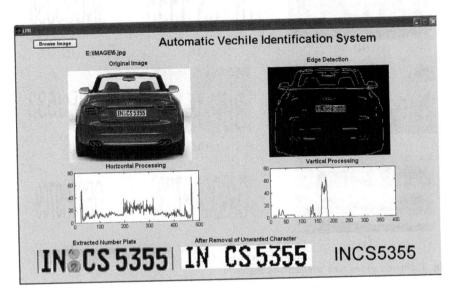

Fig. 7 Final GUI for AVI system

The Fig. 7 shows the complete Graphical User Interface. It helps to browse an input image and convert it into equivalent gray-scale image then detects the edges in the first phase. In the second phase it extracts the number plate region and in the third phase it generates a character string by recognizing the characters in the individual image segments.

The Fig. 8a–d shows the license plates images with various types of artefacts, Fig. 8g shows the license plate image with a vertical scanning angle with artefacts, Fig. 8j show the plate image with horizontal scanning angle with artefacts, Fig. 8b, e, h, k represents the number plate regions with artefacts and the Fig. 8c, f, i, l shows the vehicle number for its identification.

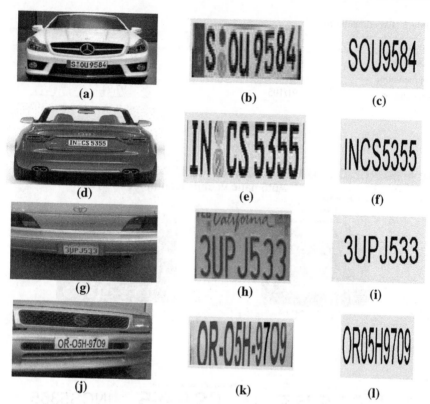

Fig. 8 Number plates with artefacts and angles

4 Conclusion

The proposed model works well for wide verity of images taken with different illuminations as well for the varied angular images. It successfully removes the additional artefacts from the plate region. This recognizes the characters irrespective of the font size and recognizes many popular font types. The algorithm has been tested with 230 images having different background and lighting condition, with varied angle and fonts, for partially damaged characters. It is observed that it works well for a wide variety of images available on Indian roads, which was the main motivation of this work. This work can be further enhanced by considering the blurred image or the video frames, which are captured when the vehicle is on motion. Again the plate regions extraction technique from multiple frames can be used for recognizing the license number.

References

1. Bailey, D.G., Irecki, D., Lim, B.K., Yang, L.: Test bed for number plate recognition applications. In: Proceedings of First IEEE International Workshop on Electronic Design, Test and Applications (DELTA'02), IEEE Computer Society (2002)
2. Shidore, N.: Number plate recognition for indian vehicles. Int. J. Comput. Sci. Netw. Secur. (IJCSNS), **11**(2), 143–146 (2011)
3. Ozbay, S., Ercelebi, E.: Automatic vehicle identification by plate recognition. World Acad. Sci. Eng. Technol. **9** (2005)
4. Chang, S.-L., Chen, L.-S., Chung, Y.-C., Chen, S.-W.: Automatic license plate recognition. IEEE Trans. Intell. Transp. Syst. **5**(1) (2004)
5. Kasaei, H., Kasaei, M., Kasaei, A.: New morphology-based method for robust iranian car plate detection and recognition. Int. J. Comput. Theory Eng. **2**(2) (2010)
6. Senthilkumaran, N., Rajesh, R.: Edge detection techniques for image segmentation—a survey of soft computing approaches. Int. J. Recent Trends Eng. **1**(2) (2009)
7. Kulkarni, P., Khatri, A., Banga, P. Shah, K.: Automatic number plate recognition system for indian conditions. IEEE Trans. (2009)
8. Zakaria, M.F., et al.: Malaysian car number plate detection system based on template matching and colour information. Int. J. Comput. Sci. Eng. (IJCSE) **2**(4), 1159–1164 (2010)
9. Sharma, C., Kaur, A.: Indian vehicle license plate extraction and segmentation. Int. J. Comput. Sci. Commun. **2**(2), 593–599 (2011)
10. Soni, S., Khan, M.I.: Content based retrieval for number plate extraction of vehicle. Int. J. Adv. Electron. Eng. **1**(1) ISSN 2278-215X
11. Wang, J.-X., Zhou, W.-Z., Xue, J.-F., Liu, X.-X.: The research and realization of vehicle license plate character segmentation and recognition technology. In: Proceedings of the 2010 International Conference on Wavelet Analysis and Pattern Recognition, Qingdao, pp. 11–14 (2010)
12. Kumar, N., Kumar, B., Chattopadhyay, S., Palai, C., Jena, P.K.: Optical character recognition using ant miner algorithm: a case study on oriya character recognition. Int. J. Comput. Appl. **57**(7) (2012)
13. Yetirajam, M., Jena, P.K.: Enhanced colour image segmentation of foreground region using particle swarm optimization. Int. J. Comput. Appl. **57**(8) (2012)
14. Horowitz, M.: Efficient use of a picture correlator. J. Opt. Soc. Am. **47** (1957)

GLCM Based Texture Features for Palmprint Identification System

Y.L. Malathi Latha and Munaga V.N.K. Prasad

Abstract In this paper, a new Palmprint recognition technique is introduced based on the gray-level co-occurrence matrix (GLCM). GLCM represents the distributions of the intensities and the information about relative positions of neighboring pixels of an image. GLCM matrices are calculated corresponding to different orientation (0, 45, 90, 135) with four different offset values. After the calculation of GLCMs, each GLCM is divided into 32 × 32 sub-matrices. For each such sub-matrix four Haralick features are calculated. The performance of the proposed identification system based on Haralick features is determined using False Acceptance Rate (FAR) and Genuine Acceptance Rate (GAR).

Keywords Palmprint recognition · Haralick features · GLCM · Gabor filter

1 Introduction

Biometric based recognition is the most popular human recognition by their biological features, inherent in each individual. Palmprint based biometric approach have been intensively developed over the past decade because they possess several advantages such as a rich set of features, high accuracy, high user friendly and low cost over other biometric systems. Palmprint recognition has five stages, palmprint acquisition, preprocessing, feature extraction, enrollment (database) and matching.

Y.L. Malathi Latha (✉)
CSE Department, Swami Vivekananda Institute of Technology (SVIT),
Secunderbaad, India
e-mail: malathilatha_99@yahoo.com

M.V.N.K. Prasad
Institute for Development and Research in Banking Technology (IDRBT),
Castle Hills, Masab Tank, Hyderabad, India
e-mail: mvnkprasad@idrbt.ac.in

© Springer India 2015
L.C. Jain et al. (eds.), *Computational Intelligence in Data Mining - Volume 1*,
Smart Innovation, Systems and Technologies 31, DOI 10.1007/978-81-322-2205-7_15

The major approach for palmprint recognition is to extract feature vectors corresponding to individual palm image and to perform matching based on some distance metrics. Palmprint research employs high resolution or low resolution images. Principle lines, wrinkles and texture-based features can be extracted from low resolution images. More discriminate features such as ridges, singular points and minutiae can be extracted using high resolution palm images. In our present work, we have used low resolution images to extract texture features.

Texture based feature extraction methods are widely adopted for palmprint identification because of their high performance. In the literature, numerous texture based approaches for palmprint recognition have been proposed. The palmprint textures can be obtained using techniques, such as Gabor wavelets [20], Fourier transformation [9], Cosine transformation [3, 9], Wavelet transformation [12] and Standard Deviation [6].

In [21], fingerprint image is represented by co-occurrence matrices. Features are extracted based on certain characteristics of the co-occurrence matrix and then fingerprint classification is done using neural networks. Rampun et al. [16] proposed new texture based segmentation algorithm which uses a set of features extracted from Gray-Level Co-occurrence Matrices. Principal Component Analysis is used to reduce the dimensionality of the resulting feature space. Gaussian Mixture Modeling is used for the subsequent segmentation and false positive regions are removed using morphology.

An effective Iris recognition system is proposed by Chen et al. [2]. Gray Level Co-occurrence Matrix (GLCM) and multi-channel 2D Gabor filters are adopted to extract iris features. The combined features are in the form of complementary and efficient effect. Particle Swarm Optimization (PSO) is employed to deal with the parameter optimization for Support Vector Machine (SVM), and then the optimized SVM is applied to classify Iris features.

Palmprint recognition based on Haralick features was proposed by Ribaric and Lopar [17]. Haralick features are extracted from a sub-image and the matching process between the live template and the templates from the system database is performed in N matching modules. Fusion at the matching-score level is used and the final decision is made on the basis of the maximum of the total similarity measure. The experiments are performed on small databases (1,324 hand images). The work in [13] extracts Haralick features along the principal lines and experiments were evaluated on small part of polyU database and shows poor performance (EER above 14 %). An optimal thenar palmprint classification model is proposed by Zhu et al. [22]. Thirteen textural features of gray level co-occurrence matrix (GLCM) are extracted and support vector machine is used for classification. To the best of our knowledge, only few papers on palmprint identification using GLCM were reported in the literature. Most of them have used support vector machine and k-neural network classifiers and experiments have been performed on small databases and results reported in the literature were not promising. In this paper, we have proposed GLCM based palmprint identification using euclidean distance classifier. The aim of this paper is to investigate and analyse the behaviour of the texture features. The proposed method is evaluated by comparing it with other texture based methods in the literature.

2 Gray Level Co-occurrence Matrix and Haralick Features

2.1 Gray-Level Co-occurrence Matrix

Gray-Level Co-occurrence Matrix or GLCM [1, 5] is a matrix that contains information about the distribution of intensities and information about the relative position of neighborhood pixels. As name suggests, it uses Grey-Scale images. Given a Grey-Scale Image I, the GLCM matrix P is defined as [1]:

$$P(i, \Delta x, \Delta y) = WQ(i, j)|\Delta x, \Delta y) \tag{1}$$

where $W = 1/(M\Delta x)(N - \Delta y) \quad \sum_{n=1}^{N-\Delta y} \sum_{m=1}^{M-\Delta x} A$

$$A = \begin{cases} 1 & \text{if } f(m, n) = i \text{ and } f(m + \Delta x, n + \Delta y) \\ 0 & \text{elsewhere} \end{cases}$$

where f(m, n) be the intensity at sample m, line n of the neighborhood, (i, j) is the index and $(\Delta x, \Delta y)$ denotes the offset and the orientation respectively. Offset represents the distance between the interested neighborhood pixels and orientation represents the angle between interested neighborhood pixels. After calculation of GLCM we do the normalization by performing a divide operation on the product of M and N, where M and N are the dimensions of the Grey Scale image.

In our proposed methodology the local Haralick features are calculated from normalized gray level co-occurrence matrices. Haralick introduced 14 statistical features [18] which are basically texture features which can be extracted from a GLCM matrix. The texture features [4, 7, 8] are Angular Second Moment, Contrast, Inverse Difference moment, Entropy, Correlation, Variance, Sum, Average, Sum Entropy, etc. In our investigation four features that can successfully characterize the statistical behavior (experimentally determined) are:

Contrast: The relative difference between light and dark areas of an image. Contrast is how dark to how light something is. The contrast makes the lighter colors more lighter, and the darker colors darker.

$$\text{Contrast} = \sum_{n=0}^{G-1} \left\{ \sum_{i=1}^{G} \sum_{j=1}^{G} p(i, |i - j| = n \right. \tag{2}$$

where P is GLCM matrix and G is Grey Scale value.

Entropy: It can be described as a measure of the amount of disorder in a system. In the case of an image, entropy is to consider the spread of states which a system can adopt. A low entropy system occupies a small number of such states, while a high entropy system occupies a large number of states. For example, in an 8-bit pixel there are 256 such states. If all such states are equally occupied, as they are in the case of an image, which has been perfectly histogram equalized, the spread of states is a maximum, as is the entropy of the image. On the other hand, if the image has been threshold, so that only two states are occupied, the entropy is low. If all of the pixels have the same value, the entropy of the image is zero.

$$\text{Entropy} = \sum_{i=0}^{G-1} \sum_{j=0}^{G-1} p(i,j) \times \log(P(i)) \tag{3}$$

where P is GLCM matrix and G is Grey Scale value.

Variance: The variance is a measure of how far a set of numbers is spread out. It is one of several descriptors of a probability distribution, describing how far the numbers lie from the mean (expected value).

$$\text{Variance} = \sum_{i=0}^{G-1} \sum_{j=0}^{G-1} (i - \mu P(i,j)) \tag{4}$$

where P is GLCM matrix and G is Grey Scale value, μx is the mean value.

Correlation: Measure that determines the degree to which two pixel values are associated.

$$\text{Correlation} = \sum_{i=0}^{G-1} \sum_{j=0}^{G-1} \{i \times j\} p(i,j) - \{\mu x \times \mu y\} / \sigma x \times \sigma y \tag{5}$$

where P is GLCM matrix and G is Grey-Scale value, μx, μy are mean values and σx, σy are standard deviations along X and Y axis.

$$\mu x = \sum_{t=0}^{G-1} i \sum_{j=0}^{G-1} P(i,j)$$

$$\mu y = \sum_{i=0}^{G-1} i \sum_{j=0}^{G-1} jP(i,j)$$

$$\sigma x^2 = \sum_{i=0}^{G-1} (i - \mu) \sum_{i=0}^{G-1} P(i,j)$$

$$\sigma y^2 = \sum_{j=0}^{G-1} (j - \mu) \sum_{j=0}^{G-1} P(i,j)$$

GLCM matrices are calculated corresponding to different orientations (0, 45, 90, 135) with four different offset values. The above mentioned four Haralick features are obtained from GLCM matrix created on sub-images of palmprint's ROI. By calculating this four texture features it is possible to see how they behave for different textures. The size of feature vector for a biometric template is Mn—component feature vectors where M is the number of sub images defined by sliding window on the palmprint's ROI and n is the number of local Haralick features.

We have used the following parameters for our experiment: $M \times M = 128 \times 128$ dimensions of palmprint ROI, $g = 256$ number of grey levels, offset value $\delta = 1, 2, 3$ and 4, $d \times d = 8 \times 8$ dimension of sliding window, $t = 4$ sliding window translation step and $\emptyset = 0°, 45°, 90°$ and $135°$.

3 Computation of Feature Vector and Matching

3.1 Haralic Features Extraction

For the computation of the GLCM not only the displacement (offset value δ), but also the orientation between neighbor pixels must be established. The orientations can be horizontal (0°),Vertical (90), Right Diagonal (45) and Left Diagonal (135) degree respectively. GLCM matrices for each plamprint's ROI are calculated corresponding to different orientation (0, 45, 90, 135) with four offset values. The local Haralick features contrast, entropy, variance and correlation are obtained from normalized gray level co-occurrence matrices.

After the calculation of GLCMs, each GLCM is divided into 32×32 sub-matrices. For each such sub-matrix Haralick features are calculated. There will be 64 such sub-matrices for each such GLCM. There are 4 GLCMs for each image. So total 256 such sub-matrices will be there for each image. So the size of feature vector for each image is: 4 (offset values) × 4 (four Haralick features) × 256 (subimages). Therefore palmprint ROI is represented by 4,096 features.

3.2 Matching

In matching process, the comparison is performed between template and query. In this paper, Euclidean distance similarity method is used to calculate the matching score. The Matching Score [19] between the template and query given by

$$\text{Matching Score} = 1 - \frac{||X - Y||2}{||X||2 + ||Y||2} \tag{6}$$

where X is the feature vector of template and Y is the feature vector of query palmprint calculated using Haralick formulas. Norm is the Euclidean Norm which is the square root of the summation of square of the values i.e. If Vector P is [x y z] then Euclidean Norm [19] is:

$$||Norm|| = \sqrt{x2 + y2 + z2}$$

The range of matching score is between 0 and 1. If the matching score is greater than a reference threshold, the user is considered as genuine otherwise imposter.

4 Experimental Result

We experimented our approach on PolyU Palmprint database [15]. PolyU database consists of 8,000 grayscale palm print images from 400 users. Each ROI image is 128 × 128 in dimension and sample ROI of PolyU database shown in Fig. 1.

The biometric system classifies an individual either as a genuine or as an imposter. The system may make two types of recognition errors, either falsely recognizing an imposter as genuine or by rejecting a genuine user as an imposter resulting to False acceptance rate and false rejection rate. The false acceptance rate is a fraction of imposter score greater than the threshold and false rejection rate is portion of genuine score less than threshold. Both FAR and FRR are functions of system threshold, if the threshold is increased then FAR will decrease, but FRR will increase and vice versa. So, for a given biometric system both the errors cannot be decreased simultaneously by varying the threshold.

The performance of the proposed identification system based on Haralick features is determined using False Acceptance Rate (FAR) and Genuine Acceptance Rate (GAR). The results of Haralick Features are shown in Table 1. Genuine Acceptance Rate (GAR) is percentage of genuine users will be authorized during authentication and False Acceptance Rate(FAR) is percentage of fake/imposter user will be authorized during authentication. The GAR/FAR value of PolyU database is 98.46/0.8873 for 0.58 threshold. The plot between FAR and FRR at various thresholds is shown in Fig. 2. It is observed that FAR reduces as threshold increases and FRR increases with threshold. Table 2 presents the performance comparison of

Fig. 1 Sample ROI polyU database

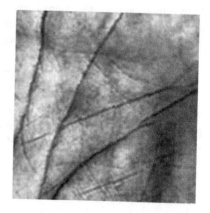

Table 1 GAR, FAR and FRR rates for different thresholds values using Haralick features

Threshold	GAR (%)	FAR	FRR
0.58	98.46	0.8873	0.02
0.6	98.12	0.8673	0.02
0.7	96.97	0.7066	0.04
0.72	96.66	0.6583	0.04
0.8	92.17	0.3917	0.08
0.84	83.82	0.2397	0.17
0.88	62.319	0.097	0.38
0.9	52.59	0.0474	0.5

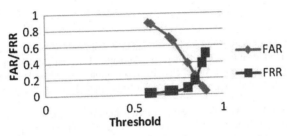

Fig. 2 FAR and FRR at various thresholds for different thresholds using Haralick features

Table 2 Performance comparison of proposed approach with existing approaches

Approaches		Database	Recognition rate (%)
Ribaric and Lopar [17]	Co-occurrence matrix and k-NN classifier	243 users, 1,874 images	98.91
Meiru et al. [14]	Local binary pattern and k-NN classifier	146 users 1,460 images	92.72
Khalifa et al. [11]	Co-occurrence matrix and SVM classifier	PolyU 2D, 400 people, 8,000 images	91.19
Proposed method	GLCM and Euclidean distance	PolyU 2D, 400 people, 8,000 images	98.46

our proposed approach with existing approaches. In this paper, we have proposed new palmprint recognition technique using Euclidean distance and experiment is performed on PolyU database.

5 Conclusion

We have developed a new palmprint recognition technique based on gray-level co-occurrence matrix (GLCM). The efficiency of texture based features using GLCM are investigated and analyzed. For each ROI image, GLCM matrix and Haralick features are extracted. The efficiency of this feature was tested using GAR /FAR and obtained recognition results for PolyU—GAR/FAR as 98.46/0.88. In this paper, we have used Euclidean classifier and achieved better recognition accuracy when compared with the existing palmprint recognition systems [11, 14, 16]. Further, the GAR can be increased by using optimization techniques such as Genetic algorithm or Particle Swarm techniques.

References

1. Albregtsen, F.: Statistical texture measures computed from gray level co-ocurrence matrices. Image Processing Laboratory, Department of Informatics, University of Oslo, pp. 1–14 (2008)
2. Chen, Y., Yang, F., Chen, H.: An effective iris recognition system based on combined feature extraction and enhanced support vector machine classifier. J. Inf. Comput. Sci. 5505–5519 (2013)
3. Dale, M.P., Joshi, M.A., Gilda, N.: Texture based palmprint identification using DCT features. In: International Conference on Advances in Pattern Recognition, vol. 42, pp. 221–224 (2009)
4. Du Buf, J.M.H., Kardan, M., Spann, M.: Texture feature performance for image segmentation. Pattern Recogn. 291–309 (1990)
5. Eleyan, A., Demirel, H.: Co-occurrence based statistical approach for face recognition. Comput. Inf. Sci. IEEE 3, 611–615 (2009)
6. Gonzalez, R.C., Woods, R.E.: Digital Image Processing. Pearson Education, London (2009)
7. Haralick, R.M., Shanmugam, K., Dinstein, I.: Textural features for image classification. IEEE Trans. Syst. Man Cybern. 6, 610–621 (1972)
8. Haralick, R.M., Shanmugam, K.: Its' Hak Dinstein: Textural features for image classification. IEEE Trans. Syst. Man Cybern. 6, 610–621 (1973)
9. Imatiaz, H., Fattah, S.A.: A spectral domain dominant feature extraction algorithm for palmprint recognition. Int. J. Image Process. (IJIP) 5, 130–144 (2011)
10. Kekre, H.B., et al.: Image retrieval using texture features extracted from GLCM, LBG and KPE. Int. J. Comput. Theory Eng. 2, 1793–8201 (2010)
11. Khalifa, A.B., Rzouga, L., BenAmara, N.E.: Wavelet, Gabor filters and co-occurrence matrix for palmprint verification. Int. J. Image Graph. Signal Process. 5, 1–8 (2013)
12. Khanna, P., Tamrakar, D.: Analysis of palmprint verification using wavelet filter and competitive code. In: International Conference on Computational Intelligence and Communication Network, pp. 20–25 (2010)
13. Martins, D.S.: Biometric recognition based on the texture along palmprint principal lines. Faculdade de Engenharia da Universidade do Porto (2013)
14. Meiru, M., Ruan, Q., Shen, Y.: Palmprint recognition based on discriminative local binary patterns statistic feature. In: International Conference on Signal Acquisition and Processing, pp. 193–197 (2010)
15. Poly palmprint database available at http://www4.comp.polyu.edu.hk/~biometrics/
16. Rampun, A., Strange, H., Zwiggelaar, R.: Texture segmentation using different orientations of GLCM features. In: 6th International Conference on Computer Vision/Computer Graphics Collaboration Techniques and Applications. MIRAGE, pp. 519–528 (2013)

17. Ribaric, S., Lopar, M.: Palmprint recognition based on local Haralick features. In: Proceedings of the IEEE MELECON, vol. 18, pp. 657–660 (2012)
18. Shen, L., Ji, Z., Zhang, L., Guo, Z.: Applying LBP operator to gabor response for palm print identification. In: Proceedings of the International Conference on Information Engineering and Computer Science, vol. **24**, 1–3 (2009)
19. Wang, S., Hu, J.: Alignment-free cancelable fingerprint template design: a densely infinite-to-one mapping (DITOM) approach. Pattern Recogn. 4129–4137 (2012)
20. Xuan, W., Li, L., Mingzhe, W.: Palmprint verification based on 2D Gabor wavelet and pulse coupled neural network. Knowl.-Based Syst. **27**, 451–455 (2012)
21. Yazdi, M., Gheysari, K.: A new approach for the fingerprint classification based on gray-level co-occurrence matrix. Int. J. Comput. Inf. Sci. Eng. 171–174 (2008)
22. Zhu, X., Liu, D., Zhang, Q.: Research of thenar palmprint classification based on GLCM and SVM. J. Comput. **6**, 1535–1541 (2011)

Virtual 3D Trail Mirror to Project the Image Reality

Mattupalli Komal Teja, Sajja Karthik, Kommu Lavanya Kumari
and Kothuri Sriraman

Abstract Magical virtual mirror, is the proposal that is going to present the virtual mirror which is used to display the jewelleries, clothes, footwear's etc. in the imaginary way means that instead of not wearing physically that in reality. Virtual Mirror uses highly sophisticated 3D image processing techniques to visualize the look of new garments and all other accessories without any need to actually put them on. By using the barcode tag on the clothes or footwear, can easily get the sizes, color, etc. features of that particular item. Previous methods usually involve motion capture, 3D reconstruction; another method based on combining all image-based renderings images of the user and previously recorded garments to save the time processing. Many several existing methods are developed for knowing the outlook of the garment is looked like. No method focusing on exactly fitting of that particular costume or footwear that is actually user interested.

Keywords Mixed reality · Augmented reality · Image-based rendering · Virtual mirror · Intellifit body scanner · Wireless barcode scanner · Camera

M.K. Teja (✉) · S. Karthik · K.L. Kumari · K. Sriraman
Department of CSE, Vignan's Lara Institute of Technology and Science,
Vadlamudi, Guntur, Andhra Pradesh, India
e-mail: komalteja@gmail.com

S. Karthik
e-mail: karthik.sajja@gmail.com

K.L. Kumari
e-mail: kommulavanya6@gmail.com

K. Sriraman
e-mail: ksriraman@hotmail.com

© Springer India 2015
L.C. Jain et al. (eds.), *Computational Intelligence in Data Mining - Volume 1*,
Smart Innovation, Systems and Technologies 31, DOI 10.1007/978-81-322-2205-7_16

1 Introduction

Now-a-days it became most popular by achieving a most popular and benefits one that mostly for Shoppe's. In the past there is only process of wearing each and every garment manually with getting more physical stress. Mainly having the problem of security when physically wearing a garment or anything inside a trail room will have commonly the doubt of whether anybody recording anything. So, the usage of trail rooms are less in number and most of them want to check the look and feel of the accessory (Garment, Jewellery, Footwear, etc.) how it look actually to them. This is the way that the base paper for this proposed system came actually from. Now many virtual trail rooms came into existence but will not fulfillment of all the needs of the customer. May be coming up society will make use of this applications to make a better system with having more secured. A fit technology may be categorised according to the problem that it resolves (size, fit or styling) or according to the technological approach. There are many different types of technological approach, of which the most established and credible are:

- Size recommendation services, Body scanners, 3D solutions, 3D customer's model, Fitting room with real 3D simulation, Dress-up mannequins/mix-and-match, Photo-accurate virtual fitting room, Augmented reality, Real models.

2 Existing System

In this existing work we improve all stages of our previous augmentation pipeline. We contribute efficient methods for GPU-based silhouette matching by line-based and area-based silhouette sampling and evaluate their performance. Moreover, a novel method to extend the pose space by combining multiple matches is introduced. They introduce an extended non-rigid registration formulation that adapts the garment shape by silhouette and depth data terms. We evaluated all improvements for their impact on the visual quality by comparing them to ground truth images. There are certain modules in the existing system that is going to show in the output process. The extension of the work is going to be done on fitting of costumes to the user is proposed in this paper.

3 Limitations in the Existing Paper

The suggested clothes augmentation pipeline transfers the appearance of garments from one user to another. It uses captured images for rendering, which inherently contain features like realistic light-cloth interaction and wrinkles. It is suitable for a wide range of garments, like shirts, sweaters, trousers, dresses and accessories like

ground truth   skeleton   silhouette
           tracking   matching

Fig. 1 A frame of the evaluation data set of users

scarf's, sun-glasses etc. Moreover, recording the garment data is quick and cheap compared to manual 3D modeling. However, the proposed pipeline stages also have limitations that can reduce the visual quality that the system achieves.

Figure 1 shows a frame of the evaluation data set of user where skeleton- and silhouette-based matching strongly differ and also the visual impact of a missing pose in the garment database is given.

Matching algorithms that are based on silhouette shapes assume a certain shape similarity between images of the model- and the end user. This is not always given. For example, when the desired garment is a dress and the current user is wearing trousers, then the shapes are not similar in the leg region. While silhouette matching is quite reliable for small dissimilarities, it loses precision with, for example, longer dresses. In practice, we achieved good matching results even for dresses and coats, but we cannot generally guarantee that it works for arbitrary garment and user shapes. Figure 1 shows the sample of adjusting the garments which are retrieved from the database and adjusting with the balancing of motion recorded by the users.

4 Introduction to the Proposed System

In our proposed system we are not physically wear a dress to analyze whether it fits or not. In this, the trailer selects a dress and standee in front of the virtual mirror. Based on that dress id, the virtual mirror takes the measurements of that dress and takes the measurements of the person who standee in front of the virtual mirror. Based on the person's body measurements and dress measurements the virtual mirror concluded whether that dress is perfectly fits to that person or not. It saves some time and avoids the security problems as there is no problem in wearing dress virtually. In this system, virtuosity avoids so many problems and there is no problem to wear the dress virtual in outside places rather than wears in real. It is like

Fig. 2 Showing sample virtual mirror

as existing system in displaying images but the difference here is that the person doesn't wear the dress physically.

Having the curiosity in checking about how it looks and very much anxiety to wear the dresses. It became the most common problem for all the persons for that; a virtual mirror is the most important one for visually displaying the accessories. It gives the more security than existing system and more efficient. The sample of showing virtual mirror in Fig. 2.

For jeweler it is used better for the designers to design by replacing the different stones colors, structure, and features of the jeweler. Person can also wear the footwear magically by not wearing it physically. It can be seemed like in the below image. There does visually seem in the below a person trying different types of the footwear and want to take one of them which is more suitable for their foot. This is a trending Shoppe because with in less time can buy more items on exactly based on user's interest. Every Showroom tries to make the customer to be satisfied. That customer satisfaction is received whenever they felt happy with the shopping.

4.1 Motion Capture

The person who is using the virtual mirror is identified and later by using of the Intellifit body Scanner we make use of reading the persons sizes and stored in the virtual database. Motion capture, or human pose tracking [1], is the task of determining the user's body poses. It usually involves a pose and shape model that is fitted to sensor data and therefore comprises a model-based tracking problem. The more recent Microsoft Kinect device allows for real-time recording of color and depth images at very low cost, as well as high-quality real-time human pose estimation. It plays a major role in defining the motion of the pose which is placed behind the mirror which makes to use of trail of the garments with the interaction of feeling it as real. Microsoft Kinect devices are used frequently in gaming to possess a user interaction i.e., used for capturing the same ideology to be implemented here.

4.2 Clothes Reconstruction

Initially the clothes which are having the tag consists of bar code will contains a data in the database those are having all the details of the clothes (like color, type, size, etc.). The measurements which are taken from the person's size will be taken as one of the input. The clothes which are selected based on their interest will be stored and makes visible on the output device. Many approaches use markers on the cloth for capturing [2], which makes them less suitable for our method. More recent approaches do not require markers. However, all approaches that rely on point correspondences that are computed from the image data assume a certain text redness of the garment. By using the light dome of or a laser scanner this limitation can be removed, but such hardware is expensive and processing cannot be performed in real time. Once the shape of a garment is digitized it needs to be fitted to the user's body model. This is a complex problem that is usually not handled in real time system that is actually in.

4.3 Virtual Try-on

Previous methods work by finding the best matching dataset in a previously recorded database that contains all possible poses of the user. These systems first learn and then search a database of poses by using a pose similarity metric. The best match is used to deform a texture to fit the user. However, like many other re-texturing approaches they operate in 2D [3, 4] and therefore do not allow the user to view him or her from arbitrary viewpoints. In Fig. 3 the image-based visual hull rendering is shown that is helpful in projecting the cloth. The Virtual Try on project offers a set of applications for various tailoring, modeling and simulation tools. 3D

Fig. 3 Illustrating the concept of image-based visual hull rendering. For every pixel its viewing ray (*red arrow*) is projected onto each camera plane (*red line*) (color figure online)

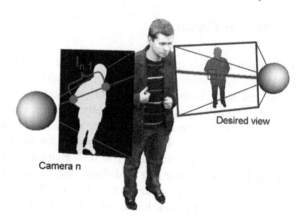

scans of real garments are acquired by color-coded cloth. MIRACloth [5] is a clothes modeling application which can create garments, fit them to avatars and simulate them. Kinect-based body scanning enables virtual try-on applications at low costs but systems with a single sensor require multiple views.

4.4 Prerequisites

The virtual dressing room that is used for this work consists of a 2×3 m footprint cabin with green walls in the existing papers. Ten cameras are mounted on the walls: two in the back, two on the sides and six in the front. The cameras are synchronized and focused at the center of the cabin, where the user is allowed to move freely inside a certain volume. All cameras are calibrated intrinsically and extrinsically and connected to a single PC. The mirror also offered accessory options and suggested different pieces that could be worn together for fashion coordination.

4.5 The Augmentation Process

Similar to [6] our clothes augmentation process has an offline phase for recording garments and an online phase where users can be augmented. The stages of the recording

1. A user wears a garment which should be used for future augmentations. He or she enters the dressing room and performs a short series of different poses while being recorded.
2. Garments are segmented and features are extracted from the recorded video streams. Results are stored in a garment database. This phase can be controlled: it is possible to recapture the scene when incomplete, or switch segmentation strategies. From now on, users who enter the dressing room can be augmented with previously recorded garments. We call this the runtime phase:
3. Users can move freely inside the room while being captured.
4. Features from the captured images are extracted.
5. The best fitting pose from the garment database is selected.
6. The selected pose of the desired garment and the captured user are rendered from the same viewpoint using image-based rendering.
7. Small pose mismatches are compensated for by rigid and non-rigid registration.

- Offline: garment database construction
  A user puts on one or multiple garments which should be transferred to other users. The other clothing should be selected to allow for easy segmentation. The model-user enters the dressing room and performs a series of different poses [7] while being recorded. Each garment database contains a single recorded sequence and therefore a single piece of clothing. Each garment is stored in the database with the proper data relevantly need to be stored. Another database is also maintained to store the measurements of the end user.

- At runtime: clothes augmentation
  Once one or more garment databases have been created, users can enter the dressing room and watch themselves wearing different clothes on the display in front of them. First, the same features as in the offline phase are extracted from the current camera images. These features are transformed to PCA space by applying the precomputed transformation matrix from the offline stage. The result is used to find the garment database entry where the model-user pose is closest to the pose of the current user. We use a simple Euclidean distance measure to find the closest feature vector during the search. The camera images [8] associated with this entry show the best match for the desired garment.

- Image-based rendering
  In this phase, the matched garment dataset and the current user are rendered. Both rendering tasks need to quickly generate an output image from a set of camera images with silhouette information. We employ the image-based visual hull (IBVH) algorithm to compute depth maps of novel viewpoints directly from the segmented camera images [9, 10]. It bypasses the

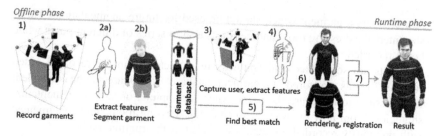

Fig. 4 Overview of our pipeline. In the offline phase, garments are recorded and processed to create a database. At runtime, the database can be queried for a record that matches the current user's pose. The recorded data is used to augment the user with a garment

computation of an explicit representation, such as a voxel grid or a mesh. Therefore, it is beneficial to the overall performance to directly derive an output image from the camera images.

5 Composing and Display

Parts of the model-user are still visible in the garment buyer, but with alpha values that indicate which pixels have to be removed. These regions were required as a good optimization target during the non-rigid registration procedure. After optimization the unwanted pixels are removed. In a final step, the garment buyer and the current user's buyer are composed. In these regions, the current user's body replaces the model user according to the previous work [5, 10]. This occlusion effect is correct, because the model-user occluded the same regions. The overview of the offline phase in shown in Fig. 4. When multiple garments are augmented, the items are consecutively composed with the previous result.

6 Implementation Process of the Proposed System

The process of the proposed system shown in Fig. 5 involves where the user interacts and they trail with the garments and also shows the user interacts with the system and quits from this process.

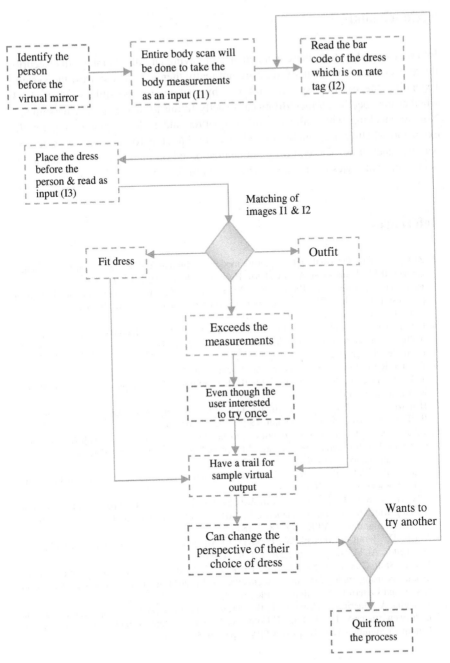

Fig. 5 Internal and external process of the virtual trail room

7 Conclusion

In our proposed system, there is no chance of having any disadvantage like security threat because of the virtually wearing not really then anyone can trust the system. Extending the existing system proposal by making the possible way to check whether the specific garment/footwear exactly fitting to that particular personality. Not by checking with different colors, textures and styles implementing for the extension whether the suitable one is there in that showroom makes easy for not waiting much time. The end user can change based on their interest and can try as many different styles of materials which are provided with.

References

1. Zhang, X., Fan, G.: Dual gait generative models for human motion estimation from a single camera. IEEE Trans. Syst. Man Cybern. B Cybern. **40**(4), 1034–1049 (2010)
2. Kanaujia, A., Kittens, N., Ramanathan, N.: Part segmentation of visual hull for 3D human pose estimation. In: 2013 IEEE Conference on Computer Vision and Pattern Recognition Workshops (CVPRW), pp. 542–549 (2013)
3. Liu, Y., Gall, J., Stoll, C., Dai, Q., Seidel, H.-P., Theobalt, C.: Markerless motion capture of multiple characters using multiview image segmentation. IEEE Trans. Pattern Anal. Mach. Intell. **35**(11), 2720–2735 (2013)
4. Sandhu, R., Dambreville, S., Yezzi, A., Tannenbaum, A.: Non-rigid 2D-3D pose estimation and 2D image segmentation. In: 2009 IEEE Conference on Computer Vision and Pattern Recognition (CVPR), pp. 786–793 (2009)
5. Hauswiesner, S., Straka, M., Reitmayr, G.: Virtual try-on through image-based rendering. IEEE Trans. Visual. Comput. Graph. **19**(9), 1552–1565 (2013)
6. Tanaka, H., Saito, H.: Texture overlay onto flexible object with PCA of silhouettes and k-means method for search into database. In: Proceedings of the IAPR Conference on Machine Vision Applications, Yokohama, Japan (2009)
7. Mansur, A., Makihara, Y., Yagi, Y.: Inverse dynamics for action recognition. IEEE Trans. Cybern. **43**(4), 1226–1236 (2013)
8. Elhayek, A., Stoll, C., Hasler, N., Kim, K.I., Seidel, H., Theobalt, C.: Spatio-temporal motion tracking with unsynchronized cameras. In: 2012 IEEE Conference on Computer Vision and Pattern Recognition (CVPR), pp. 1870–1877
9. Vlasic, D., Baran, I., Matusik W., Popovic, J.: Articulated mesh animation from multi-view silhouettes. ACM Trans. Graph. **27**(3), 1–9 (2008)
10. Liu, Y., Stoll, C., Gall, J., Seidel, H.-P., Theobalt, C.: Markerless motion capture of interacting characters using multi-view image segmentation. In: 2011 IEEE Conference on Computer Vision and Pattern Recognition (CVPR), pp. 1249–1256
11. Kanaujia, A., Haering, N., Taylor, G., Bregler, C.: 3D Human pose and shape estimation from multi-view imagery. In: 2011 IEEE Computer Society Conference on Computer Vision and Pattern Recognition Workshops (CVPRW), pp. 49–56

Data Aggregation Using Dynamic Tree with Minimum Deformations

Ashis Behera and Madhumita Panda

Abstract In Wireless Sensor Network, energy efficient routing is a challenging matter. In addition to this the loss of data packets during transmission should be handled carefully. Data aggregation is a technique which can be used to eliminate the unnecessary data or information and transmit only required data to the sink node. To maximize the lifetime of the sensor network, protocols like LEACH, PEGASIS had been proposed, where aggregation of data are done at the cluster head of each cluster and at the sink node. This scheme gives better performance than the conventional protocol of Direct Transmission. In this paper a new scheme has been proposed where the sink node is mobile in nature and the data are forwarded to the sink node through a tree which is constructed on the cluster head of each cluster. To maintain the dynamic structure of the tree during motion of the sink node, a minimum deformation is carried out on the tree so that total energy consumption is less which enhance the life time of the sensor network.

Keywords Wireless sensor network · Sensor nodes · Cluster · Dynamic tree · Data aggregation

1 Introduction

A Wireless Sensor Network (WSN) consists of spatially distributed autonomous sensors to monitor physical or environmental conditions, such as temperature, sound, pressure, etc. and to cooperatively pass their data through the network to a main location, generally to Base Station or Gateway Sensor Node [1]. The sensors are the

A. Behera (✉)
Department of Computer Science and Engineering, PKACE, Bargarh, India
e-mail: ashisbeherasuiit@gmail.com

M. Panda
SUIIT, Sambalpur University, Jyotivihar, Burla 768018, India
e-mail: mpanda.suiit@gmail.com

© Springer India 2015
L.C. Jain et al. (eds.), *Computational Intelligence in Data Mining - Volume 1*,
Smart Innovation, Systems and Technologies 31, DOI 10.1007/978-81-322-2205-7_17

small electronic devices which are equipped with a transceiver, for transmitting and receiving signal, a tiny memory for storing data, a tiny processor, and battery for power back up. Due to its low cost and adaptability nature it has been used in so many areas like in disaster management, combat field reconnaissance, border protection and in security surveillance [2]. Though WSN has a wide range of application, sensors have severe resource constraints in terms of power processing capabilities, memory and storage. WSN nodes are generally severe energy constrained due to the limitation of batteries. Due to its limited power source it will become dead once its power goes down. So hardware improvements in battery design and energy harvesting are the different challenges to improve the lifetime of the network and could make the network operational for a long time even without replacing the batteries. Data Aggregation is a technique which can be used to reduce the energy consumption of nodes and enhance the network lifetime [1, 3]. It tries to alleviate the localized congestion problem. It attempts to collect useful information from the sensors surrounding the event. It then transmits only the useful information to the end point thereby reducing congestion and its associated problems [3]. Data aggregation is used to aggregate the data before transmission so that redundancy data can be eliminated. It also reduces the traffic or congestion occurs at the base station hence the problem of packet loss can somehow be reduced.

2 Related Works

In a paper Fukabori et al. [4] suggested a scheme where the nodes in the sensing area are grouped into 'k' clusters by using EM algorithm and the mobile sink traces a trajectory of TSP through these cluster centroids. To aggregate the data efficiently, the mobile sink and nodes uses cluster adapted directed diffusion. To reduce the traffic and unwanted message transmission, each node may choose to transmit or not to transmit based on its degree of dependence ratio. This also gives better performance of the sensor network and also increase the lifetime of the network. Kamat et al. [2] in his paper has proposed a fine grained location based cluster with no cluster heads. In this paper [5], the authors have proposed a novel energy efficient clustering scheme (EECS) for single-hop wireless sensor networks, which better suits the periodical data gathering applications. Their approach elects cluster heads with more residual energy in an autonomous manner through local radio communication with no iteration while achieving good cluster head distribution. Furthermore, this paper introduces a novel distance-based method to balance the load among the cluster heads. Simulation results have showed that EECS prolongs the network lifetime significantly against the other clustering protocols such as LEACH and HEED. Virmani et al. in AIEEDA [6] proposed a method where the formation of clusters in the sensor networks can be done based on close proximity of nodes. The data transfer among the nodes is done with a hybrid technique of both TDMA/FDMA which leads to efficient utilization of bandwidth and maximizing throughput. It utilizes the energy and bandwidth by minimizing the distance between the nodes and schedules the clusters. Ozgur et al. [7] proposed PEDAP and PEDAP-PA techniques which are based on power consumption of individual node in a spanning tree which is constructed on the sensor nodes of the network. The power consumption of each node, thus network, is reduced by scheduling the transmission of each node of the tree. Aiming at the problem of limited energy of sensors in Wireless Sensor Network, based on the classic clustering routing algorithm LEACH, a distance-energy cluster structure algorithm [8] considering both the distance and residual energy of nodes is presented in the dissertation" which improves the process of cluster head election and the process of data transmission. It reduces the adverse effect on the energy consumption of the cluster head, resulting from the non-uniform distribution of nodes in network and avoids the direct communication between the base station and cluster head, which may has low energy and far away from base station. The results of simulation indicate that the improved algorithm effectively balances the energy consumption, prolongs 31 % of the lifetime, reduces 40 % of the energy consumption and has a better performance than the original LEACH protocol. In paper [6], the authors have investigated the reduction in the total energy consumption of wireless sensor networks using multi-hop data aggregation by constructing energy-efficient data aggregation trees. They have proposed an adaptive and distributed routing algorithm for correlated data gathering and exploit the data correlation between nodes using a game theoretic framework. Routes are chosen to minimize the total energy expended by the network using best response dynamics to local data. The cost function that is used for the proposed routing algorithm takes into account energy, interference and in-network data aggregation. The iterative algorithm is shown to converge in a finite number of steps. Simulations results show that multi-hop data aggregation can significantly reduce the total energy consumption in the network. The authors of [7] have intended to propose a new protocol called Fair Efficient Location-based Gossiping (FEL Gossiping) to address the problems of Gossiping and its extensions. They show how their approach increases the network energy and as a result maximizes the network life time in comparison with its counterparts. In addition, they also show that the energy is balanced (fairly) between nodes. Thus saving the nodes energy leads to an increase in the node life in the network, in comparison with the other protocols. Furthermore, they also show that the protocol reduces the propagation delay and loss of packets.

3 Problem Statement

According to Low Energy Adaptive Clustering Hierarchy (LEACH), the data are transmitted by each cluster head to the sink node directly. But this will consume more energy when the sink is far away from the site. In LEACH the sink node is stationary. So the energy consumption is more to transmit the aggregated data by the cluster heads to the sink node directly. So this paper proposed an in-network data aggregation scheme where the clustering technique is used to aggregate the data and used a dynamic tree to route the aggregated data to the sink node. Here the sink is mobile in nature to reduce further the transmitted cost. The dynamic nature of the tree should be maintained with minimum possible changes on the tree so that the total energy consumption will be less. It has been assumed that the sensor nodes in each cluster know their location at priori by implementing some network localization scheme.

4 Design Methodology

This paper has been organized into four parts.

(a) Formulation for Cluster and Cluster Head
(b) Construction of tree
(c) Formation of trajectory path
(d) Maintenance of the dynamic tree

4.1 Formation of Clusters and Cluster Head

Since the required power of wireless transmission is proportional to the square of the transmission distance the transmission distance of individual node should be reduced as well as the balancing the load or traffic in the network also be maintained. The following algorithm divides the nodes of WSN into k numbers of clusters.

Algorithm for formation of clusters

1. Divide the network region of size into K numbers of small regions of same size.
2. Each sensor node has the pre-knowledge of the mid-points of each the regions.
3. Let N number of sensor nodes that are randomly distributed in the network region.
4. All sensor nodes know their location in the network relative to the sink node, which is GPS enabled, by applying some network localization algorithm.
5. Each sensor can determine the nearest sub-region by calculating Euclidian distance of itself from mid-point of all other regions.
6. A node can associate with that region whose calculating distance from that node is minimum. The node stores the mid-point of its associated region.
7. For any node, if more than one regionis found to be shortest distance and same, then association of that node should be done one region.

Fig. 1 Association of nodes
to a region

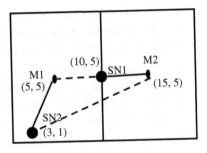

The association of each node to a certain region is illustrated as below:

Suppose two nodes SN1 and SN2 have been deployed in the regions shown above. The coordinates of SN1 and SN2 are (10, 5) and (3, 1) respectively. Let the mid-point of two regions are (5, 5) and (15, 5) respectively as shown in Fig. 1.

Now the two nodes SN1 and SN2 determine the shortest distance from M1 and M2 respectively.

For SN2:

$$D1 = \left(|(3-5)|^2 + |(1-5)|^2 \right)^{1/2} = 4.472$$

$$D2 = \left(|(3-15)|^2 + |(1-5)|^2 \right)^{1/2} = 12.649$$

So SN2 choose M1 as the closest region and associated with that region.

For SN1:

$$D1 = \left(|(10-5)|^2 + |(5-5)|^2 \right)^{1/2} = 5,$$

$$D2 = \left(|(10-15)|^2 + |(5-5)|^2 \right)^{1/2} = 5$$

Since both the mid-points are equidistance from SN1, SN1 choose any one let M2 as the nearest region and associated with that region.

Algorithm for selection of Cluster head in each cluster

1. Each node maintains a list of IDs and residual energy of all other nodes including itself in a descending order of their residual energy.
2. Initially, each node stores the ID and residual energy of itself.
3. Each node sends a packet to its neighbors who contain the residual energy of itself, its ID and its associated region's mid-point.

4. Repeat steps 5, 6, 7 for N/K times where N = Total number of sensor nodes and K = Total number of clusters or regions.
5. On receiving a packet from its neighbor, a node can do the followings:

   a. Compare the mid-point which is present in the packet with the mid-point of itself. If matching is found then go to next step otherwise go to step 7.
   b. Update its list by comparing the residual energy present in the packet with the residual energies present in its list.

6. If a node updates its list, then it will retransmit its updated list along with its associated region's mid-point to its neighbors. Go to step 4.
7. If a node does not update its list then discard the packet.

After M/K number of iterations, all nodes of each cluster have the same node IDs and in same order. The node ID placed at the top of the list will be declared as a cluster head.

So after creation of cluster head, all other nodes in that cluster send their data to their cluster head. This is the responsibility of the cluster head to transmit the collected data to the sink node after aggregation. To maintain the uniformity, the cluster head in each cluster is changed in a regular interval among their members. When a cluster head has the remaining residual energy less than 25 % of the residual energy of the last node in the list, then it will quit from the leadership by sending its residual energy and unwillingness to all other nodes in its associated region after updating its list. After receiving the unwillingness signal, all other nodes within that region update their list and the node placed at the top of the list declare itself as a cluster head.

4.2 Construction of Tree

After formation of cluster head, a tree will be constructed which takes the cluster heads as nodes. The root of the tree will be the node closest to the sink.

Algorithm for formation of tree:

1. All cluster heads send their locations to the sink nodes.
2. The sink node will store a copy of that and calculate the distance of itself from all other cluster heads and designate the closest cluster head as root of the tree.
3. The sink node stores the location of the root node as recipient and destination address and sends a message to the designated root node which contains the location of all other nodes except the location of the designated root node.
4. If more than one heads are equidistant from the sink nodes, then the sink node can choose any one of them as root node.
5. Repeat the steps 6, 7, 8 until all the cluster heads are exhausted.

6. After receiving the message, the root node stores recipient location as recipient address and calculate the distance of itself from other heads.

7. It then designate the closest head as the root of the remaining tree, store its location as destination address and sends the message to the designated root after removing the location of the designated root node from the message.

8. It should be ensured that each node should not receive the message from more than one neighbor.

After this algorithm, a tree will be constructed (Fig. 2) taking the cluster heads as its nodes. The child nodes are connected to their parent node through edges. The directions of these edges are from parent node to child node. Now make this tree as converge-cast tree (Fig. 3), by reversing the directions of each edge i.e. from child node to parent node. This is required to reduce the transmission cost and reduce the duplicity of data.

It found the single neighbor and connects it with an edge. Then remove that connected node from the pool. This process continued until the pool is empty. And finally a tree is constructed among the cluster heads.

Fig. 2 Construction of tree

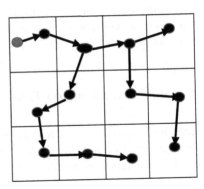

Fig. 3 Converge-cast tree

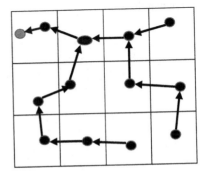

4.3 Formation of Trajectory Path

Since the root node is responsible for data transmission, it must be the closest node to the sink node hence reduction of energy consumption by reducing the transmission distance. The deformation of the tree occurs due to mobility of the sink node. Hence new root should be chosen by the sink node while it is moving into its trajectory path. The following algorithm selects new root nodes in the deformation tree:

Algorithm for finding the new root nodes during the motion of sink node

1. The sink node will choose number of rest points in the network.
2. For each rest point

    a. The sink node calculates the distance from each node location to that rest point.
    b. Select the node which has the shortest distance from that rest point and keep it in the list. If more than one node exists, then select any one node.

3. The node listed in the list of the sink node may serve as intermediate root nodes when sink node moves.

As shown in the Fig. 4, the location of the sink node is near the root. After the sink shown as square size block, determine its two rest points, it will determine the new root nodes for the tree which is shown in faded circle.

4.4 Maintenance of the Tree

After the construction of the tree, a converge-cast tree is to be constructed, where all the edge are directed towards the root. By doing this we can have the maximum number of outgoing edge from a node is 1 but the number of incoming edge to a

Fig. 4 Formation of trajectory path

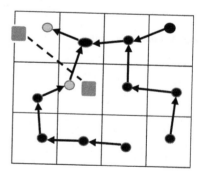

node may be more than one. The maintenance of the tree is essential when the root of the tree changes due to the mobility of the sink. The algorithm is given below.

Algorithm for maintaining the tree

1. The sink node after positioning itself to a new position, it sends a packet to the root node along with the node ID of next root node.
2. The root node keeps the location and ID of the next root node.
3. When the sink node selects a new root node the following condition should be checked.

    i. An edge is to be added from the previous root node to the current root node, keeping other edges of the tree unchanged.
    ii. Each node must have at most one out degree. If a node has more than one outgoing edges then select the nearest node edge and delete other outgoing edges of that node.

4. The root node after receiving the packet from the sink, it will forward the query packet to the other nodes in the tree.
5. The node which has the data related to the query responds by forwarding the data to the root node.
6. The root node in turn transmits the data to the sink node.

Let the sink node is at its first rest point top left corner of the grid. It will send a query packet to the root node. The root node forwards the packet to other nodes as shown in the Fig. 5.

When the intended node (mark as faded dotted circle in Fig. 6), which has the data related to the query, receive the query packet, it forward the data to the root node which forward the data to the sink node which is shown in dotted lines. Let the sink node is at new rest point as shown in Fig. 7. A new edge is added from previous root node to the new root node. When it is added, it has been found that the new root has more than one out-degree, so it deletes the outgoing edge of the

Fig. 5 Sink node forward the request to the root and the root send it to the intended node

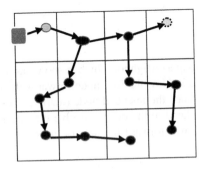

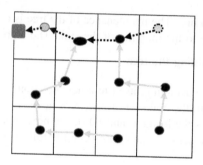

Fig. 6 After receiving the request the intended node forward the message to the skin node

Fig. 7 Sink node forward the request to the new root and the new root send it to the intended node

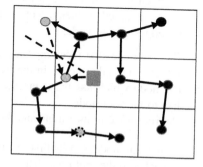

Fig. 8 After receiving the request the intended node forward the message to the sink node

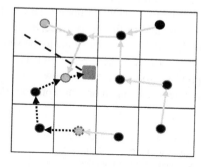

longest neighbor which is shown as dotted lines in Fig. 7. It sends the query packet to it. Now the root node forwards the packet to other nodes as shown in the Fig. 7. When the intended node (mark as faded dotted circle in Fig. 8), which has the data related to the query, receive the query packet, it forward the data to the root node which forward the data to the sink node which is shown in dotted lines.

5 Conclusion and Future Works

Wireless Sensor Networks are powered by the limited capacity of batteries. Due to the power management activities of these sensor nodes, the network topology changes dynamically. These essential properties pose additional challenges to communication. In-network data aggregation using cluster is a good technique to reduce the transmission of redundancy data, hence, a reduction of the energy consumption. Energy consumption further reduced by forwarding the data through a tree. As explained, the cost incurred in maintaining the tree, due to mobility of the sink node, is somehow minimized by reducing the number of deformations in the tree. But this incurred some delay as compared to direct transmission. The task of implementation and simulation of this new module is to be done in a suitable network simulator. Also the presence of the malicious node in the network has not been considered which has been kept for future works.

References

1. Vidyanathan, K., Sur, S., Narravula, S., Sinha P.: Data aggregation techniques in sensor network. OSU-CISRC-11/04-TR60, The Ohio State University, OSU (2004)
2. Kamat, M., Ismail, A.S., Olari S.: Efficient in-network aggregation for WSN with fine grain location based cluster. In: Proceeding of International Conference on MOMM, Jan 2011, pp. 66–71 (2011)
3. Akkaya, K., Ari I.: In-network data aggregation in WSNs
4. Fukabori, T., Nakayamaetall H.: An efficient data aggregation scheme using degree of dependence on cluster in WSNs. In: IEEE ICC Proceedings (2010)
5. Ye, M., LI, C., Chen, G., Wu, J.: An energy efficient clustering scheme in wireless sensor networks. Ad Hoc Sens. Wirel. Netw. 3, 99–119
6. Zeydan, E., Kivanc, D., Comaniciu, C., Tureli U.: Networks Ad Hoc Networks. Elsevier, New York (2012)
7. Norouzi, A., Babamir, F.S., Zaim, A.H.: A novel energy efficient routing protocol in wireless sensor network. Wirel. Sens. Netw. 3, 341 (2011)
8. Yong, Z., Pei, Q.: A energy-efficient clustering routing algorithm based on distance and residual energy for wireless sensor networks. Procedia Eng. 29, 1882–1888 (2012) (Elsevier)
9. Heinzelman, W.R., Chandrakasan, A., Balakrishnan H.: Energy-efficient communication protocol for wireless microsensor networks. In: Proceeding of the HICSS (2000)
10. Nakayama, H., Ansari, N., Jamalipour, A., Kato, N.: Fault-resilient sensing in WSN. Comput. Commun. Arch. 30, 2375–2384 (2007)
11. Virmani, D. Singhal, T., Ghanshyam, Ahlawat, K. Noble: Application independent energy efficient data aggregation in WSN. IJCSI 9(2) (2012)
12. Ozgur, H., Korpeoglu I.: Power efficient data gathering and aggregation in WSN. SIGMOID Rec. 32(4), 66–71 (2003)
13. Enachescu, M., Goel, A., Govindan R.: Scale free aggregation in sensor networks. Elsevier Science, New York (2006)
14. Zhao, J., Govindan, R., Estrin D.: Computing Aggregates for Monitoring WSN (2003)

15. Zhu, X., Tang, B., Gupta H.: Delay efficient data gathering in sensor networks. Lecture Note in Computer Science, vol. 3794, pp. 380–389. Springer, Heidelberg (2005)
16. Al-Yasiri, A., Sunley A.: Data aggregation in WSN using SOAP protocol. J. Phys. Conf. Ser. **76**, 012039 (2007)

An Improved Cat Swarm Optimization Algorithm for Clustering

Yugal Kumar and G. Sahoo

Abstract Clustering is an efficient technique that can be put in place to find out some sort of relationship in the data. Large number of heuristic approaches have been used for clustering task. The Cat Swarm Optimization (CSO) is the latest meta-heuristic algorithm which has been applied in clustering field and provided better results than K-Means and Particle Swarm Optimization (PSO). However, this algorithm is suffered with diversity problem. To overcome this problem, an improved version of CSO method using Cauchy mutation operator is proposed. The performance of improved CSO is compared with the existing methods like K-Means, PSO and CSO on several artificial and real datasets. From the simulation study, it came to revelation that the improved CSO algorithm gives better quality solution than others.

Keywords Cat swarm optimization · Cauchy mutation operator · Clustering and particle swarm optimization

1 Introduction

Clustering is a process to find out the groups of similar objects in a given dataset and it can be applied in many areas like image analysis, pattern recognition, data mining, medical science etc. In clustering, a similarity criterion function is defined to search out the resemblance between objects. In literature, it is found that

Y. Kumar (✉) · G. Sahoo
Department of Computer Science and Engineering,
Birla Institute of Technology, Mesra, Ranchi, Jharkhand, India
e-mail: yugalkumar.14@gmail.com

G. Sahoo
e-mail: gsahoo@bitmesra.ac.in

© Springer India 2015
L.C. Jain et al. (eds.), *Computational Intelligence in Data Mining - Volume 1*,
Smart Innovation, Systems and Technologies 31, DOI 10.1007/978-81-322-2205-7_18

similarity criterion functions for clustering are non-convex and nonlinear in nature. It becomes a NP-Hard problem when the number of clusters increases more than three. The study of relevant literature suggests that the heuristic approaches are more suitable and gain wide popularity to solve large size of problems even when it becomes NP hard. In clustering field, the K-Means is one of the oldest and well known approach which has been developed in [1]. But, this algorithm has several shortcomings- stuck in local optima, lack of information to treat inappropriate and clatter attributes [2]. Thus many researchers have proposed hybrid and heuristic approaches to conquer these shortcomings. Several heuristic approaches which have been applied for clustering came to knowledge after extended literature survey such as genetic algorithm [3, 4], simulated annealing [5, 6], tabu search [7], ACO [8], PSO [9], ABC [10], CSS [11, 12], TLBO [13, 14] and many more.

Conversely, the Cat swarm optimization (CSO) is the latest meta-heuristic algorithm developed by Tasi et al. [15] modeled on the behavior of cats and applied to solve optimization problems. Santosa et al. [16] have applied cat swarm optimization (CSO) algorithm to find out optimal cluster centers and showed that the CSO provided more accurate results than others. Nevertheless, this algorithm is suffered with population diversity problem. Hence to address the above mentioned shortcoming, a mutation operator is introduced in the CSO algorithm to maintain and enhance the population diversity. Besides this, it is also captured in local optima as there is no predefined method to deal with data objects which cross the boundary values. Therefore, some modifications are also proposed in the original CSO to stick out the local optima problem and get an improved version of CSO (ICSO) algorithm.

2 Cat Swarm Optimization

Chu and Tsai have introduced a new meta-heuristic algorithm based on the natural behavior of cats and named it cat swarm optimization [15]. Cats have two distinct features that make them distinct from other species. These features are intense curiosity of moving objects and outstanding hunting skill. Cats always stay alert but change their positions very slowly. This behavior of cats is characterized as seeking mode. When cats sense the presence of a target, they trace it very quickly. This behavior of cats is presented as tracing mode. Thus, a mathematical model is constituted by combining these two modes to solve optimization problems. In a random search space, each cat is represented using position and velocity. A problem specific fitness function is used to direct the next step of the search. Along these, a flag is used in determining whether the cat is in seeking mode or tracing mode.

2.1 Seeking Mode

The seeking mode of CSO algorithm can be viewed as a global search of solution in the random search space of the optimization problem. Few terms related to this mode are reproduced below.

- Seeking Memory Pool (SMP): Number of copies of cat in seeking mode.
- Seeking Range of selected Dimension (SRD): It is the maximum difference between the new and old values in the dimension selected for mutation.
- Counts of Dimension to Change (CDC): It is the number of dimensions to be mutated.

The steps involved in this mode are:

1. Make "i" copies of cat_j, where "i" equal to the seeking memory pool of cat_j, if "i" is one of candidate solutions then "i" = SMP-1 Else "i" = SMP
2. Determine the shifting value for each "i" copies using (SRD*position of cat_j)
3. Determine the number of copies undergo mutation (randomly add or subtract the shifting value to "i" copies)
4. Evaluate the fitness of all copies
5. Pick the best candidate from i copies and place it in the position of jth cat.

2.2 Tracing Mode

The tracing mode of CSO algorithm is the same as a local search technique for the optimization problem. In this mode, cats update their velocities due to target the object with high speed. Thus, enormous differences between the positions of cats are occurred. So, the position (X_j) and velocity (V_j) of cat_j in the d-dimensional space can be described as $X_j = \{X_{j,1}, X_{j,2}, X_{j,3} \ldots X_{j,D}\}$ and $V_j = \{V_{j,1}, V_{j,2}, V_{j,3} \ldots V_{j,D}\}$ where $D(1 \leq d \geq D)$. The global best position of the cat_j is represented as $P_g = \{P_{g,1}, P_{g,2}, P_{g,3} \ldots P_{g,D}\}$ and the velocity and position are updated using the Eqs. 1 and 2 respectively.

$$V_{jd} = w * V_{jd} + c * r * (X_{gd} - X_{jd}) \tag{1}$$

$$X_{jd,new} = X_{jd} + V_{jd} \tag{2}$$

3 Improved Cat Swarm Optimization (ICSO)

In CSO method, when a data instance crosses the boundary of the dataset, it will replace the value of data instances by the value near to the boundary of dataset. So every time, a value that is close to the boundary of dataset will be assigned to the data instances. Hence, there is a possibility that the algorithm will be stuck in local optima if the solution lies near to the boundary and also lost its diverse nature. Therefore, to address these short-comings, we are hereby reporting few improvements. Thus, to overcome sticking in local optima, two modifications are done—one for the seeking mode and another for the tracing mode of CSO method and Cauchy mutation operator is used to handle the diversity problem.

In the seeking mode, the movement (position) of cats is obtained by randomly adding or subtracting the shifting value from cluster centers and as a result of this (SMP*K) number of positions are obtained. Thus, addition and subtraction of shifting value from cluster centers may lead the data vectors to cross the boundary of dataset. Finally, a mechanism is introduced to deal with such data vectors. The proposed mechanism can be described as follows.

If the data vector $X_j(d) <$ Min (d), then Eq. 3 is used to obtain the new position of data instance. In Eq. 3, "a" is a variable which is used to escape the data vectors stuck in local optima near the boundary of data set and it is calculated using the Eq. 4.

$$X_{jd} = D_{min}^m + rand(0, 1) * \left(D_{max}^m - D_{min}^m\right) * a \qquad (3)$$

$$a = (1 + iteration/iteration\ max.) \qquad (4)$$

If the data vector $X_j(d) >$ Max (D), then Eq. 5 is used to reallocate the position of data instances and the value of "a" is calculated using given Eq. 6.

$$X_{jd} = D_{max}^m - [rand(0, 1) * \left(D_{max}^m - D_{min}^m\right) * a] \qquad (5)$$

$$a = (1 - iteration/iteration\ max.) \qquad (6)$$

Another modification is made in tracing mode. The tracing mode of CSO algorithm seems to be a local search technique for an optimization problem. In tracing mode, a cat traces its target with high speed. Mathematically, it can be achieved by defining the positions and velocities of cats in d-dimensional search space. Thus, the new position of jth cat is obtained by Eq. 2. Hence, there is also a chance that data vectors may go beyond the boundary limits. Thus, to deal with such data vectors, another method is described. The proposed method can be summarized as follows. Thus, the Eqs. 7 and 8 are used to handle the boundary constraints in tracing mode.

When any data vector $X_j(d) <$ Min (D) then

$$X_{j,d} = \min(D), \quad V_{j,d} = \text{rand}(0,1) * |V_{j,d}| \tag{7}$$

When any data vector $X_j(d) <$ Min (D) then

$$X_{j,d} = \max(D), \quad V_{j,d} = -\text{rand}(0,1) * |V_{j,d}| \tag{8}$$

The other issue related to CSO algorithm is to maintain the population diversity especially in tracing mode. Numerous researchers have introduced the concept of mutation for this [17, 18]. The idea behind the inclusion of the mutation operator with heuristic approaches is to increase the probability to escape from local optima. Thus, a Cauchy mutation operator is applied to maintain the population diversity. The Cauchy mutation operator is described as follows.

$$W(i) = \left(\sum_{i=1}^{D} V[j][i] \right) / D \tag{9}$$

where, $V[j][i]$ represents the ith velocity vector of the jth cat, D represents the dimension of jth particle and $W(i)$ is a weight vector in the range of $[-W_{min}, W_{max}]$ whose values vary in between $[0.5, -0.5]$.

$$X_{gd,new}(j) = X_{gd}(j) + W(i) * N * (X_{min} - X_{max}) \tag{10}$$

where, $X_{gd}(j)$ represents the best position of jth cat, $W(i)$ represents the Cauchy mutation and its value is calculated using Eq. 9, N represents the Cauchy distribution function and (X_{min}, X_{max}) is a minimum and maximum value of ith attribute of a dataset D.

The population diversity of CSO method can be improved using Eq. 10 and the algorithm can explore more solution space. Along this, restriction of the data vectors in random space search increases the chances of better results. The flow chart of proposed ICSO algorithm is shown in Fig. 1.

3.1 Steps of Improved CSO Algorithm

Step 1: Load the dataset; initialize the parameters for CSO method and number of cats.

Step 2: Initialize positions of cats in random fashion and the velocities of every cat.

Step 3: Determine the value of the objective function (Euclidean distance) and group the data according to the objective function value.

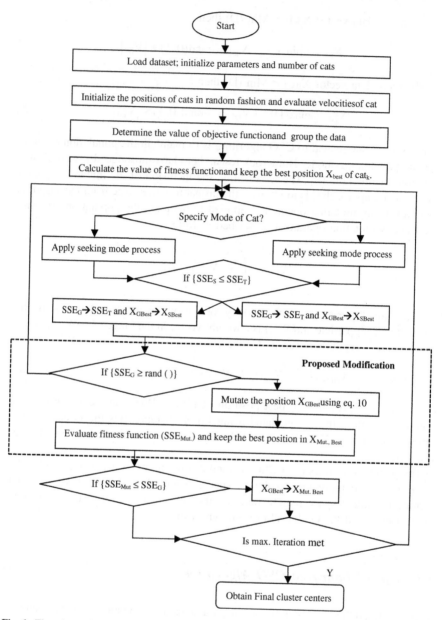

Fig. 1 Flowchart of proposed ICSO Algorithm

Step 4: Calculate the value of the fitness function (SSE) and keep the best position X_{best} of cat_k with minimum fitness value.

Step 5: For each cat_k, apply seeking mode process:

Step 5.1: Make the j copy of each cat_k.

Step 5.2: Calculate shifting value for each cat_k using (SRD*cluster center (k)).

Step 5.3: Add or subtract each cat_k to shifting value.

Step 5.4: Calculate the objective function value, group the data.

Step 5.5: Evaluate the fitness function (SSE) and keep the best position X_{Sbest} of cat_k.

Step 5.6: If $\{SSE \leq SSE_S\}$, then $SSE_S \rightarrow SSE$ and $X_{SBest} \rightarrow X_{best}$.

Step 5.7: Else, $SSE \rightarrow SSE_S$ and $X_{Sbest} \rightarrow X_{SBest}$.

Step 6:   For each cat_k, apply tracing mode process:

Step 6.1: Update the velocity of cat_k using Eq. 1.

Step 6.2: Update the position of each cat_k using Eq. 2.

Step 6.3: Calculate the objective function value and group the data.

Step 6.4: Compute the fitness function (SSE_T) and keep the best position X_{TBest} of cat_k.

Step 7:   If $(SSE_S \leq SSE_T)$ then $SSE_G \rightarrow SSE_T$ and $X_{GBest} \rightarrow X_{SBest}$, Else $SSE_G \rightarrow SSE_T$ and $X_{GBest} \rightarrow X_{SBest}$

Step 8:   If $\{SSE_G \geq rand()\}$ then, goto step 5

Step 9:   Else, mutate the position X_{GBest} using Eq. 10

Step 10:   Compute the fitness function ($SSE_{Mut.}$) and keep the best position in $X_{Mut., Best}$

Step 11:   If $\{SSE_{Mut} \leq SSE_G\}$, then, $X_{GBest} \rightarrow X_{Mut., Best}$

Step 12:   Go to step 5, until the maximum iteration reached

Step 13:   Obtain the final solutions.

where SSE_S, SSE_T, SSE_{Mut} and SSE_G represent the value of fitness function in seeking mode, tracing mode, mutation with Cauchy operator and global fitness of CSO algorithm and X_{Sbest}, X_{TBest}, $X_{Mut., Best}$ and X_{GBest} describe the best position achieved by a cat_k in seeking mode, tracing mode, mutation with Cauchy operator and global best respectively.

4 Experimental Results

This section describes the simulation study of the proposed ICSO algorithm with two artificial datasets, generated in Matlab 2010a and four real datasets which are taken from the UCI repository. The real datasets are iris, CMC, cancer and wine. The characteristics of these datasets are mentioned in Table 1. The performance of proposed algorithm is compared with K-Means, PSO and CSO via sum of intra cluster distance, standard deviation and f-measure parameters. Matlab 2010a

Table 1 Features of datasets used in experiment

Dataset	Class	Feature	Total data	Data in each classes
ART 1	3	2	300	(100, 100, 100)
ART 2	3	3	300	(100, 100, 100)
Iris	3	4	150	(50, 50, 50)
Cancer	2	9	683	(444, 239)
CMC	3	9	1,473	(629, 334, 510)
Wine	3	13	178	(59, 71, 48)

Table 2 ICSO parameters value

Parameters	Values
Max. Iter.	100
SRD	(0,1)
SMP	5
r	(0,1)
c	2
SPC	(0, 1)

environment is used to implement the proposed algorithm. The algorithm is run for a hundred times independently with randomly initialized cluster centers. Parameters of proposed method are discussed in Table 2.

4.1 Datasets

ART1: A two dimensional dataset consist of 300 instances with the two attributes and three classes to validate the proposed algorithm. Classes in dataset are circulated using μ and λ where μ is the mean vector and λ is the variance matrix and values of $\mu 1 = [1, 3]$, $\mu 2 = [0, 3]$, $\mu 3 = [1.5, 2.5]$ and $\lambda 1 = [0.3, 0.5]$, $\lambda 2 = [0.7, 0.4]$, $\lambda 3 = [0.4, 0.6]$.

ART2: It is three dimensional data which consist of 300 instances with three attributes and three classes. The data has created using $\mu 1 = [10, 25, 12]$, $\mu 2 = [11, 20, 15]$, $\mu 3 = [14, 15, 18]$ and $\lambda 1 = [3.4, -0.5, -1.5]$, $\lambda 2 = [-0.5, 3.2, 0.8]$, $\lambda 3 = [-1.5, 0.1, 1.8]$.

Table 3 indicates the simulation results of improved CSO algorithm with K-Means, PSO and CSO methods. From this table, it is concluded that improved CSO algorithm provides more accurate results than others. It is also pointed out that performance of CSO algorithm is not good for ART1 and cancer dataset as compare to PSO algorithm.

Table 3 Performance comparison of ICSO algorithm with different techniques

Dataset	Parameters	K-means	PSO	CSO	ICSO
ART 1	Best	157.12	154.06	154.26	154.13
	Average	161.12	158.24	159.06	158.17
	Worst	166.08	161.83	164.56	162.08
	Std.	0.34	0	0.292	0.14
	F-Measure	99.14	100	100	100
ART2	Best	743	740.29	740.18	740.08
	Average	749.83	745.78	745.91	745.74
	Worst	754.28	749.52	749.38	748.24
	Std.	0.516	0.237	0.281	0.247
	F-Measure	98.94	99.26	99.32	99.35
Iris	Best	97.33	96.89	96.97	96.78
	Average	106.05	97.23	97.16	97.08
	Worst	120.45	97.89	98.18	97.83
	Std.	14.631	0.347	0.192	0.156
	F-Measure	0.782	0.782	0.782	0.783
Cancer	Best	2,999.19	2,973.5	2,992.45	2,967.07
	Average	3,251.21	3,050.04	3,109.14	3,036.49
	Worst	3,521.59	3,318.88	3,456.63	3,291.16
	Std.	251.14	110.801	132.47	43.56
	F-Measure	0.829	0.819	0.831	0.834
CMC	Best	5,842.2	5,700.98	5,696.23	5,685.76
	Average	5,893.6	5,820.96	5,778.12	5,756.31
	Worst	5,934.43	5,923.24	5,908.32	5,917.21
	Std.	47.16	46.959	41.33	35.04
	F-Measure	0.334	0.331	0.336	0.341
Wine	Best	16,555.68	16,345.96	16,331.56	16,317.46
	Average	18,061	16,417.47	16,395.18	16,357.89
	Worst	18,563.12	16,562.31	16,548.54	16,534.76
	Std.	793.213	85.497	57.34	40.73
	F-Measure	0.521	0.518	0.523	0.524

However, modification of CSO method not only improves the results with ART1 and cancer datasets but also improved the results with all other datasets using all of parameters. The improved CSO algorithm achieves minimum intra cluster distance for all datasets among all techniques being compared. Figures 2 and 3 shows the convergence of the intra cluster distance parameter and objective function (f-measure) for wine dataset using all methods.

Fig. 2 Convergence of sum of intra cluster distance parameter of wine data set

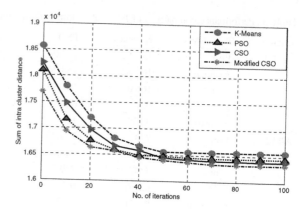

Fig. 3 Convergence of objective function (F-measure) of wine data set

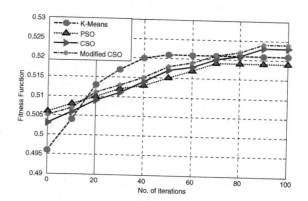

5 Conclusion

In this work, Cauchy mutation operator based CSO algorithm is proposed to improve the population diversity of CSO algorithm. Along this, some heuristic approaches are also proposed to deal with the data vectors which cross the boundary of dataset and to overcome the fall in local optima if solution exists near the boundary values. Conclusively, some amendments are here by proposed to sort out these problems in the CSO method and also to enhance its performance. To ascertain the efficiency of the proposed method, it is compared with K-Means, PSO and CSO algorithms utilizing six datasets and the resulted outcomes are in favor of improved CSO algorithm.

References

1. MacQueen, J.: On convergence of k-means and partitions with minimum average variance. Ann. Math. Statist **36**, 1084 (1965)
2. Jain, A.K.: Data clustering: 50 years beyond K-means. Pattern Recog. Lett. **31**(8), 651–666 (2010)
3. Bandyopadhyay, S., Maulik, U.: Genetic clustering for automatic evolution of clusters and application to image classification. Pattern Recogn. **35**(6), 1197–1208 (2002)
4. Krishna, K., Narasimha, M.: Murty: genetic K-means algorithm. IEEE Trans. Syst. Man Cybern. B Cybern. **29**(3), 433–439 (1999)
5. Selim, S.Z., Alsultan, K.L.: A simulated annealing algorithm for the clustering problem. Pattern Recogn. **24**(10), 1003–1008 (1991)
6. Maulik, U., Mukhopadhyay, A.: Simulated annealing based automatic fuzzy clustering combined with ANN classification for analyzing microarray data. Comput. Oper. Res. **37**(8), 1369–1380 (2010)
7. Sung, C.S., Jin, H.W.: A tabu-search-based heuristic for clustering. Pattern Recogn. **33**(5), 849–858 (2000)
8. Shelokar, P.S., Jayaraman, V.K., Kulkarni, B.D.: An ant colony approach for clustering. Anal. Chim. Acta **509**(2), 187–195 (2004)
9. Kao, Y.T., Zahara, E., Kao, I.W.: A hybridized approach to data clustering. Expert Syst. Appl. **34**(3), 1754–1762 (2008)
10. Zhang, Changsheng, Ouyang, Dantong, Ning, Jiaxu: An artificial bee colony approach for clustering. Expert Syst. Appl. **37**(7), 4761–4767 (2010)
11. Kumar, Y., Sahoo, G.: A charged system search approach for data clustering. Prog. Artif. Intell. **2**, 153–166 (2014)
12. Kumar, Y., Sahoo, G.: A chaotic charged system search approach for data clustering. Informatica (accepted, in press)
13. Satapathy, S.C., Naik, A.: Data clustering based on teaching-learning-based optimization. In: Swarm, Evolutionary, and Memetic Computing, pp. 148–156 (2011)
14. Sahoo, A.J., Kumar, Y.: Improved teacher learning based optimization method for data clustering. In Advances in Signal Processing and Intelligent Recognition Systems, pp. 429–437. Springer International Publishing, Heidelberg (2014)
15. Chu, S.C., Tsai, P.W., Pan, J.S..: Cat swarm optimization. In: PRICAI 2006: Trends in Artificial Intelligence, pp. 854–858. Springer, Heidelberg (2006)
16. Santosa, B., Ningrum, M.K.: Cat swarm optimization for clustering. In: International Conference of Soft Computing and Pattern Recognition, SOCPAR'09, IEEE, 54–59 (2009)
17. Hu, X., Eberhart, R. C., and Shi, Y.: Swarm intelligence for permutation optimization: a case study on n-queens problem. In: Proceedings of IEEE Swarm Intelligence Symposium, pp. 243–246 (2003)
18. Yao, Xin, Liu, Yong, Lin, Guangming: Evolutionary programming made faster. Evol. Comput. IEEE Trans. **3**(2), 82–102 (1999)

Watermarking Techniques in Curvelet Domain

**Rama Seshagiri Rao Channapragada
and Munaga V.N.K. Prasad**

Abstract This paper proposes four different methods for embedding and extraction of the watermark into the cover image based on Curvelet Transform Technique. Magic Square Technique was used in the algorithms for spreading the watermark and embedding into the curvatures of original image. The Curvelet transform is a type of the Wavelet transform technique designed to represent images in sparse mode consisting of all objects having curvature information taken in higher resolution even for lower resolution content. The experiments indicated that these algorithms embedded the watermark efficiently such that the images have possessed robust watermark on extraction after the image compression like JPEG, GIF, scaling, rotation and noise attacks.

Keywords Digital watermarking · Magic square · Curvelet transform · Peak signal to noise ratio

1 Introduction

Watermarking is a process of inserting authentication details into digital contents. This ensures that the authentication proof cannot be easily separated from the watermarked content. The first paper watermarking technique was appeared around 1282 in Italy [1, 2]. After discovery of the digital camera by Steven Sasson in 1975 and due to the increase in WiFi technology usage in digital image processing

R.S.R. Channapragada (✉)
Department of CSE, Geethanjali College of Engineering and Technology,
Cheeryal, Rangareddy 501301, Telangana, India
e-mail: crsgrao@yahoo.com

M.V.N.K. Prasad
Department of CSE, IDRBT, Hyderabad, Telangana, India
e-mail: mvnkprasad@idrbt.ac.in

© Springer India 2015
L.C. Jain et al. (eds.), *Computational Intelligence in Data Mining - Volume 1*,
Smart Innovation, Systems and Technologies 31, DOI 10.1007/978-81-322-2205-7_19

devices, the digital image sharing and accessing has become simple and convenient [3]. This has increased the vulnerability of protection towards the digital images. This has triggered the importance of research for protecting owners copy rights.

The wavelets in Discrete Wavelet Transform (DWT) are discretely sampled with which it gains temporal resolution advantage. The Belgian mathematician Ingrid Daubechies has invented Daubechies wavelets in 1988, which belongs to the family of the discrete wavelet transforms. The Hungarian mathematician Alfréd Haar has discovered the first DWT [4]. The transformation functions mainly useful to find self-similarity properties of a signal or fractal problems, signal discontinuities, etc.

Curvelet Transform belongs to the family of DWT. Curvelets represent images with objects having minimum length scale bounded curvatures. It is important for graphical representation of cartoons, images etc. Curvelet represent these objects by considering higher resolution curvelets for handling lower resolutions, which appear straight as it gets zoomed. However, photographs will not possess this property. Curvelets provide the information in spare matrix format [5]. The DWT and Curvelet techniques have motivated many researchers to develop robust digital watermarking techniques.

Zhang and Hu [6] have discussed a watermarking technique based on curvelet transform domain. In this technique [6], the scrambled watermark was embedded to the coarse coefficients of original image, which are larger than some threshold values in the curvelet transform domain. Xiao et al. [7] have presented a Human Visual System (HVS) model in curvelet domain. The method embeds the watermark into the cover image curvelet coefficients. This paper also has discussed blind watermark extraction technique. Leung et al. [8] have discussed a method using selective curvelet coefficients. The given technique has encoded the binary watermark repeatedly to obtain redundant data, which will be embedded into the selective curvelet coefficients of the original image. The results obtained in the method were improved by the authors by extracting more rules in HVS model [9]. Hien et al. [10] have developed a technique in which the watermark is embedded into the selective curvelet coefficients of cover image. The paper [10] has discussed the extraction technique for extracting the watermark from the altered curvelet coefficients. Zhang et al. [11] have presented a technique based on curvelet domain. In the technique, the watermark was scanned both vertically and horizontally to form a sequence and then it was encoded with a key to produce pseudo random key. This was XOR with the watermark to produce a sequence to embed into curvelet coefficients of the image obtained through unequally spaced fast curvelet transform procedure. Tao et al. [12] have presented a technique in which the cover image was segmented into number of small blocks. The curvelet transformation is applied to those blocks, which were consisting of strong edges. The watermark was converted to pseudo random sequence and then embedded into the significant curvelet coefficients.

Zhao and Wu [13] have discussed a technique in which the watermark was initially scrambled with Arnold technique and then embedded into the curvelet coefficients of original image. Lari et al. [14] have discussed a method in which the amplitude modulation technique was applied for watermarking. In the technique, to obtain robust watermarking the watermark was not embedded in pixels identified through curvelets. Qian et al. [15] have discussed a technique in which the watermark was initially scrambled with Arnold technique and then embedded using quanta method according to mean value of the coefficients into the curvelet coefficients of original image.

The techniques discussed in [7, 9, 10, 12] have complexities due to the size of the watermark. In this paper, Discrete Wavelet Transform and Curvelet Transform Techniques are used in embedding and extraction procedures to achieve robustness and efficiency [6]. The rest of this paper is organized as follows. Section 2 discusses four proposed watermarking techniques. Section 3 gives the results obtained through the proposed techniques and Sect. 4 discusses the conclusion.

2 Proposed Watermarking Techniques

This section presents four techniques based on DWT and curvelet transform techniques. The first two techniques embed watermark into curvelets of the original image, where as in the other two techniques the spread spectrum of the given watermark is embedded into the original image. The following assumptions are considered for all the proposed methods. The cover image (CI) is read in spatial domain by using RGB color space. The watermark is embedded in blue component as eye is less sensitive for blue and yellow component [16]. The blue component of image is segmented into a number of small blocks and the curvelet procedure [17] is applied on each block to obtain curvelet coefficients (CC). The gray scale watermark image (W) is read in spatial domain and partitioned into number of small blocks equal to the size of cover image blocks.

2.1 Method-1

Each block of watermark obtained is added to the respective curvelet coefficients (CC) of cover image blocks using Eq. 1. The inverse curvelet transform procedure is applied to obtain coefficients for watermarked image (CC1).

$$CMm, n = Wm, n + CCm, n \qquad (1)$$

While extracting, the watermarked image, WMI, is read by using RGB colour space The blue component of watermarked image (WMI) is partitioned into a number of small blocks. The curvelet procedure is applied on each block to obtain curvelet coefficients (CW). The curvelet coefficients of each block of cover image (CC) are subtracted from CW to obtain watermark image, EW using Eq. 2. The extracted watermark (EW) can be compared to W.

$$EWm, n = CWm, n - CCm, n \qquad (2)$$

2.2 Method-2

The discrete wavelet transform procedure is applied on each block of the watermark image, W, to obtain transform coefficients (WD) using Eq. 3. The transform coefficients (WD) are added to the respective curvelet coefficients (CC) of each block, using Eq. 4 to obtain watermarked image, WMI.

$$WDm, n = DWT(W) \qquad (3)$$

$$CMm, n = WDm, n + CCm, n \qquad (4)$$

For extraction of watermark, the watermarked image (WMI) is read by using RGB colour space. The blue component of watermarked image (WMI) is divided into small blocks. The curvelet procedure is applied on each of these blocks to obtain curvelet coefficients (CW). The curvelet coefficients of cover image (CC) are subtracted from watermarked image curvelet coefficients (CW). The inverse discrete wavelet transform procedure is applied on the resultant blocks using Eq. 5. The output generates the extracted watermark (EW) which can be compared with original watermark (W).

$$EWm, n = InverseDWT(CWm, n - CCm, n) \qquad (5)$$

2.3 Method-3

The watermark image, W, is resized by using magic square spread spectrum technique as discussed in [18] to produce resized watermark image IM. The spread spectrum technique is an effective technique in spatial domain for embedding, in which not only the watermark spreads across the cover image but also it survives against many signal processing attacks like, compressing, noising etc. [19, 20]. For simplicity the watermark is added in continuous locations of cover image in all the procedures. The results obtained through these techniques are proven to be acceptable. A Magic Square is a square matrix consisting of n2 distinct numbers, such that the sum of numbers of any row, any column or any diagonal will be the same constant [21]. The magic square is part of Indian culture from the times of Vedic days for example a 4 × 4 magic square appearing in Khajuraho in the Parshvanath Jain temple, India [22]. Example Magic Square looks as in Fig. 1.

34	34	34	34	34	34
34	1	13	12	8	
34	2	14	7	11	
34	15	3	10	6	
34	16	4	5	9	

Fig. 1 4 × 4 magic square

Magic square procedure discussed in [18] is used to resize the watermark image (W) of size m × m pixels to the equivalent cover image (CI) size of N × N. An adjustment value array, AD, is also generated based on the given procedure.

A block of IM is considered and divided into small blocks and added to the respective curvelet coefficients (CC) using Eq. 6. The inverse curvelet transform procedure is applied on individual blocks after addition, which together obtains the third dimension of watermarked image (CC1). The first two dimensions of cover image and CC1 are used to produce the watermarked image (WMI).

$$CM_{m,n} = IM_{m,n} + CC_{m,n} \tag{6}$$

The watermark extraction procedure reads the watermarked image (WMI) by using RGB colour space. The blue component of watermarked image (WMI) is segmented into a number of small blocks. The curvelet procedure is applied on each block to obtain curvelet coefficients (CW). The curvelet coefficients of cover image (CC) are subtracted from curvelet coefficients of watermarked image (CW) to obtain resultant array (EW) using Eq. 7. The adjustment array (AD) is added to the resultant array (EW) to obtain extracted watermark with cover image size (EMW). By applying magic square principle on EMW watermark will be regenerated and compared with original watermark (W).

$$EW_{m,n} = CW_{m,n} - CC_{m,n} \tag{7}$$

2.4 Method-4

A block of IM obtained in method-3 is taken and segmented into a number of small blocks. Discrete wavelet transform procedure is applied on each block to obtain wavelet transform coefficients which are added to the respective curvelet coefficients (CC) using Eq. 8. The inverse curvelet transform procedure is applied on individual blocks and combined to form third dimension of the watermarked image (CC1). The first two dimensions of cover image and CC1 are combined to produce the watermarked image (WMI).

$$CMm, n = DWT(IMm, n) + CCm, n \qquad (8)$$

The watermarked image, WMI, is read by using RGB colour space for extracting watermark. The blue component is partitioned into small blocks and curvelet procedure is applied on each block to obtain curvelet coefficients (CW). The curvelet coefficients of original image (CC) are subtracted from CW and the inverse discrete wavelet transform procedure is applied to obtain extracted intermediate watermark (EMW) as given in Eq. 9. By applying magic square principle on EMW, the watermark, EW, is regenerated and compared with original watermark (W).

$$EMWm, n = Inv.DWT(CWm, n - CCm, n) \qquad (9)$$

3 Results

The quality of image is measured with the comparative study between the original and the affected image. Normalized Correlation (NC) is applied to compare the extracted watermark with the embedded watermark to prove the authentication [23]. The value of NC ranges between −1.0 and 1.0. Peak Signal to Noise Ratio (PSNR) is applied to measure the watermarked image quality through the equation Eq. 10 [23].

$$PSNR = 10 \cdot \log_{10}\left(\frac{MAX^2}{MSE}\right) \qquad (10)$$

In the equation MAX represents maximum value of a pixel in the respective colour component and MSE stands for Mean Square error given by Eq. 11.

$$MSE = \frac{1}{mn}\sum_{i=0}^{m-1}\sum_{j=0}^{n-1}[I_{ij} - K_{ij}]^2 \qquad (11)$$

For testing the proposed methods, a Lenna colour image (Fig. 3) of size 512×512 pixels was taken as cover image (CI) and read using RGB colour space to obtain a three dimensional array. The third dimension of the image (CI) is partitioned into 64×64 blocks, where each block size is 8×8 pixels. Each block is individually transformed by using curvelet transform procedure to obtain curvelet transform coefficients (CC). A gray scale image with 64×64 pixels size was considered as watermark, W (Fig. 2). These CI and CC values are common for the four proposed methods.

In Method-1, the watermark image of size 16×16 pixels (W) was segmented into number of blocks, where each block size was 8×8 pixels. These individual blocks were added to CC as per said procedure. The resultant individual blocks were applied with inverse Curvelet Transform function and added back to the third dimension of original image to obtain watermarked image (WMI). The comparison between

Fig. 2 Original watermark

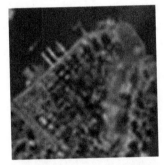

Fig. 3 Original lenna image

original image and watermarked image was done with respect to PSNR and found to be 43.63 db. The technique was repeated with 32×32 pixel and with 128×128 pixel watermark and the PSNR between original and watermarked images was obtained as 37.48 and 31.49 db respectively. The results are comparable with other discussed methods [9, 12].

In Method-2, the segmented blocks of watermark image (W) of size 16×16 pixels were transformed with discrete wavelet transform function and added to CC as stated in the procedure. The resultant blocks were individually inverted by using inverse Curvelet Transform function and combined to form third dimension of the watermarked image, which was then added with the first two dimensions of cover image to form watermarked image. The PSNR between original image and watermarked image was calculated and found to be 44.29 db. The technique was repeated with 32×32 pixel and 128×128 pixel watermark images and the PSNR results were obtained as 38.15 and 32.35 db respectively. The results are comparable with other discussed methods [7, 9, 12].

For the method-3 and method-4, the watermark, W, is resized by using Magic Square procedure [18]. The cover image CI size is of 512×512, so for every pixel of the watermark image an 8×8 magic square is generated as per the procedure given in [18] and adjustment array AD is updated. The obtained magic square consists of 64 elements, which will be added in the respective pixel location of 64 blocks to generate the resized image. The resized image consists of 64 blocks of images with varied intensities of the original watermark, as shown in Figs. 4 and 5.

Fig. 4 Resized watermark

Fig. 5 Resized watermark
scaled with 10 factor

In Method-3, the watermark image (W) of size 64 × 64 pixels was first resized to 512 × 512 pixels size by using magic square procedure. A block of 16 × 16 pixels was considered. This block was segmented into a number of small blocks with size of 8 × 8 pixels. These values were added to CC as per method 3 procedure. The individual blocks were transformed back by using inverse Curvelet Transform function. They were added as the third dimension of original image to obtain watermarked image. The watermarked image and original image were compared and the obtain PSNR is 64.86 db. The technique was repeated by using 32 × 32 pixel block of resized watermark and 128 × 128 pixel block of resized watermark. The PSNR obtained is 58.8 and 42.09 db respectively. The results were comparable to [7, 9, 10, 12].

In Method-4, the watermark image, W, was first resized using magic square procedure. 16 × 16 pixels block of the resized image was partitioned into 8 × 8 pixel sized blocks and each block was transformed with discrete wavelet transform function. These transform coefficients were added to curvelet coefficients (CC) of original image. The inverse Curvelet Transform technique was applied and added to obtain the third dimension of original image. The other two dimensions were added to obtain the watermarked image as shown in Fig. 6. The extracted watermark from the watermarked image is shown in Fig. 7. The watermarked image and the original image were compared to prove the quality of the watermark. The obtained PSNR is

Fig. 6 Watermarked BMP image

Fig. 7 Extracted watermark from BMP image

64.91 db. The technique was repeated by using 32 × 32 pixel and 128 × 128 pixel blocks of resized watermark and the PSNR obtained is 59.89 and 42.10 db respectively. The results were comparable to [7, 9, 10, 12].

The Table 1 provide the PSNR results after comparing the compressed watermarked images with respect to their original images. Xiao et al. [7] and Leung et al. [9] have used 16 × 16 pixel binary watermark image in their developed digital watermarking technique. Hien et al. [10] has used 32 × 32 pixel watermark image and Tao et al. [12] has used 6,144 × 49 elements of watermark for obtaining the experimental results. These techniques were compared with the obtained results of the proposed methods.

The watermarked images are attacked with various image compression, scaling, rotation and noise attacks. The watermark is extracted by applying the respective extraction procedures. The Table 2 has given the PSNR and correlation results by comparing the extracted watermark images with respect to original watermark image. The observations show that the Method-3 and Method-4 are possessing better results when compared to [7, 12, 9]. At the same time the observations also reveal that the watermark can be extracted efficiently even after various image manipulation attacks.

Table 1 PSNR results for different BMP images

S. No.	Name (Size 512 × 512)	Method 1			Method 2			Method 3			Method 4			Xiao et al. [7]	Leung [9]	Hien [10]	Tao et al. [12]
		C1	C2	C3	C1	C2	C3	C1	C2	C3	C1	C2	C3	C5	C5	C2	C4
1	Lenna	31.49	37.48	43.63	32.35	38.15	44.29	42.09	59.89	64.86	42.10	59.89	64.91	44.47	42.86	48.73	39.73
2	Baboon	31.17	37.21	43.49	31.50	37.23	43.43	36.19	51.79	57.41	36.21	51.69	57.39		42.84		
3	Barbara	31.30	37.56	43.80	31.63	37.67	43.99	39.68	52.88	56.54	39.68	53.05	56.57				
4	Pepper	31.20	37.13	43.31	31.52	37.25	43.55	39.14	50.76	55.70	39.23	50.78	55.85		42.71		

Table 2 PSNR results after attacks on watermarked image

S. No.	Type of attack on 24 bit color lenna image (Size 512 × 512)	Method 1 PSNR/NC			Method 2 PSNR/NC			Method 3 PSNR/NC			Method 4 PSNR/NC			Xiao et al. (NC) [7]	Leung (PSNR/NC) [9]	Hien et al. (NC) [10]
		C1	C2	C3	C1	C2	C3	C1	C2	C3	C1	C2	C3	C5	C5	C2
1	Watermark extracted from BMP image	25.69/ 0.9390	19.27/ 0.7076	23.61/ 0.8163	15.84/ 0.5959	15.48/ 0.3882	19.75/ 0.3400	34.65/ 0.9943	40.74/ 0.9984	41.39/ 0.9988	29.76/ 0.9816	40.47/ 0.9982	41.35/ 0.9988			0.9063
2	Bmp converted to GIF format	21.41/ 0.8355	19.28/ 0.7056	23.61/ 0.8163	15.65/ 0.4457	15.49/ 0.3867	16.78/ 0.3342	33.27/ 0.9911	40.37/ 0.9982	41.29/ 0.9987	29.48/ 0.9810	40.20/ 0.9981	41.28/ 0.9987			
3	Bmp converted to JPG format	20.37/ 0.7804	18.68/ 0.6505	22.64/ 0.7700	14.89/ 0.2310	12.65/ 0.04	13.21/ 0.05	32.76/ 0.9917	40.19/ 0.9981	41.22/ 0.9987	29.61/ 0.9775	39.93/ 0.9979	41.24/ 0.9987	~0.9	32.3/ 0.9448	0.826
4	Bmp converted to PNG format	21.41/ 0.8363	19.27/ 0.7076	23.61/ 0.8163	15.64/ 0.4426	15.48/ 0.3882	19.75/ 0.3400	34.65/ 0.9943	40.74/ 0.9984	41.39/ 0.9988	29.76/ 0.9816	40.47/ 0.9982	41.35/ 0.9988			
5	Scaled to 1024 × 1024 pixels	15.66/ 0.2964	15.17/ 0.1554	17.76/ 0.3050	9.77/ 0.0961	9.52/ 0.0196	10.0/ 0.089	20.46/ 0.9133	35.51/ 0.9947	38.14/ 0.9972	18.52/ 0.7177	35.65/ 0.9944	38.57/ 0.9974			
6	Rotated by 90°	25.69/ 0.9390	19.27/ 0.7074	23.60/ 0.8158	16.84/ 0.5959	15.48/ 0.3881	16.75/ 0.3399	34.65/ 0.9943	40.74/ 0.9984	41.39/ 0.9988	29.76/ 0.9816	40.47/ 0.9982	41.35/ 0.9988			
7	Salt and pepper noise	23.81/ 0.8990	19.01/ 0.6756	21.85/ 0.6984	16.29/ 0.5617	15.03/ 0.3630	16.07/ 0.3236	26.51/ 0.9537	37.37/ 0.9963	39.69/ 0.9981	26.16/ 0.9437	36.91/ 0.9959	40.13/ 0.9983			

(C1)—128 × 128 pixel gray scale watermark image

(C2)—32 × 32 pixel gray scale watermark image

(C3)—16 × 16 pixel gray scale watermark image

(C4)—6144 × 49 elements watermark

(C5)—16 × 16 pixel binary watermark image

4 Conclusion

This paper has explained four watermark embedding and extraction algorithms based on Curvelet Transform technique. It was shown that the curvelet application on cover image results equally for both spatial and transform domain embedding procedures. Also this is robust against regular image compression techniques due to the watermark get embedded on smooth curvatures of the cover image. The method 3 has proven to be acceptable for both embedding and extraction processes.

References

1. Philip, B.M.: A History of Graphic Design. Wiley, 3rd edn. 58 (1998)
2. Hunter, D.: Handmade Paper and its Watermarks: A Bibliography. Burt Franklin publisher, New York (1967)
3. Tekla, S.P.: Digital Photography: The Power of Pixels. IEEE Spectrum's Special Report: Top 11 Technologies of the Decade (2011)
4. Steele, L.D.: Review and description of the MacTutor at the CM magazine, vol. III(17). The Manitoba Library Association, Winnipeg (1997)
5. Donoho, D.L., Flesia, A.G.: Digital ridgelet transform based on true ridge functions. Int. J. Stud. Comput. Math. **10**, 1–30 (2003)
6. Zhang, C.J., Hu, M.: Curvelet image watermarking using genetic algorithms. Congr. Image Signal Process. **1**, 486–490 (2008)
7. Xiao, Y., Cheng, L.M., Cheng, L.L.: A robust image watermarking scheme based on a novel hvs model in curvelet domain. In: International Conference on Intelligent Information Hiding and Multimedia Signal Processing, pp. 343–346 (2008)
8. Leung, H.Y., Cheng, L.M., Cheng, L.L.: A robust watermarking scheme using selective curvelet coefficients. In: International Conference on Intelligent Information Hiding and Multimedia Signal Processing, pp. 465–468 (2008)
9. Leung, H.Y., Cheng, L.M., Cheng, L.L.:Digital watermarking schemes using multi-resolution curvelet and HVS model. In: 8th International Workshop on Digital Watermarking, vol. 5703, pp. 4–13 (2009)
10. Hien, T.D., Kei, I., Harak, H., Chen, Y.W., Nagata, Y., Nakao Z.: Curvelet-domain image watermarking based on edge-embedding. In: 11th International Conference on Knowledge-Based Intelligent Information and Engineering Systems, vol. 4693, pp. 311–317 (2007)
11. Zhang, Z., Huang, W., Zhang, J., Yu, H., Lu, Y.: Digital image watermark algorithm in the curvelet domain. In: International Conference on Intelligent Information Hiding and Multimedia Signal Processing, pp. 2015–108 (2006)
12. Tao, P., Dexter, S., Eskicioglu, A.M.: Robust digital image watermarking in curvelet domain. In: International conference on Security. Forensics. Steganography and Watermarking of Multimedia Contents X, vol. 6819, pp. 68191B1–68191B12 (2008)
13. Zhao, Xiuling, Aidi, Wu: Image digital watermarking techniques based on curvelet transform. Recent Adv. Comput. Sci. Inf. Eng. **128**, 231–236 (2012)
14. Lari, M.R.A., Ghofrani, S., McLernon, D.: Using curvelet transform for watermarking based on amplitude modulation. Int. J. Sign. Image Video Process. **8**(4), 687–697 (2014)
15. Qian, Z., Cao, L., Su, W., Wan, T., Yang, H.: Image digital watermarking techniques based on curvelet transform. Recent Adv. Comput. Sci. Inf. Eng. **124**, 231–236 (2014)
16. Wandell, B.A : Foundations of Vision. Sinauer Associates Publisher, Sunderland (1995)

17. Starck, J.L., Candes, E.J., Donoho, D.L.: The curvelet transform for image denoising. IEEE Trans. Image Process. **11**(6), 670–684 (2002)
18. Channapragada, R.S.R., Prasad, M.V.N.K.: Digital watermarking based on magic square and ridgelet transform techniques. Intell. Comput. Netw. Inf, **243**, 143–161 (2014)
19. Cox, J., Kilian, J., Leighton, T., Shamoon, T: Secure spread spectrum watermarking for images, audio and video. International Conference on Image Processing, vol. 3, pp. 243–246
20. Cox, J., Kilian, J., Leighton, T., Shamoon, T.: Secure spread spectrum watermarking for multimedia. IEEE Trans. Image Process. **6**(12), 1673–1687 (1997)
21. Xie, T.: An evolutionary algorithm for magic squares. In: The 2003 Congress on Evolutionary Computation, vol. 2, pp. 906–913 (2003)
22. Andrews, W.S., Frierson, L.S. Browne, C.A.: Magic Squares and Cubes. Open court publish company (1908)
23. Eratne, S., Alahakoon, M. : Fast predictive wavelet transform for lossless image compression. In: Fourth International Conference on Industrial and Information Systems, pp. 365–368 (2009)

A Method for the Selection of Software Testing Techniques Using Analytic Hierarchy Process

Mohd Sadiq and Sahida Sultana

Abstract For the development of high quality systems, software testing has been used as a way to help software engineers. Selection of a software testing technique is an important research issue in software engineering community. Verification and validation activities are conducted to enhance the software quality throughout the entire life cycle of software development. The success rate of software system depends upon the type of software testing techniques that we employ at the time of software testing process. In literature, we have identified different types of software testing techniques like, black box techniques, white box techniques, and gray box techniques etc.; and choosing one of them is not an easy task according to need/ criteria of the software projects because each technique pursues a specific objective or goal; and is less suitable for specific kind of software system. Therefore, in order to address this issue, we present a method for the selection of software testing techniques using analytic hierarchy process (AHP) by considering the following criteria: New or modified system (NMS), Number of independent paths (NIP), Number of test cases (NTC), and Cost of requirements (CoR). Finally, the utilization of the proposed approach is demonstrated with the help of an example.

Keywords Software testing techniques · Decision making process · AHP

M. Sadiq (✉)
Software Engineering Laboratory, Lab. no. 305, Computer Engineering Section,
University Polytechnic, Faculty of Engineering and Technology,
Jamia Millia Islamia (A Central University), New Delhi 110025, India
e-mail: sadiq.jmi@gmail.com

S. Sultana
Department of Computer Science and Engineering, Al-Falah School
of Engineering and Technology, Dhauj, Faridabad, Haryana, India
e-mail: sahida.sultana3@gmail.com

© Springer India 2015
L.C. Jain et al. (eds.), *Computational Intelligence in Data Mining - Volume 1*,
Smart Innovation, Systems and Technologies 31, DOI 10.1007/978-81-322-2205-7_20

1 Introduction

Software testing identifies defect, flows or errors in the software. In literature, we have identified various definitions of software testing. Few of them are given below [1–3]: (i) Testing is the process of demonstrating that errors are not present (ii) The purpose of testing is to show that a program performs its intended functions correctly. The three most important techniques that are used for finding errors are functional testing, structural testing and gray box testing [2–5]. Functional testing is also referred to as black box testing in which contents of the black box are not known. Functionality of the black box is understood on the basis of the inputs and outputs in software. There are different methods which are used in black box testing methods like boundary value analysis, robustness testing, equivalence class partitioning, and decision table testing. White box testing or structural testing is the complementary approach of functional testing or black box testing. White box testing permits us to examine the internal structure of the program. In functional testing, all specifications are checked against the implementation. This type of testing includes path testing, data flow testing, and mutation testing. In white box testing there are various applications of graph theory which is used to identify the independent path in a program or software like decision to decision (DD) flow graph, Cyclomatic complexity [1] etc.

Gray box testing is the testing of software application using effective combination of white box testing, black box testing, mutation, and regression testing [4]. This testing provides a method of testing software that will be both easy to implement and understand using commercial of the shelf (COTS) software [4]. In the Gray box testing, tester is usually has knowledge of limited access of code and based on this knowledge the test cases are designed; and the software application under test treat as a black box and tester test the application from outside. Gray box software testing methodology is a ten steps process for testing computer software. The methodology starts by identifying all the inputs and output requirements to computers systems. This information is captured in the software requirements documentation. The steps are given as follows: (1) Identify inputs (2) Identify outputs (3) Identify major paths (4) Identify sub-function (SF) X (5) Develop inputs for SF X (6) Develop outputs for SF X (7) Execute test cases for SF X (8) Verify correct results for SF X (9) Repeat steps from 4 to 8 for other SF X and (10) Repeat steps 7 to 8 for regression; where X is the name of a sub-function [4]. Testing is a vital part of software development, and it is important to start it as early as possible, and to make testing a part of the process of deciding requirements. In literature, we have identified various studies for the selection of testing techniques. For example, in 2013, Cotroneo et al. [14] proposed a method for the selection of software testing techniques based on orthogonal defect classification (ODC) and software metrics. This method is based on the following steps: (i) construct a model to characterize software to test in-terms of fault types using empirical studies and (ii) characterizing testing techniques with respect to fault types. In 2013, Farooq and Quadri [6] proposed the guidelines for software testing evaluation.

In literature, we have identified different types of software testing techniques like, black box techniques, white box techniques, and gray box techniques etc.; and choosing one of them is not an easy task according to need/criteria of the software projects because each technique pursues a specific objective or goal; and is less suitable for specific kind of software system. Therefore, in order to address this issue, we present a method for the selection of software testing techniques using analytic hierarchy process (AHP) by considering the following criteria, i.e., New or modified system (NMS), Number of independent paths (NIP), Number of test cases (NTC), and Cost of requirements (CoR). Several researchers advocate the use of AHP method in banking system, manufacturing system, drug selection etc. [7, 8]. In 2009 and 2010, we proposed an approach for eliciting software requirements and its prioritization using AHP [9, 10]. In a similar study, in 2014, we proposed a method for the selection and prioritization of software requirements using fuzzy analytic hierarchy process [11, 12].

The paper is organized as follows: In Sect. 2, brief introduction about AHP is given. We present the proposed method for the selection of software testing techniques in Sect. 3. Case study is given in Sect. 4. Finally, we conclude the paper in Sect. 5.

2 Analytic Hierarchy Process

In 1972, Saaty [13] proposed the analytic hierarchy process. It is a multi-criteria decision (MCDM) making method. AHP helps decision maker facing a complex problem with multiple conflicting and subjective criteria [7, 13]. This process permits the hierarchical structure of the criteria or sub-criteria when allocating a weight. AHP involves following steps: (a) problem definition (b) pair-wise comparisons (c) compute the eigenvector of the relative importance of the criteria (d) check consistency. Once we have identified the criteria or sub-criteria according to the need of the problem or **problem definition**, then the next step is to express the decision makers opinion on only two alternatives than simultaneously on all the alternatives. On the basis of the pair wise comparison with all the alternatives, we construct the **pair-wise comparison matrix** on the basis of the following rating scale (**Judgment scale**). Table 1 presents the rating scale proposed by Saaty.

There are several methods or algorithm for the calculation of eigenvector. In this paper, we adopt the following algorithm:

Algorithm:

Step 1 : Multiplying together the entries in each row of the matrix and then take the nth root of the product.

Step 2 : Compute the sum of nth root and store the result in SUM.

Step 3 : The value of SUM would be used to normalize the product values and the resultant would be the eigenvector.

Table 1 The Saaty rating scale

Intensity of importance	Definition
1	Equal importance
3	Somewhat more importance
5	Much more important
7	Very much important
9	Absolutely more important
2, 4, 6, 8	Intermediates values (when compromise is needed)

Saaty argues that a Consistency Ratio (CR) > 0.1 indicates that the judgment are at the limit of consistency, where as CR = 0.9 would mean that the pair wise judgment are random and are completely untrustworthy [7, 13].

3 Proposed Method

This section presents a method for the selection of software testing techniques (STT) using AHP. The proposed method is presented simply in the following:

(i) Identify the criteria
(ii) Construct the hierarchical structure of STT
(iii) Construct the decision matrix
(iv) Calculate the ranking values
(v) Selection of STT

(i) Identify the criteria

Before the selection of any STT, software tester should identify the criteria's for the selection of STT. On the basis of our literature review, we have identified the following factors which influence the decision of choosing a STT:

(a) New or modified system (NMS),
(b) Number of independent paths (NIP),
(c) Number of test cases (NTC), and
(d) Cost of requirements (CoR).

(ii) Construct the hierarchical structure of STT

As the STT selection decision requires a systematic approach to help integrate different attributes or criteria into software project development. Therefore, it is essential to break down the problem into more manageable sub-problems. As illustrated in Fig. 1, the problem studied here has three level of hierarchy. The first level, i.e., the overall objective, is the selection of a STT model. Level two contains

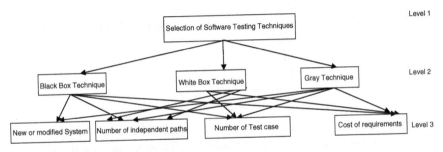

Fig. 1 Hierarchical structure of the SDLC selection problem

three different STT like BBT, WBT, GBT, and at level three following decision criteria is given: New or modified system (NMS), Number of independent paths (NIP), Number of test cases (NTC), and Cost of requirements (CoR).

(iii) Construct the decision matrix

We will create the decision matrix using AHP method [7, 9, 10]. Detailed description for the construction of decision matrix is given in Sect. 4.

(iv) Calculate the ranking values

Ranking values will be obtained after computing the eigenvector values from the pair wise comparison matrix [7, 9, 10].

(v) Selection of STT

Construct the binary search tree of the ranking values (BSRTV) that we have obtained in previous step. Apply in-order tree traversal technique on BSTRV and as a result, we will get the prioritized list of STT. The model which has highest priority will be selected for the testing of the project.

4 Case Study

This section presents a case study of our work. In-order to test any software, it is indispensible to select the software testing technique (STT) according to the need of our project. There are various STT which are available in the literature. In this paper, we have considered the project, developed by our students, i.e., "Institute Examination System (IES)". We have identified the following criteria for the selection of STT: New or modified system (NMS), Number of independent paths (NIP), Number of test cases (NTC), and Cost of requirements (CoR). The hierarchical structure of the STT selection problem is given in Fig. 1 (Step second). For the third step, we have defined the initial matrix for the pair wise comparison (see Table 2). In this matrix, the principal diagonal matrix contains entries of 1 because each factor is important as itself.

Table 2 Initial matrix

Criteria	NMS	NIP	NTC	CoR
NMS	1			
NIP		1		
NTC			1	
CoR				1

To make the pair wise comparison among all the criteria, we decide that NIP is more important than NMS. In the next matrix, i.e., Table 3, that is rated as 7 in the cell NIP and NMS; and 1/7 in NMS and NIP. We also decide that CoR is more important than NMS. Therefore, in Table 4, we put 9 in CoR and NMS; and 1/9 in NMS and CoR. In a similar way, we complete the matrix, that we call the "*Overall Preference Matrix (OPM)*".

The eigenvector or relative value vector (RVV) corresponding to each criterion is calculated by the algorithm, given in Sect. 2. Therefore, as a result, we have identified the following values: (0.269, 1.495, 0.588, 4.21). These four values correspond to the relative value of NMS, NIP, NTC, CoR. The value 4.21 means that CoR is an important criterion. 1.495 shows that NIP is also an important parameter for the selection of STT. The remaining two data represent that NMS and NTC are least considerable parameters.

After this, we evaluate different STT on the basis of the given parameters, i.e., NMS, NIP, NTC, and CoR. Tables 5, 6, 7 and 8 are created according to the NMS, NIP, NTC, and CoR; and it ranks the three STT, i.e., Black box technique (BBT), White box technique (WBT), and Gray box technique (GBT).

On the basis of our analysis, we identify that GBT is important for testing the IES. In our case study, we identify that Gray Box Testing (GBT) is important for the testing of Institute Examination System because it has highest priority having priority vector = 11.536. WBT and BBT have second and third priority respectively (see Table 9).

Table 3 Initial overall preference matrix

Criteria	NMS	NIP	NTC	CoR
NMS	1	1/7		
NIP	7	1		
NTC			1	
CR				1

Table 4 Overall preference matrix

Criteria	NMS	NIP	NTC	CoR
NMS	1	1/7	1/3	1/9
NIP	7	1	5	1/7
NTC	3	1/5	1	1/5
CoR	9	7	5	1

Table 5 Pair-wise comparison matrix w.r.t. NMS

Criteria	BBT	WBT	GBT	Priority vector
BBT	1	1/7	1/9	0.252
WBT	7	1	1/3	1.325
GBT	9	3	1	2.996

Table 6 Pair-wise comparison matrix w.r.t. NIP

Criteria	BBT	WBT	GBT	Priority vector
BBT	1	1/9	1/7	0.252
WBT	9	1	5	3.552
GBT	7	1/5	1	1.119

Table 7 Pair-wise comparison matrix w.r.t. NTC

Criteria	BBT	WBT	GBT	Priority vector
BBT	1	9	7	3.97
WBT	1/9	1	1	0.481
GBT	1/7	1	1	0.523

Table 8 Pair-wise comparison matrix w.r.t. CoR

Criteria	BBT	WBT	GBT	Priority vector
BBT	1	1	1/9	0.481
WBT	1	1	1	1
GBT	9	1	1	2.078

Table 9 Final decision matrix

Criteria	BBT	NIP	NTC	CoR	Priority vector
STT	0.269	1.495	0.588	4.21	
BBT	0.252	0.252	3.97	0.481	4.8
WBT	1.325	3.552	0.481	1	10.159
GBT	2.996	1.119	0.523	2.078	11.536

5 Conclusion

This paper presents a method for the selection of software testing techniques using AHP. Proposed method is a five step process, namely, (i) identify the criteria, (ii) construct the hierarchical structure of STT, (iii) construct the decision matrix, (iv) calculate the ranking values, and (v) the selection of a STT. This method would be used for the selection of STT according to the need of the project. In this paper, we have considered four criteria's for the selection of STT, i.e., NMS, NIP, NTC, and CoR.; and as a result we select GBT method for the testing of IES. Future research agenda includes the following:

1. To propose a fuzzy decision making approach or the selection of STT.
2. To propose a method for the selection of STT using hybrid techniques like fuzzy AHP and fuzzy ANP.

References

1. Sommerville, I.: Software engineering, 5th edn. Addison Wesley, New York (1996)
2. Khan, MA., Sadiq, M.: Analysis of black box software testing techniques: A case study. In: IEEE International Conference and Workshop on Current Trends in Information Technology, pp. 1–5. Dubai, UAE, Dec 2011
3. Khan, MA., Bhatia, P., Sadiq, M.: BBTool: a tool to generate the test cases. Int. J. Recent Technol. Eng. 1(2), 192–197 (2012)
4. Coulter, A.: Gray box software testing methodology. White paper, Version 0.8
5. Coulter, A.: Gray box software testing methodology-embedded software testing technique. In: 18th IEEE Digital Avionics Systems Conference Proceedings, p. 10.A.5-2 (1999)
6. Farooq, S.U., Quadri, S.M.K.: Empirical evaluation of software testing techniques—need, issues and mitigation. Softw. Eng.Int. J. 3(1), 41–51 (2013)
7. Ishizaka, A., Labib, A.: Review of the main development in the analytic hierarchy process. Experts Syst. Appl. 38(11), 14336–14345 (2011)
8. Kaa et al.: Supporting decision making in technology standards battles based on a fuzzy analytic hierarchy process. IEEE Trans. Eng. Manage. 61(2) (2014)
9. Sadiq, M., Ahmad, J., Asim, M., Qureshi, A., Suman, R.: More on elicitation of software requirements and prioritization using AHP. In: IEEE International Conference on Data Storage and Data Engineering, pp. 232–236. Bangalore, India (2010)
10. Sadiq, M., Ghafir, S., Shahid, M.: An approach for eliciting software requirements and its prioritization using analytic hierarchy process. In: IEEE International Conference on Advances in Recent Technologies in Communication and Computing, pp. 790–795. Kerala, India (2009)
11. Sadiq, M., Jain, S.K.: A fuzzy based approach for the selection of goals in goal oriented requirements elicitation process. Int. J. Syst. Assur. Eng. Maint (2014). ISSN No.: 0975-6809 (Print Version) ISSN No.: 0976-4348 (Electronic Version) (Springer)
12. Sadiq, M., Jain, S.K.: Applying fuzzy preference relation for requirements prioritization in goal oriented requirements elicitation process. Int. J. Syst. Assur. Eng. Maint (2014). ISSN No.: 0975-6809 (Print Version) ISSN No.: 0976-4348 (Electronic Version) (Springer)
13. Saaty, T.: The Analytic Hierarchy Process. McGraw-Hill, New York (1980)
14. Cotroneo, D., et al.: Testing techniques selection based on ODC fault types and software metrics. J. Syst. Softw. 86, 1613–1637 (2013)
15. Sadiq, M., Jain, S.K.: A fuzzy based approach for requirements prioritization in goal oriented requirements elicitation process. In: 25th International Conference on Software Engineering and Knowledge Engineering, Boston, USA, 27–29 June 2013
16. The Standish Group International, CHAOS Summary (2009)
17. Viera, M., et al.: Automation of GUI testing using model-driven approach. ACM-AST, China (2006)

Implementation of an Anti-phishing Technique for Secure Login Using USB (IATSLU)

Amit Solanki and S.R. Dogiwal

Abstract In the area of computer security, phishing is the criminally fraudulent process of attempting to acquire sensitive information such as usernames, passwords and credit card details by masquerading as a trustworthy entity in an electronic communication. Phishing is typically carried out by e-mail, and it often directs users to enter details at a fake website which almost identical to original one. Phishing filters help Internet users avoid scams that create fake websites. Such sites ask for personal information, including banking passwords or offer software downloads. This paper concerned with anti-phishing techniques with the help of hardware device. Anti phishing software is designed to track websites and monitor activity, any suspicious behavior can be automatically reported and even reviewed as a report after a period of time.

Keywords Phishing · Anti-phishing technique · Phishing software · Filters

1 Introduction

The word 'Phishing' initially emerged in 1990s. The early hackers often use 'ph' to replace 'f' to produce new words in the hacker's community since they usually hack by phones. Phishing is a new word produced from 'fishing' it refers to the act that the attacker allure users to visit a faked Web site by sending them faked e-mails (or instant messages) and stealthily get victim's personal information such as user name, password and national security ID, etc. This information then can be used for future target advertisements or even identity theft attacks (e.g., transfer money from

A. Solanki (✉) · S.R. Dogiwal
Department of Computer Science and Engineering, Swami Keshvanand
Institute of Technology, Jaipur, India
e-mail: amit.solanki48@gmail.com

S.R. Dogiwal
e-mail: dogiwal@gmail.com

© Springer India 2015
L.C. Jain et al. (eds.), *Computational Intelligence in Data Mining - Volume 1*,
Smart Innovation, Systems and Technologies 31, DOI 10.1007/978-81-322-2205-7_21

221

victims' bank account). The frequently used attack method is to send e-mails to potential victims, which seemed to be sent by legitimate online organizations. In these e-mails, they will make up some causes. The style, the functions performed, sometimes even the URL of these faked Web sites is similar to the real Web site. The attackers successfully collect the information at the server side, if you give input the account number and password and with that information it is able to perform their next step actions [1].

2 Problem Statement

Today a wide range of Anti phishing software and techniques are available but the most important problem is to find the best anti-phishing technique which solves the problem faced by the user, and also which is compatible with the runtime environment and easily modified as per need. Anti-phishing techniques provides the best solution and different technique can be implemented that will be more efficient.

3 Classification Techniques and Attacks

3.1 Impersonation Attack

Impersonation is the simplest and the popular method of deception. It consists of a completely fake website that receiver is deceived to visit [2]. This fake site contains images from the real Web site and might even be associated to the real site [3].

3.2 Forwarding Attack

In the forward phishing technique the model approach is to gather the data and forward the victim to the actual site. This is one of the further sophisticated types of phishing attack since there is no collection Web page, no images and the only server is concerned it has just a redirect script. The user is prompted for his or her information inside the e-mail itself.

3.3 Popup Attack

The Popup phishing technique introduces a pop-up window on the actual site that will forward the intended victim to the objective phishing server. This approach is the most unusual type of attack today because popup blockers are broadly used and included as default setting by numerous browsers existing in the market which lowers the success rate of this process [4].

4 Existing Anti-phishing Toolbars

There are a range of methods that can be used to recognize a page as a phishing site, together with white lists (lists of known safe sites), blacklists (lists of recognized fraudulent sites), different heuristics to see if a URL is approximating to a well-known URL and community ratings [5].

4.1 Cloudmark Anti-fraud Toolbar

When visiting a site user have the choice of reporting the site as superior or bad. Therefore the toolbar will display a colour icon for every site visited. Green icons shows that the site has been rated as legitimate, red icons point out that the site has been determined to be fake and yellow icons specify that not sufficient information is known about the site to construct a determination. In addition the users themselves are rated according to their record of accurately identifying phishing sites [5].

4.2 EarthLink Toolbar

The EarthLink Toolbar, rely on a grouping of heuristics user ratings and manual verification. Little information is showed on the EarthLink website however we used the toolbar and observed how it functions. The toolbar allows users to report suspected phishing sites to EarthLink. These sites are then verified and added to a blacklist [6].

4.3 eBay Toolbar

The eBay Toolbar, uses a combination of heuristics and blacklists. The Account Guard indicator has three modes: green, red and gray. The icon is indicated with a green background when the user looks a site known to be operated by eBay (or PayPal). The icon is indicated with a red background when the site is a recognized phishing site [7].

4.4 Geo Trust Trust Watch Toolbar

Geo Trust's Trust Watch Toolbar, labels sites as green (confirmed as trusted), yellow (not verified) or red (verified as fraudulent). Geo Trust's web site provides no information regarding how Trust Watch determines if a site is fake, however we suppose that the company compiles a blacklist that includes sites reported by users through a button showed on the toolbar.

5 Proposed Anti-phishing Technique

This section describes the different steps in process for implementation of my proposed anti-phishing technique as follows.

5.1 Registration Process

The registration process used by the bank requires the following field to authorize the given client during login such as Username (Account Number), Password, Full name, Email address, Secret Code, Serial Number of USB.

5.2 Detail Steps for Registration

Step 1: For the registration process bank requires the serial number of USB. To get the serial number of USB bank will choose the USB in random fashion and through the GET_USB method and then find out the serial number of USB after plug in the USB. This serial number is unique for each USB.

Figure 1 shows various serial no. of the external drives, that are present in the system. When we begins the process named as GET USB, it will shows the serial no. of all the drives, by selecting the appropriate drive where the USB is connected, and show the drive name along with the serial no. of USB drive in the text box in the bottom of the diagram as shown "Drive Name :E:\ || Serial No: 842117". The current drive where USB is connected is drive 'E' and the serial no. of the of the USB is '842117' as shown in Fig. 1. After getting the serial number of USB which is unique for each USB bank will save this serial number for the future reference.

Step 2: Now after the success of step 1 bank will require the account number of the user who applied for the online facility. The Bank also having the account number of user, so bank will save the account no. in the text file and copied it to the USB in the step 1 of registration process. For the reference we

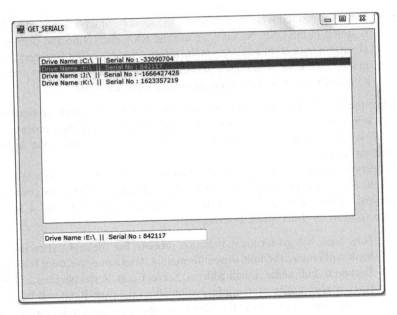

Fig. 1 Detecting serial number of USB

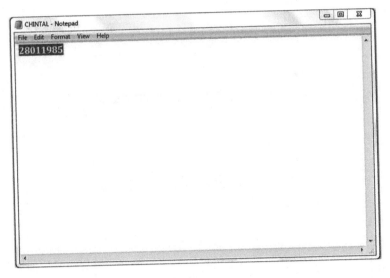

Fig. 2 Display of user account number in text format

used the name of text file as CHINTAL.TXT here. Figure 2 gives the details of the text file which contains the account no. of the user that is '28011985' which is saved in the USB with the file name CHINTAL.TXT.

Step 3: For enhanced the security user wants the security of the information which is stored in the USB. So here we use the RSA encryption and decryption algorithm to encrypt and decrypt the information (Account Number) stored in the USB. RSA algorithm will encrypt the information (account number) so nobody can read the information saved into the USB.

When the process is started, Fig. 3 shows the bank has to select various fields such as select drive, select folder and select file. For the Encryption process, the drive 'E' is selected, also select the folder and the file which contain the account information of user.

To encrypt the account no. of user, click on the encrypt button present in Fig. 4 to start the encryption process, after clicking the account number is now converted to encrypted form, and the status of the encryption is shown in Fig. 4 by the message window as 'Encryption Is Complete'.

Step 4: Now bank will go for the registration process. For the registration process bank will require the following information, Username (Account Number), Password, Full name, Email address, Secret Code, Serial Number of USB. Bank is also having all the information about the user. Bank will have the user ID, password, Email ID, Secret Number (to reset the password) and the Serial Number is also which is saved in the step 1. After filling all the

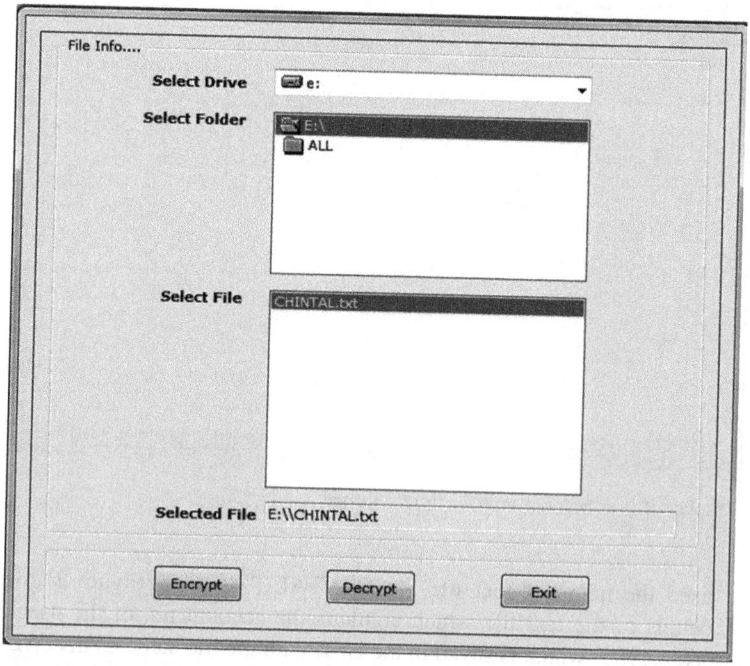

Fig. 3 Encryption process using RSA algorithm

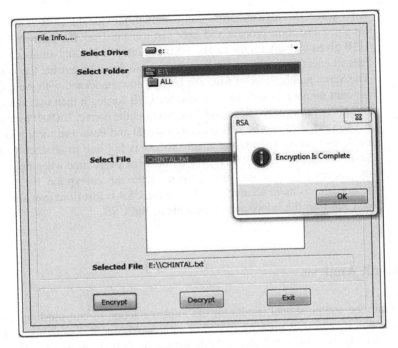

Fig. 4 Successful completion of encryption process

information necessary for the registration process bank will click on the proceed button shown in Fig. 5 to confirm the registration of the user on this serial number of USB and the Account number.

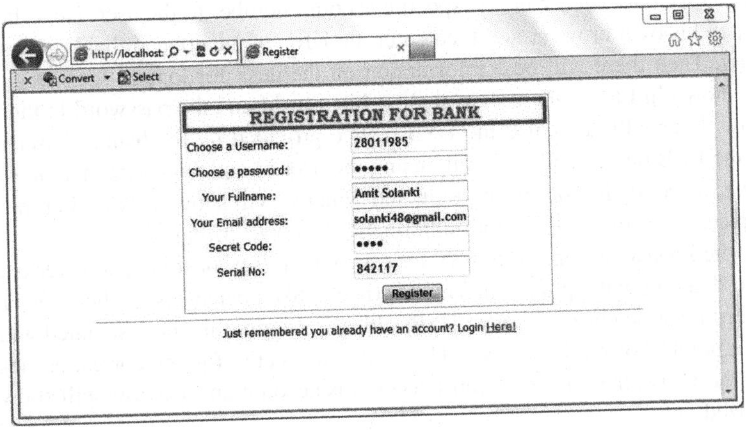

Fig. 5 Registration process for bank

Step 5: After successful registration of user bank will provide the USB to the user for the transaction process and user can use the User ID, password and USB given by bank for the Login. USB contains the account number of the user in encrypting way and it is in unreadable form for outside the world. The Serial number extract from the USB is now associated with the user's account number. If user will use another USB for login then user will not be able to achieve the successful login without the correct USB. If user will not use the USB and he will use only user ID and Password for the login, he cannot proceed. Now the login process is fail due to absence of the USB. So user needs the USB given by the bank every time when he wants the successful login into his account. After the encryption the user's information is more secured and even if the USB is lost from user side, no one can read the information saved inside the USB.

6 Result Analysis

To obtain the computational results for the implementation of anti phishing technique which is named as implementation of an anti-phishing technique for secure login using USB, which includes a USB device to provide more security and authentication to any system? I have studied various techniques for detection and prevention of phishing and they have defined and implemented various anti phishing techniques. I described the anti-phishing technique that is developed by summarizing our findings and offer observations about the usability and overall effectiveness of these techniques. To design and implementation of this anti phishing technique i have used the Microsoft .net frame work 3.5 for coding as a front end and for the back end data base SQL server 2008 is used. Initially after completion of registration process the bank will hand over the User_Id, password and the USB associated with the specific account number to the user. Now the user will open his own browser and then enter the URL of the concern bank for internet banking. Then there will be a information on the desk for login.

As shown in Fig. 6, the user will enter his user-id and the password in the login window. User will have also the USB device provided by the bank. If user is not using the USB device and he is only using the user-id and password then he will be not able to access the successful login and window will show the warning message like "please plugin the USB for authorization" shown in Fig. 7.

For the successful login it is very necessary that user must have to enter user-id, password along with plugin the appropriate USB device also. The user-id and password must be correct and the USB which is plugin must be associated with the account number of particular user. User must take care of given correct entries and plugin the USB otherwise user can't access his account and window will show "un authorized entry".

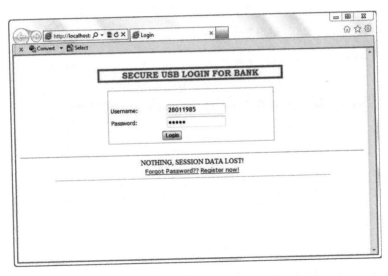

Fig. 6 Secure USB login process with user's information

Fig. 7 Invalid user authorization

And if user will enter correct entries along with the correct USB which is provided by the bank then he will achieve the successful login given in the Fig. 8. Through the IATSLU phishing technique user can achieve the authentication and confidentiality to enhance the security.

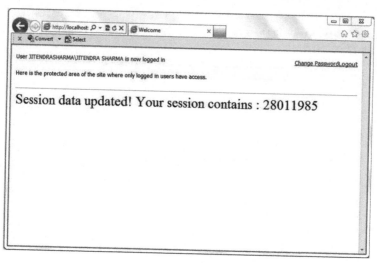

Fig. 8 Successful login of user

7 Conclusion

The proposed anti-phishing technique gives detailed experienced towards its simplicity and it also setting up the various test cases. The extensive experimentation is performed for substantiation of implementation of all the components of the techniques work together. For this technique, the proposed work successfully implemented the various steps for anti-phishing technique such as Username (Account Number), Password, Full name, Email address, Secret Code, Serial Number of USB. The GET USB method is to get the serial number of USB, and is executed successfully by providing the serial number to the user screen and this serial number is used for the registration process in future. The user account or user id in anti-phishing technique is securely encrypted using the RSA algorithm and copied to the respective USB which was issued to the customer for secure login. The main advantage of encrypting the user account is to protect the data from the other user, if in case USB is lost, other person get the USB and try to access the information contained within the USB, but now the data contained within the USB is in encrypted form or in more general we can say it is in non-readable format.

References

1. Madhuri, M., Yeseswini, K., Vidya Sagar, U.: Intelligent phishing website detection and prevention system by using link guard algorithm. Int. J. Commun. Netw. Secur. **2**, 9–15 (2013)
2. Carmona, P.L., Sánchez, J.S., Fred, A.L.N.: Pattern Recognition—Applications and Methods, pp. 149–158. Springer, New York (2013)

3. Padmos, A.: A Case of Sesame Seeds Growing and Nurturing Credentials in the Face of Mimicry, pp. 33–45. Royal Holloway, University of London, Egham (2011)
4. James, L.: Phishing Exposed, pp. 50–56. Syngress Publishing, USA (2005). ISBN 159749030X
5. Cranor, L., Egelman, S., Hong, J., Zhang, Y.: Phinding Phish: An Evaluation of Anti-Phishing Toolbars. Pittsburgh (2006), pp. 1–17
6. EarthLink Inc.: EarthLink toolbar. Accessed 9 Nov 2006
7. eBay Inc.: Using eBay toolbar's account guard. Accessed 13 June 2006

Clustering Based on Fuzzy Rule-Based Classifier

D.K. Behera and P.K. Patra

Abstract Clustering is the unsupervised classification of patterns which has been addressed in many contexts and by researchers in many disciplines. Fuzzy clustering is recommended than crisp clustering when the boundaries among the clusters are vague and uncertain. Popular clustering algorithms are K-means, K-medoids, Hierarchical Clustering, fuzzy-c-means and their variations. But they are sensitive to number of potential clusters and initial centroids. Fuzzy rule based Classifier is supervised and is not sensitive to number of potential clusters. By taking the advantages of supervised classification, this paper intended to design an unsupervised clustering algorithm using supervised fuzzy rule based classifier. Fuzzy rule with certainty grade plays vital role in optimizing the rule base which is exploited in this paper. The proposed classifier and clustering algorithm have been implemented in Matlab R2010a and tested with various benchmarked multidimensional datasets. Performance of the proposed algorithm is compared with other popular baseline algorithms.

Keywords Clustering · Classification · Fuzzy clustering · Fuzzy rule-based classifier

1 Introduction

This report contributes to the major subject and discussion area of Data Clustering and mainly focuses on clustering by the use of supervised classifier algorithm in an unsupervised manner. Important survey papers on clustering algorithm and their performance exist in the literature [1].

D.K. Behera (✉)
Trident Academy of Technology, Bhubaneswar, India
e-mail: dayal.behera@tat.ac.in

P.K. Patra
College of Engineering and Technology, Bhubaneswar, India
e-mail: pkpatra@cet.edu.in

© Springer India 2015
L.C. Jain et al. (eds.), *Computational Intelligence in Data Mining - Volume 1*,
Smart Innovation, Systems and Technologies 31, DOI 10.1007/978-81-322-2205-7_22

In supervised classification a collection of labeled patterns are given and the problem is to label a newly encountered, unlabeled pattern. In the case of unsupervised clustering, the problem is to group a given collection of unlabeled patterns into meaningful clusters. But the work of this paper has been motivated by the desire to improve and develop clustering methods by the use of classifier so that better learning agents can be built.

In the past, algorithms like K-mean [2], K-medoids [3], Mountain Clustering [4] and Fuzzy C-Mean (FCM) gain popularity. In K-means, if an instance is quite far away from the cluster centroid, it is still forced into a cluster and thus, results distort cluster shape. The two algorithms ISODATA [1] and PAM [1, 5] consider the effect of outliers in clustering. Unlike K-mean [1], K-Medoids algorithm [3] utilizes medoids (real data points) as the cluster centroid and overcomes the effect of outliers to some extent. Both K-means and K-medoids are not appropriate to handle vague boundaries of clusters. Fuzzy C-Means (FCM) [5, 6] is a simplification of ISODATA was realized by Bezdek [1]. FCM is suitable to handle vague boundaries of cluster by partition based on fuzzy cluster. In FCM, dataset is grouped into n number of clusters with every data point in the dataset belonging to every cluster to a certain degree. By iteratively updating the cluster centers and the membership grades for each data point, it moves the cluster centers to the appropriate location within a dataset. It also suffers from presence of noise and outliers and the difficulty to identify the initial partition. Many extension of FCM has been proposed in the literature [6–10].

The main challenge for most of clustering algorithms is to know the number of clusters for which to look. This issue of obtaining the clusters that better fit a dataset, as well as their evaluation, has been the subject of almost all research efforts in this field. In the paper [11], automatically determination of the clusters in unlabeled dataset is proposed. In the paper [12], a fuzzy Association Rule based classification model for high dimensional problem is proposed.

In this paper a novel clustering approach has been proposed where advantages of fuzzy rule with certainty grade is taken for classification. And then fuzzy classifier is used to do clustering in an unsupervised manner.

The following sections are organized as follows: Sect. 2 focuses on general idea of fuzzy rule based system. Section 3 furnished with proposed model of fuzzy classifier, training and testing phase of classification are also presented. Section 4 furnished with proposed model of fuzzy rule based clustering. Experimental results are illustrated in Sects. 5 and 6 focused with conclusion.

2 Fuzzy Rule Based System

Fuzzy Rule Based Systems (FRBS) [13, 14] are intelligent systems those are based on mapping of input spaces to output spaces where the way of representing this mapping is known as fuzzy linguistic rules [15]. These intelligent systems provide a framework for representing and processing information in a way that looks like human communication and reasoning process.

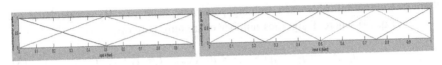

Fig. 1 Member function representation for P = 3 and P = 5

Generally fuzzy rule based system possesses a fuzzy inference system [16] composed of four major modules: Fuzzification module, Inference Engine, Knowledge Base and Defuzzification module [17].

One of the biggest challenges in the field of modeling fuzzy rule based systems is the designing of rule base as it is characterized by a set of IF–THEN linguistic rules. This rule base can be defined either by an expert or can be extracted from numerical data. Fuzzy IF–THEN rules for a classification problem with n attributes can be written as follows:

$$\text{If } x_1 \text{ is } L_{j1} \text{ and } \ldots \quad \text{and } x_n \text{ is } L_{jp} \text{ Then class } C_j \tag{1}$$

where $X(x_1, x_2, \ldots x_n)$ is an n-dimensional pattern vector; L_{ji} $(i = 1, \ldots, p)$ is an antecedent linguistic value; C_j is the consequent class.

To generate rule base [4], first each attribute is rescaled to unit interval [0, 1]. Then, the problem space is partitioned into different fuzzy subspaces, and each subspace is identified by a fuzzy rule, if there are some patterns in that subspace. To do partitioning, each attribute is represented by P number of membership functions. Figure 1 shows membership functions for P value 3 and 5 of an attribute.

3 Design of Fuzzy Rule Based Classifier

The fuzzy IF–THEN rule for classification in the proposed algorithm is taken as

$$\text{If } x_1 \text{ is } L_{j1} \quad \text{and } \ldots \text{ and } x_n \text{ is } L_{jp}, \text{ then Class } C_j \text{ with } CG = CG_j \tag{2}$$

where certainty grade (CG) of rule j is CG_j. CG signifies the impact factor of a rule in determining the consequent class label. It plays an important role in optimizing the rule base. Rules with higher certainty grade are selected for the classification.

3.1 Training Phase

Classification involves mainly two phases, training and testing. Different steps in training phase of the proposed algorithm are

Step 1: Input space is divided into different fuzzy regions.
Step 2: Possible fuzzy rules are generated.

Step 3: Compatibility of each training pattern X_t with rule R_j is calculated by Eq. (4) where T signifies number of training patterns.

$$\mu_{R_j}(X_t) = \mu_{L_{j1}}(x_{t1}) \times \cdots \times \mu_{L_{jp}}(x_{tm}) \quad t = 1, 2, \ldots T \tag{3}$$

Step 4: Sum of compatibility grades for each class is calculated as follows

$$S_{class\,c}(R_j) = \sum_{X_t \in Class\,c} \mu_{R_j}(X_t), \quad c = 1, 2, \ldots m \tag{4}$$

Step 5: Among C classes of the problem, class having maximum value of $S_{class\,c}(R_j)$ is chosen as class for Rule R_j.

$$S_{Class\,C_j}(R_j) = \max\{S_{class\,1}(R_j), \ldots, S_{Class\,m}(R_j)\} \tag{5}$$

Step 6: The certainty grade of R_j can be calculated as follows:

$$CG_j = \frac{S_{Class\,C_j}(R_j) - \lambda}{\sum_{c=1}^{m} S_{Class\,c}(R_j)} \quad where\ \lambda = \frac{\sum_{c \neq c_j} S_{Class\,c}(R_j)}{m - 1} \tag{6}$$

3.2 Testing Phase

For testing phase we have taken fuzzy winner rule reasoning of cluster analysis [18]. In the testing phase class level is not used but it is calculated. To calculate the class level from the rule list obtained in training phase, a winner rule is selected by using Eq. (8) and then consequent class of winner rule is taken as the class level of testing instance.

The winner rule of the rule-set is the rule R_j for which:

$$\mu_{R_j}(X_t) = \max\left\{\mu_{R_j}(X_t) \cdot CG_j \quad where\ R_j \in R\right\} \tag{7}$$

In the above equation R is rule base. If R_j is the winner rule then it says that the pattern is in class C_j with the degree of CG_j.

4 Fuzzy Rule-Based Clustering

The algorithm CBFC proposed here take the advantages of supervised classification. Randomly generated uniform patterns called auxiliary data are added to the main data and taken as Class 2 where as main data is taken as class 1.

To estimate the number of auxiliary data patterns the summation of a within-cluster distance is used. This value for original dataset is defined as

$$d = \sum_{i<j} D(X_i, X_j) \tag{8}$$

where X_i is one of the data pattern. D is the distance metric. Similarly for auxiliary data pattern X'.

$$d' = \sum_{i<j} D(X'_i, X'_j) \tag{9}$$

By the use of d and d', the auxiliary random patterns are added incrementally until d' exceeds d. Hence, distribution of original data pattern determines the size of the auxiliary data pattern. After the preparation of the two-class problem that consists of original and auxiliary data patterns, the CBFC uses fuzzy rule base classifier to classify this two-class problem and a fuzzy rule base is generated. From the rule base the best rule is determined based on certainty grade. The instances of the original pattern that are covered by the best rule are chosen as members of first cluster and those instances are removed from the dataset to not be reconsidered for other clusters.

After removal of instances from the dataset; remaining instances are taken as class 1 and newly generated random instances are taken as class 2 and above procedure is repeated until a stopping criteria reached.

After extracting major clusters from the dataset, there may be some instances remain in the dataset which could not help to shape another cluster. This situation is taken as stopping criteria in the algorithm. Mathematically this is measured by rule fitness and defined as

$$Fit(R_j) = \left(\frac{m_{rj}}{n_p}\right) \cdot \left(\frac{l_{rj}}{l_{max}}\right) \cdot (1 - l_{rj}S_{rj}) \tag{10}$$

where n_p is the number of problem patterns, m_{rj} is the number of problem patterns that are covered by rule R_j, l_{rj} is the length of R_j, l_{max} is the length of rule that represents C1 and S_{rj} is the rule subspace size of R_j. After revealing all potential clusters in this manner, the CBFC assigns distinct class labels to the consequent of fuzzy rules that represent the discovered clusters. It then uses the fuzzy rules simultaneously to classify the main data patterns and, therefore, identify the clusters' boundaries. Since the CBFC uses the single winner-rule reasoning method, a data pattern that is covered by more than one rule is assigned to the cluster that is identified by a fuzzy rule, which has the highest certainty grade as in (6). However, some of the main data patterns might not be covered by any fuzzy rule. To include these patterns, as well as to increase the clustering accuracy, the centroids of explored clusters are computed using Eq. (11) and are used to group the main data patterns according to the nearest centroid [9].

$$c_j = \frac{\sum_{X_t \in G_j} \mu_j(X_t) \cdot X_t}{\sum_{X_t \in G_j} \mu_j(X_t)} \qquad (11)$$

4.1 Proposed Clustering Algorithm

Step 1: Randomly generated patterns are appended to the problem space to enumerate unsupervised clustering problem as supervised Classification one until d' exceeds d as in Eqs. 9 and 8 respectively.

Step 2: Main data is taken as Class 1 and Auxiliary data as Class 2.

Step 3: Set of rules from numerical data are extracted by using Fuzzy Rule Based Classifier.

Step 4: Among the rules generated for class 1, the best one is chosen and named R_j by using Eq. 6.

Step 5: Rule effectiveness of R_j using Eq. 10 is calculated and if it is less than threshold value i.e. 0.1, step 7 is traced.

Step 6: Cluster members identified by rule R_j from the problem space are removed.

Step 7: All above steps are repeated to generate all possible clusters.

Step 8: By the assignment of some distinct labels to the obtained fuzzy rules, which represent the clusters, the actual boundaries of the clusters can be identified.

Step 9: Centroid of the identified cluster is calculated using Eq. 11.

Step 10: Regrouping is done on unlabeled patterns according to nearest centroid of the clusters.

5 Experimental Results

We have implemented the related programs for the proposed method on a system with MATLAB. We have tested the result with one artificial dataset and six real multidimensional datasets collected from UCI repository (iris, lense, thyroid, haberman, balance scale, hayes-roth).

5.1 Clustering Based on Classifier

In this section, the CBFC is evaluated via experiments on two sets of data. First, the main steps of the CBFC's algorithm are pictorially explained by the use of artificial data. Next, it is applied to other datasets mentioned in Table 1 in order to examine its clustering accuracy.

Table 1 Datasets taken for evaluation

Dataset	Dim (n)	Classes (m)	Instances (N)
Artificial	2	3	1,200
Iris	4	3	150
Lenses	4	3	24
Thyroid	5	3	215
Haber Man	3	2	306
Balance scale	4	3	625
Hayes Roth	5	3	132

In order to visually analyze the operation of the CBFC to explore clusters, one artificial dataset is used in the simulation which is shown in Fig. 2h. To identify the first cluster in artificial dataset with N = 1,200 main data patterns, about A = 1,147 auxiliary instances are added in Step 1 of the CBFC's algorithm as shown in Fig. 2a. In step 4, the best generated rule is: "R1: IF x1 is L3 and x2 is L5 THEN C1 with CG = 0.7636" Fig. 2b shows the fuzzy subspace region of this rule, which assigns 546 main data patterns for cluster C1.

After the removal of the members of cluster C1, the new space of the problem is shown in Fig. 2c, which includes 654 remained main data and 566 newly generated auxiliary patterns. Figure 2d shows the fuzzy subspace of the rule R2 representing cluster C2. This rule is expressed as "R2: IF x1 is L5 and x2 is L2 THEN C2 with CG = 0.8541." and sets aside 263 main patterns for cluster C2. After removal of cluster C2, the new problem space is shown in Fig. 2e. By the repetition of this

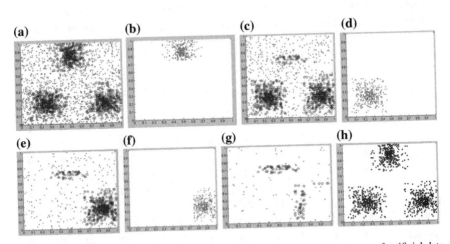

Fig. 2 **a** Main and auxiliary data patterns of artificial data. **b** Main data patterns of artificial data when exploring 1st cluster. **c** Main and aux. data patterns of artificial data when identifying 1st cluster. **d** Main data patterns of artificial data when identifying 2nd cluster. **e** Main and aux. data patterns of artificial data when identifying 2nd cluster. **f** Main data patterns of artificial data when identifying 3rd cluster. **g** Remaining data after major clusters identified. **h** Clustering results of CBFC on artificial dataset

process, the next rule, i.e., "R3: IF x1 is L2 and x2 is L2 THEN C3 with CG = 1.8956," and assigns 313 main data patterns to cluster C3 as shown in Fig. 2f.

After the removal of the main data patterns of three obvious clusters from the problem space, as shown in Fig. 2g, there is no meaningful cluster in 78 remaining patterns. By the repetition of the CBFC, the fuzzy rules will be generated to assign

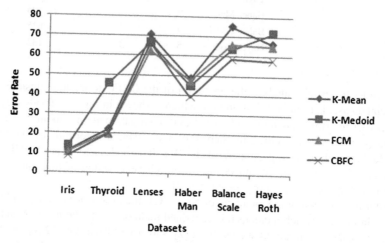

Fig. 3 Performance comparison of different clustering algorithms

Table 2 Confusion matrix of different clustering algorithms

Dataset and actual class level		CBFC			FCM			K-mean			K-medoid		
		No. of cluster			No. of cluster			No. of cluster			No. of cluster		
		1	2	3	1	2	3	1	2	3	1	2	3
Iris	1	50	0	0	50	0	0	50	0	0	50	0	0
	2	0	38	12	0	47	3	0	35	15	0	33	17
	3	0	1	49	0	13	37	0	2	48	0	1	49
Lenses	1	2	1	1	2	1	1	0	4	0	2	0	2
	2	2	1	2	2	1	2	2 1	1	2	3	0	2
	3	4	6	5	4	5	6	5	4	6	5	4	6
Thyroid	1	150	0	0	150	0	0	150	0	0	76	60	14
	2	0	16	19	0	16	19	0	10	25	19	16	0
	3	24	0	6	24	0	6	0	23	07	0	5	25
Haber Man	1	163	62		118	107		120	105		142	83	
	2	58	23		38	43		44	37		55	26	
Balance scale	1	15	20	14	18	6	25	23	19	7	24	17	8
	2	135	105	48	64	84	140	184	51	53	185	94	9
	3	25	125	138	159	18	111	94	114	80	91	89	108
Hayes Roth	1	23	13	15	13	15	23	15	12	24	16	22	13
	2	12	22	17	17	22	12	17	21	13	17	11	23
	3	8	11	11	11	8	11	11	11	8	12	8	10

some more main patterns to another cluster. However, according to the rule fitness measure in (10), the stopping condition of the CBFC is reached, and the algorithm must be continued from Step 8. Computation of the rule fitness measure for these six potential clusters obtains the values 0.2244, 0.1694, 0.1313, 0.0158, 0.0156, and 0.0079, respectively. The sudden drop in these measures after identifying cluster C3 reveals that the stopping condition of the CBFC is reached. Accordingly, the threshold in Step 5 of the algorithm is set to 0.1 in all these experiments.

By continuing the CBFC from Step 9 for these three identified clusters, their final members for artificial dataset is depicted in Fig. 2h. Clearly, the identified fuzzy rules with certainty grade are interpretable and readable by human users.

6 Conclusion

This report investigated K-Mean, K-Medoids and FCM algorithm and devised a new algorithm, named CBFC, based on Fuzzy rule based classifier to address the behavioral shortcoming of K-Mean, K-Medoids and FCM. The shortcomings are like, sensitive to initial parameters, can't explore number of cluster automatically and not user understandable. The generated fuzzy rules with certainty grade, which represent the clusters, are human understandable with reasonable accuracy.

The Proposed algorithm when tested in Matlab (R2010a) with different datasets listed in Table 1, shows that it classifies with adequate accuracy. On datasets like Iris, Haber Man, Hayes Roth, Balance Scale and Thyroid, CBFC performs better as compared to K-Mean, K-Medoids and FCM which has been depicted in Fig. 3 and Table 2.

Furthermore, to achieve high accuracy, fuzzy member function parameter can be adjusted by using neural network machine learning algorithm but this may reduce interpretability to some extent.

References

1. Xu, R., Wunsch, D.: Survey of clustering algorithms. IEEE Trans. Neural Netw. **16**, 645–678 (2005)
2. Likas, A., Vlassis, N., Verbeek, J.: The global K-means clustering algorithm. Pattern Recogn. **36**, 451–461 (2003)
3. Park, H.S., Jun, C.H.: A simple and fast algorithm for K-medoids clustering. Expert Syst. Appl. **36**, 3336–3341 (2009)
4. Yager, R., Filev, D.: Generation of fuzzy rules by mountain clustering. J. Intel. Fuzzy Syst. **2**, 209–211 (1994)
5. Baraldi, A., Blonda, P.: A survey of fuzzy clustering algorithms for pattern recognition—parts I and II. IEEE Trans. Syst., Man, Cybern. B, Cybern. **29**, 778–801 (1999)
6. Yuan, B., Klir, G.J., Stone, J.F.: Evolutionary fuzzy c-means clustering algorithm. In: Fuzzy-IEEE, pp. 2221–2226 (1995)

7. Pal, N.R., Pal, K., Keller, J.M., Bezdek, J.C.: A possibilistic fuzzy c-means clustering algorithm. IEEE Trans. Fuzzy Syst. **13**, 517–530 (2005)
8. Kolen, J., Hutcheson, T.: Reducing the time complexity of the fuzzy c-means algorithm. IEEE Trans. Fuzzy Syst. **10**, 263–267 (2002)
9. Zhu, L., Chung, F.L., Wang, S.: Generalized fuzzy c-means clustering algorithm with improved fuzzy partitions. IEEE Trans. Syst. Man, Cybern. **39**, 578–591 (2009)
10. Wikaisuksakul, S.: A multi-objective genetic algorithm with fuzzy c-means for automatic data clustering. Appl. Soft Comput. **24**, 679–691 (2014)
11. Wang, L., Leckie, C., Ramamohanarao, K., Bezdek, J.: Automatically determining the number of clusters in unlabeled data sets. IEEE Trans. Knowl. Data Eng. **21**, 335–350 (2009)
12. Alcala-Fdez, J., Alcala, R., Herrera, F.: A fuzzy association rule-based classification model for high-dimensional problems with genetic rule selection and lateral tuning. IEEE Trans. Fuzzy Syst. **19**, 857–872 (2011)
13. Sugeno, M., Yasukawa, T.: A fuzzy-logic-based approach to qualitative modeling. IEEE Trans. Fuzzy Syst. **1**, 7–31 (1993)
14. Setnes, M., Babuska, R., Verbruggen, B.: Rule-based modeling: precision and transparency. IEEE Trans. Syst. Man Cybern. Part C: Appl. Rev. **28**, 165–169 (1998)
15. Zadeh, L.: Fuzzy sets. Inf. Control **8**, 338–353 (1965)
16. Klir, G.J., Yuan, B.: Fuzzy Sets and Fuzzy Logic: Theory and Applications. In Pattern Recognition. Prentice-Hall, Englewood Cliffs (1995)
17. Mamdani, E.H., Assilian, S.: An experiment in linguistic synthesis with a fuzzy logic controller. Int. J. Man-Mach. Stud. **7**, 1–13 (1975)
18. Kaufman, L., Rousseeuw, P.J.: Finding Groups in Data: An Introduction to Cluster Analysis. Wiley, Hoboken (1990)

Automatic Contrast Enhancement for Wireless Capsule Endoscopy Videos with Spectral Optimal Contrast-Tone Mapping

V.B. Surya Prasath and Radhakrishnan Delhibabu

Abstract Wireless capsule endoscopy (WCE) is a revolutionary imaging method for visualizing gastrointestinal tract in patients. Each exam of a patient creates large-scale color video data typically in hours and automatic computer aided diagnosis (CAD) are of important in alleviating the strain on expert gastroenterologists. In this work we consider an automatic contrast enhancement method for WCE videos by using an extension of the recently proposed optimal contrast-tone mapping (OCTM) to color images. By utilizing the transformation of each RGB color from of the endoscopy video to the spectral color space La*b* and utilizing the OCTM on the intensity channel alone we obtain our spectral OCTM (SOCTM) approach. Experimental results comparing histogram equalization, anisotropic diffusion and original OCTM show that our enhancement works well without creating saturation artifacts in real WCE imagery.

Keywords Contrast enhancement · Wireless capsule · Endoscopy · Contrast tone mapping · Spectral

V.B. Surya Prasath (✉)
Department of Computer Science, University of Missouri-Columbia,
Columbia 65211, USA
e-mail: prasaths@missouri.edu
URL: http://goo.gl/66YUb

R. Delhibabu
Cognitive Modeling Lab, IT University Innopolis, Kazan, Russia
e-mail: delhibabur@ssn.edu.in

R. Delhibabu
Artificial Consciousness Lab, Kazan Federal University, Kazan, Russia

R. Delhibabu
Department of Computer Science Engineering, SSN Engineering College,
Chennai 603110, India

© Springer India 2015
L.C. Jain et al. (eds.), *Computational Intelligence in Data Mining - Volume 1*,
Smart Innovation, Systems and Technologies 31, DOI 10.1007/978-81-322-2205-7_23

1 Introduction

Wireless capsule endoscopy (WCE) is a novel imaging technique that provides an inner view of the gut without much hindrance and discomfort to the patient. Each endoscopy exam obtains multiple hours of color (RGB) video data, for example the Pillcam® Colon2 capsule (Given Imaging, Yoqnem, Isreal) has 8 h of video with around 55,000 frames. Sifting through the video imagery places a burden on gastroenterologist's time and effort in identifying important frames. Hence automatic computer-aided diagnosis (CAD) tools can greatly aid in making diagnosis using the big data available through WCE imagery.

Recent years saw various image processing, computer vision tasks performed for WCE images for automatic CAD purposes [1]. For example, polyp detection/classification [2, 3], mucosa-lumen surface reconstruction/segmentation [4–6] and contrast enhancement [7], see [8] for a recent review.

1.1 Contrast Enhancement—A Brief Review

In this work, we consider the problem of image enhancement for WCE videos. With respect to enhancing the contrast in WCE images we mention the work of Li and Meng [7] who used an inverse diffusion [9] based method. Among a plethora of generic image contrast enhancement techniques we mention the following well-known and widely utilized methods. Histogram equalization (HE) is one of the classical algorithm which relies on the first order statistics namely the histogram computed from the given image to be mapped into a uniform histogram [10]. To avoid the drawbacks of HE, such as the exaggeration of in smooth regions, contrast adaptive histogram equalization (CLAHE) is proposed as an improvement [11].

In a rigorous study of HE based methods Wu [12] exposed the intrinsic weakness of HE in particular and histogram based image enhancement techniques in general. In [12] the gray scale contrast enhancement problem is posed as an optimal allocation of dynamic range to maximize contrast gain and the optimization problem is solved using linear programming principles. The method known as optimal contrast-tone mapping (OCTM) is shown to obtain visually more pleasing results when compared with other histogram based methods.

In our work we adapt the OCTM and extend it to the color images using the RGB to La*b* color transformation and applying it to the luminance channel. Note that utilizing the OCTM independently on the RGB channels does not take into account the color discontinuities. Thus, the color space transformation to a more spectral-spatial separation aides in improving the method. We compare the spectral OCTM (named SOCTM hereafter) with HE, CLAHE, OCTM in each channel on WCE images. Our results indicate that we obtain contrast enhancement without saturation artifacts associated with other methods.

1.2 Organization

The rest of the paper is organized as follows. In Sect. 2 we provide a brief introduction of OCTM from [12] along with the La*b* transformation based SOCTM with illustrations. Section 3 contains experimental results on various WCE images depicting different scenarios of varying light, contrast conditions. Finally, Sect. 4 concludes the paper.

2 Spectral Optimal Contrast-Tone Mapping Based Enhancement

2.1 OCTM

The problem of maximizing the contrast gain for optimal allocation of output dynamic range can be posed as a constrained linear program. The objective function and the constraints are written as follows and we refer to [12] for more details:

$$
\begin{aligned}
\text{maximize} \quad & \sum_{i=1}^{L-1} p_i s_i \\
\text{subject to} \quad & \mathbf{1}^T s \leq L \\
& 0 \leq s \leq u \\
& \sum_{j=0}^{d-1} s_{i+j} \geq 1, \quad i = 1, \ldots, L-d,
\end{aligned}
\tag{1}
$$

where $p = (p_1, p_2, \ldots, p_{L-1})$ is the histogram vector of the input image with L gray values. The variable s_i is known as the *context-free* contrast at gray level i which is computed as the unit rate of change from level i to level $i + 1$ in the output image. Note that this is independent of pixel locations hence the name context-free and this variable uniquely determines the following increasing transfer function

$$
T(i) = \left\lfloor \sum_{t=1}^{i} s_t \right\rfloor, \quad 0 \leq i \leq L-1
\tag{2}
$$

which maps i gray level to $T(i)$. The objective function in the maximization problem (1) is seen as

$$
\sum_{i=1}^{L-1} p_i s_i = \sum_{i=1}^{L-1} p_i \cdot [T(i) - T(i-1)]
\tag{3}
$$

Thus, the OCTM maximization (1) is based on the expected context-free contrast achieved by the transfer function $T(i)$ in (2). Note that OCTM approach uses the histogram, which is a first order statistics computed from the given input image.

2.2 Spectral OCTM

To motivate the spectral transformation we show an example result of OCTM on an RGB image by applying the contrast enhancement on each channel independently before combining in the output image. Figure 1a shows the input RGB WCE image (from ileum region of the gut, image courtesy of RapidAtlas, Given Imaging, image taken using a Pillcam® COLON2 capsule) and Fig. 1b–d show the Red, Green, Blue channels separately. Figure 1f–h show independent OCTM enhanced results using the maximization scheme in (1). As can be seen from Fig. 1e, the final combined OCTM result, saturates some of the regions due to strong homogenous regions in a particular channel (in this case in the saturation occurred in Red channel, see Fig. 1g middle region).

To avoid this saturation issue, we first transform the given input RGB image onto the La*b* space, where L denotes the luminance and a*, b* are the spectral channels. This space is based on the concept of color-opponent and includes all

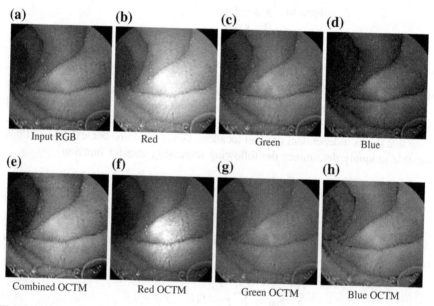

Fig. 1 Applying OCTM enhancement to each channel in an RGB image gives saturation artifacts, as the cross-channel spectral information is not utilized. **a** Example RGB frame from ileum using Pillcam® COLON2 capsule vide data, it contains villus structure. **b–d** RGB channels **e** OCTM applied RGB image and its RGB channels **f–h**. Clearly OCTM channel wise inherits the illumination problems. In particular the specular reflection in the Red channel is amplified

perceivable colors [13]. Due to its perceptual uniformity this space can be used for gamut mapping and efficient segmentation [13]. This space, also known as the CIELAB is widely used in the vision and computer graphics community. The nonlinear relations between L, a*, and b* mimic nonlinear response of the human eye to various colors.

Since the wireless capsule imaging involves multiple near lights it is paramount that we account for illumination changes. Converting RGB to La*b* provides a spectral separation and the OCTM introduced in Sect. 2.1 can then be applied effectively. The transformation from RGB to La*b* is done through RGB-to-XYZ and XYZ-to-La*b* transformations with D50 reference white. Thus, our proposed SOCTM consist of the following steps:

Step 1   Transform the RGB input image into the La*b* space with D50 reference white

Step 2   Apply the OCTM (1) to the luminance (L) channel alone and keep the a*, and b* channels intact

Step 3   Convert the modified (contrast enhanced) luminance channel along with original a*, and b* channels to RGB space.

Note that these RGB → La*b* transformations are available in all the standard color imaging references, see for example the book [13] and thus omitted here for brevity.

3 Experimental Results

We present WCE image frames taken from different parts of the gastrointestinal tract to illustrate the advantage of our SOCTM against other histogram equalization approaches [10, 11] and anisotropic diffusion method from [7].

First in Fig. 2a–c we show the L, a*, and b* channels for the same RGB WCE image shown in Fig. 1a. Next, we use the OCTM only on the luminance channel (L) and keep a* and b* spectral channels intact. Figure 2d shows the OCTM based contrast enhancement applied to the luminance channel L. As can be seen the contrast mapping is better (compare to Fig. 2a). The final output image is obtained by doing a reverse transformation of this modified L channel and combining it with a* and b* channels and doing a reverse transformation to the RGB space. Figure 2e shows this combined final results using our SOCTM result and it is devoid of the saturations issues, compare Fig. 1e with Fig. 2e.

To highlight the difference between, RGB, OCTM and SOCTM, we show the differences between them (pixels where the intensity values are changed using L_1 differences) in Fig. 2f–h. As can be seen, the difference between OCTM and RGB

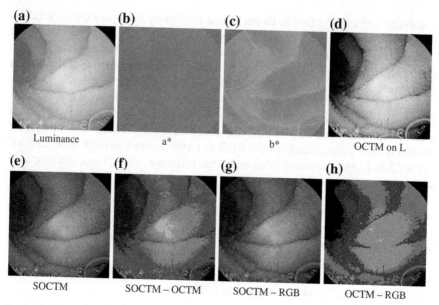

Fig. 2 Switching to the La*b* color space helps in differentiating the spectral part of the RGB (shown in Fig. 1a) and an application of SOCTM to the luminance provides a visually pleasing result. **a** L channel of the RGB image **b** a* and **c** b* channel **d** OCTM applied to the L channel **e** SOCTM result. We show the difference of different contrast mapping using the L_1 differences (residual) **f** SOCTM and OCTM **g** SOCTM and RGB **h** OCTM and RGB. As can be seen proposed SOCTM keeps the RGB pixel numerical range and OCTM dramatically changes the values. Better viewed online and zoomed in

(Fig. 2h) shows the specularity effect at the center that corresponds to the result in Fig. 1e. On the other hand, the difference between SOCTM and RGB (Fig. 2g) shows the lack of specularity at the center. This can further be seen in by comparing SOCTM and OCTM (Fig. 2f), where the difference highlights the advantage of using the proposed modification from La*b* space.

Next Fig. 3 shows a series of comparisons on different WCE image frames taken at different parts of the colon between various schemes. As can be seen, the HE [10], CLAHE [11] provide spurious saturation due to the lack of spatial context, inverse diffusion [7] propagates artifacts to darker regions and the proposed SOCTM obtains most visually pleasing results without these drawbacks. Currently, we are working on evaluating the effect of final contrast enhanced images in various CAD tasks, for example, in polyp segmentation and image classification [3].

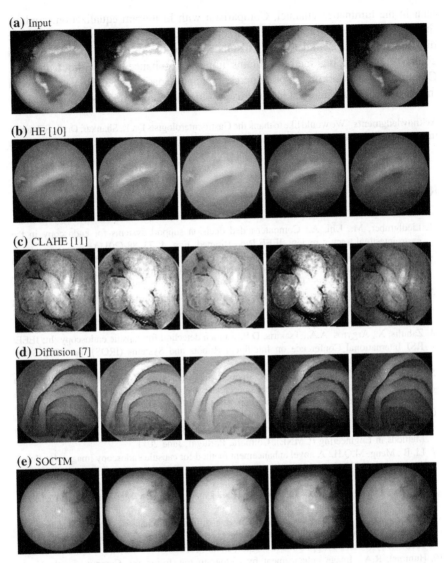

Fig. 3 Comparison results with other contrast enhancement techniques. **a** Input frames from WCE, different parts of the colon. Results of: **b** histogram equalization (*HE*) [10] **c** CLAHE [11] **d** inverse diffusion [7] **e** proposed SOCTM result. Better viewed online and zoomed in. Our proposed SOCTM obtains better results without flare exaggeration and color change

4 Conclusions

We presented a spectral optimal contrast-tone mapping based enhancement approach for wireless capsule endoscopy. By utilizing a reversible transformation from RGB to La*b* we preserve spectral information and utilize the maximization of contrast gain in the luminance channel. Comparison with histogram equalization, contrast limited adaptive histogram equalization, optimal contrast-tone mapping, anisotropic diffusion methods is undertaken. Results indicate the proposed spectral optimal contrast-tone mapping method obtains improved enhancement without artifacts. Currently we are working on other perceptual color spaces (XYZ, HSV) along with improved tone mapping geared towards capsule endoscopy images.

Acknowledgments We would like to thank the Gastroenterologists Dr. R. Shankar, Dr. A. Sebastian from Vellore Christian Medical College Hospital, India for their help in interpreting WCE imagery.

References

1. Liedlgruber, M., Uhl, A.: Computer-aided decision support systems for endoscopy in the gastrointestinal tract: a review. IEEE Rev. Biomed. Eng. **4**, 73–88 (2011)
2. Figueiredo, P.N., Figueiredo, I.N., Prasath, S., Tsai, R.: Automatic polyp detection in PillCam COLON 2 capsule images and videos: preliminary feasibility report. Diagn. Ther. Endosc. **2011**, Article ID 182435, 16 p (2011)
3. Prasath, V.B.S., Pelapur, R., Palaniappan, K.: Multi-scale directional vesselness stamping based segmentation for polyps from wireless capsule endoscopy. In: 36th IEEE EMBS International Conference, Chicago, USA (2014)
4. Zabulis, X., Argyros, A.A., Tsakiris, D.P.: Lumen detection for capsule endoscopy. In: IEEE/RSJ International Conference on Intelligent Robots and Systems (IROS), pp. 3912–3926, Nice, France (2008)
5. Prasath, V.B.S., Figueiredo, I.N., Figueiredo, P.N., Palaniappan, K.: Mucosal region detection and 3D reconstruction in wireless capsule endoscopy videos using active contours. In: 34th IEEE EMBS International Conference, pp. 4014–4017, San Diego, USA (2012)
6. Prasath, V.B.S., Figueiredo, I.N., Figueiredo, P.N.: Colonic mucosa detection in wireless capsule endoscopic images and videos. In: Proceedings of the Congress on Numerical Methods in Engineering (CMNE), Coimbra, Portugal, June 2011
7. Li, B., Meng, M.Q.H.: A novel enhancement method for capsule endoscopy images. Int. J. Inf. Acquisition **4**, 117–126 (2007)
8. Karargyris, A., Bourbakis, N.: A survey on wireless capsule endoscopy and endoscopic imaging: a survey on various methodologies presented. IEEE Eng. Med. Biol. Mag. **29**, 72–83 (2010)
9. Prasath, V.B.S., Singh, A.: Controlled inverse diffusion models for image restoration and enhancement. In: IEEE International Conference on Emerging Trends in Engineering and Technology (ICETET), pp. 90–94, Nagpur, India (2008)
10. Hummel, R.A.: Image enhancement by histogram transformation. Comput. Graph. Image Process. **6**, 184–195 (1977)
11. Zuiderveld, K.: Contrast limited adaptive histogram equalization. In: Heckbert, P. (ed.) Graphics Gems IV, pp. 474–485. Academic Press, Waltham (1994)
12. Wu, X.: A linear programming approach for optimal contrast-tone mapping. IEEE Trans. Image Process. **20**, 1262–1272 (2011)

Cuckoo Search Algorithm Based Optimal Tuning of PID Structured TCSC Controller

Rupashree Sethi, Sidhartha Panda and Bibhuti Prasad Sahoo

Abstract Cuckoo Search Algorithm (CSA) is one of the new nature inspired meta-heuristic algorithms for solving many engineering optimization problems. This paper investigates the application of CSA technique for the tuning of proportional integral derivative (PID) structured Thyristor Controlled Series Compensator (TCSC) based controller to improve damping of the power system when subjected to different disturbances at various loading conditions. The dynamic performances of proposed approach are analyzed by taking an example of Single Machine Infinite Bus (SMIB) power system in MATLAB/Simulink environment. The superiority of the proposed approach is demonstrated by comparing the simulation results with previously published technique such as Non-dominated Sorting Genetic Algorithm-II (NSGA-II) based TCSC controller for the same power system.

Keywords Cuckoo search algorithm · Thyristor controlled series compensator · PID controller · Power system stability · Single-machine infinite-bus system · Non-dominated sorting genetic algorithm-II

1 Introduction

Optimization is the process of adjusting the inputs to or characteristics of a device, to find the minimum error or maximum performance improvement. The Cuckoo Search Algorithm (CSA) developed in 2009 by Yang and Deb [1, 2], is one of the latest nature inspired meta-heuristic algorithms for solving optimization problems,

R. Sethi (✉) · S. Panda · B.P. Sahoo
Department of Electrical Engineering, Veer Surendra Sai
University of Technology, Burla 768018, Odisha, India
e-mail: rupashree.sethi21@gmail.com

S. Panda
e-mail: panda_sidhartha@rediffmail.com

B.P. Sahoo
e-mail: bibhutivssut@gmail.com

© Springer India 2015
L.C. Jain et al. (eds.), *Computational Intelligence in Data Mining - Volume 1*,
Smart Innovation, Systems and Technologies 31, DOI 10.1007/978-81-322-2205-7_24

which is based on the brood parasitism of some cuckoo species. This algorithm is enhanced by Levy flights, rather than by simple isotropic random walks.

Within the FACTS initiative, it has been demonstrated that variable series compensation is highly effective in both controlling power flow in the lines and in improving stability. Thyristor controlled series capacitor (TCSC) is one of the important member of FACTS family that came into existence from their conventional parents i.e. fixed series capacitor [3] whose effective fundamental equivalent reactance can be regulated periodically to cancel a portion of the reactive line impedance and thereby increase the transmittable power. As a novel method for electrical network control, TCSC can be utilized in the power system oscillation damping, the sub synchronous resonance mitigation and load flow control, scheduling power flow, decreasing unsymmetrical components, reducing net loss, limiting short-circuit currents, power system transient stability enhancement [4].

Evolutionary algorithms such as genetic algorithms [5], particle swarm optimization (PSO) algorithm [6], Differential Evolution (DE) algorithm [7], Non-dominated Sorting Genetic Algorithm (NSGA)-II [8] have been successfully applied for optimizing the parameters of TCSC based controller with different control structure. This paper introduces a new evolutionary optimization algorithm i.e. Cuckoo Search Algorithm (CSA) that has a huge prospective to be as an effective alternative to other evolutionary algorithm in solving many optimization problems. CSA was previously applied to solve several engineering design optimization problems, such as the design of springs and welded beam structures [2], optimal location and sizing of distributed generation on a radial distribution system [9], and also applied for solving multi-objective optimization problems [10]. But the design of an optimal PID controller requires optimization of multiple performance parameters.

In this paper CSA is successfully applied for searching the optimal parameters of PID structured TCSC based controller for the dynamic performance improvement under various disturbances for different loading conditions in MATLAB/Simulink environment. Simulation results show the advantages of using the modeling and tuning method when performing control and stability analysis in a power system including a CSA tuned TCSC controller over NSGA-II tuned TCSC controller.

2 TCSC and PID Structure of TCSC Based Controller

TCSC constitutes three components—a capacitor bank C, a bypass inductor L and the bidirectional thyristor. The firing angle (α) of the thyristor is controlled to adjust the TCSC reactance. TCSC can be controlled to work in capacitive zone. The equation of reactance which is function of (α) is represented by the following equation [5]:

$$X_{TCSC}(\alpha) = X_C - \frac{X_C^2}{(X_C - X_L)} \frac{(\sigma + \sin \sigma)}{\pi}$$
$$+ \frac{4X_C^2}{(X_C - X_L)} \frac{\cos^2(\sigma/2)}{(k^2 - 1)} \frac{[k \tan(k\sigma/2) - \tan(\sigma/2)]}{\pi} \tag{1}$$

where, X_C is the nominal reactance of the fixed capacitor, X_L the inductive reactance of inductor connected in parallel with fixed capacitor, $\sigma = 2(\pi - \alpha)$, the conduction angle of TCSC controller and $k = \sqrt{(X_C/X_L)}$, the compensation ratio. TCSC is modeled here as a variable capacitive reactance within the operating region defined by the limits imposed by α. Thus $X_{TCSCmin} \leq X_{TCSC} \leq X_{TCSCmax}$, with $X_{TCSCmax} = X_{TCSC}(\alpha_{min})$ and $X_{TCSCmin} = X_{TCSC}(180°) = X_C$. In this study, the controller is assumed to operate only in the capacitive region, i.e., $\alpha_{min} > \alpha_r$, where α_r corresponds to the resonant point, as the inductive region associated with $90° < \alpha < \alpha_r$ induces high harmonics that cannot be properly modeled in stability studies.

The structure of TCSC-based damping controller, to modulate the reactance offered by the TCSC, $X_{TCSC}(\alpha)$ is shown in Fig. 1. The structure consists of a PID controller with proportional gain K_p, integral gain K_i, and derivative gain K_d, a first order lag and a limiter. The details of TCSC based controller modeling is explained in [8]. The input signal of the proposed controller is the speed deviation ($\Delta\omega$) error and the output signal is the reactance $X_{TCSC}(\alpha)$. According to the speed error, the parameters of PID controller (K_p, K_i, K_d) are to be optimized so that the TCSC reactance $X_{TCSC}(\alpha)$ can be effectively modulated to cancel some portions of line reactance and thereby improve the damping of power oscillations. The effective reactance is given by:

$$X_{Eff} = X - X_{TCSC}(\alpha) \tag{2}$$

where,

$X_{TCSC}(\alpha)$ is the reactance offered by TCSC at firing angle α.
X is the equivalent line reactance.

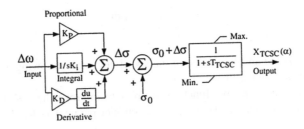

Fig. 1 PID structure of TCSC based controller

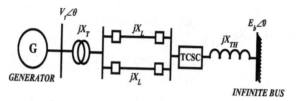

Fig. 2 Single machine infinite bus power system with TCSC

3 Modeling of Power System with TCSC

The single machine infinite bus (SMIB) power system with TCSC (shown in Fig. 2), is considered in this study where a synchronous generator propagates power to the infinite-bus via double circuit transmission line and a TCSC. In Fig. 2, V_t and E_b are the generator terminal voltage and infinite bus voltage respectively. X_T, X_L and X_{TH} represent the reactances of the transformer, transmission line per circuit and the Thevenin's impedance of the receiving end system respectively.

The synchronous generator is represented by model 1.1 [5], i.e. with field circuit and one equivalent damper winding on q-axis, comprising of the electromechanical swing equation and the generator internal voltage equation. The system data's are represented in [8]. Using state equations [11], the modeling of power system with TCSC in MATLAB/Simulink environment is explained in [5, 8].

4 Objective Function

The rotor of the generator machine experiences a deviation in its motion when it gets a fault in any of the phase which leads the power system oscillations. TCSC-based controller is designed to minimize the power system oscillations after a disturbance so as to improve the system stability.

In the present study, an integral time absolute error (ITAE) of the speed deviations ($\Delta\omega$) is taken as the objective function J which is defined as follows:

$$J = \int_0^{t_{sim}} t|\Delta\omega(t)|dt \tag{3}$$

where, $|\Delta\omega(t)|$ is the absolute value of the speed deviation following a disturbance and t_{sim} is the time range of simulation.

To minimize this objective function in order to improve the system response, CSA is used to optimize the TCSC based controller parameters K_p, K_i, K_d.

5 Cuckoo Search Algorithm

Cuckoo search algorithm (CSA) is one of the new optimization techniques. This algorithm is more computationally efficient (uses less number of parameters) than the Particle Swarm Optimization [12]. In CSA, special lifestyle of cuckoos and their characteristics in egg laying and breeding has been the basic motivation for development of this new evolutionary optimization algorithm. Cuckoos search for the most suitable nest to lay eggs in order to maximize their eggs survival rate. Each egg in a nest represents a solution, and a cuckoo egg represents a new solution. The aim is to employ the new and potentially better solutions (cuckoos) to replace not-so-good solutions in the nests. In the simplest form, each nest has one egg. The CSA is based on three idealized rules [1, 2]:

- Each cuckoo lays one egg at a time and dumps it in randomly chosen nest;
- The best nests with high quality of eggs (solutions) will carry over to the next generations;
- The number of available host nests is fixed, and a host can discover an alien egg with probability $p_a \in [0, 1]$. In this case, the host bird can either throw the egg away or abandon the nest to build a completely new nest in a new location.

For simplicity, the last assumption can be approximated by a fraction p_a of the n nests being replaced by new nests, having new random solutions.

5.1 Cuckoo Search Via Levy Flights

In nature, animals search for food in a random or quasi-random manner. In general, the foraging path of an animal is effectively a random walk. Which direction it chooses depends implicitly on a probability which can be modeled mathematically. Various studies have shown that the flight behavior of many animals and insects has demonstrated the typical characteristics of Lévy flights [1]. A Lévy flight is a random walk in which the step-lengths are distributed according to a heavy-tailed probability distribution. According to Yang [13], Lévy flights are more efficient for searching than regular random walks or Brownian motions. Equation (4) shows how a new solution $X_i^{(t+1)}$ for ith cuckoo is generated using a Lévy flight.

$$x_i^{(t+1)} = x_i^{(t)} + \alpha \oplus \text{Lévy}(\lambda) \tag{4}$$

where $\alpha > 0$ is the step size of random walk. The above equation is the schotastic equation for a random walk. A random walk is a Markov chain whose next status or location $\left(X_i^{(t+1)}\right)$ only depends on the current location $\left(X_i^{(t)}\right)$ and the transition probability $(\alpha \oplus \text{Lévy}(\lambda))$. Product $\oplus$ means entry wise multiplications. Here random walk via Lévy flight is more efficient in exploring the search space, as its step length is much longer in the long run [13].

The simple way to calculate step size (α) of random walk is as the Eq. (5).

$$\alpha = 0.01^*(s) \oplus \left(x_i^{(t)} - x_{best}\right) \tag{5}$$

where 0.01 is a factor for controlling step size of cuckoo walk/flights which comes from the facts that L/100 should the typical step size of walk/flights where L is the typical length scale; otherwise, Lévy flights may become too aggressive/efficient, which makes new solutions jump outside the design domain and thus wasting the evaluations.

$X_i^{(t)}$ is the current solution of iteration t and x_{best} is the global best solution. S is the step length which is distributed according to the most efficient Mantegna algorithm for a symmetric Lévy stable distribution [13]. For generating new solution $X_i^{(t+1)}$ this step size (α) is taken.

5.2 Mantegna's Algorithm

Mantegna's algorithm produces random numbers according to a symmetric Levy stable distribution. It was developed by R. Mantegna. In Mantegna's algorithm, the step length s can be calculated by [13]:

$$s = \frac{u}{|v|^{1/\beta}} \tag{6}$$

where u and v are random value which is drawn from normal distributions i.e. $u \sim N\left(0, \sigma_u^2\right)$, $v \sim N\left(0, \sigma_v^2\right)$,

and

$$\sigma_u = \left\{ \frac{\Gamma(1 + \beta) \sin(\pi\beta/2)}{\Gamma[(1 + \beta)/2]\beta\, 2^{(\beta-1)/2}} \right\}^{1/\beta}, \tag{7}$$

$$\sigma_v = 1 \tag{8}$$

Γ is gamma function which is extended from factorial function of positive real number. β is the variable which controls distribution by $0 < \beta < 2$. σ_u^2 is the variance. This variance, hence, is used to calculate the step size of random walk. The algorithmic control parameters of the CS algorithm are the scale factor β and mutation probability value P_a.

Fig. 3 Flowchart of Cuckoo
search optimization algorithm

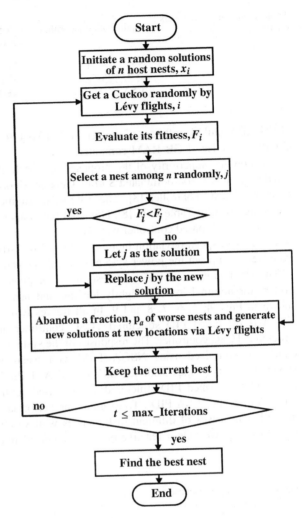

In this work, $\beta = 1.5$ and $P_a = 0.25$ have been taken. The flow chart of the CSA used in this work is shown in Fig. 3.

6 Simulation Results, Discussion and Comparison of CSA with NSGA-II

The CSA tuned PID structured TCSC based controller is tested in Single Machine Infinite Bus power system to damp out power system oscillations when the system is subjected to various disturbances. The simulation test is carried out by using

Table 1 Optimal parameters for PID controller

Optimization algorithm	TCSC-based controller parameters		
	K_p	K_i	K_d
NSGA-II [8]	29.5477	2.4866	0.1533
CSA	48.7238	4.1930	0.0010

MATLAB/Simulink software of the version 7.10.0.499 (R2010a), on an Intel Core i3 CPU, 2.3 GHz, 4 GB RAM computer.

To show the superiority of the proposed approach, the performances are compared with that of Non-dominated Sorting Genetic Algorithm (NSGA)-II tuned PID structured TCSC controller [8]. The optimized parameters of TCSC-based controller are given in Table 1. In [8], NSGA-II based approach for the tuning of PID structured TCSC based controller was proposed in which the author has proved that NSGA-II with speed deviation input signal (ω-NSGA-II) tuned PID controller can effectively damp out the power system oscillations of the same system.

The simulation results of time responses computed with the alternative controllers for nominal loading, lightly loading and heavy loading respectively are shown in Figs. 4, 5, 6, 7, 8 and 9. In all figures two cases are considered; with NSGA-II [8] tuned PID structured TCSC controller and with CSA tuned PID structured TCSC controller. The responses of the system with PID structured TCSC controller, optimized using NSGA-II [8] with speed deviation input signal are shown by dashed lines with legend 'ω NSGA-II PID' and the responses with proposed CSA tuned PID structured TCSC based controller are shown by solid lines with legend 'CSA PID'. To examine the performance of proposed approach, the following cases of different disturbances with various loading conditions [8] (Pe = active power, Qe = reactive power) are considered:

Fig. 4 Power angle response

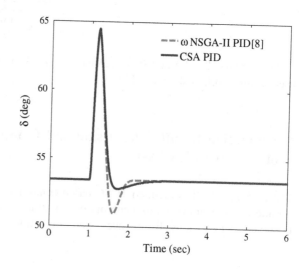

Fig. 5 Rotor speed deviation response

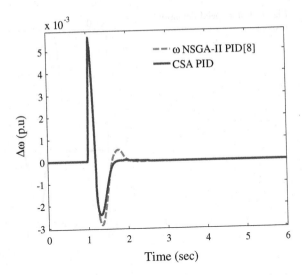

Fig. 6 Power angle response

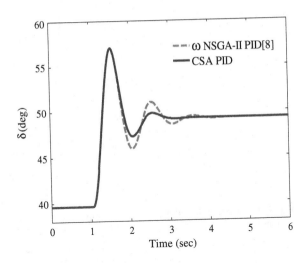

6.1 Three-Phase Fault Disturbance at Nominal Loading (Pe = 0.8, Qe = 0.017)

The fault, that is considered here, is a three phase fault which is applied at the generator terminal at t = 1 s and cleared after 3 cycles. The original system is restored upon the fault clearance. The responses of system are shown in Figs. 4 and 5 for two different system conditions; namely, for the system with NSGA-II [8] tuned controller, and for the system with proposed CSA tuned controller.

Fig. 7 Rotor speed deviation
response

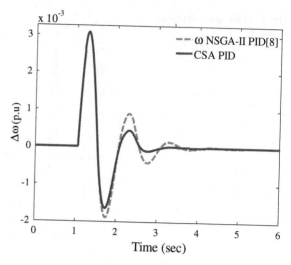

Fig. 8 Power angle response

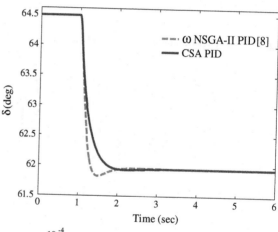

Fig. 9 Rotor speed deviation
response

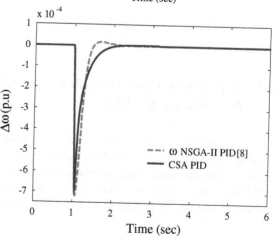

From the simulation results it is observed that a well optimized TCSC based controller significantly improves the damping of the subsequent power swings. From the dynamic response of power angle variation in Fig. 4, it is cleared that with NSGA-II tuned controller, though the system is stable following a disturbance, there is still some oscillations or variation in angle but by using CSA tuned controller the angle deviation is minimized with minimization of undershoots and settling time. Figure 5 shows the rotor speed deviation when three phase fault is applied. Here it can be seen that with CSA tuned PID structured TCSC based controller the deviation in speed is minimized very quickly with minimum overshoot and undershoot values than that of with NSGA-II [8] PID structured TCSC based controller. Hence in this paper it is proved that CSA tuned PID structured TCSC based controller performs better than NSGA-II tuned PID structured TCSC based controller under the most severe fault.

6.2 Permanent Line Outage Disturbance at Light Loading (Pe = 0.5, Qe = 0.006)

In order to test the performance of the proposed approach, another severe disturbance is considered here. One of the transmission line is permanently tripped out at $t = 1$ s. The system responses of power angle variation and rotor speed deviation are shown in Figs. 6 and 7 respectively.

The reactance of the transmission line increases in the post-fault steady-state period due to permanent line outage. Assuming that the mechanical input power remains constant during the disturbance period, to transmit the same power, the power angle increases in the post-fault steady state period as shown in Fig. 6. It is cleared from the above figures that the system becomes stable following a disturbance with both NSGA-II tuned and CSA tuned TCSC controller but the CSA tuned controller provides better act than NSGA-II tuned controller in minimizing both speed deviation and angle variation.

6.3 Small Disturbance at Heavy Loading (Pe = 1.2, Qe = 0.038)

In this case the proposed approach is examined under a small fault at heavy loading condition. The input mechanical power is decreased by 0.1 p.u at $t = 1.0$ s. The system responses under this small disturbance contingency are shown in Figs. 8 and 9.

From the simulation results of the above case study, it is observed that, the optimized TCSC controller with cuckoo search algorithm (CSA) improves the stability performance of the power system and performs satisfactorily under this small disturbance at heavy loading condition. From Figs. 8 and 9, it can be seen that

Table 2 Error comparison

Optimization algorithm	ITAE		
	Nominal loading (three phase fault)	Light loading (permanently line outage)	Heavy loading (small disturbance)
NSGA-II [8]	19.28×10^{-4}	36.17×10^{-4}	18.31×10^{-5}
CSA	16.32×10^{-4}	26.48×10^{-4}	17.73×10^{-5}

the CSA tuned controller provides good damping characteristics to low frequency oscillations and quickly stabilizes the system under a disturbance when compared with Non-dominated Sorting Genetic Algorithm (NSGA)-II [8] tuned TCSC Controller.

The error comparison between NSGA-II [8] technique and the proposed approach for above three cases of loading conditions is shown in Table 2. It is cleared from Table 2 that CSA tuned controller reduces integral time absolute error (ITAE) value effectively compared to NSGA-II [8] tuned controller, hence proving here that proposed approach is better than previous technology.

7 Conclusion

In this paper the parameter tuning of PID structured TCSC controller is presented by a new evolutionary nature inspired meta-heuristic algorithm i.e., cuckoo search algorithm (CSA). The performance of proposed approach is examined by taking an example of single machine connected to infinite bus power system with TCSC-based controller. To analyze performance of proposed controller it is tested under various disturbances at different loading conditions. The results are obtained and compared with the previous publication i.e. NSGA-II [8] tuned PID structured TCSC controller. From system responses of CSA tuned PID controller and NSGA-II tuned PID controller, it can be observed that the disturbance is effectively damped out by CSA with a faster response. Hence it is proved that the optimal solutions for PID controller obtained by CSA provide better performance than previous technology.

References

1. Yang, X.-S., Deb, S.: Cuckoo search via levy flights. In: Proceedings of World Congress on Nature and Biologically Inspired Computing (NaBIC 2009), India, pp. 210–214. IEEE Publications, USA (2009)
2. Yang, X.-S., Deb, S.: Engineering optimization by Cuckoo search. Int. J. Math. Model. Numer. Opt. 1(4), 330–343 (2010)
3. Hingorani, N.G., Gyugyi, L.: Understanding FACTS: Concepts and Technology of Flexible AC Transmission Systems. IEEE Press, Piscataway (2000)

4. Kimbark, E.W.: Improvement of system stability by switched series capacitors. IEEE Trans. Power Apparatus Syst. **PAS-85**(2), 180 (1966)
5. Panda, S., Padhy, N.P.: MATLAB/Simulink based model of single-machine infinite-bus with TCSC for stability studies and tuning employing GA. Int. J. Electr. Electron. Eng. 1–5 (2007)
6. Baliarsingh, A.K., Dash, D.P., Samal, N.R., Panda, S.: Design of TCSC-based controller using particle swarm optimization algorithm IOSR. J. Eng. **2**(4), 810–813 (2012)
7. Panda, S.: Differential evolutionary algorithm for TCSC-based controller design. Int. J. Simul. Model. Pract. Theory **17**, 1618–1634 (2009)
8. Panda, S.: Multi-objective PID controller tuning for a FACTS-based damping stabilizer using non-dominated sorting genetic algorithm-II. Electr. Power Energy Syst. **33**, 1296–1308 (2011)
9. Tan, W.S., Hassan, M.Y., Majid, M.S., Rahman, H.A.: Allocation and sizing of DG using Cuckoo search algorithm. In: IEEE International Conference on Power and Energy (PECon), Kota Kinabalu, Sabah (2012)
10. Coelho, L.S., Guerra, F.A., Batistela, N.J., Leite, J.V.: Multiobjective Cuckoo search algorithm based on Duffing's oscillator applied to Jiles-Atherton vector hysteresis parameters estimation. IEEE Trans. Magn. **49**(5), 1745–1748 (2013)
11. Padiyar, K.R.: Power System Dynamics Stability and Control, 2nd edn. BS Publications, Hyderabad (2002)
12. Adnan, M.A., Razzaque, M.A.: A comparative study of particle swarm optimization and Cuckoo search techniques through problem-specific distance function. In: International Conference of Information and Communication Technology (2013)
13. Yang, X.S.: Nature-Inspired Metaheuristic Algorithms, 2nd edn. Luniver Press, UK (2010)

Design and Simulation of Fuzzy System Using Operational Transconductance Amplifier

Shruti Jain

Abstract In this paper a well defined method for the design of fuzzy system using operational transconductance amplifier was discussed. This paper also presents the various types of operational transconductance amplifier: simple OTA, fully differential OTA and Balanced OTA. Proposed model of fully differential OTA and balance OTA is illustrated, out of which balance OTA is the best. Fuzzy system includes fuzzification of the input variables, application of the fuzzy operator (AND or OR) in the antecedent, implication from the antecedent to the consequent, aggregation of the consequents across the rules, and defuzzfication. The fuzzy system (including all blocks) for estimating risk involved in an engineering project has been designed, exhibits a gain of 36.24 dB, input resistance with 44.13 kΩ, output resistance with 2.163 kΩ, CMRR with 47.68 dB, slew rate with 0.6 V/µs and power dissipation with 197 W.

Keywords Operational transconductance amplifier · Fuzzy system · Analog circuit · Electrical parameters · SPICE

1 Introduction

An operational amplifier (op-amp) is a direct coupled high gain amplifier usually consisting of one or more differential amplifiers and usually followed by a level translator and an output stage. The output stage is generally a push pull or push pull complementary symmetry pair [1, 2]. Op-amp circuits are the key components of analog processing systems and are widely used in electronic and communication systems such as differential amplifier, negative feedback amplifier, isolation amplifier, comparators, oscillators, filters, sensors, instrumentation amplifier, biomedical amplifier etc. [3].

S. Jain (✉)
Jaypee University of Information Technology, Waknaghat, Solan, Himachal Pradesh, India
e-mail: jain.shruti15@gmail.com

© Springer India 2015
L.C. Jain et al. (eds.), *Computational Intelligence in Data Mining - Volume 1*,
Smart Innovation, Systems and Technologies 31, DOI 10.1007/978-81-322-2205-7_25

Two-stage op-amp mainly consists of a cascade of voltage to current and current to voltage stages. The first stage consists of a differential amplifier converting the differential input voltage to differential currents. These differential currents are applied to a current mirror load recovering the differential voltage. The second stage consists of common source MOSFET (transconductance ground gate) converting the second stage input voltage to current. This transistor is loaded by a current sink load, which converts the current to voltage at the output [4, 5].

Operational Transconductance Amplifier (OTA) is a voltage controlled current source (VCCS) [5]. The input stage will be a differential amplifier similar to CMOS differential amplifier using a current mirror load. Since the output resistance of the differential amplifier is reasonably high, a simple differential amplifier may suffice to implement the VCCS [5].

If more gain is required, a second stage consisting of an inverter can be added. If both higher output resistance and more gain are required, then the second stage could be a cascade with a cascade load. The proposed circuit diagram of VCCS based on the block diagram is shown in Fig. 1a. The output is current; to convert current to voltage we can use current mirror circuit shown in Fig. 1b is used at the output side of VCCS circuit. Current mirror circuit increases the gain of the circuit.

There are some electrical parameters which I have calculated for fuzzy system which are as follows [1]: differential input resistance, output resistance, large signal voltage gain (20 $\log_{10} V_o/V_{id}$ in dB), common mode rejection ratio (CMRR) [20 log (A_d/A_{cm}) in dB], slew rate (SR) [max (dV_o/dt) in V/µs].

Fuzzy logic is a powerful problem solver methodology with wide range of applications in industrial control, consumer electronics, management, medicine, expert system and information technology [6–8]. There are five parts of the fuzzy system [9]. Fuzzification of the input variables, application of the fuzzy operator

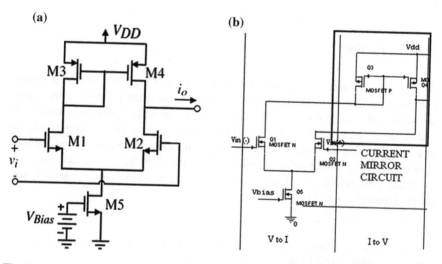

Fig. 1 **a** Proposed circuit diagram of VCCS, **b** current mirror circuit

(AND or OR) in the antecedent, implication from the antecedent to the consequent, aggregation of the consequents across the rules, and defuzzfication [10–13].

In this paper I will discuss regarding the designing of various types of OTA and electronic implementation of fuzzy system using OTA and simulating it with PSPICE software [14, 15].

2 Operational Transconductance Amplifier

The operational transconductance amplifier (OTA) is an amplifier whose differential input voltage produces an output current. Thus, it is a voltage controlled current source (VCCS) [5]. There is usually an additional input for a current to control the amplifier's transconductance. The OTA is similar to a standard operational amplifier in that it has a high impedance differential input stage and that it may be used with negative feedback [16].

2.1 Types of Operational Transconductance Amplifier (OTA)

There are three types of OTA based on the structure: Simple Differential OTA, Fully Differential OTA and Balanced OTA.

2.1.1 Simple Operational Transconductance Amplifier

It is the basic operational transconductance amplifier with only a current-mirror circuit, a differential input and a current sink inverter [17]. Although simple OTA's are single input these are unsymmetrical and are not considered for analysis. We have used differential amplifier as current mirror load as simple OTA (Fig. 1a). The parameters calculated are presented in Table 1.

Table 1 Comparison of three types of OTA's

	Simple	Fully differential OTA [17]	Fully differential OTA (proposed)	Balanced OTA [17]	Balanced OTA (proposed)
Voltage gain (dB)	2.38	1.93	10.8	2.5	55.1
Output resistance (kΩ)	0.0139	0.01	0.8	0.035	8.87
CMRR (dB)	2.2	1.78	10	2.31	51.01
Slew rate (V/μs)	1.5	1.49	2	2.32	2.7
Power dissipation (mW)	0.159	0.132	6.5	57.2	65.2

For parameters: V_{DD} = 4 V, k'_N = 110 μA/V^2, k'_p = 50 μA/V^2, V_{TN} = 0.7 V, V_{TP} = −0.7 V, λ_N = 0.04 V^{-1}, λ_P = 0.05 V^{-1}, −1.5 V < ICMR < 2 V, I have calculated aspect ratio for every transistor. Aspect ratio for differential input [i.e. $(W/L)_1$ and $(W/L)_2$] is 27.6, for current mirror [i.e. $(W/L)_3$ and $(W/L)_4$] circuit is 0.75, for current sink [i.e. $(W/L)_5$] is 0.232.

2.1.2 Fully Differential Operational Transconductance Amplifier

Fully differential operational transconductance amplifier contains simple inversely connected current-mirror pairs. These current mirror pairs are used as active load in the circuit [17]. Along with these current mirror pair we have a differential circuit for which we have differential inputs. This amplifier is fully symmetrical and their main advantage is due to these characteristics only.

The circuit illustrates in Fig. 2 is the **proposed circuit** in which there are few changes compared to circuit and is simple and suitable for implementation. In this circuit instead of current source I1, we have used current sink load. The other difference is using of 2 PMOS current mirror pairs. Different parameters are calculated for both these circuits and are presented and compared in Table 1.

For parameters: V_{DD} = 4 V, k'_N = 110 μA/V^2, k'_p = 50 μA/V^2, V_{TN} = 0.7 V, V_{TP} = −0.7 V, λ_N = 0.04 V^{-1}, λ_P = 0.05 V^{-1}, −1.5 V < ICMR < 2 V, I have calculated aspect ratio for every transistor. Aspect ratio for differential input [i.e. $(W/L)_5$ and $(W/L)_6$] is 27.6, for current mirror [i.e. $(W/L)_1$ and $(W/L)_2$, $(W/L)_3$ and $(W/L)_4$] circuit is 1.5, for current sink [i.e. $(W/L)_9$] is 0.232.

Fig. 2 Proposed fully differential OTA

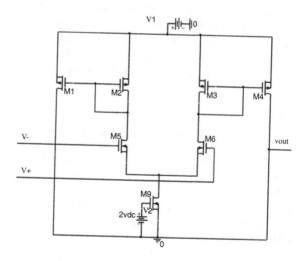

2.1.3 Balanced Operational Transconductance Amplifier (BOTA)

BOTA is the modified circuit of the two discussed above. It contains three current mirrors, a differential circuit and a current sink inverter [17]. These are symmetrical too and have better parameters than the two discussed above. The parameters are shown in the Table 1. The main advantage of BOTA is that filters realized using BOTA provide more simplify structures and perform better performance in higher frequency range than the single output OTA'S. The proposed circuit of BOTA is shown in Fig. 3 that's adding the current sink load and removing current source I1. The other difference is using of 2 PMOS current mirror pairs and 1 pair of NMOS.

For parameters: V_{DD} = 4 V, k'_N = 110 μA/V^2, k'_p = 50 μA/V^2, V_{TN} = 0.7 V, V_{TP} = −0.7 V, λ_N = 0.04 V^{-1}, λ_P = 0.05 V^{-1}, −1.5 V < ICMR < 2 V, I have calculated aspect ratio for every transistor. Aspect ratio for differential input [i.e. $(W/L)_5$ and $(W/L)_6$] is 27.6, for pMOS current mirror [i.e. $(W/L)_1$ and $(W/L)_2$, $(W/L)_3$ and $(W/L)_4$] circuit is 1.5, for current sink [i.e. $(W/L)_7$] is 0.232, for nMOS current mirror [i.e. $(W/L)_1$ and $(W/L)_{2,}$] circuit is 0.48.

I have calculated the electrical parameters of all circuits of OTA and compared it with Sinenchio et al. [17] paper. As observed from the Table 1 the Balanced OTA is the best out of the three in all terms i.e. high CMRR, high voltage gain, high output impedance. Now we implement all the structures using Balanced OTA.

3 Methodology

Our aim is to design and implementation of fuzzy system using OTA. The problem is to estimate risk involved in an engineering project. Let's assume inputs as project funding and project staffing and output as risk. Then, define linguistic variables to

Fig. 3 Proposed balanced OTA

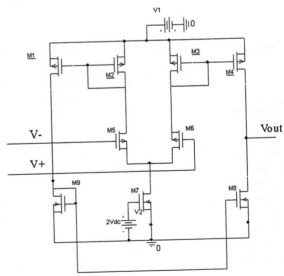

all input and output. The linguistic variables defined for *project funding* are **inadequate**, **marginal** and **adequate**, for *project staffing* are **small**, **medium** and **large**, while, for *risk* are **low**, **medium** and **high** [9]. Now membership functions (S, Z, triangular and trapezoidal) are assigned to every linguistic variable [10]. I have designed all the functions using OTA.

Let's assume the sigma (S) function for all linguistic variables. Then we define the range of each linguistic variable: inadequate as 0–4, marginal as 3–7 and adequate as 6–10. Similarly, for every input and output function. Let's define these ranges in volts so as to implement electronically: inadequate as 1 V, marginal as 2 V and adequate as 3 V. Similarly, I have defined the ranges in volts for rest linguistic variables in 1–3 V range. I have assumed reference voltage as 4 V and the OTA works at 4 V.

Second step was rule composition i.e. IF–THEN statement. There are different operators like AND (min), OR (max) and inverter.

Let's find the electrical parameters of all functions i.e. S, Z, triangular, trapezoidal, MIN and MAX. Table 2 shows the different electrical values of different functions. For parameters: $V_{DD} = 4$ V, $k'_N = 110$ $\mu A/V^2$, $k'_p = 50$ $\mu A/V^2$, $V_{TN} = 0.7$ V, $V_{TP} = -0.7$ V, $\lambda_N = 0.04$ V^{-1}, $\lambda_P = 0.05$ V^{-1}, -1.5 V $<$ ICMR < 2 V, I have calculated aspect ratio for every transistor. Aspect ratio for differential input [i.e. $(W/L)_5$ and $(W/L)_6$] is 27.6, for pMOS current mirror [i.e. $(W/L)_1$ and $(W/L)_2$, $(W/L)_3$ and $(W/L)_4$] circuit is 1.5, for current sink [i.e. $(W/L)_7$] is 0.232, for nMOS current mirror [i.e. $(W/L)_1$ and $(W/L)_2$,] circuit is 0.48.

The similar work has been done using two stage operational amplifiers. Table 3 shows the electrical parameters of all functions i.e. S, Z, triangular, trapezoidal, MIN and MAX. For parameters: $V_{DD} = 5$ V, $k'_N = 110$ $\mu A/V^2$, $k'_p = 50$ $\mu A/V^2$, $V_{TN} = 0.7$ V, $V_{TP} = -0.7$ V, $\lambda_N = 0.04$ V^{-1}, $\lambda_P = 0.05$ V^{-1}, -1.5 V $<$ ICMR < 2 V, I have calculated aspect ratio for every transistor. Aspect ratio for differential input

Table 2 Electrical parameters of S, Z, triangular, trapezoidal, MIN and MAX function using OTA

	S member-ship function	Z-member-ship function	Trapezoidal membership function	Triangular membership function	MAX operator	MIN operator
Voltage gain (dB)	48.3	51.2	57.3	58.5	67.6	13.9
Input resistance (kΩ)	450	289	450	450	999	1,000
Output resistance (kΩ)	8.87	8.77	8.77	8.77	14.7	30.68
CMRR (dB)	66.35	67.7	75.39	76.97	88.9	18.28
Slew rate (V/μs)	2.7	2.5	2.4	2.4	0.6	1.25
Power dissipation (mW)	0.652	0.652	0.72	0.82	198	0.995

Table 3 Electrical parameters of S, Z, triangular, trapezoidal, MIN and MAX function using 2 stage CMOS op-amp

	S member-ship function	Z-member-ship function	Trapezoidal membership function	Triangular membership function	MAX operator	MIN operator
Voltage gain (dB)	37.61	29.54	46.48	46.76	46.48	36.77
Input resistance (kΩ)	240	240	240	240	999	1,000
Output resistance (mΩ)	9.9	9.9	9.9	9.9	0.0177	0.127
CMRR (dB)	34.64	20	40.42	40.66	45.34	29.24
Slew rate (V/µs)	1.5	2.8	2.8	2.8	0.5	0.98
Power dissipation (mW)	82	98.5	82	82	7.23	8.16

[i.e. $(W/L)_1$ and $(W/L)_2$] is 18.4, for current mirror [i.e. $(W/L)_3$ and $(W/L)_4$] circuit is 0.33, for current sink [i.e. $(W/L)_5$ and $(W/L)_7$] is 0.16.

The rules are:

- IF project funding is inadequate *AND* project staffing is medium THEN risk is high.
- IF project funding is marginal *AND* project staffing is large THEN risk is low.
- IF project funding is adequate *AND* project staffing is small THEN risk is medium.

Before THEN and after IF is known as antecedent part and after THEN is consequent part. Antecedent part is rule composition part and consequent part is *implication process* [9]. Third step implication process is of different types but in this paper I have used mamdani implication style. Electronically rules are used as

- IF project funding is inadequate (1 V) *AND* project staffing is medium (2 V) THEN risk is high (3 V).
- IF project funding is marginal (2 V) *AND* project staffing is large (3 V) THEN risk is low (1 V).
- IF project funding is adequate (3 V) *AND* project staffing is small (1 V) THEN risk is medium (2 V).

I have implemented all these rules electronically and, get their output. Fourth step is to aggregate (add) the every output after implication process and then final step is to defuzzify it. In this paper I have used Maximum defuzzification techniques [18].

Figure 4 shows the output after every step. V(32), V(62) and V(89) are the output after mamdani implication process, V(36) is output after aggregating all the rules after implication process. V(65) is the final output i.e. defuzzified output.

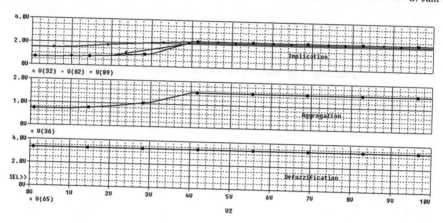

Fig. 4 Final output of fuzzy system

Table 4 Comparison table of the electrical parameters of fuzzy system

	Balanced OTA	Two stage CMOS op-amp
Voltage gain (dB)	36.24	41.47
Input resistance (kΩ)	44.13	6.522
Output resistance (kΩ)	5.163	0.239
CMRR (dB)	47.68	38.39
Slew rate (V/µs)	0.6	1
Power dissipation (W)	197	0.402

The electrical parameter for fuzzy system using balanced OTA and two stage CMOS op-amp is shown in Table 4. As the OTA is a current source, the output impedance of the device is high, in contrast to the op-amp's very low output impedance. Input impedance and CMRR is high. Comparing the Table 4 it is noticed that fuzzy system using OTA is the best.

4 Conclusion

In this paper an attempt has been made for designing the different types of OTA circuits out of which balanced OTA is the best. The paper also gives the designing of the fuzzy system and its electronic implementation using OTA. I have successfully designed and implemented the corresponding functions like S, Z, triangular, trapezoidal, MIN and MAX functions. Along with this I have clubbed all the steps of fuzzy system i.e. fuzzification, rule composition, implication, aggregation and defuzzification process for various rules. I have also calculated its various parameters like slew rate, CMRR, power dissipation, gain, input resistance and output resistance. In the last I have calculated the electrical parameters of fuzzy system using OTA and fuzzy system using two stage CMOS op-amp out of which fuzzy system using OTA is best in all terms except voltage gain. In this paper S membership function is used for all the rules. In future I'll try different membership functions for the rules.

References

1. Gayakwad, R.A.: Op-Amps and Linear Integrated Circuits, 3rd edn. Prentice Hall of India Pvt. Ltd., New Delhi (2002)
2. Sedra, S.S.: Microelectronic Circuits, 5th edn. Oxford University Press, Oxford (2004)
3. Mahalingam, S.A.P., Mamun, M., Rahman, L.F., Zaki, W.M.D.W.: Design and analysis of a two stage operational amplifier for high gain and high bandwidth. Aust. J. Basic Appl. Sci. **6** (7), 247–254 (2012)
4. Patel, P.D., Shah, K.A.: Design of low power two stage CMOS operational amplifier. Int. J. Sci. Res. **2**(3), 432–434 (2013)
5. Allen, P.E., Holberg, D.R.: CMOS Analog Circuit Design. International Student Edition, Oxford (2011)
6. Berkan, R.C., Trubatch, S.L.: Fuzzy System Design Principles, 1st edn. Wiley-IEEE Press, Hoboken-Piscataway (1997)
7. Drankov, D., Hellendoorn, H., Reinfrank, M.: An Introduction to Fuzzy Control. Springer, New York (1993)
8. Yen, J., Langari, R.: Fuzzy Logic: Intelligent Control and Information. Prentice Hall, Upper Saddle River (1998). (United States Edition)
9. Jain, S.: Design and simulation of fuzzy implication function of fuzzy system using two stage CMOS operational amplifier. Int. J. Emerg. Technol. Comput. Appl. Sci. (IJETCAS) **7**(2), 150–155 (2014)
10. Jain, S.: Design and simulation of fuzzy membership functions for the fuzzification module of fuzzy system using operational amplifier. Int. J. Syst. Control Commun. (IJSCC) **6**(1), 69–83 (2014)
11. Zadeh, L.A., Berkeley, C.A.: Online Database. http://radio.feld.cvut.cz/matlab/toolbox/fuzzy/fuzzytu3.html
12. Pant, S.N., Holbert, K.E.: A Fuzzy Logic in Decision Making and Signal Processing [online database]. http://enpub.fulton.asu.edu/powerzone/fuzzylogic/chapter%205/chapter5.html
13. Pant, S.N., Holbert, K.E.: A Fuzzy Logic in Decision Making and Signal Processing [online database] http://enpub.fulton.asu.edu/powerzone/fuzzylogic/chapter%206/frame6.htm
14. Nguyen, M.A.: PSPICE Tutorial, Class: Power Electronic 2. Colorado State University Student
15. Rashid, M.H.: Introduction to PSICE Using OrCAD for Circuits and Electronics, 3rd edn. Prentice-Hall, Inc., Upper Saddle River (2009)
16. Sorkhabi, M.M., Toofan, S.: Design and simulation of high performance operational transconductance amplifier. Can. J. Electr. Electron. Eng. **2**(7), 275–281 (2011)
17. Sinenchio, E.S., Martinez, J.S.: CMOS transconductance amplifiers, architectures and active filters: a tutorial. IEEE Proc. Circuits Devices Syst. **147**(1), 3–12 (2000)
18. Jain, S.: Implementation of fuzzy system using operational transconductance amplifier for ERK pathway of EGF/Insulin leading to cell survival/death. J. Pharm. Biomed. Sci. **4**(8), 701–707 (2014)

DFT and Wavelet Based Smart Virtual Measurement and Assessment of Harmonics in Distribution Sector

Haripriya H. Kulkarni and D.G. Bharadwaj

Abstract This paper presents MATLAB based VI measurement for harmonics using DFT and WT. Results of both the techniques are compared and validated against the standard power analyzer. It is proved that DWT based VI have promising performance. For Data collection fifteen field visits are done to IT Industries, Steel Industries, Paper Industries, workshops etc. Sample of the readings are considered for analysis purpose. It is observed that the current harmonics generated are beyond the tolerable limits and many literatures are available in the area of Harmonic measurements but they do not have the capacity to identify the limits of harmonics as per the standards. In this paper GUI is developed in connection with MATLAB simulation to assess the harmonics and identify the normal and abnormal presence of harmonics with green and red colors respectively. Through this colour coding technique an unskilled worker can easily identify the objectionable presence of harmonics. It creates the awareness in the utility for measurement and control of harmonics which in turn helps to improve the PQ.

Keywords Discrete Fourier transform · Wavelet transform · Power quality · Graphical user interface · Virtual instrument

1 Introduction

In the past few years the use of nonlinear devices which are prone to draw harmonic current has been increased exponentially. Few examples are Computers, Induction furnace, cranes, welding machines, power electronics driven convertors [1–3] etc. Out of these loads Computers, Fluorescent lamps, UPS etc. are rich in third

H.H. Kulkarni (✉) · D.G. Bharadwaj
B.V.D.U.C.O.E., Pune, Maharashtra, India
e-mail: electricalmcoe@yahoo.com

D.G. Bharadwaj
e-mail: dgbharadwaj@gmail.com

© Springer India 2015
L.C. Jain et al. (eds.), *Computational Intelligence in Data Mining - Volume 1*,
Smart Innovation, Systems and Technologies 31, DOI 10.1007/978-81-322-2205-7_26

harmonics [1, 2]. Some other loads such as transformer, Induction motor or conventional machines give rise to generation of triplen harmonics due to iron saturation [3]. Presence of harmonics have many detrimental effects such as extra power loss in Distribution transformer, Reduced life expectancy, loss of reliability, increased operating cost, overheating, machine failures, maloperation of switchgear and inaccurate power metering [4].

The Assessment of harmonics is done by referring to IEEE 519 standard [5–7], IEC 61000-2-2 [8] or EN 50160 standard [9]. Till today the assessment is done by an expert experienced person who observes the harmonics using standard equipment available for measurement of harmonics and compares it with the harmonic standard. In fact it is difficult to identify the abnormality just by visual inspection. In this paper the harmonic assessment is proposed through a smart concrete solution which uses the threshold value prescribed by the standards and represents the results in the form of Bar chart with colour code. Colour represents the presence of harmonics in safe or below the threshold zone where as Red color is for danger or abnormal zone. In preparation of this paper actual site visits are done at Steel Industries are considered for analysis which are rich in fifth and seventh harmonics for multiple periods over peaks and low load times within 24 h of a day. The current pattern observed is mathematically generated in MATLAB then it is simulated using DFT and DWT. The results of Individual harmonic and THD are compared with each other and also with Actual measurements. It is found that proposed (Db40) DWT based VI gives accurate results with Assessment of harmonics as per the selected standard.

2 Field Measurements at Steel Industry

Measurement of harmonics has been done by using Advanced Harmonic Analyzer "Yokogawa" (model CW240). Wiring connections for this harmonic analyzer is 3 phase 4 wire. C.T. Ratio is 50/5.

2.1 Load Current in Three Phases

Table 1 lists Average currents in all the three phases i.e. R, Y and B of steel industry over an interval of 2 h.

Readings were taken for 24 h with a step size of 10 min. This steel industry is located in Maharashtra (India). Figure 1 shows variations in Average current in all 3 phases with respect to time. From this graph nature of load pattern of the steel industry is observed which is highly fluctuating. Switching is the main reason for generation of harmonics in the furnace. This switching takes place at very high frequencies; It leads to generation of harmonics.

Table 1 Average currents in all the three phases

Time (h)	I_R_AVE (amp)	I_Y_AVE (amp)	I_B_AVE (amp)
15:06:50	0.54	0.51	0.39
17:06:50	1.21	1.19	0.94
19:06:50	1.90	1.78	1.34
21:06:50	2.00	1.95	1.86
23:06:50	2.32	1.81	1.82
01:06:50	2.30	1.80	1.78
03:06:50	1.98	1.51	1.87
05:06:50	1.92	1.49	1.84
07:06:50	0.93	0.96	0.78
09:06:50	1.72	1.65	1.24
11:06:50	1.96	1.86	1.47
13:06:50	1.54	1.46	1.10

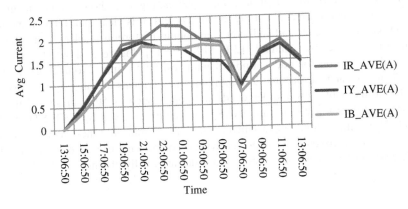

Fig. 1 Graph of average currents versus time

Table 2 shows the Active power consumption in steel Industry and it is found that due to the nature of non-linear load it is varying widely. Out of this total active power the power consumed due to Harmonic content is unbilled power.

Table 3 lists % THD in current in all three phases.

Table 2 Average power consumption in steel industry

Time (h)	13:06:50	15:06:50	17:06:50	19:06:50	21:06:50	11:06:50 PM
P_{avg} (W)	–	80	170	300	320	340
Time (h)	03:06:50	05:06:50	07:06:50	09:06:50	11:06:50	13:06:50
P_{avg} (W)	320	320	120	260	310	240

Table 3 % I_{THD} in all three phases

Time (h)	I_{RTHD} (%)	I_{YTHD} (%)	I_{BTHD} (%)
13:06:50	30.6	22.7	0.00
15:06:50	48.6	53.1	70.2
17:06:50	44.8	47.6	58.6
19:06:50	42.9	45.4	44.4
21:06:50	42.6	42.5	46.8
23:06:50	45.0	46.1	44.4
01:06:50	45.8	46.2	45.4
03:06:50	46.3	43.6	37.6
05:06:50	44.9	41.7	36.8

2.2 I_{THD} in All the Three Phases

Table 3 lists % THD in current in all three phases. Figure 2 shows Variations in % I_{THD} with respect to time. From Fig. 3 and Table 4, it is clear that, total harmonic distortion Current level and Individual Harmonic Distortion level exceeds above standard limits set by standard IEEE 519-1992 [5]. Figure 3 shows Individual harmonic distortion for odd harmonics with respect to time. From Table 4 and Fig. 4, it is clear that, 3rd and 9th harmonics i.e. Triplens are absent in observed steel industry. Only lower ordered odd harmonics such as 5th, 7th and 11th are present and their magnitudes are above tolerance limits given by IEEE STD 519-1992 [5]. So filters should be installed to eliminate or minimize these harmonics, so that their ill effects on various power system components can be eliminated. All higher odd ordered harmonics above 9th order are totally absent.

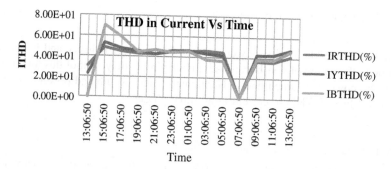

Fig. 2 Graph of % I_{THD} in all three phases versus time

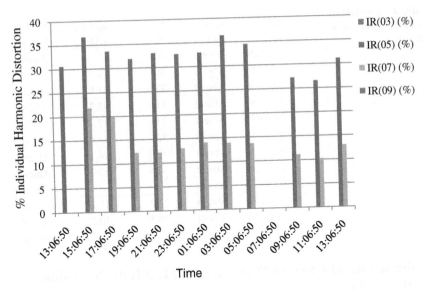

Fig. 3 Graph of % individual harmonic distortion in currents versus time

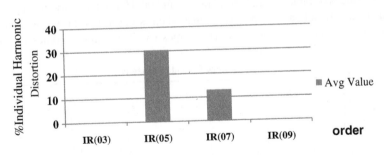

Fig. 4 Individual harmonic distortions versus order

2.3 Individual Harmonic Distortion in Currents in Three Phases

Table 4 lists percentage individual harmonic distortion in current, for 3rd, 5th, 7th, 9th, 11th harmonics in R Phase.

Figure 4 shows Individual harmonic distortion for odd harmonics with respect to time. From Table 4 and Fig. 4, it is clear that, 3rd and 9th harmonics i.e. Triplens are absent in observed steel industry. Only lower ordered odd harmonics such as 5th, 7th and 11th are present and their magnitudes are above tolerance limits given by IEEE STD 519-1992 [5]. So filters should be installed to eliminate or minimize these harmonics, so that their ill effects on various power system components can be eliminated. All higher odd ordered harmonics above 9th order are totally absent.

Table 4 % individual harmonic distortion in currents

Time (h)	I_R (03) (%)	I_R (05) (%)	I_R (07) (%)	I_R (09) (%)
13:06:50	Absent	30.6	–	Absent
15:06:50	Absent	36.7	21.8	Absent
17:06:50	Absent	33.6	19.9	Absent
19:06:50	Absent	31.9	12.3	Absent
21:06:50	Absent	33.0	12.2	Absent
23:06:50	Absent	32.7	13.0	Absent
01:06:50	Absent	32.9	14.1	Absent
05:06:50	Absent	34.5	13.7	Absent
07:06:50	Absent	–	–	Absent
09:06:50	Absent	27.2	11.1	Absent
11:06:50	Absent	26.5	10.2	Absent
13:06:50	Absent	31.1	13.0	Absent

3 Simulation of Current Waveform in Steel Industry Using DFT in MATLAB

Average values of Fundamental, 3rd, 5th, 7th and 9th Harmonics currents in phase R are calculated. Resultant current waveform is taken as a distorted sine wave function x (t) containing Fundamental +0 % 3rd Harmonics +32.75 % 5th Harmonics +14.02 %

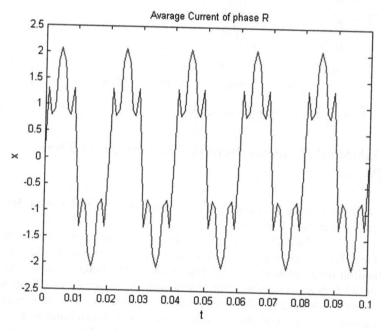

Fig. 5 Average current of phase R

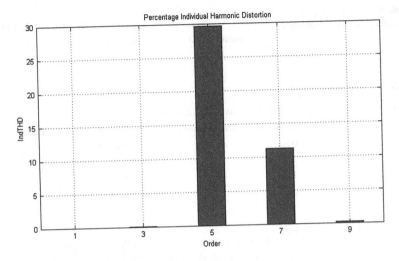

Fig. 6 Simulation result in MATLAB

7th Harmonics +0 % 9th Harmonics as per IEC STD 61000 [6]. Here 0, 32.75, 14.02 and 0 % are average values of 3rd, 5th, 7th and 9th Harmonics currents actually present in a steel industry under study. Similar analysis can be done on currents present in the remaining phases i.e. and B. When x(t) is plotted in the MATLAB it is totally distorted as shown in Fig. 5.

X(t) is simulated in the MATLAB using DFT. Result of simulation is shown in Fig. 6.

Results of Actual field measurements shown in Fig. 5 and simulation shown in Fig. 7 are compared and tabulated in Table 5.

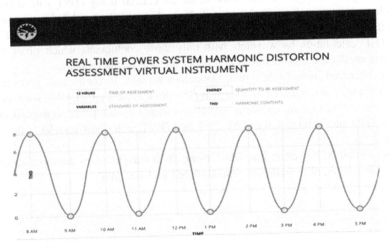

Fig. 7 Graphical user interface developed

Table 5 Average value of individual harmonic distortion

Order	I_R (03)	I_R (05)	I_R (07)	I_R (09)
Avg. value	0	30.3	13	0

Table 6 Comparative results

Harmonic order	% individual harmonic distortion	
	Actual field measurements	Simulation in MATLAB
3	0	0.0181
5	30.3	29.62
7	13	11.25
9	0	0.163

Table 7 THD calculation by various wavelets

Wavelet	Db40	Sym40	Coif5	Bior6.8	Rbio	Dmey
THD (%)	35.03	34.64	27.92	32.28	33.52	31.85

Table 8 Comparison of DFT

Actual field measurement	DFT	DWT With db40 wavelet
35.68	31.69	35.03

4 Simulation of Current Waveform in a Steel Industry Using DWT in MATLAB

X(t) shown in the Fig. 6 is simulated in the MATLAB using DWT with different wavelets such as such as db, sym, coif, bior, rbio and dmey. All these wavelets with its all available coefficients are used for the harmonic analysis using DWT [6]. But THD calculation by wavelets with only those coefficients which gives most accurate results are considered in this paper and are tabulated below.

It is observed from Table 6 that, Dabuchies wavelet 40 coefficients gives more accurate results as compared to other wavelets. So it is the most suitable wavelet for harmonic analysis of the nonlinear signal considered in this paper. Comparison of actual THD and THD calculated by DFT and DWT with db40 wavelet is done in Table 7.

From Table 8, it is clear that, DWT using db40 wavelet gives the most accurate results as compared to tradition computational tool i.e. DFT.

Fig. 8 Harmonic assessment
by colour-code

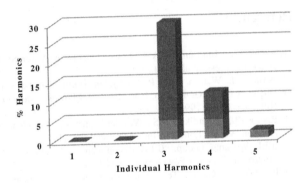

5 Graphical User Interface

This is developed to access the harmonics observed in VI using DFT and DWT as shown in Fig. 7. The result of this is representation of individual harmonics and total harmonic distortion shown below with a colour code. The comparative bar chart shows the identification of abnormal and normal harmonic level presence with RED and GREEN colour respectively as shown in Fig. 8.

6 Conclusion

It is observed through site visits that the current harmonics generated are beyond the tolerable limits. Due to harmonic presence true power factor gets reduced and power consumption only due to harmonics is up to 20 % in Harmonic rich loads. Hence the proposed solution of measurement will definitely inculcate the awareness regarding the measurement and controlling of harmonics.

The proposed Discrete Wavelet based Virtual measurement and its Assessment gives simple and accurate results. It is an easy way of identifying the presence of harmonics in Distribution sector. In addition to it any of the harmonic standard IEC 61000-2-2 or IEEE 519-1996 or EN 50160 IEEE can be selected for its assessment as per users wish. Hence the proposed DWT based VI is the offline but cost effective solution in Distribution sector for measurement of harmonics and its assessment. Developed Virtual Instrument will cater the need of accurate measurement and its assessment. Through this colour coding technique an unskilled worker can easily identify the objectionable presence of harmonics. Advanced tools like WPT [8], ANN, and Hanning window framework can also be used for performance optimization [10, 11]. Harmonics can be mitigated by use of appropriate Active or Passive filter.

Acknowledgments The project work is being carried out in Bharti Vidyapeeth Deemed University COE, Pune. The Authors wish to thank authorities of BVDUCOE, Pune for granting permission to publish the work. Special Thanks to MSEDCL and MSETCL for collection of data at Consumer end.

References

1. Fuchs, E.F., Masoum, M.A.S.: Power Quality in Power Systems and Electrical Machines. Academic Press, Waltham (2011)
2. Arrilaga, J., Watson, N.R.: Power System Harmonics, 1st edn. Wiley, Chichester (2004)
3. Dugan, R.C., McGrangahan, F.M.: Electrical Power System Quality, 2nd edn. Tata McGraw-Hill, New York (1996)
4. Rao, R.M., Bopardikar, A.S.: Wavelet Transforms-Introduction to Theory and Applications. Addison-Wesley Longman Inc., Massachusetts (1998)
5. IEEE Working Group.: IEEE Recommended Practices and Requirements for Harmonic Control in Electrical Power Systems (1992)
6. IEC Working Group.: Testing and Measurement Techniques—General Guide on Harmonics and Interharmonics Measurement and Instrumentation, for Power Supply Systems and Equipment Connected Thereto, 2002. IEC 61000-4-7 (2002)
7. Waje, V.B., Gudaru, U.: Analysis of harmonics in power system using wavelet transform. In: IEEE Students Conference on Electrical, Electronics and Computer Science, pp. 1–5 (2012)
8. Chen, C., Yang, X., Xie, W.: A wavelet packet transform power harmonic detection method based on DSP and LabVIEW. In: IEEE Information and Automation (ICIA), pp. 2219–2214 (2010)
9. Alles, M.L., Schrimpf, R.D., Fleetwood, D.M., Pasternak, R., Tolk, N.H., Standley, R.W.: Experimental evaluation of second harmonic generation for non-invasive contamination detection in SOI wafers. In: Advanced Semiconductor Manufacturing Conference (ASMC), pp. 1–5 (2006)
10. Li, T., Chen, Y., Li, G.M.: An optimized method for electric power system harmonic measurement based on back-propagation neural network and modified genetic algorithm. In: Power Electronics Systems and Applications, pp. 1–5 (2009)
11. Xie, W., Yang, X.: A power harmonic measurement system based on wavelet packet transform and ARM9. In: Advanced Computer Control (ICACC), vol. 3, pp. 445–450 (2010)
12. Koch, A.S., Myrzik, J.M.A., Wiesner, T., Jendernalik, L.: Harmonic measurement and modeling for mass implementation of nonlinear appliances. In: IEEE Power and Energy Society General Meeting, pp. 1–6 (2012)

Addressing Object Heterogeneity Through Edge Cluster in Multi-mode Networks

Shashikumar G. Totad, A. Smitha Kranthi and A.K. Matta

Abstract The oceans of data generated by social media have become a goldmine to researchers in the data mining domain. Discovering actionable knowledge by extracting latent patterns has many advantages. One of the utilities of mining social data is learning collective behavior which helps in taking well informed decisions pertaining to humanitarian assistance, disaster relief and such real world applications. In multi mode while studying the collective behavior using edge centric approach, object heterogeneity is a problem. In this paper, we propose a scheme temporal regularized evolutionary multimode clustering algorithm which can address object heterogeneity in social media with multi-mode more effectively. With this the prediction performance of collective behavior is improved further. We built a prototype application to demonstrate the proof of perception. The empirical results are encouraging and our approach can be used in real world applications that mine social media data explicitly.

Keywords Social networking · Data mining · Social dimensions · Collective behavior

1 Introduction

Due to the advancements in technologies, virtual communities have been realized as a new phenomenon that empowers people to get together online. Social media provides plethora of opportunities to gain business intelligence by studying human

S.G. Totad (✉) · A. Smitha Kranthi
Department of Computer Science and Engineering, GMRIT, Razam, Andhra Pradesh, India
e-mail: shasikumar.gt@gmrit.org

A. Smitha Kranthi
e-mail: alapatismitha1@gmail.com

A.K. Matta
Department of Mechanical Engineering, GMRIT, Razam Andhra Pradesh, India
e-mail: anilnatas@rediffmail.com

L.C. Jain et al. (eds.), *Computational Intelligence in Data Mining - Volume 1*,
Smart Innovation, Systems and Technologies 31, DOI 10.1007/978-81-322-2205-7_27

interactions and obtaining collective behavior of people of various walks of life who participate in virtual computing. Social network analysis has become important in many fields such as targeted marketing, intelligent analysis, epidemiology, and sociology. An important study which has given much importance in social media is to predict collective behavior of some individuals of a group provided the knowledge of some people of the same group. In social networking sites behavior of one actor is usually influenced by interest of other actor. In this paper, we focus on understanding actor's behavior in multimode networks. Behavior study is essential to improve actor's skill and to know what exact the interest of user is. It is very important for so many businesses to know the interest of actors for enhancing their business. This work contains different aspects for understanding actor's behavior in online social networking sites.

2 Related Works

Mining social media content has been around for some years. However, this kind of research started long back. For instance network classification was explored in [1]. In similar fashion relational learning was focused by Getoor and Taskar [1]. Conventional data mining is different from that of social data mining. The datasets used for network instances is not uniformly distributed. In such datasets objects have relationships and correlating with neighboring data objects is done with an assumption known as "Markov dependency assumption". It does mean that label of one network node relies on one of more labels of neighboring nodes. Classification is one of the data mining techniques for supervised learning which is used to classify network objects. For instance in [1] a weighted vote relational classifier [2] was built which showed good performance in classification against benchmark datasets.

A network contains heterogeneous relations and only capturing local dependency is possible with Markov assumption. For this reason in [3, 4] class labels are used with latent groups. Similar kind of research was done in [5] to explore heterogeneous relationships and differentiate the same by extracting social affiliations and dimensions from social media data. Soft clustering scheme was suggested by them in order to explore community membership in social dimensions. Social dimensions extracted from social media data are known as features and data mining technique such as Support Vector Machine (SVM) can be used to classify such data. The social dimensions approach is better than other approaches to explore social media data based on collective inference. Soft clustering techniques can be used to achieve this which is based on modularity maximization [6], spectral clustering [7], and matrix factorization. To solve the same problem other methods such as probabilistic methods came into existence [8–9]. A drawback of soft clustering is that the social dimensions are naturally dense and throw challenges pertaining to computational overhead. Palla et al. [10] proposed this method by name "clique percolation method" which is used to find dense communities which are overlapping. This method has two fundamental phases. They are finding all cliques in graph, and finding connections between cliques in order to discover various communities. Similar idea was explored in [11] where all maximal cliques are found in a network through hierarchical clustering. Overlapping communities are handled by the method proposed by Newman–Girvan [12] which is an extension to the method proposed by Gregory [13]. The Newman–Girvan method recursively removes edges in order to generate disconnected components. The method removes only edges with high between's among them. Finally it generates output consisting of non-overlapping communities. Node splitting is another feature added by Gregory besides removing edges. Their algorithm splits nodes recursively where multiple communicates reside and remove edges that are used to interconnect communities. These methods list out all possible cliques and choose the paths which are very short in the network where computational cost is very high in case of large scale networks.

For finding overlapping communities, graph partition algorithms were explored in [14, 15] that work on line graphs. However, just construction of ling graph is not sufficient as it prohibits functional with large scale networks. Scalable approaches are required in order to deal with huge number of networked objects present in the data of social media. Recently in [16] K-means was used to achieve partitioning of edges for disjoint sets. They also proposed a variant of K-means to hand scarcity of data in order to handle huge number of edges effectively. In order to accelerate the process more advanced data structures can be exploited [17, 18]. When the data loaded into RAM is very high, other variants of K-means such as distributed k-means [19], scalable k-means [20] and online k-means [21] can be used.

3 Preliminaries

This section familiarizes the reader about preliminaries required to understand the problem solved in this paper. The details provided here include social networking, social dimension, affiliations, communities, collective behavior, sparse social dimensions, edge clusters, and object heterogeneity. Social networking refers to online or virtual community including friends, relatives, classmates, family members, researchers and so on who can have a platform to get together and exchange views. Social dimension refers to the relationship an actor has with others. Affiliation refers to a group of nodes in social network to which an actor belongs. One actor may belong to multiple affiliations. Community refers to a set of edges in the network. Collective behavior refers to the result of a process where the behavior of some objects is known based on the other objects in the same affiliation. Sparse social dimension refers to the social dimension where density is very low. Edge cluster refers to a cluster of objects that is connected to another in the network. Sample edge clusters are presented in Fig. 1 with a toy example.

Each object in the network is associated with other objects. One actor can have multiple affiliations. The affiliations can be represented with modularity

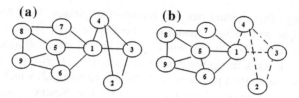

Fig. 1 **a** Sample network, **b** edge clusters

maximization. The actors, modularity maximization and edge partition details are presented in Table 1. It does mean that the table shows social dimensions in the Toy example.

Multi-mode network is the network which essentially contains multiple and heterogeneous social actors. There are interactions among these actors through which communications can be identified or evaluated over a period of time. Figure 2 shows a sample multi-mode network which is based on online marketing scenario. Such multi-mode networks exhibit object heterogeneity.

From Fig. 2, multiple modes are involved in the same network. The resultant network obtained is known as multi-mode network. The queries, users, and ads are intervened with seemingly perfect coexistence. This kind of network shows more

Table 1 Social dimensions of Toy example

Actors	Modularity maximization	Edge	Partition
1	−0.1185	1	1
2	−0.4043	1	0
3	−0.4473	1	0
4	−0.4473	1	0
5	0.3093	0	1
6	0.2628	0	1
7	0.1690	0	1
8	0.3241	0	1
9	0.3522	0	1

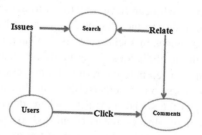

Fig. 2 Illustrates a multi-mode network with three actors

object heterogeneity which has to be handled. Object heterogeneity is explored in [16] but the approach is sensitive to number of social dimensions. Handling object heterogeneity has many real world utilities social mining domain.

4 Proposed Scheme to Address Object Heterogeneity

Scalable learning of collective behavior is explored in [16] in which a variant of K-means is sued for Edge Clustering. The generated clusters and used to mine the collective behavior. Given knowledge of some actors in a group predicting the behavior of other actors in the same group is known as learning collective behavior. Scalability of this approach is achieved in [16] by using sparse social dimensions. However their solution is sensitive to number of social dimensions. To overcome this drawback, in this paper, we proposed a new scheme that handles object heterogeneity more gracefully. The proposed scheme is presented as pseudo code in Fig. 3. In multimode network with m types of actors $X_i = \{x_1^i, x_2^i, x_3^i, \ldots, x_n^i\}$ $i = 1, 2, \ldots, m$ where n_i is the number of actors for X_i. The interaction between the actors is approximated by the interaction between groups in the form of latent cluster membership for a group interaction and transpose of a matrix. A d-dimensional network is represented as $R = \{R_1, R_2, \ldots, R_d\}$ where R_i represents the interactions among actors in the ith dimension. Ideally, the interaction between actors can be approximated by the interactions between groups in the following form:

$$R_{i,j}^t \sim C^{(i,t)} A_{i,j}^t \left(C^{(j,t)} \right)^T$$

where $C^{(i,t)} \in \{0, 1\}^{n_i \times k_i}$ denotes latent cluster membership for X_i.

Fig. 3 Learning collective behavior and addressing object heterogeneity

Algorithm Name: Learning Collective Behavior
Inputs: multimode networks, labels of some nodes, social dimensions
Outputs: Labels of unlabeled nodes
Step1:
Convert multimode network into edge centric view
Step 2:
Perform clustering using K-means
Step 3:
Construct social dimensions
Step 4:
Construct classifier that can handle object heterogeneity
Step 5:
Use the classifier that can predict the dimensions which have not been labeled

As can be seen in Fig. 3, the algorithm is used to cluster objects in multi-mode network. Temporal information has been for analyzing the multi-mode network. The results revealed that the algorithm is scalable and solves problems with complex correlations in the social media data. The Learning collective behavior is used to access files more frequently for a user.

5 Experimental Results

We built a prototype application to demonstrate the efficiency of the proposed system. It is a web based application which facilitates social networking thus the synthetic data is generated. Thus the generated data is used for experiments. It also supports real time data sets which are compatible or tailored to meet the requirements of the application. The environment used to build the application is a PC with 4 GB RAM, core 2 dual processor running Windows 7 operating system. The experiments are made in terms of both rating of collective behavior of users in social networking and also the sensitivity to object heterogeneity.

As can be seen in Fig. 4, the overall collective behavior of actors involved in multimode network is presented. Audio files usage is highest and the file usage is least. The text file usage is second highest. Video file usage is the third highest. The second least is the image file usage as per the interactions and dimensions available in the dataset. More importantly the experimental results of sensitivity to object heterogeneity are presented in Fig. 5. The results of our approach are also compared with that of Edge clustering approach explored in [16].

As can be seen in Fig. 4, it is evident that the existing Edge Clustering algorithm explored in [16] has more sensitivity to dimensionality. The proposed algorithm outperforms it as it is suitable for multi-mode networks where object heterogeneity is high.

Fig. 4 Overall collective behavior of actors

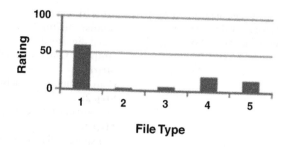

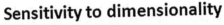

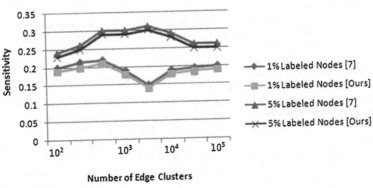

Fig. 5 Sensitivity to dimensionality

6 Conclusion

In this paper we study the object heterogeneity problem in social networking. Learning collective behavior has very important utility in real world applications. However, the existing schemes for extracting collective behavior are not scalable. To overcome the problem recently Tang et al. [16] proposed an Edge Clustering algorithm that demonstrated the scalable learning of collective behavior. However, this algorithm is sensitive to object heterogeneity or dimensionality. Addressing this problem will give more meaningful results that can be used in various real world business applications. As object heterogeneity results in multi-mode network, in this paper we focus on multi-mode networks in order to extract collective behavior from social interactions with multiple modes. We built a prototype application to demonstrate the proof of concept. The empirical results revealed that the proposed scheme is very useful in extracting useful knowledge from social mining.

References

1. Getoor, L., Taskar, B. (eds.): Introduction to Statistical Relational Learning. MIT Press, Cambridge (2007)
2. Macskassy, S.A., Provost, F.: A simple relational classifier. In: Proceedings of Multi-Relational Data Mining Workshop (MRDM) at the Ninth ACM SIGKDD International Conference on Knowledge Discovery and Data Mining (2003)
3. Xu, Z., Tresp, V., Yu, S., Yu, K.: Nonparametric relational learning for social network analysis. In: KDD'08: Proceedings of Workshop Social Network Mining and Analysis (2008)
4. Neville, J., Jensen, D.: Leveraging relational auto correlation with latent group models. In: MRDM'05: Proceedings of Fourth International Workshop Multi-Relational Mining, pp. 49–55 (2005)

5. Tang, L., Liu, H.: Relational learning via latent social dimensions. In: KDD'09: Proceedings of 15th ACM SIGKDD International Conference on Knowledge Discovery and Data Mining, pp. 817–826 (2009)
6. Newman, M.: Finding community structure in networks using the eigenvectors of matrices. Phys. Rev. E (Stat. Nonlin. Soft Matter Phys). **74**(3), 036104. http://dx.doi.org/10.1103/PhysRevE.74.036104 (2006)
7. Luxburg, U.V.: A tutorial on spectral clustering. Stat. Comput. **17**(4), 395–416 (2007)
8. Airodi, E., Blei, D., Fienberg, S., Xing, E.P.: Mixed membership stochastic blockmodels. J. Mach. Learn. Res. **9**, 1981–2014 (2008)
9. Fortunato, S.: Community detection in graphs. Phys. Rep. **486**(3–5), 75–174 (2010)
10. Palla, G., Derényi, I., Farkas, I., Vicsek, T.: Uncovering the overlapping community structure of complex networks in nature and society. Nature **435**, 814–818 (2005)
11. Shen, H., Cheng, X., Cai, K., Hu, M.: Detect overlapping and hierarchical community structure in networks. Phys. A: Stat. Mech. Appl. **388**(8), 1706–1712 (2009)
12. Newman, M., Girvan, M.: Finding and evaluating community structure in networks. Phys. Rev. E **69**, 026113. http://www.citebase.org/abstract?id=oai:arXiv.org:cond-mat/0308217. 2004. 1090 IEEE Trans. Know. Data Eng. **24**(6) (2012)
13. Gregory, S.: An algorithm to find overlapping community structure in networks. In: Proceedings of European Conference on Principles and Practice of Knowledge Discovery in Databases (PKDD), pp. 91–102. http://www.cs.bris.ac.uk/Publications/pub_master.jsp?id=2000712 (2007)
14. Evans, T., Lambiotte, R.: Line graphs, link partitions and overlapping communities. Phys. Rev. E **80**(1), 16105 (2009)
15. Ahn, Y.Y., Bagrow, J.P., Lehmann S.: Link communities reveal multi-scale complexity in networks. http://www.citebase.org/abstract?id=oai:arXiv.org:0903.3178 (2009)
16. Tang, L., Wang, X., Liu, H.: Scalable learning of collective behavior. IEEE Trans. Knowl. Data Eng. **24**(6) (2012)
17. Kanungo, T., Mount, D.M., Netanyahu, N.S., Piatkom, C.D., Silverman, R., Wu, A.Y.: An efficient k-means clustering algorithm analysis and implementation. IEEE Trans. Pattern Anal. Mach. Intell. **24**(7), 881–892 (2002)
18. Bentley, J.: Multidimensional binary search trees used for associative searching. Comm. ACM **18**, 509–175 (1975)
19. Jin, R., Goswami, A.Y., Agrawal, G.: Fast and exact out-of-core and distributed k-means clustering. Knowl. Inf. Syst. **10**(1), 17–40 (2006)
20. Bradley, P., Fayyad, U., Reina, C.: Scaling clustering algorithms to large databases. In: Proceedings of ACM Knowledge Discovery and Data Mining (KDD) Conference (1998)
21. Sato, M., Shii, S.: On-line EM algorithm for the normalized Gaussian network. Neural Comput. **12**, 407–432 (2000)
22. Wasserman, S., Faust, K.: Social Network Analysis Methods and Applications. Cambridge University Press, Cambridge (1994)
23. Baumes, J., Goldberg, M., Ismail, M.M., Wallace, W.: Discovering hidden groups in communication networks. In: 2nd NSF/NIJ Symposium on Intelligence and Security Informatics (2004)
24. Meyers, M.N.L.A., Pourbohloul, B.: Predicting epidemics on directed contact networks. J. Theor. Biol. **240**, 400–418 (2006)
25. Tang, L., Liu, H.: Toward predicting collective behavior via social dimension extraction. IEEE Intell. Syst. **25**(4), 19–25 (2010)
26. Tang, L., Liu, H., Zhang, J., Nazeri, Z.: Community evolution in dynamic multi-mode networks. KDD'08 (2008)
27. Yu, K., Yu, S., Tresp, V.: Soft clustering on graphs. In: Proceedings of Advances in Neural Information Processing Systems (NIPS) (2005)

Microstrip Patch Antenna with Defected Ground Plane for UWB Applications

**Bharat Rochani, Rajesh Kumar Raj, Sanjay Gurjar
and M. SantoshKumar Singh**

Abstract In this paper a compact microstrip antenna with substrate 22 mm × 27 mm using microstrip line feed is proposed. Two stubs are introduced in half ground plane to enhance the bandwidth of the antenna. Antenna covers the frequency range from 1.65 to 10.68 GHz. The antenna is proposed to achieve broadband characteristics. The antenna is fabricated on PCB and tested on Vector Network Analyzer. This antenna can be used for Ultra Wide Band applications.

Keywords UWB · Defected ground plane · VNA · PCB · VSWR

1 Introduction

In 2002, the release of the Ultra Wide Band (UWB) for commercial communications by the Federal Communication Commission (FCC, USA) excited interests in the construction of UWB communication system, and since then, continuous efforts have been made to find the good property of UWB antennas [1]. In this article, an UWB microstrip patch antenna [2–4] with defected ground plane [5–7] has been presented. The proposed antenna is designed for Ultra Wide Band (UWB)

B. Rochani (✉) · R.K. Raj · M.S. Singh
Department of Electronics and Communication Engineering,
Government Engineering College, Ajmer 305001, Rajasthan, India
e-mail: bharat.rochani@gmail.com

R.K. Raj
e-mail: raj_raj2002@rediffmail.com

M.S. Singh
e-mail: eceait.2011@gmail.com

S. Gurjar
Department of Electronics and Communication Engineering,
Bhagwant University, Ajmer 305001, Rajasthan, India
e-mail: ersanjay86@yahoo.in

© Springer India 2015
L.C. Jain et al. (eds.), *Computational Intelligence in Data Mining - Volume 1*,
Smart Innovation, Systems and Technologies 31, DOI 10.1007/978-81-322-2205-7_28

applications. The proposed antenna having a rectangular top patch and fed by microstrip line feed technique to increase the bandwidth ground is defected. The antenna is well matched within the UWB frequency band. Parametric study of the proposed antenna parameters has also been done.

For the design studied here, the radiator and ground plane are etched on the opposite sides of a Printed Circuit Board (PCB) made of material FR4_epoxy with dielectric constant of 4.4 and substrate thickness of 1.6 mm. The size of substrate and ground considered as 27 mm × 22 mm. The antenna has been feeded with microstrip line and has width and length of 1.9 mm × 8 mm.

2 Parametric Study of the Proposed Antenna Design

The microstrip patch antenna with defected ground plane has been designed in three steps shown in Fig. 1. In Step 1, simple rectangular patch antenna has been designed to resonate at 5.5 GHz by using the standard equations [1]. Ground plane has the dimensions of 2 mm × 22 mm. In Step 2, stub is introduced along feed side to increase the current path in the ground plane. As a result bandwidth is improved. The length and width of stub is 3 mm × 12 mm. In the final step, two rectangular stubs have been introduced to increase the excitation of resonant modes. Result of return loss plots for various steps shown in Fig. 2 and Table 1.

2.1 Design of Optimized Patch Antenna with Defected Ground Plane at 1.65–10.68 GHz

The final antenna geometry with all design parameters has been shown in Fig. 3. The antenna design parameters to resonate at 1.65–10.68 GHz have been shown in Table 2.

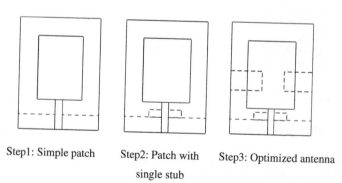

Step1: Simple patch   Step2: Patch with   Step3: Optimized antenna
                         single stub

Fig. 1 Development of the design

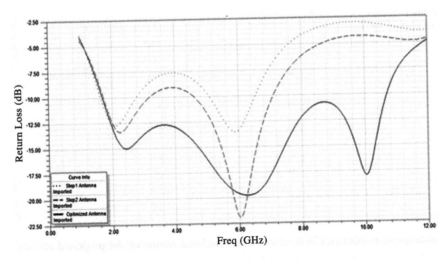

Fig. 2 Return loss plots for various steps in development in the design

Table 1 Results of return loss plots for development of the design

Design steps	f_L (GHz)	f_H (GHz)	Bandwidth (GHz)
Step 1	1.66	2.81	1.15
	5.125	6.55	1.43
Step 2	1.66	3.14	1.48
	4.63	7.10	2.47
Step 3	1.71	10.68	8.97

Fig. 3 Dimensions of the patch and defected ground

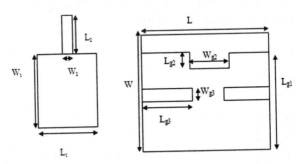

3 Simulated Results of the Optimized Antenna

The software used to designed and simulate the proposed antenna is Ansoft High Frequency Structure Simulator (HFSS). The HFSS uses the Finite Element Method (FEM). The HFSS can be used to calculate the return loss plot, radiation pattern,

Table 2 Dimensions of patch and feed of the antenna

Parameters	Dimensions (mm)
W_1	15.5
L_1	10.5
W_2	1.9
L_2	8.0
W	27
L	22
L_{g1}	25
L_{g2}	3.0
W_{g2}	12
L_{g3}	8.0
W_{g3}	3.0

smith chart, E-field and H-field etc. The simulated results of the proposed antenna are presented in Figs. 4, 5, 6, 7 and 8.

3.1 Return Loss Plot and Bandwidth

Return loss is the measure of the effectiveness of electrical energy delivery from feed to an antenna. For maximum energy transfer the return loss should be minimum. Figure 4 shows the S_{11} parameters (Return loss) for the proposed antenna. The bandwidth of the antenna considers those ranges of frequencies over which the return loss is greater than -10 dB (around VSWR of 2). Thus from the Fig. 4, the graph shows that the return loss below the -10 dB is started from 1.71 to 10.68 GHz which covers the entire UWB applications. The bandwidth of the proposed antenna is 8.97 GHz.

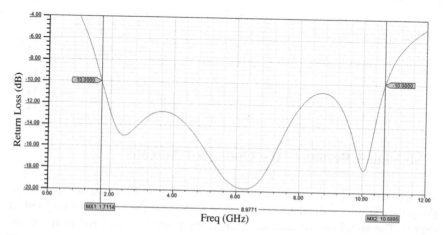

Fig. 4 Return loss plot for optimized antenna

3.2 VSWR

Voltage Standing Wave Ratio (VSWR) can be derived from the level of reflected and incident waves shown in Fig. 5. It is also an indication of how closely or efficiently, an antenna's terminal input impedance is matched to the characteristic impedance of the transmission line. The VSWR is always a positive and real number. Increase in VSWR indicates an increase in the mismatching between the antenna and the transmission line. Practically, the VSWR must lie between 1 and 2.

3.3 Radiation Pattern

The radiation patterns of an antenna provide the information that describes how the antenna directs the energy it radiates. All antennas, if are 100 % efficient, will radiate the same total energy for equal input power regardless of pattern shape. Radiation patterns are generally presented on a relative power dB scale. Figure 6 presents the E-plane and H-plane radiation pattern at 5.5 GHz.

3.4 3D Polar Plot

The 3D polar plot can be shown in Fig. 7 at 5.5 GHz frequency as top view and side view respectively. The red color shows the maximum field intensity in the broadside direction.

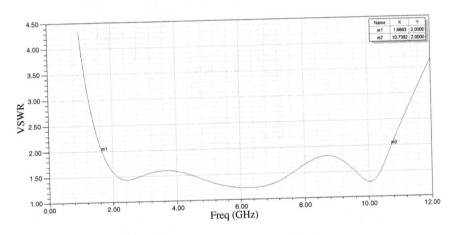

Fig. 5 VSWR versus frequency curve

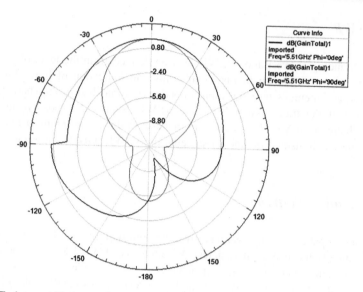

Fig. 6 E-plane and H-plane radiation pattern at 5.5 GHz

Fig. 7 3D polar plot at 5.5 GHz

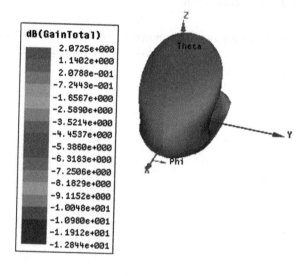

Fig. 8 Smith chart for
optimized antenna

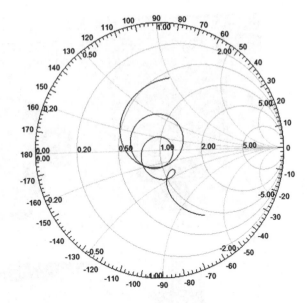

3.5 Smith Chart

Figure 8 shows that antenna behaves resistive as we know upper part shows inductive nature and lower part shows capacitive nature.

4 Fabricated Antenna at Frequency 5.5 GHz

Figure 9 shows the front and back view of the fabricated antenna.

4.1 Fabricated Result of the Optimized Antenna

Antenna is tested on Vector Network Analyzer (VNA) and covers Ultra Wide Band frequency range. Figure 10 shows the return loss of fabricated antenna.

Figure 11 shows the VSWR plot at 5.5 GHz. VSWR is between 1 and 2 dB in Ultra Wide Band frequency range.

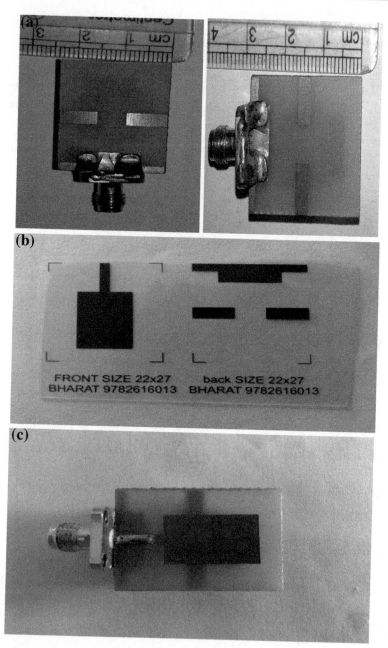

Fig. 9 **a** Bottom view of fabricated antenna along width and length **b** negative of front and back view of fabricated antenna **c** front view of fabricated antenna

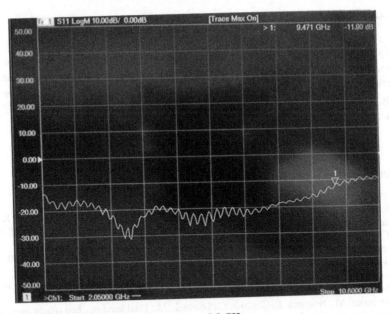

Fig. 10 Return loss plot of fabricated antenna at 5.5 GHz

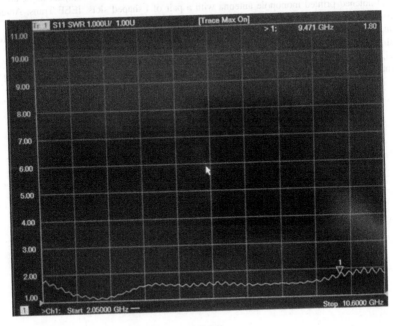

Fig. 11 VSWR plot of fabricated antenna at 5.5 GHz

5 Conclusion

In this article, defected ground structure technique is used to improve the bandwidth of microstrip patch antenna. The proposed antenna have a compact size, stable radiation pattern, constant group delay and return loss below -10 dB over the whole desirable band. It is a good antenna candidate for personal and mobile UWB applications due to the features described above.

References

1. Federal Communications Commission (FCC), Revision of Part 15 of the Commission's Rules Regarding Ultra-Wideband Transmission Systems. First Report and Order, ET Docket 98-153, FCC 02–48 (2002)
2. Taheri, M.M.S., Hassani, H.R., Nezhad, S.M.A.: UWB printed slot antenna with bluetooth and dual notch bands. IEEE Antennas Wirel. Propag. Lett. **10** (2011)
3. Foudazi, A., Hassani, H.R., Nezhad, S.M.A.: Small UWB planar monopole antenna with added GPS/GSM/WLAN bands. IEEE Trans. Antennas Propag. **60**(6) (2012)
4. Mehranpour, M., Nourinia, J., Ghobadi, C., Ojaroudi, M.: Dual band-notched square monopole antenna for ultra wide band applications. IEEE Antennas Wirel. Propag. Lett. **11** (2012)
5. Zaker, R., Ghobadi, C., Nourinia, J.: Bandwidth enhancement of novel compact single and dual band-notched printed monopole antenna with a pair of l-shaped slots. IEEE Trans. Antennas Propag. **57**(12) (2009)
6. Liu, W.C., Wu, C.M., Dai, Y.: Design of triple-frequency microstrip-fed monopole antenna using defected ground structure. IEEE Trans. Antennas Propag. **59**(72457) (2011)
7. Pei, J., Wang, A.G., Gao, S., Leng, W.: Miniaturized triple-band antenna with a defected ground plane for WLAN/WiMAX applications. IEEE Antennas Wirel. Propag. Lett. **10** (2011)

An Enhanced K-means Clustering Based Outlier Detection Techniques to Improve Water Contamination Detection and Classification

S. Visalakshi and V. Radha

Abstract In many data mining applications, the primary step is detecting outliers in a dataset. Outlier detection for data mining is normally based on distance, clustering and spatial methods. This paper deals with locating outliers in large, multidimensional datasets. The k-means clustering algorithm partitions a dataset into a number of clusters, and then the results are used to find out the outliers from each cluster, using any one of the outlier's detection methods. The k-means clustering algorithm is enhanced in three manners. The first is by using a different distance metric. The second and third enhancements are brought forward by automating the process of estimating 'k' value and initial seed selection using the enhanced clustering algorithm. Outliers are detected in the drinking water dataset after the clustering process is over. The results show that classification accuracy, speeds are improved and normalized root mean square error is reduced.

Keywords K-means · Similarity matrix · Dissimilarity co-efficient · Fixed-width clustering · Distance-based · Density-based

1 Introduction

An 'Outliers' is defined as an examination that is radically different from the other data in its set. Outliers are also referred as abnormalities, discordant, deviants or anomalies in data mining. According to Mendenhall et al. the term "Outliers" is "that lies very far from the middle of the distribution in either direction". Outlier detection must be performed during the preprocessing step for locating whether the data pre-

S. Visalakshi (✉) · V. Radha
Department of Computer Science, Avinashilingam Institute for Home Science and Higher Education for Women, Coimbatore 641042, India
e-mail: visaraji@gmail.com

V. Radha
e-mail: radhasrimail@gmail.com

© Springer India 2015
L.C. Jain et al. (eds.), *Computational Intelligence in Data Mining - Volume 1*,
Smart Innovation, Systems and Technologies 31, DOI 10.1007/978-81-322-2205-7_29

303

sented in the drinking water dataset are normal or abnormal. It has become an active research issue in data mining, which has important applications in the field of medical care, public safety and security, image processing, sensor/video network related surveillance, intrusion detection, monitoring criminal activities in e-commerce, monitoring water quality etc.

In my research, the main goal is to identify the contamination from the drinking water dataset. The basic event detection framework utilizes a data mining algorithm for studying the interactions between multivariate water quality parameters and detecting possible outliers. The classifier SVM is used for studying the interplay between multivariate water quality parameters and detecting possible outliers. The SVM make use of several models to detect complex outliers based on the characteristics of the support vectors obtained from SVM-models. SVM is an iterative approach and remove severe outliers in the first iteration itself. So from the next iteration it starts to learn from "cleaner" data and thus reveals outliers that were masked in the initial models [1].

The outlier detection method can be divided into uni-variant and multi-variant. Uni-variate means, considering only one variable or one parameter and multi-variate means, considering more than one variable and check for the relationship between variables. Uni-variate outlier is easy for detection and correction, but multi-variate are more difficult to detect and consumes more time for detection. Another method of outliers is parametric and non-parametric. A parametric method uses statistical models and non-parametric uses some outlier detection methods which are distance based, clustering based and spatial based.

According to [2, 3], the existing method for detecting outliers is classified based on the availability of labels present in training data sets and it is categorized namely: Supervised, Semi-Supervised and Unsupervised. In principle, models belonging to supervised or semi-supervised approaches, all the data must be trained before use, while in unsupervised approach training is not required. Additionally, in the supervised approach, training set should be provided with labels for anomalous or normal. In contrast, in the training set with normal object, labels alone are needed for semi-supervised approach. In other words, the unsupervised approach does not require any object label information. In the paper [4] the classification model treats outlier detection, classification and feature selection as separate step. Thus the proposed method combines all three methods.

In this research work, the outliers are detected before going for feature selection. Clustering is used to partition data into large or small clusters. Based on the classification, the small cluster is considered as outliers and removed safely from the dataset. The main goal of this paper is to remove the outlier effectively. For the experiment, the real time drinking water dataset is used for evaluation. Section 2 analyses the concepts of outlier detection and clustering. In Sect. 3 the proposed techniques are explained in detail. Section 4 reports the experimental results of enhanced technique. Finally conclusion is presented in Sect. 5.

2 Outlier Detection (OD) and Clustering

Clustering analysis and Outlier Detection are two related tasks which will go hand in hand. Clustering finds the major patterns in a dataset and sorting will be performed according to the data, whereas outlier detection aims at capturing the exceptional cases which deviate from the majority patterns. Taking binary decision on whether the object present in the dataset is an outlier or not are becoming a challenging task for the real time dataset. Some of the general terms used in clustering are Cluster (ordered list of objects, which will have common characteristics), Distance (calculating the distance between two points or two elements), Similarity (similarity between two documents SIMILAR (D_i, D_j), Average Similarity (similarity measure will be computed for all the documents (D_i, D_j), except $i = j$ an average value will be obtained), Threshold (finding out the lowest possible input value of similarity which is required to join two objects in one cluster), Similarity Matrix (to find out the similarity between two objects, the similarity function SIMILAR (D_i, D_j) will be used and result will be represented in the form of matrix), Dissimilarity Co-efficient (distance between two clusters) and Cluster seed (first object or first point of cluster is defined as initiator of that cluster and this initiator is known as cluster seed) [1]. Clustering is an important and popular tool for outlier analysis. Most of the techniques presented in outlier will rely on the key assumption that normal objects belong to either large/dense clusters, while small clusters are considered as outliers [5, 6]. Many researchers argue that clustering algorithm is not an appropriate choice for outlier detection. There is no single and specific algorithm for detecting outlier. Therefore, many approaches have been proposed and existing algorithms are also enhanced to improve the outlier detection metrics.

The approaches are classified into four major categories, namely, Distance-based approach, Density-based approach, Distribution-based approach, Deviation-based and clustering-based approach. Distance-based approach, outliers are detected, based on the distance between two points. Density-based approach helps to find out non-linear shapes and structure based on the density. Distribution-based approach helps to detect clusters with arbitrary shape and it does not require any input parameters. It can also handle large amount of spatial data. Clustering-based approach considers small sizes of clusters as outliers. Deviation-based approach helps to identify outliers which deviate from the selected objects [4, 7, 8].

In this paper, enhancement of k-means is proposed in the first phase and the second phase analyses detection of outliers. The main focus of this work is to identify outlier using distance-based clustering, which results in discovering normal and abnormal clusters. Classic k-means algorithm is very sensitive in nature. The selection of initial cluster prototypes will converge to suboptimal solution, only when the initial prototypes are chosen properly and the value of k must be specified in advance. While solving the real-world applications fixing the value of k will be difficult. For such real-time applications [8], first-width clustering algorithm is used to perform clustering process, and classifies outliers as erroneous value or interesting event. The main disadvantage of this algorithm is that it will not consider the intra clusters. The width must be specified before processing starts. For high-dimensional real time data, this algorithm will not work. The required number of distance calculation is high. The selection of centroid is important and if it is not initialized correctly, the classified cluster might have outlier. To overcome this, the Bayesian Information Criterion is included with Modified Dynamic Validity Index (MDVI). Generally, the k-means clustering is used to cluster and classify normal and abnormal clusters. The process of computation is high when k-means is applied to high dimensional dataset. To overcome this issue, a different distance calculation mechanism is applied. In the proposed algorithm all issues are handled and solved. The k-means is enhanced and there is no need to specify the k value; it is automatically assigned to the variable k and the centroid seed selection is estimated automatically. Proposed research is discussed in the following section.

3 Proposed Work

One of the best top ten algorithms in data mining is k-means, which is simple and scalable in nature. Clustering algorithm partitions the dataset into k clusters and it has the two main objectives of making each cluster as compact as much as possible [9]. This paper proposes a new cost function, and distance measure, based on the values present in the dataset. The traditional k-mean algorithm divides the data set X into k clusters and calculates the centroid of each cluster. k value must be assigned before the clustering process. The distance must be calculated with each instance and each instance is to be assigned to the cluster with the nearest seed. Finally threshold % for each cluster, and the distance between each point of cluster from centroid are calculated. When the distance is greater than the threshold value it will considered as "outlier" [9, 10]. The main objective of this paper is to enhance the k-mean clustering algorithm, handling the issues of first width clustering algorithm and evaluating the proposed algorithm with the real time application.

In this work, k-means clustering algorithm is used to cluster the dataset into k clusters. In the proposed algorithm the k-value is generated automatically by using enhanced Bayesian Information Criterion (BIC) along with Modified Dynamic Validity Index (MDVI). The Akaike Information Criterion (AIC), Bayesian Information Criterion (BIC) or Deviation Information Criterion (DIC) are used to determine the number of clusters and the distance metric used for this work is Euclidean distance. The distance from centroid to cluster will be processed until convergence is achieved. The below algorithm describes the process and steps to perform k-mean clustering.

Input: Dataset D with x_i (i = 1...n) data points **Output**: Clusters (C_1, ..., C_k)

Step 1: Feature selection is applied.
Step 2: Estimation of K is automatic and the procedure is given below:

    I. Pre-cluster real time dataset (drinking water dataset) using Birch algorithm.

    II. BIC is computed for each cluster using Eq. (1), where J is a cluster.

$$BIC(J) = -2\sum_{j=1}^{J} \xi_j + m_J \log(N). \tag{1}$$

    III. The ratio of change in BIC at each successive merging relative to the first merging determines the initial estimate and is calculated using Eq. (2).

$$dBIC(J) = BIC(J) - BIC(J+1). \tag{2}$$

From these initial estimates the change ratio of the J cluster is calculated using Eq. (3) as the ratio of dBIC(J) to the dBIC(1) of the first cluster.

$$R_1(J) = \frac{dBIC(J)}{dBIC(1)}. \tag{3}$$

If dBIC(1) < 0, then $K_T = 1$ and go to step 8 else calculate inter-cluster ratio and K_T = number of cluster for which the recorded ratio is minimum of all and repeat steps V, VI and VII for all K_T.

    IV. Calculate modified inter and intra cluster ratio between cluster C_k and C_{k+1} using Eq. (4).

$$\text{IntraRatio(k)} = \frac{\text{Intra(k)}}{\text{MaxIntra}} \quad \text{InterRatio(k)} = \frac{\text{Inter(k)}}{\text{MaxInter}}. \tag{4}$$

K is the pre-defined upper bound number of the clusters.

    V. Calculate the modified dynamic validity index using Eq. (5).

$$MDVI = \min_{k=1,...k} \{\text{IntraRatio(k)} + \gamma * \text{InterRatio(k)}\}. \tag{5}$$

    VI. K_T = Number of clusters for which the dynamic validity index is maximum Optimal k = K_T.

Step 3: Estimation of K initial seeds (C_j) is automatically generated.

I. Calculate K.

II. Compute the distances between objects in D is calculated using Eq. (6).

$$N(X_i) = \{any\, x_j : d(x_i, x_j) < x_j, D, i, j\}. \tag{6}$$

where $d(x_i, x_j)$ is the distance between x_i and x_j calculated using DLG and the average distance between all objects is calculated using the following equation.

III. Compute the average distance between all objects using Eq. (7).

$$\varepsilon = \frac{1}{n(n-1)} \sum_{i=1}^{n-1} \sum_{j=i+1}^{n} d(x_i, x_j). \tag{7}$$

IV. Find neighborhood of objects in D

The coupling degree between neighborhoods of objects x_i, x_j is the ratio of number of objects neighbor to both x_i and x_j is calculated using Eq. (8).

$$\text{Coupling}(N(x_i), N(x_j)) = \frac{|N(x_i) \cap N(x_j)|}{N(x_i) \cap N(x_j)}. \tag{8}$$

V. If Coupling $(N(x_i), N(x_j)) < \varepsilon$ (average distance between all objects), then next centroid is found

Add to C(Next Centroid)

VI. If |No of centroids| < k, Go to Step 6, otherwise go to Step 9.

VII. End

Step 4: Steps 5–9 for each point x_i in D′ are repeated.

Step 5: Distance between each data point x_i and all k cluster centre is calculated using Eq. (9).

$$\text{DLG}_{ij} = \min \left(\sum_{e=1}^{p} L(p_e, p_{e+1}) \right). \tag{9}$$

where $e \in E$ and ranges from 1...p. Thus, DLG_{ij} satisfies the four conditions for a metric, that is, $D_{ij} = D_{ji}$; $D_{ij}\ 0$; $D_{ij}\ D_{ie} + D_{ej}$ for all x_i, x_j, x_p and $D_{ij} = 0$ iff $x_i = x_j$. Thus the new measure considers both global and local consistency and can adapt itself to the data structure. L the density length is computed using Eq. (10).

$$L(x_i, x_j) = \rho^{dist(x_i, x_j)} - 1. \tag{10}$$

where $dist(x_i, x_j)$ is the Euclidean distance between x_i and x_j and >1 is the flexing factor (the value 8 is used during experimentation).

Step 6: The closest centre c_j is found and assigned x_i to cluster j.

Step 7: Label of cluster centre j along with the distance of x_i to c_j and stored in array Cluster[] and Dist[] respectively

Set Cluster[i] = j (j is the nearest cluster)
Set Dist[i] = DLG_{ij} (distance between x_i to closest cluster centre c_j)

Step 8: Cluster centres are recalculated.

Step 9: DLG_{new} of x_i is computed to new cluster centres until convergence

If DLG_{new} is less than or equal to DLG_{ij}, then x_i belongs to the same cluster j
else
DLG is computed with every other cluster centre and assign x_i to the cluster whose DLG is minimum
Set Cluster[i] = j and Set Distance[i] = DLG_{new}

Step 10: Output clustered results.

Features selection chooses distinctive features from a set of dataset. Selection of features helps to reduce the size of dataset and make the process simpler for all subsequent design [11].

Once the feature selection is over, pre-cluster the dataset using Balanced Iterative Reducing and Clustering using Hierarchies (BIRCH) algorithm which can handle large datasets. The ratio of change in cluster is calculated in BIC at each consecutive merging. The intra and inter cluster relationship are evaluated using the formula. To improve the performance of algorithm and to find the better cluster number the MDVI is used. Next process of algorithm is to select the initial seed selection for the assignment of data to cluster. The performance of initial seed selection will be based on the sum of square, difference between members of cluster, cluster center and normalized data size [8, 9]. The inter and intra cluster distance validity measure allows to determine the number of clusters automatically. For initial centroid selection, enhanced distance measure, Reverse Neighbor Node and coupling degree are used. The coupling degree is used to measure the similarity between two objects.

Thus conventional k-means algorithm begins with a decision on the value of k. Any initial partition which classifies the data into k clusters is assigned. The problem of Euclidean distance is overcome in the proposed work. Hence the Euclidean distance of both x_i and x_j is calculated with the threshold value of 8 is used for experiment. Thus the new measure considers both global and local consistency and can adapt itself to the data structure. The distance from centroid to cluster will process until convergence is achieved. The above algorithm outlines the process of automatic key generation of k value and initial seed selection for clustering.

3.1 Outlier Detection (OD)

The second phase of the work is outlier detection. Detection of OD is basic issue in data mining. Outlier detection will remove 'noise' or 'unwanted' data from the dataset. Once the cluster formation has taken place with the help of enhanced k-mean clustering, then the output will be given as input for outlier detection. The dataset is partitioned into small and large clusters and the resultant cluster will be checked for anomalies, and if anomalies are present then those anomalies are safely removed from the whole dataset. For each cluster c_i in the cluster set C, a set of inter cluster distances $Dc_i = \{d(c_i, c_j): j = 1... (|C| - 1), j = i\}$ is computed. Here, $d(c_i, c_j)$ is the Euclidean distance between centroids of c_i and c_j, and $|C|$ is the number of clusters in the cluster set C. Among the set of inter-cluster distances Dc_i for cluster c_i, the shortest K (parameter of KNN) distances are selected and using those, the average inter-cluster distance ICD_i of cluster c_i is computed using Eq. (11).

$$ICD_i = \begin{cases} \frac{1}{K} \sum_{j=1, \neq i}^{K} d(c_i, c_j) & K \leq |C| - 1 \\ \frac{1}{|C|-1} \sum_{j=1, \neq i}^{|C|-1} d(c_i, c_j) & K > |C| - 1 \end{cases}. \tag{11}$$

The average inter-cluster distance computation is enhanced. Instead of using the whole cluster set C to compute the average inter-cluster distance ICD_i for a cluster c_i, the presented algorithm uses the K-Nearest Neighbor (KNN) for cluster c_i. The advantage of this approach is that clusters at the edge of a dense region are not considered compared to clusters in the centre of the region. A cluster is identified as anomalous $Ca \subset C$ are defined as $C_a = \{c_i \in |C| \; ICD_i > AVG (ICD) + SD (ICD)\}$, where ICD is the set of average inter-cluster distances.

Once the anomaly is identified and removed then the two clusters are merged into single cluster, if it satisfies the rules which are given below. A pair of clusters c1 and c2 are similar if the inter-cluster distance d(c1, c2) between their centers is less than the width w. If c1 and c2 are similar, then a new cluster c3 is produced. The centre of c3 is the mean of the centers of c1 and c2 and whose number of data vectors is the sum of those in c1 and c2. In the proposed system, the merging procedure compares each cluster c_i with clusters $\{c_{i+1}, c_{i+2}, ...\}$, and merges c_i with the first cluster c_j such that $d(c_i, c_j) < T$ and $j > i$. The value of T is set to 0.38 after experimentation with different values ranging between 0.01 and 2.02 in steps of 4. For detecting outlier, the k-means clustering algorithm is enhanced and KNN is used to find out the nearest neighbor. The DLG is used to consider both intra and inter distance between data points. For automatic estimation, BIC is enhanced using

MDVI. The centroid selection enhanced distance measure is combined with RNN and coupling degree. Finally, for dimensionality reduction, Principal Component Analysis is used. This section concludes that outliers are detected effectively with the help of the proposed technique. The enhanced k-means clustering will be able to discover clusters with correct arbitrary shape, it works well for large databases efficiently and lot of heuristics to determine the parameters. The enhanced technique helps to detect outliers efficiently and accurately.

4 Experimental Results

Clustering-based method determine cluster with shape, efficient to handle large database and determines the number of input parameters. Based on clustering, the exception will be considered as "noise", where it is bearable in some cases and sometimes that leads to inaccurate results. The two different season (summer and winter) real time datasets are collected from TWAD Board, Coimbatore, Tamil Nadu, India. The dataset contains various parameters which brief about the characteristics of drinking water. Some of the parameter used for research are Turbidity, EC, TDS, Ph, Ca etc. The SVM and BPNN classifiers are used in this experiment for training and testing. Enhanced k-means and outlier detection are evaluated with various metrics namely, Accuracy, Normalized Root Mean Square Error and Speed.

$$Accuracy = \frac{Number\ of\ correct\ predictions}{Total\ number\ of\ predictions} \times 100$$

$$NRMSE = \frac{\sqrt{mean[(y_{true} - y_{imp})^2]}}{variance\ y_{true}}$$

Table 1 presents the classification accuracy of before and after OD to various % of outliers for two different seasons. It is obvious that the classification accuracy is improved compared with BPNN.

The normalized root means square error before and after outlier detection values are presented in Table 2 for two different seasons and it is evident that the normalized root mean square error value is minimized compared with BPNN.

The execution speed of before and after outlier detection is listed in the above Table 3 for summer and winter seasons and it is clear that the execution speed is reduced compared with BPNN. Thus the experimental results shows that proposed algorithm works effectively for detecting outlier in contaminated drinking water dataset.

Table 1 Classification accuracy (%)

Datasets	Outliers	BPNN before OD	BPNN after OD	SVM before OD	SVM after OD
Summer	0	79.80	81.23	83.79	84.66
	10	80.52	85.34	88.92	90.45
	20	81.27	84.34	87.17	87.46
	30	84.12	85.64	86.22	87.61
	40	80.97	82.34	84.97	85.93
Winter	0	79.83	81.76	82.87	84.16
	10	81.73	83.45	89.51	91.49
	20	82.24	85.54	86.42	87.17
	30	80.12	83.45	85.61	86.00
	40	79.34	80.16	84.73	85.72

Table 2 Normalized root mean square error

Datasets	Outliers	BPNN before OD	BPNN after OD	SVM before OD	SVM after OD
Summer	0	0.8976	0.8765	0.4451	0.4389
	10	0.7921	0.7832	0.4049	0.3990
	20	0.8012	0.7986	0.4103	0.4035
	30	0.8123	0.7989	0.4282	0.4213
	40	0.9013	0.8967	0.4251	0.4190
Winter	0	0.8876	0.8675	0.4726	0.4664
	10	0.8012	0.7954	0.4410	0.4348
	20	0.8234	0.8123	0.4573	0.4512
	30	0.8100	0.8024	0.4415	0.4350
	40	0.9876	0.9765	0.4548	0.4486

Table 3 Execution speed (seconds)

Datasets	Outliers	BPNN before OD	BPNN after OD	SVM before OD	SVM after OD
Summer	0	6.69	6.11	5.13	5.07
	10	6.76	6.34	5.23	5.17
	20	7.23	7.05	5.41	5.35
	30	6.98	6.87	6.28	6.24
	40	7.45	7.25	7.12	7.07
Winter	0	6.89	6.50	6.37	6.32
	10	6.54	6.50	6.41	6.36
	20	7.15	7.10	7.05	6.98
	30	9.05	8.79	8.26	8.22
	40	9.34	9.10	8.97	8.93

5 Conclusion

This paper focused to detect outlier from dataset and aims to find objects which are different or contradictory from other data. An outlier detection method is proposed. The issues in k-means clustering algorithm are handled to enhance its clustering operations and to identify the outliers from the dataset are proposed. The first step is grouping of data into a number of clusters for this the average of inter cluster distance is used. Next outlier is identified from the resultant cluster. The experimental results show that the classification accuracy is improved and normalized root mean square error is minimized and it is evident that the proposed algorithm is efficient in identifying outliers to detect the contamination quickly. In future, Feature Selection algorithm can be combined with outlier detection to improve contamination detection.

Acknowledgments The authors express their gratitude to TWAD Board for their whole hearted support in providing dataset for research.

The author expresses their gratitude to Avinashilingam Institute for Home Science and Higher Education for Women, Coimbatore, Tamil Nadu, India for the progress of research work.

References

1. Cateni, S., Colla, V., Vannucci, M.: Outlier detection methods for industrial applications. Advances in robotics. In: Automation and Control, pp. 274–275 (2008)
2. Ahmad, A., Dey, L.: A k-mean clustering algorithm for mixed numeric and categorical data. Data Knowl. Eng. **63**, 502–527 (2007)
3. Hodge, V.J., Austin, J.: A survey of outlier detection methodologies. Artif. Intell. Rev. **22**(2), 85–126 (2004)
4. Fawzy, A., Mokhtar, H.M.O., Hegazy, O.: Outliers detection and classification in wireless sensor networks. Egypt. Inf. J. **14**, 157–164 (2013)
5. Khan, F.: An initial seed selection algorithm for k-means clustering of geo-referenced data to improve replicability of cluster assignments for mapping application. Appl. Soft Comput. **12**, 3698–3700 (2012)
6. http://members.tripod.com/asim_saeed/paper.htm
7. Chandola, V., Banerjee, A., Kumar, V.: Anomaly detection: a survey. ACM Comput. Surv. **41**(3), 1–58 (2009)
8. Pachgade, S.D., Dhande, S.S.: Outlier detection over data set using cluster-based and distance-based approach. Int. J. Adv. Res. Comput. Sci. Soft. Eng. **2**(6), 12–16 (2012)
9. Zhu, C., Kitagawa, H., Papadimitriou, S., Faloutsos, C.: Outlier detection by example. J. Intell. Inf. Syst. **36**, 217–247 (2011)
10. Shi, Y., Zhang, L.: COID: a cluster–outlier iterative detection approach to multi-dimensional data analysis. Knowl. Inf. Syst. **28**, 710–733 (2010)
11. Indira Priya, P., Ghosh, D.K.: A survey on different clustering algorithms in data mining techniques. Int. J. Mod. Eng. Res. **3**(1), 267–274 (2013)
12. Zhang, T., Ramakrishnan, R., Livny, M.: BIRCH: a new data clustering algorithm and its applications. Data Min. Knowl. Discov. **1**, 141–182 (1997)

A System for Recognition of Named Entities in Odia Text Corpus Using Machine Learning Algorithm

Bishwa Ranjan Das, Srikanta Patnaik, Sarada Baboo
and Niladri Sekhar Dash

Abstract This paper presents a novel approach to recognize named entities in Odia corpus. The development of a NER system for Odia using Support Vector Machine is a challenging task in intelligent computing. NER aims at classifying each word in a document into predefined target named entity classes in a linear and non-linear fashion. Starting with named entity annotated corpora and a set of features it requires to develop a base-line NER System. Some language specific rules are added to the system to recognize specific NE classes. Moreover, some gazetteers and context patterns are added to the system to increase its performance as it is observed that identification of rules and context patterns requires language-based knowledge to make the system work better. We have used required lexical databases to prepare rules and identify the context patterns for Odia. Experimental results show that our approach achieves higher accuracy than previous approaches.

Keywords Support vector machine · Name entity recognition · Part of speech tagging · Root word

B.R. Das (✉) · S. Patnaik
Department of Computer Science and Information Technology,
Institute of Technical Education and Research, SOA University,
Bhubaneswar, India
e-mail: biswadas.bulu@gmail.com

S. Patnaik
e-mail: baboosarada@gmail.com

S. Baboo
Department of Computer Science and Application,
Sambalpur University, Burla, India
e-mail: patnaik_srikanta@yahoo.co.in

N.S. Dash
Linguistic Research Unit, Indian Statistical Institute, Kolkata, India
e-mail: ns_dash@yahoo.com

© Springer India 2015
L.C. Jain et al. (eds.), *Computational Intelligence in Data Mining - Volume 1,*
Smart Innovation, Systems and Technologies 31, DOI 10.1007/978-81-322-2205-7_30

1 Introduction

Named Entity Recognition (NER) is a technique to identify and classify named entities for particular domain of a piece of text. It is an important task as it is directly related to applications like Information Extraction, Question Answering, and Machine Translation, Data Mining, and other Natural Language Processing (NLP) applications. This paper proposes a novel NER system for Odia—one of the Indian national languages. It performs NER act on three types of named entity—person names, location names, and organization names. These named entities are addressed because identification of these is the most challenging task in the whole scheme of NER. For our task, suitable set features are first identified for the named entities in Odia. The feature list includes orthography features, suffix and prefix information, morphological information, part-of-speech information as well as information about the neighbouring words and their POS tags, which are combined together to develop the Support Vector Machine (SVM)-based NER System of the language. Some rules are defined for classification of person, location, organization names based on certain criteria, which are made available to the system through gazetteers-based identification for person, location, and organization names.

There are several named entity classification methods which may be successfully applied on this task. Kudo and Matsumoto [1] have used the Support Vector Machine in chunking which may also help in our proposed work. Biswas et al. [2] have used the Max Entropy model for hybrid NER for classification. Their approach can achieve higher precision and recall, if it is provided with enough training data and appropriate error correction mechanism. Ekbal and Bandyopadhyay [3] have used the SVM for classification of Bengali named entities with 91.8 % accuracy. Saha et al. [4] have described the development of Hindi NER system by using ME approach with 81.51 % accuracy. Their system is tested with a lexical database of 25k words having 4 classes of named entities. Goyal [5] has also developed a system for NER for South Asian Language. Saha et al. [4] have identified suitable features for Hindi NER task that are used to develop an ME based Hindi NER system. Two-phase transliteration methodology has been used to make the English lists useful in the Hindi NER task. This system gives the accuracy with 81.2 %.

Various approaches that are used in NER system include Rule Based, Handcrafted Approach, Machine Learning, Statistical Approach, and Hybrid Model [6]. In Rule-Based approaches, a set of rules or patterns is defined to identify the named entities in a text. For instance, while pre-tags like 'sri', 'sriman', 'srimati' etc. are used to identify person names, forms like 'nagar', 'sahara', 'vihar' etc. are used to identify place names, and the forms like 'vidyalaya', 'karjyalaya' etc. are used to identify organization names. Chieu and Ng [7] have used Maximum Entropy Model to find NE (Global information) with just one classifier. In another work ("A survey of named entity recognition and classification"), they have presented a survey of 15 years of research (1991–2006) in NERC field.

The introduction of this paper (Sect. 1) describes in some details about the early works on NER system development in other languages; Sect. 2 describes the composition and content of the Odia newspaper text corpus; Sect. 3 describes the Support Vector Machine which is used for classification of named entities; Sect. 4 describes the training data, how it is specially used, and data mapping with the test dataset; Sect. 5 presents the evaluation results to show how our proposed system works; and Sect. 6 describes the conclusion part of the paper.

2 The Odia Text Corpus

An Odia newspaper text corpus is recently developed to describe in details the form and texture of the Odia language used in the present data Odia newspapers. Following some well-defined strategies and methods [8] this Odia corpus is designed and developed in a digital with texts obtained from Odia newspapers. The corpus is developed with sample news reports produced and published by some major Odia news papers published from Bhubaneswar and neighboring places [9]. We have followed several issues relating to text corpus design, development and management, such as, size of the corpus with regard to number of sentences and words, coverage of domains and sub-domains of news texts, text representation, question of nativity, determination of target users, selection of time-span, selection of texts, amount of sample for each text types, method of data sampling, manner of data input, corpus sanitation, corpus file management, problem of copy-right, etc. [8]. Since this corpus is very much rich with data relating to named entities of various types, we have been using it to perform linear and nonlinear classification of named entities in which we prepared our own digital corpus from various Odia news papers. In essence, we are using this corpus to identify and classify Odia person names, place names and organization names along with some miscellaneous named entities.

3 Support Vector Machine

The Support Vector Machines is a binary learning machine with some highly elegant properties that are used for classification and regression. It is a well known system for good generalization performance and it is used for pattern analysis. In NLP, it is applied to categorize the text, as it gives high accuracy with a large number of features set. We have used this machine to defining a very simple case—a two class problem where the classes are nonlinearly separable. Let the data set D be given as $(X_1, y_1), (X_2, y_2)...(X_D, y_D)$, where X_i is the set of training tuples with associated class labels y_i. Each y_i can take one of two values, either $+1$ or -1 (i.e., $y_i \in \{+1, -1\}$.

A separating hyperplane equation can be written as $wx + b = 0$, where x is an input vector, w is the adjustable weight, and b is the bias. Training tuples are 2-D, e.g. $x = \{x_1, x_2\}$, where x_1, x_2 are the values of attributes A_1 and A_2 respectively for x.

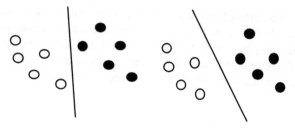

Fig. 1 Classification of textual data

It finds an optimal hyperplane which separates the training data as well as the test data into two classes. It find separating hyperplane which maximizes its margin. Two parallel lines and margin M can be expressed as wx + b = +1, M = 2/||w||. To maximizes this margins r = 1/||w|| and Minimize ||w|| = ||w||²/2, Subject to $d_i(w \cdot x_i + b) \geq 1$, where i = (1, 2, 3,..., 1). Any training tuples that falls on either side of the margins are called support vector. It has strength to carry out the nonlinear classification. The optimization problem can be written usual form, where all feature vectors appear in their dot products. By simple substituting every dot product of x_i and x_j in dual form with a certain Kernel function $K(x_i, x_j)$. SVM can handle nonlinear hypotheses. Among these many kinds of Kernel function available. We shall focus on the polynomial kernel function with degree d such as $K(x_i, x_j) = (x_i * x_j + 1)^d$. Here d degree polynomial kernel function helps us to find the optimal separating hyperplane from all combination of features up to d. The hypothesis space under consideration is the set of functions. The linear separable case is almost done. The non linear SVM classifier gives a decision making function f(x). Figure 1 shows the classification of textual data.

$$f(x) = \sum_{i=1}^{m} w_i K(x, z_i) + b, \quad g(x) = \text{sign}(f(x)) \tag{1}$$

If g(x) is +1, x is classified as class C_1 and −1 x is classified as class C_2. z_i are called support vectors and representative of training examples, m is the number of support vectors is a kernel that implicitly maps vectors into a higher dimensional space and can be evaluated efficiently. The polynomial kernel $K(x, z_i) = (x \cdot z_i)^d$.

4 Training Data

We used our own training data set that was developed by ourselves in Odia. It gives the 100 % correct result for our system.

$$T = \{x_k, d_k\}_{k=1}^{Q} \tag{2}$$

where $X_k \in \mathbb{R}^n$, $d_k \in \{-1, +1\}$.

4.1 Features

It is mentioned the following set of features that have been applied to the NER task.

(i)   After POS tagging, the nominal word or surrounding word is set to be +1 otherwise it is set to −1. This binary value used to all POS feature.
(ii) Person prefix word, if the prefix belongs to 'sriman', 'srimati' etc. then set to +1.
(iii) If middle names like 'kumar', 'ranjan', 'prasad' etc. appear inside the person name, then it is set to be +1.
(iv) If surnames like 'Das', 'Mishra', 'Sahoo' etc. appear set to be +1.
(v)   Location name with suffix 'nagar', 'sahara', 'podaa', 'vihar' etc. is set to be +1.
(vi) Organization name with suffix 'mahabidyalaya', 'karjyalaya', 'bidyalaya' etc. is set to be +1, otherwise set to be −1.

All positive words used in the training set are considered as +1 and rest of the words are considered as −1.

It is identified that various features may be considered to find out NE in Odia language as mentioned below. Following the features many place names, person names, and organization names are identified. Also some rules are mentioned in this paper that is used for such purpose, as summarized below.

(a) A Odia word which is associated with its prefix or suffix word and its surrounding words i.e., *desha* "country", *rajya* "state", *anchala* "area", *jilla* "district", *sadar mahakumaa* "dist. head quarter", *grama* "village", *panchayata* "panchayata", *pradesh* "state", *sahara* "town", are treated as place names. Some other words which belong to *nagara, vihara, pura, podaa* also used to identify place name.
(b) An Odia word which is associated with *sriman, srimati, kumara, kumari, ranjan,* etc. are used to identify person names. Some of the bivokti or markers are also used in Odia to identify person names, e.g., *-ku, -re, -ro*.
(c) An Odia word which is associated with forms like *bidyalaya* "school", *mahabidyalaya* "college", *vishwabidyalaya* "university", *karjyalaya* "office", is used to identify organization name.

Fig. 2 Flowchart of finding
NER

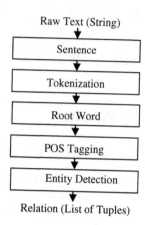

Raw Text (String)

Sentence

Tokenization

Root Word

POS Tagging

Entity Detection

Relation (List of Tuples)

The flowchart to find NE (Fig. 2).

4.2 Suffix and Prefix

Some suffix and prefix alphabets are used to identify NE, which are mentioned in the features. Firstly a fixed length word suffix of the current and surrounding words are used as features.

4.3 Part of Speech Tagging

POS tagging is used to find out noun and verb as POS information of the current word and the surrounding words are useful features for NER. For this purpose an Odia POS tagger using ANN is used here. The tagset of the tagger contains 28 tags. The POS values of the current and surrounding tokens as features is used here.

4.4 Root Word

Morphological analyzer is used to find the root words by stripping suffix-prefix from a word.

4.5 Algorithm

The proposed algorithm is used for finding the NE in the Odia corpus data. First, the entire Odia text corpus is entered by user in our proposed system, and then the process of NER is divided into seven steps which are described in the following algorithm.

Step 1:   Enter a text.
Step 2:   Convert entire text into token by tokenization.
Step 3:   Find root word using morphological analysis.
Step 4:   Compare each word with our valid features.
Step 5:   Extract the features from each and every word.
Step 6:   Compare each word with the training data set.
Step 7:   Find the exact Name Entity.

5 Result Evaluation

Odia news corpus is used to identify the test set for NER experiment. Out of one lakh word forms, a set of one thousand word forms has been manually annotated with the 10 tags initially. In our system we have used several important features to find NE and these are already described in the earlier sections. The general result obtained from our experiment is presented below (Fig. 3).

For classification of NE, the SVM technique is used. A baseline model is defined where the NE tag probabilities depend only the current word.

$$P(t_1, t_2, t_3 \ldots t_n | w_1, w_2, w_3 \ldots w_n) = \prod_{i=1 \ldots n} P(t_i, w_i) \tag{3}$$

The test data is assigned to a particular NE tags POS tags that occur in the training data after some empirical analysis. The combination of words from a set 'F' gives the best features for Odia NER. The given set 'F' mentioned below.

Fig. 3 Analysis of the proposed system

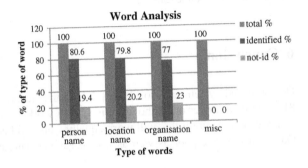

$F = \{w_{i-4}, w_{i-3}, w_{i-2}, w_{i-1}, w_i, w_{i+1}, w_{i+2}, w_{i+3}, |prefix| <= 3, |suffix| <= 3, NE$ information, POS information of current word, digit features$\}$.

Some experimental notations are used in this work as follows: pw (previous word), cw (current word), nw (next word), pp (POS tag of previous word), cp (POS tag of the current word), np (POS tag of the next word). The cardinality of the prefix, suffix length is measured up to 3 characters.

The Precision, Recall, and F_Score formula are used for measuring the level of accuracy of results. Mathematical equations, which get from SVM, are giving proper classification. Construction of SVM, taking training set in the Eq. (2) Minimize, $\Phi(w) = 1/2 \parallel w \parallel^2$, subject to the constraints $d_i(w^T x_{i+} b) - 1 \geq 0, i = 0,$ 1, 2...N. The objective was to maximize the margin $1/ \parallel W \parallel$. Since the square root is monotonic function, one can switch to $\parallel w \parallel^2$ instead of $\parallel w \parallel$, and in order to minimize $1/2 \parallel w \parallel^2$. To solve this optimization problem, the technique of language multiplier is used to turn here. It is used because it is easy to handle. Also to find the accuracy, we use the mathematical formula of precision, recall, F_score. POS information helps to fine the accuracy. Most of the words are tagged with appropriate tagset. From the tagged word, named entities can find easily.

$$Precision = |ANE \cap ONE|/|ONE|$$
$$Recall = |ANE \cap ONE|/|ANE|$$
$$F\_Score = 2(Recall * Precision)/(Recall + Precision)$$

Here ANE—Actual named entity, ONE—Obtained named entity. Precision means how many correct entities from whatever has been obtained are. Recall means out of the correct once how many have been obtained named entities. Here accuracy is calculated through F_Score in percentage. With the help of harmonic mean (HM) more accurate result also calculated.

Let us consider some instances to know how it works. For instance, let us consider a sentence: *Sriman Hariprasad jone volo gayaka* "Sriman Hariprasad is a good singer". Here the term Hariprasad is Person Name Entity, because it contains the middle name 'prasad'. Similarly, consider this sentence *Hariprasadnko ghara Bhubaneswar* "Hariprasad's home is at Bhubaneswar". Here the term *Bhubaneswar* is a Location Name Entity. Similarly, in the sentence *Revenshaw mahabidyalayare se patho podhithile* "He was studying at Revenshaw college" the term *Reveshaw* is an Organization Name Entity.

6 Conclusion

Our proposed system tries to identify NE nearly accurately with a success rate of 81 % without any error. Although this system worked fine on the Odia newspaper text, we are not sure if this will work equally well in other types of Odia text. Since Odia is a resource-poor as well as less-researched language, it is obvious that we need more exhaustive research in this direction before we can claim appreciable success in recognition and identification of named entities used in Odia written texts. The performance of this system has been compared with the existing one Odia NER [2] system and one Bengali NER [3] system. There are many linguistic and stylistic issues (e.g., agglutinative nature and different writing style, etc.) that also need careful attention for developing NER system for the Odia language. Definitely, the availability of an Odia text corpus of only five lakh words collected from Odia newspapers cannot be the benchmark trial database for systems like this, even if SVM system works fine on our database. With this limited success we propose to move further as application relevance of NER is approved in many domains of NLP: parsing, word sense disambiguation, information retrieval, question answering, machine learning—to mention a few.

References

1. Kudo, T., Matsumoto, Y.: Chunking with support vector machine. In: Proceedings of NAACL, pp. 192–199 (2001)
2. Biswas, S., Mishra, S.P., Acharya, S., Mohanty, S.: A hybrid Oriya named entity recognition system: harnessing the power of rule. Int. J. Artif. Intell. Expert Syst. 1(1), 639–643 (2010)
3. Ekbal, A., Bandyopadhyay, S.: Bengali named entity recognition using support vector machine. In: Proceedings of the IJCNLP-08 Workshop on NER for South and South East Asian Languages, pp. 51–58 (2008)
4. Saha, S.K., Sarkar, S., Mitra, P.: A hybrid feature set based maximum entropy hindi named entity recognition. In: Proceedings of the 3rd International Joint Conference on NLP, Hyderabad, India, pp. 343–349, Jan 2008
5. Goyal, A.: Named entity recognition for South Asian languages. In: Proceedings of the IJCNLP-08 Workshop on NER for South and South-East Asian Languages, Hyderabad, India, pp. 89–96, Jan 2008
6. Sasidhar, B., Yohan, P.M., Babu, A.V., Govardhan, A.: A survey on named entity recognition in Indian languages with particular reference to Telugu. Int. J. Comput. Sci. 8(2). ISSN 1694-0814. www.IJCSI.org (2011)
7. Chieu, H.L., Ng, H.T.: Named entity recognition: a maximum entropy approach using global information. In: 19th International Conference on Computational Linguistics (COLING 2002), 24 Aug–1 Sept 2002
8. Dash, N.S.: Indian scenario in language corpus generation. In: Dash, N.S., Dash, P.D., Sarkar, P. (eds.) Rainbow of Linguistics, vol. I, pp. 129–162. T Media Publication, Kolkata (2007)
9. Das, B.R., Patnaik, S., Dash, N.S.: Development of Odia language corpus from modern news paper texts: some problems and issues. In: Proceedings of the International Conference on Intelligent Computing, Communication and Devices (ICCD 2014). SOA University, Bhubaneswar, India, Springer Book Series on AISC, pp. 88–94 (2014)
10. Sharma, P., Sharma, U., Kalita, J.: Named entity recognition: a survey for the Indian languages. Language in India. Special Volume: Problems of Parsing in Indian Languages 11 (5). www.languageinindia.com, May 2011
11. Ekbal, A., Bandyopadhyay, S.: Named entity recognition using support vector machine: a language independent approach. Int. J. Electr. Electron. Eng. 4(2), 155–170 (2010)
12. Saha, S.K., Ghosh, P.S., Sarkar, S., Mitra, P.: Named entity recognition in Hindi using maximum entropy and transliteration. Res. J. Comput. Sci. Comput. Eng. Appl. 33–41 (2008)

13. Bharati, A., Sangal, R., Chaitnya, V.: Natural language processing—a Paninian perspective. Prentice Hall-India, New Delhi (1995)
14. Ray, P.R., Harish, V., Sarkar, S., Basu, A.: Part of speech tagging and local word grouping techniques for natural language parsing in Hindi. In: Proceedings of the International Conference on Natural Language Processing (ICON 2003), pp. 118–125 (2003)
15. Satish, K.: Neural Network Book: A Classroom Approach, 10th edn. TMH Publication, New Delhi (2010)
16. Mahapatra, D.: Adhunika Odia Byakarana (Modern Odia Grammar), 5th edn. Kitab Mahal, Cuttack (2010)

A New Grouping Method Based on Social Choice Strategies for Group Recommender System

Abinash Pujahari and Vineet Padmanabhan

Abstract Recommender System is a software or tool that helps users to select items or things according to their preferences. These are used in almost every web sites today. A lot of research is going on, how to produce efficient recommendations for individuals. Even more, today the recommendation of items/things are for a group of users where there is more than one user in a group and each user have their own preferences. Group Recommender System recommends items or things for a group of users based on their individual preferences. There are many social choice grouping strategies available. We proposed a new grouping algorithm which will first generate homogeneous groups and then generate recommendation of items for them. In this paper we followed the rule based approach to learn the user's preferences. All the results of our approach is validated with the movie lens data set which is the bench mark data set for recommender system testing.

Keywords Recommender system · Rule learning · Predictive rule mining

1 Introduction

The Internet, nowadays, is overloaded with information related to books, movies, music etc. and choosing items that suits one's interest has become a difficult task. So it is reasonable to think of building a system that can recommend items according to our interest. Such systems are commonly known as Recommender

A. Pujahari (✉)
Institute of Information Technology, Sambalpur University,
Jyoti Vihar, Burla 768019, India
e-mail: abinash.pujahari@gmail.com

V. Padmanabhan
School of Computer and Information Sciences, University of Hyderabad,
Hyderabad 500046, India
e-mail: vcpnair73@gmail.com

325

© Springer India 2015
L.C. Jain et al. (eds.), *Computational Intelligence in Data Mining - Volume 1*,
Smart Innovation, Systems and Technologies 31, DOI 10.1007/978-81-322-2205-7_31

Systems in the Machine Learning community. Recommender systems [1, 2] are very popular because it reduces the overhead of the user by providing the recommendations of their interest from a large volume information. These days almost every web site uses a recommender system. For instance, in on-line shopping websites when we select an item similar types of items are shown that we'll probably like. But the problem with most recommender systems is that they are built for individuals or are personalized to suit an individual's interest. In this paper we aim to build a Group Recommender System that makes recommendations to a group of users within the problem domain of a Movie recommender system. We need to aggregate all the users' preferences from the group and then recommend movies for that group of users.

Lots of research work is taking place on how to build efficient Group Recommender System [3, 4]. Here efficient means more accurate recommendations. Every recommender system uses some kind of machine learning algorithms like Decision Tree (ID3, C4.5), etc., to learn from users past behavior in order to know his/her preferences. In this paper we have followed the rule based approach for learning rules from users past experiences and then while generating group recommendation we have used our proposed algorithm which is based on other social choice aggregation strategies.

2 Group Recommender System

Recommending items or things to a group of users is much more difficult than recommending items to individuals. Because each users of the group has individual preferences. These days, almost all shopping mall, shopping sites are using group recommender system. Because the customers, who come to their premise have their own preferences and the shop owner can't stick to one's preferences. For example, the music being played at a shopping mall is applicable to group of user. The owner of the mall can't stick to individual's preferences. Hence recommending items or things for a group of users is a difficult task. There are many social choice strategies available for aggregating users' preferences and recommend items that are suitable for a group of users.

2.1 Recommendation Procedures

Whether recommender system or a group recommender system, they generally recommends items/things by using two techniques i.e. Collaborative Filtering and Content Based Filtering. The Collaborative filtering methods are based on collecting and analyzing a large amount of information on users behaviors, activities or preferences and predicting what users will like based on their similarity to other users. This procedure requires large volume of data. The idea behind the collaborative filtering is that, to get the recommendation from someone with similar liking to the current user. This filtering technique generally uses KNN, Bayesian Network and other algorithms to find out the similarity of preferences between two users. In this technique we need any machine analyzable content because it do not consider the properties of items/things while generating recommendation.

Content Based filtering methods recommends items that are similar to the items that the user liked in the past. So these type of recommender system generally study the users past behaviors and preferences and generate recommendation of items that are similar to those past items/things. In order to learn users past preferences the system has to learn it by using any machine learning algorithm which can later be used for predicting new items/things.

2.2 Aggregation Strategies

The main problem of group recommendation is that, how to adapt to the group as a whole based on information about individual users' like or dislike. In the movie recommender system, a lot of users rated some movies in the part. Based on their previous ratings we have to learn their preferences. After learning their preferences we have to aggregate their rating information to provide group recommendation. Some of the aggregation strategies [2, 5] are discussed below:

Utilitarian Strategy: Here utility values are taken into consideration for group recommendation. The utility values are of two types i.e. additive or multiplicative. For example, consider the movie recommender system. Here, we first add/multiply all the ratings of each movies separately for a group of users. Then we take the highest value of the aggregated value of the movies and list out those movies, those have equal utility value with the highest value, as recommended items.

Most Pleasure Strategy: Here we make the group ratings list based on the maximum of individual ratings. Those movies, whose have the highest rating values in common, will be added to recommended list.

Least Misery Strategy: Here we make a group ratings based on the minimum of the given individual ratings. Then the item with large minimum individual rating will be recommended. The idea behind this strategy is that, a group is as happy as its least happy member.

Vineet et al. implemented a new strategy called RTL strategy [5] which is the combination of all the above three strategy. Here the least ratings values of a movie are removed and the recursively and the maximum of the utility values are taken into consideration. This algorithm is performing better in terms of accuracy in group recommendation as compared to other three grouping strategies. But this algorithm do consider any method to create a better group. So in this paper we will make a new grouping strategy that will also consider how to make a better group while listing group recommendation.

3 Learning Users Preferences

In order to provide group recommendation, we have to first learn individual user's preferences. To achieve this end, we need to follow some machine learning algorithm. Using these algorithm and the users' past experience we will learn rule. Based on that learned rule we predict the new items that arise in the future. This is the most important part of any recommender system. Because when the learning of users' preferences is more accurate, the accuracy of the recommended items also increases. In this paper we followed the rule based approach to learn the user preferences. The algorithm used for learning users' preferences is described below.

3.1 Learning Algorithm

There are many machine learning algorithms available for rule learning. In this paper we followed a decision list based approach to Predictive Rule Mining [6] algorithm, for learning users' preferences. It is a multiclass rule learner, since our problem is a multiclass problem i.e. each user can rate a movie from 1 to 5. The rules which we learn are represented in DNF format. The rules are represented in a sequence of if-else if-else statements. It also comes with a default rule, means whenever a new instance do not satisfy any of the rules present in the list then the instance is assigned with default class associated with the rule.

Algorithm Used for Classification:

Input: Train data, decay factor (α),totalWeightFactor(δ),mingain, k $-$ value
Output: Set of Rules

Rule Set R $= \phi$
Find the class with maximum number of examples and set it as default class
Assign each example with a weightage 1.0
For eachclassdo
       Create the positive (P) and negative(N) examples array for the current class
       minimumTotalWeightThreshold $= \delta \times$ totalWeight(P)
       While totalWeight(P) $>$ minimumTotalWeightThreshold**do**
           $P' \leftarrow P , N' \leftarrow N$
           Make an empty rule r
           If bestgainaccordingto$P'$and$N' <$ mingain
                break
           While (bestgain $>$ mingainandrulelength $<$ maxlength)
                Find literal l with highest gain according to P'andN'
                Add literal l to rule r
                Remove examples from P'andN' that do not satisfyr
           Add rule r to the rule set R
           Decrease the weight of each example by a factor α from P
    Return Rule Set R

The above written algorithm is based on Predictive Rule Mining with a decision list based approach for a multiclass problem. Here we store the training examples in two arrays i.e. positive examples array and negative examples array for a particular class. Suppose we are learning rules for a particular class, then the training examples belong to that particular class will be treated as positive examples and the rest will be treated as negative examples. While the total weight of the positive examples is above a certain threshold value we continue to learn rule for that class. Unlike other sequential covering algorithms i.e. FOIL and RIPPER, this algorithm does not remove the positive examples after learning a rule. Hence we do not miss out any important rule from being generated due to removal of examples. After learning a rule we decrease the weight of the positive examples by a factor'α' that are covered by the current rule. While learning a rule the literals with best gain are added to the rule antecedent until the best gain falls below a certain minimum value or we reach the maximum rule length. The gain of a literal can be found out by the following formula:

$$\text{Gain(l)} = WP' \times \left(\log_2 \left(\frac{WP'}{WP' + NP'} \right) - \log_2 \left(\frac{WP}{WP + NP} \right) \right) \quad (1)$$

where WP' is the total weight of positive examples according to $l + r$, NP' is the total weight of negative examples according to $l + r$, WP is the initial total weight of the positive examples and NP is the initial total weight of negative examples for a particular class.

4 Proposed Aggregation Strategy

After learning rules for each user individually, we need to provide recommendation for a group of users. We discussed some of the aggregation strategies in Sect. 2.2 for group recommendation. We followed a different approach for group recommendation which is written below.

4.1 Proposed Algorithm

Input: test instances, users' ratings in vector format
Output: Recommended List

Classify all the test instances according to the learned rules.
Create a duplicate set of test instances as testmovies.
For eachinstanceioftestmovies**do**
    For eachuserj**do**

$$\text{sum}[i] = \text{sum}[i] + \text{userrating}[i][j]$$

Find the max value in the sum array
Store the movies whose sum values is equal to or differ at most by 1 % from max Value into a separate movie array.
Find the users who have given highest rating i.e. 5 to any of the movies present in Movie array and make them one group.
Remove the movies from test movies that are present in the movie array.
Similarly find other homogeneous groups by considering the rating of 4 from all the users by repeating the above steps.
Now ask the user to choose any of the created homogeneous groups.
For eachinstanceioftestinstance**do**

    $\text{MPS}[i]$ = highestratedvalueamongalltheusersinthegroup

List out the movies those have equal MPS value for the group.

Group recommendation will be effective when the members of the group have similar kind of preferences. The idea behind the above algorithm for group recommendation is, first we make some homogeneous groups and then generate recommendation for a particular group. While making groups we have consider the utilitarian strategy and find the similar ratings by a set of users and make them one group. In this process we will we will make a group of users those have similar kind of preferences. After generating homogeneous groups then we follow the most pleasure strategy approach among the group members as described in Sect. 2.2. Doing this, we ensure that most of the group member is happy with the recommendation because we will consider the maximum ratings given by the users in the group.

Table 1 Precision comparison

| Algorithms | Precision of grouping strategies (in %) | | | |
	Utilitarian	Most pleasure	RTL	Proposed method
FOIL	40.32	38.40	42.40	42.92
RIPPER	43.89	44.32	46.90	48.32
PRM	44.12	44.05	47.67	51.76

4.2 Experimental Verification

The group recommender system we built is a group movie recommender system. There are total 1,682 no of movies available in the train data set which contains the movie information. It also contains a user ratings file which contains 100,000 user ratings to all those movies by total 943 number of users. Each user has rated at least 20 movies. For learning rules we have taken the genre attributes of movies for classification. Each movie is categorized with 19 different genres. Users' ratings are from 1 to 5 where 1 stands for bad, 2 stands for average, 3 stands for good, 4 stands for very good and 5 stands for excellent. First we learned each user's preferences using the learning algorithm (Sect. 3.1) stated above. Then we tested our group recommendation using our proposed group recommendation algorithm as stated in previous section. There is no standard formula for evaluating [7] a group recommender system. For evaluating our group recommender system we use the following formula:

$$\text{Precision}(\text{singleuser}) = \frac{R \cap U}{R} \tag{2}$$

where the R is the set of recommended movies, U is the set of movies used by a particular user and $R \cap U$ is the set movies common in R and U. For calculating precision for a group we take the average of precision of all the users in the group. We have followed three different algorithms for learning users' preferences as written in and also followed the grouping strategies to generate group recommendation. After generating all the recommendation the average precision we get using different strategies along with our proposed strategy is given in (Table 1).

5 Conclusion

As we can see from the experimental result the proposed algorithm for group recommendation is performing better in terms of precision of the recommended movies, for the data set we described above. Also the learning algorithm we used is performing better for all most all the grouping strategies in comparison with other learning algorithm stated in the precision table. The proposed algorithm can be applied to other data set where ever group recommendations is required. Further we are working on how to assign weightage to some of the members in the group. Because this may be a case also that there are some member whose preference is influential to a group, so that we can generate better recommendation wherever applicable.

References

1. Basu, C., Hirsh, H., Cohen, W., et al.: Recommendation as classification: using social and content-based information in recommendation. In: AAAI, pp. 714–720. (1998)
2. Ricci, F., Rokach, L., Shapira, B.: Group recommender systems: combining individual models. Recommender Systems Handbook, pp. 677–702. Springer, Heidelberg (2011)
3. McCarthy, K., Salam, M., Coyle, L., McGinty, L., Smyth, B., Nixon, P.: Group recommender systems: a critiquing based approach. In: Proceedings of the 11th International Conference on Intelligent User Interfaces, pp. 267–269. ACM (2006)
4. Jameson, A.: More than the sum of its members: challenges for group recommender systems. In: Proceedings of the Working Conference on Advanced Visual Interfaces, pp. 48–54. ACM (2004)
5. Padmanabhan, V., Seemala, SK., Bhukya, WN.: A rule based approach to group recommender systems. In: Multi-Disciplinary Trends in Artificial Intelligence, pp. 26–37. Springer, Heidelberg (2011)
6. Yin, X., Han, J.: CPAR: Classification based on predictive association rules. In: SDM SIAM, vol. 3, pp. 369–376. (2003)
7. Zaier, Z., Godin, R., Faucher, L.: Evaluating recommender systems. In: Automated Solutions for Cross Media Content and Multi-channel Distribution, AXMEDIS'08 International Conference on IEEE, pp. 211–217. (2008)

An Improved PSO Based Back Propagation Learning-MLP (IPSO-BP-MLP) for Classification

D.P. Kanungo, Bighnaraj Naik, Janmenjoy Nayak,
Sarada Baboo and H.S. Behera

Abstract Although PSO has been successfully used in much application, the issues of trapping in local optimum and premature convergence can be avoided by using improved version of PSO (IPSO) by introducing new parameter called inertia weight. The IPSO is based on the global search properties of the traditional PSO and focuses on the suitable balance of the investigation and exploitation of the particles in the swarm for effective solution. During IPSO iterations, with increase in possible generations, the search space is decreased. Motivated from successful use of IPSO in many applications, in this paper, it is an attempt to design a MLP classifier with a hybrid back propagation learning based on IPSO. The proposed method has been tested using benchmark dataset from UCI machine learning repository and performances are compared with MLP, GA based MLP and PSO based MLP.

Keywords Data mining · Classification · Improved particle swarm optimization · Particle swarm optimization · Multilayer perceptron · Back propagation learning

D.P. Kanungo (✉) · B. Naik · J. Nayak · H.S. Behera
Department of Computer Science Engineering and Information Technology,
Veer Surendra Sai University of Technology,
Burla, Sambalpur, Odisha 768018, India
e-mail: dpk.vssut@gmail.com

B. Naik
e-mail: mailtobnaik@gmail.com

J. Nayak
e-mail: mailforjnayak@gmail.com

H.S. Behera
e-mail: mailtohsbehera@gmail.com

S. Baboo
Department of Computer Science and Application, Sambalpur University,
Burla, Sambalpur, Odisha 768018, India
e-mail: baboosarada@gmail.com

© Springer India 2015
L.C. Jain et al. (eds.), *Computational Intelligence in Data Mining - Volume 1*,
Smart Innovation, Systems and Technologies 31, DOI 10.1007/978-81-322-2205-7_32

1 Introduction

The PSO (Kennedy and Eberhart [1, 2]) is a meta-heuristic swarm based optimization technique, which has been successfully used in much application of science and engineering. The advantages of PSO like ease of implementation, less parameter settings, fast convergence and free from mathematical computations makes it more popular among other the optimization algorithm based on swarm intelligence. During the use of PSO for many applications, researchers found that, PSO may be trapped at some local optimum which leads to premature convergence. Many researchers addressed this issue to enhance the performance of the standard PSO.

The method of improved PSO is basically based on the global search properties of the traditional PSO and focuses on the suitable balance of the investigation and exploitation of the particles in the swarm for effective solution. The common steps of PSO like updation of position, velocity and fitness terms of the swarms will remain same in IPSO. Here, a new parameter called adaptive inertia weight (λ) is added to the basic equation of PSO. With the increase number of generations and by setting the parameters for λ, the value of λ can be decreased in a gradual manner. As a result, during the search procedure of IPSO method, when the number of possible generations will increase, the search space will be decreased. Hence, during the iteration the weak particle in the current generation will make a replacement with the best particle of previous generation which will helpful to avoid the premature convergence. Also, each particle will share the information with other particles having only the global best (g_{best}) value in the search space. In the dominion of the improving properties of PSO, IPSO has been applied in various application domains. A few among them have been discussed relevant to the proposed work.

Yang et al. [3] described an improved PSO algorithm for onboard embedded applications in power-efficient wireless sensor networks (WSNs) and WSN-based security systems for significant improvement on the performance of basic PSO. The effectiveness of Vehicle routing and scheduling Problems by using the improved PSO is being realized by Zhang and Lu [4]. Wu et al. [5] has successfully used the improved PSO for optimizing the body of gravity dam and sluice gate. By improving the basic PSO, Tang et al. [6] has designed a S Curve controller for Motion control of Underwater Vehicle using IPSO technique. To optimize the authority and threshold values of back propagation nerve network, Chen et al. [7] has used the IPSO to obtain fast convergence speed. Park et al. [8] have applied the IPSO technique with chaotic sequence for Nonconvex Economic Dispatch Problems. Ran et al. [9] explained an improved PSO based amphibious mouse robot for the purpose of path planning. Chew and Zarrabi [10] have made an effort on digital speckle correlation method to measure surface displacements and strains by assuming first order linear deformation using IPSO and compared the resulting performance with PSO and GA. Ishaque et al. [11] have implemented an enhanced maximum power point tracking (MPPT) technique for the photovoltaic (PV) system using an IPSO algorithm. Qais and Wahid [12] have developed a novel method called TriPSO for the improvisation of IPSO. Geng and Zhao [13] designed a method for UAV based track planning using the IPSO algorithm. Shakiba et al. [14] have introduced a humanoid soccer playing robot by using Ferguson splines and IPSO technique. Yanqiu et al. [15] have developed an IPSO based three-axis measuring system calibration problems for solving the optimal local parameters. Qu and Yue [16] have used the IPSO algorithm to solve the constraint optimization problems and have used a new mutation operator to improve the global search ability of PSO. Barani et al. [17] have implemented a novel IPSO based chaotic cellular automata to get the high exploration capability in the randomness nature of the algorithms. An efficient classification method based on PSO and GA based hybrid ANN has been proposed by Naik et al. [18] and it is found relatively better in performance as compared to other alternatives.

In this paper, a MLP with IPSO based back propagation learning has been proposed for classification. The rest part of this paper is organized as follow: Preliminaries, Proposed Method, Experimental Setup and Result Analysis, Conclusion and References.

2 Preliminaries

2.1 Particle Swarm Optimization

Particle swarm optimization (PSO) [1, 2] is a widely used stochastic based algorithm and it is able to search global optimized solution. Like other population based optimization methods, the particle swarm optimization starts with randomly initialized population for individuals and it works on the social behavior of particle to get the global best solution by adjusting each individual's positions with respect to global best position of particle of the whole population (Society). Each individual is adjusting by changing the velocity according to its own experience and by observing the position of the other particles in search space by use of Eqs. 1 and 2. Equation 1 is for social and cognition behavior of particles respectively where $c1$ and $c2$ are the constants in between 0 and 2 and rand(1) is random function which produces random number between 0 and 1.

$$V_i(t+1) = V_i(t+1) + c_1 * rand(1) * (lbest_i - X_i) + c_2 * rand(1) * (gbest_i - X_i)$$
(1)

$$X_i(t+1) = X_i(t) + V_i(t+1)$$
(2)

Basic steps of PSO can be visualized as:

Initialize the position of particles $V_i(t)$ (population of particles) and velocity of each particle $X_i(t)$.
Do
    Compute fitness of each particle in the population.
    Generate local best particles (LBest) by comparing fitness of particles in previous population with new population.
    Choose particle with higher fitness from local best population as global best particle (GBest).
    Compute new velocity $V_i(t + 1)$ by using eq.1.
    Generate new position $X_i(t + 1)$ of the particles by using eq.2.
While *(iteration <= maximum iteration OR velocity exceeds predefined velocity range);*

2.2 Improved Particle Swarm Optimization (IPSO)

In SPSO, the basic three steps like calculation of velocity, position and the fitness value will be iterated till the required criteria of convergence is met. The ending criteria may be the maximum change in the best fitness value. However, if the velocity of the swarm will be fixed to zero or nearer to that and the best position will have a fixed value, then the SPSO may lead to be trapped at some of local optima. This happens only due to the swarm's experience on the current and global positions. This experience is to be avoided and should be based on the mutual cooperation among all the swarms in a multidirectional manner [19].

So, in IPSO a new inertia weight factor λ is introduced to control both the local and global search behavior. The value of λ may be decreased quickly [dehuri] during the initial iterations and slowly during the optimal iterations.

The new velocity and position updation can be realized through the Eqs. 3 and 4.

$$V_i^{(t+1)} = \lambda * V_i^{(t)} + c_1 * rand(1) * \left(l_{best_i}^{(t)} - X_i^{(t)} \right)$$
$$+ c_2 * rand(1) * \left(g_{best}^{(t)} - X_i^{(t)} \right) \tag{3}$$

$$X_i^{(t+1)} = X_i^{(t)} + V_i^{(t+1)} \tag{4}$$

Basic steps of Improved PSO can be visualized as:

Improved PSO

Set the inertia weight λ, maximum number of iteration maxIter and other control parameters.
For each particle in the population P, initialize particle's velocity and position randomly.
Iter=1
While (Iter <= maxIter)
    If(Iter==1)
        Lbest=P
    else
        For each particle in the population
            Evaluate the fitness value of each particle in P
            If the current fitness value is better than the previous particle in P
                Set the current particle position as the new local best position.
            EndIf
        Endfor
    Endif
    Select particle having maximum fitness from P as global best particle.
    For each particle in the population P
        Compute particle's velocity by using eq.3
        Update particle's position using eq.4
    EndFor
    Iter = Iter +1
endWhile

Even if SPSO performs well in global space due to its capability of computing promising regions in the search space, rapid search near global optimum is very slow. If the g_{best} position cannot be improved for some consecutive generations, then the self-adaptive ES [20] is used for further improvement of g_{best} position.

2.3 Multi Layer Perceptron

MLP (Fig. 1) is the simplest neural network model which is consists of neurons called perceptrons (Rosenblatt 1958). From multiple real valued inputs, the peceptron compute a single output according to its weights and non-linear activation functions. Basically MLP network is consists of input layer, one or more hidden layer and output layer of computation perceptions.

MLP is a model for supervised learning which uses back propagation algorithm. This consists of two phases. In the 1st phase, error (Eq. 6) based on the predicted outputs (Eq. 5) corresponding to the given input is computed (forward phase) and in the 2nd phase, the resultant error is propagated back to the network based on that weight of the network are adjusted to minimize the error (Back Propagation phase).

$$y = f\left(\sum_{i=1}^{n} w_i x_i + b\right) \qquad (5)$$

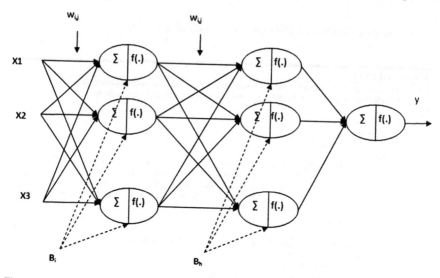

Fig. 1 MLP with input layer, single hidden layer and output layer

where w is the weight vector, x is the input vector, b is the bias and $f(.)$ is the non-linear activation function.

$$\delta_k = (t_k - y_k)f(y_{in_k}) \tag{6}$$

where t_k and y_k is the given target value and predicted output value of input kth pattern and δ_k is the error term for kth input pattern.

The popularity of MLP increases among the neural network research community due to its properties like nonlinearity, robustness, adaptability and ease of use. Also it has been applied successfully in many applications [21–29].

3 Proposed Method

In this section, we have proposed a IPSO based back propagation learning-MLP (IPSO-BP-MLP) for classification. Here basic concepts and problem solving strategy of IPSO evolutionary algorithm is used to enhance performance of MLP classifier.

Algorithm. IPSO based Back Propagation Learning- MLP (IPSO-BP-MLP) for classification

---

% **Initialization of population**

    P = round (rand(n,(c-1)*(c-1)));

    Where n is the number of weight-sets (chromosomes) in the population 'P' and 'c' is the number of attributes in dataset (excluding class label).

% **Initialization of velocity**

    v = rand (n,(c-1)*(c-1));

% **NEURAL SETUP for MLP**

    Bh =(rand(c-1,1))';   % Bh: Bias of hidden layer.

    Bo = rand(1);   % Bo: Bias of output layer.

% **IPSO Iterations**

Iter=0; % **Iter: Iteration**

while(1)

    1. Selecting local best weight-sets (lbest) by comparing with weight-sets in previous population. If it is first iteration, then initial population (P) is considered to be local best(lbest). Otherwise, new 'lbest' population is formed by selecting best weight-sets from previous population (P) and current local best (lbest).

        Iter = Iter+1;

        If (Iter == 1)

            lbest = P;

      else

            [lbest] = lbestselection (lbest, P, tdata, t);

      end

    2. Compute fitness of all weight-sets in local best 'lbest'. Each weight-sets are set individually in MLP and trained with training data 'tdata'. RMSE for each weight-sets are calculated with respect to target 't'. Based on RMSE, fitness of weight-sets are calculated by using 'fitfromtrain' procedure.

        [F] = fitfromtrain (lbest, tdata, t);

    3. Select a global best 'gbest' from local best 'lbest' based on their fitness(F) (calculated by using 'fitfromtrain' procedure) by using 'gbestselection' procedure.

        [gbest] = gbestselection (P, F, rmse);

    4. Compute new velocity 'vnew' from population (P), velocity (v), local best 'lbest' and global best 'gbest'by using 'calcnewvelocity' procedure.

        vnew = calcnewvelocity (P, v, lbest, gbest);

    5. Update next position by using current population (P) and new velocity (vnew).

        P = P + vnew;

        If (Iter == maxIter)

            break;

      end

end

function [lbest]=lbestselection (lbest, P, tdata, t)

1. Compute fitness of all weight-sets in local best 'lbest' and previous population 'P'. Each weight-sets are set individually in MLP and trained with training data 'tdata' by using 'fitfromtrain'. RMSE for each weight-sets are calculated with respect to target 't'. Based on RMSE, fitness of weight-sets is calculated.

      [F1] = fitfromtrain (lbest, tdata, t);
      [F2] = fitfromtrain (P, tdata, t);

2. Compare fitness of weight-sets in lbest and P by comparing F1 and F2 , where F1 and F2 are fitness vector of lbest and P respectively. Based on this comparison, generate new lbest for next generation.

```
for i=1:1:number of weight-sets in P or lbset.
    if(F1(i,1)<=F2(i,1))
        lbest(i,:)=P(i,:);
    else
        lbest(i,:)=lbest(i,:);
    end
end
```

end

function [F] = fitfromtrain(data, tdata, t)

Step-1: Repeat step 1 to 6 for all the weight-set 'w' in 'P'.
Here tdata is the training data without class label and it is considered to be tdata=$\{x_1,x_2,....,x_L\}$, L is the number of input patterns.
Error=0.

Step-2: Repeat step-3 to step-6 for all the patterns $x_k = (x_{k1}, x_{k2}, ..., x_{kn})$ in given dataset tdata, where n is the number of attribute value in a single input pattern.

Step-3: Feed forward stage.

$$Z_{in_j} = Bh_j + \sum_{i=1}^{n} x_{ki} w_{ij}$$

Here Z_{in_j} represents j^{th} hidden unit and i=1,2..n, where n is the number of input which are connected to a single hidden unit.

$y_{in_k} = B_{ok} + \sum_{j=1}^{m} Z_j W h_{jk}$, Where j=1,2..m, m is the number of hidden unit connected to one output unit.

$y_k = f(y_{in_k})$, for k=1,2,..p, where p is the number of output unit in output layer and $f(.)$ is the sigmoid activation function and can defined as follows.

$$f(y_{in_k}) = \frac{1}{1 + e^{-y_{in_k}}}$$

Step-4: Calculation of error terms in hidden layer.

$\delta_k = (t_k - y_k) f(y_{in_k})$, for each output unit y_k, k=1,2....m. For each hidden unit Z_j, j=1,2...n, sums of delta input is computed as follows.

$$\delta_{in_j} = \sum_{i=1}^{m} \delta_k w_{jk}$$

Error terms are calculated as: $\Delta_j = \delta_{in_j} f_1(z_{in_j})$, where $f_1(.)$ is the activation function which is defined as:

$$f_1(z_{in_j}) = f(z_{in_j})(1 - f(z_{in_j}))$$
$$= \frac{1}{1+e^{-z_{in_j}}} \left(1 - \frac{1}{1+e^{-z_{in_j}}}\right)$$

Finally error of the network in calculated as: Error = Error + Δ_j.

Step-5: if all input patterns in dataset are processed then goto step-6. Else goto step-1.
Step-6: Compute RMSE based on total error of the network and compute fitness of the weight-set as F(i)=1/RMSE. , where F(i) is the fitness of ith weight-set in P.

end

```
function vnew=calcnewvelocity (P, v, lbest, gbest)
    for i = 1:1:number of row in population 'P'.
        for j = 1:1: of column in population 'P'.
            vnew(i,j) = λ * v(i,j) + r1 * c1 * ( lbest(i,j) - P(i,j) ) + r2 * c2 * ( gbest(1,j) - P(i,j) );
        end
    end
end
```

4 Experimental Setup and Result Analysis

In this section, the classification accuracy (Eq. 7) of our proposed method has been presented and compared with other classifiers. Benchmark datasets (Table 1) from UCI machine learning repository [30] have been used for classification. Before training and testing is made, datasets are normalized using Min-Max normalization. Comparative study on classification accuracies of models have been listed in Table 2. Let 'cm' be a confusion matrix of order m × n. Then the classification accuracy of classifiers is calculated as:

$$Clssification\,Accuracy = \frac{\sum_{i=1}^{n} \sum_{\substack{j=1, \\ i==j}}^{m} cm_{i,j}}{\sum_{i=1}^{n} \sum_{j=1}^{m} cm_{i,j}} \times 100 \qquad (7)$$

4.1 Parameter Setting

During simulation, c1 and c2 constants of PSO has been set to 2 throughout the experiment. The inertia weight λ is set in between [1.8, 2]. In proposed method, one input layer, one hidden layer and one output layer for the MLP neural network has been set during training and testing.

Table 1 Data set information

Dataset	Number of pattern	Number of features (excluding class label)	Number of classes	Number of pattern in class-1	Number of pattern in class-2	Number of pattern in class-3
Monk 2	256	06	02	121	135	–
Hayesroth	160	04	03	65	64	31
Heart	256	13	02	142	114	–
New thyroid	215	05	03	150	35	30
Iris	150	04	03	50	50	50
Pima	768	08	02	500	268	–
Wine	178	13	03	71	59	48
Bupa	345	06	02	145	200	–

Table 2 Performance comparison in terms of accuracy

Dataset	Accuracy of classification in average									
	MLP		GA-MLP		PSO-MLP		IPSO-MLP	
	Train	Test	Train	Test	Train	Test	Train	Test
Monk 2	86.94648	85.27453	87.23734	87.85732	90.19375	92.44732	90.84623	93.62544
Hayesroth	83.48576	82.38657	85.43675	81.04653	88.38271	81.97365	89.28463	84.76935
Heart	82.84653	74.77453	85.44937	75.03645	86.23755	75.84651	88.57295	76.29754
New thyroid	92.03782	73.26876	92.74834	73.92756	93.02785	75.92784	93.73562	77.75562
Iris	90.87365	92.15368	92.56873	95.93158	92.51736	93.36158	93.28569	94.39826
Pima	73.73645	73.88647	76.82642	77.38275	78.17464	78.28746	78.83645	79.36244
Wine	80.69731	80.87294	88.93645	77.37565	90.75389	91.77319	91.32753	91.85934
Bupa	68.66251	69.82637	70.97485	71.27459	70.27841	71.21465	70.73923	71.72649

5 Conclusion

In this paper, the weights of MLP Artificial Neural Network have been optimized using IPSO based back propagation learning scheme. The comparison of performance analysis of the results indicates that the proposed method is better in classification accuracy than the other methods (MLP, GA-MLP, PSO-MLP). In future, our work may extend in this interest by better optimization of weights of MLP with hybridization of back propagation learning with other variants of PSO.

References

1. Kennedy, J., Eberhart, R.: Particle swarm optimization. In: Proceedings of the 1995 IEEE International Conference on Neural Networks, vol. 4, pp. 1942–1948 (1995)
2. Kennedy, J., Eberhart, R.: Swarm Intelligence Morgan Kaufmann, 3rd edn. Academic Press, New Delhi (2001)
3. Yang, E., Erdogan, A.T., Arslan, T., Barton, N.: An improved particle swarm optimization algorithm for power-efficient wireless sensor networks. In: 2007 ECSIS Symposium on Bio-inspired, Learning, and Intelligent Systems for Security. (2007). doi 10.1109/BLISS.2007.19
4. Zhixia, Z., Caiwu, L.: Application of the improved particle swarm optimizer to vehicle routing and scheduling problems. In: Proceedings of 2007 IEEE International Conference on Grey Systems and Intelligent Services, Nanjing, pp. 1150–1152, 18–20 Nov 2007
5. Wu, X., Qie, Z., Zhang, Z.Z.H.: Application of Improved PSO to optimization of gravity dam and sluice gate *. In: Proceedings of the 7th World Congress on Intelligent Control and Automation, Chongqing, pp. 6178–6182, 25–27 June 2008
6. Tang, X., Pang, Y., Chang, W., Li, Y.: Improved PSO-based S curve controller for motion control of underwater vehicle. In: Proceedings of the 7th World Congress on Intelligent Control and Automation, Chongqing, pp. 6949–6954, 25–27 June 2008
7. Chen, Q., Guo, W., Li, C.: An improved PSO algorithm to optimize BP neural network. In: 2009 5th International Conference on Natural Computation, pp. 351–360 (2009)
8. Park, J.-B., Jeong, Y.-W., Shin, J.-R., Lee, K.Y.: An improved particle swarm optimization for nonconvex economic dispatch problems. IEEE Trans. Power Syst. 25(1), 156–166 (2010)
9. Ran, M., Duan, H., Gao, X., Mao, Z. : Improved particle swarm optimization approach to path planning of amphibious mouse robot. In: 2011 6th IEEE Conference on Industrial Electronics and Applications, pp. 1146–1149 (2011). doi:978-1-4244-8756-1
10. Chew, K.S., Zarrabi, K.: Non-contact displacements measurement using an improved particle swarm optimization based digital speckle correlation method. In: 2011 International Conference on Pattern Analysis and Intelligent Robotics, Putrajaya, pp. 53–58, 28–29 June 2011
11. Ishaque, K., Salam, Z., Amjad, M., Mekhilef, S. : An improved particle swarm optimization (PSO)–based MPPT for PV with reduced steady-state oscillation. IEEE Trans. Power Electron. 27(8), 3627–3638 (2012)
12. Qais, M., AbdulWahid, Z.: A New Method for Improving Particle Swarm Optimization Algorithm (TriPSO). doi: 978-1-4673-5814-9
13. Geng, Q., Zhao, Z.: A Kind of Route Planning Method for UAV Based on Improved PSO Algorithm. doi: 978-1-4673-5534-6
14. Shakiba, R., Najafipour, M., Salehi, M.E.: An Improved PSO-Based Path Planning Algorithm For Humanoid Soccer Playing Robots. doi: 978-1-4673-6315-0

15. Yanqiu, L., Zheng, W., Dayu, Z., Yinghui, L., He, W.: Improved particle swarm optimization applied in calibrating a three-axis measuring system. In: Control Conference (CCC). 2013 32nd Chinese, pp. 3120–3123, IEEE (2013)
16. Qu, Z., Li, Q., Yue, L.: Improved particle swarm optimization for constrained optimization. In: 2013 International Conference on Information Technology and Applications, pp. 244–247 (2013)
17. Barani, M.J., Ayubi, P., Hadi, R.M.: Improved Particle Swarm Optimization Based on Chaotic Cellular Automata. doi: 978-1-4799-3351-8
18. Naik, B., Nayak, J., Behera, H.S.: A Novel FLANN with a hybrid PSO and GA based gradient descent learning for classification. In: Proceedings of the 3rd International Conference on Front of Intelligent Computing (FICTA). Advances in Intelligent Systems and Computing, vol. 1(327), pp. 745–754 (2014). doi: 10.1007/978-3-319-11933-5_84
19. Yue-bo, M.: ZouJian-hua, GanXu-sheng, ZhaoLiang: research on WNN aerodynamic modeling from flight data based on improved PSO algorithm. Neurocomputing 83, 212–221 (2012)
20. Yang, J.M., Kao, C.Y.: A robust evolutionary algorithm for training neural networks. Neural Comput. Appl. 10, 214–230 (2001)
21. Hamedi, M., et al.: Comparison of multilayer perceptron and radial basis function neural networks for EMG-based facial gesture recognition. In: The 8th International Conference on Robotic, Vision, Signal Processing and Power Applications, Springer, Singapore (2014)
22. Ndiaye, A., et al.: Development of a multilayer perceptron (MLP) based neural network controller for grid connected photovoltaic system. Int. J. Phys. Sci. 9(3), 41–47 (2014)
23. Roy, M., et al.: Ensemble of multilayer perceptrons for change detection in remotely sensed images. Geosci. Remote Sens. Lett. 49–53 IEEE 11.1, New York (2014)
24. Hassanien, A.E., et al.: MRI breast cancer diagnosis hybrid approach using adaptive ant-based segmentation and multilayer perceptron neural networks classifier. Appl. Softw Comput. 14, 62–71 (2014)
25. Aydin, K., Kisi, O.: Damage detection in Timoshenko beam structures by multilayer perceptron and radial basis function networks. Neural Comput. Appl. 24(3-4), 583–597 (2014)
26. Velo, R., López, P., Maseda, F.: Wind speed estimation using multilayer perceptron. Energy Convers. Manag. 81, 1–9 (2014)
27. Lee, S., Choeh, J.Y.: Predicting the helpfulness of online reviews using multilayer perceptron neural networks. Expert Syst. Appl. 41(6), 3041–3046 (2014)
28. Azim, S., Aggarwal, S.: Hybrid model for data imputation: using fuzzy c means and multi layer perceptron. In: IEEE Advance Computing Conference (IACC), pp. 1281–1285 (2014)
29. Chaudhuri, S., Goswami, S., Middey, A.: Medium-range forecast of cyclogenesis over North Indian Ocean with multilayer perceptron model using satellite data. Nat. Hazards 70(1), 173–193 (2014)
30. Bache, K., Lichman, M.: UCI Machine Learning Repository. [http://archive.ics.uci.edu/ml]. Irvine. CA: University of California, School of Information and Computer Science (2013)

GDFEC Protocol for Heterogeneous Wireless Sensor Network

S. Swapna Kumar and S. Vishwas

Abstract Wireless sensor networks (WSNs) in recent years shown abrupt growth in technological applications. The main research goals of WSN in the area of heterogeneity are to achieve various matrix performances such as high energy efficiency, lifetime and packet delivery nodes. Most proposed clustering algorithms do not consider the situation causes hot spot problems in multi-hop WSNs. To achieve such network the two soft computing techniques applied to energy efficient clustered heterogeneous sensor node network. In this paper proposed the implementation of the real time energy efficient clustering using a Genetic Dual Fuzzy Entropy Clustering (GDFEC) algorithm. Various matrixes of simulation carried out using MATLAB to study the performance under setup conditions. This creates a standardized power distribution among disseminated cluster nodes in the heterogeneous network. The protocol realization carried out on software simulation by different empirical test. The empirical analysis of GDFEC protocol compared with different traditional protocol to evaluate the level of resultant matrix. The protocol evaluation studies have shown that GDFEC protocol able to improve the network performance matrix under the heterogeneous distribution of network nodes.

Keywords Black hole · Clustering · Entropy-based algorithms · Fuzzy clustering · Genetic · Wireless sensor networks

S. Swapna Kumar (✉)
Department of Electronics and Communication Engineering, Vidya Academy of Science and Technology, P.O. Thalakkottukara, Thrissur, Kerala 680501, India
e-mail: drsswapnakumar@gmail.com

S. Vishwas
Department of Computer Science and Engineering, KVG College of Engineering, Sullia, Karnataka, India
e-mail: vshcool@gmail.com

© Springer India 2015
L.C. Jain et al. (eds.), *Computational Intelligence in Data Mining - Volume 1*, Smart Innovation, Systems and Technologies 31, DOI 10.1007/978-81-322-2205-7_33

1 Introduction

In the modern era the autonomous network monitoring by wireless sensor networks (WSN) proved to be every efficient. The wireless networks consist of large number of tiny low power sensor node with limited computational capability to access energy efficiently [1]. The nodes gather data from various sources and aggregate it to the destination as per the application. There are numerous application of WSN such as military surveillance, habitat monitoring, target tracking, pressure, temperature, vibration monitoring, health care monitoring, and disaster prevention monitoring etc. The data sensed by distributed sensors are randomly deployed in the network depend upon application is pre-processed to send aggregated data to the Base Station (BS) is normally called as Sink. This data processing requires numerous computation leads power constraint to standby for loner period [2]. Thus several research works towards sensor is carried out in order to operate in hostile environment to be capable of fault tolerance and processes data for energy efficiency routing in WSN [3].

Different techniques such as Reinforcement learning [5], Neural networks [6], Fuzzy logics [7, 8] and Genetic algorithms [9] have been able to optimize the problems. Here we applied the approach of Fuzzy and Genetic.

Genetic algorithms produce a robust optimization technique because they ensure a gradual improvement in the solution optimization. In this paper we study the performance implantation of Genetic Dual Fuzzy Entropy Clustering (GDFEC) algorithm in heterogeneous WSN protocol and compared with existing protocols under different matrix. The empirical study discriminate each protocol on the basis of energy efficient network lifetime, packet delivery and lifetime of nodes under numerous rounds.

The remaining part of the paper is planned as follows: Sect. 2 refers the related work done. The Sect. 3 considers the assumption and properties, Sect. 4 discuss on proposed work. Further the Sect. 5 presented the simulation and its analysis. Finally concluding Sect. 6 makes remarks and future scope.

2 Related Work

WSN Heinzelman [10] proposed Low Energy Adaptive Cluster Hierarchy (LEACH) in a distributed clustering algorithm for homogenous sensor network. LEACH protocol uses single hop routing to transmit data to sink by TDMA mechanism. Fu et al. proposed Fuzzy approach clustering is applied with crisp cluster [11]. Fuzzy Clustering Method (FCM) is one of the most popular fuzzy clustering techniques proposed by DUNN [6]. However FCM algorithm approach has showed some problems, due to the complexity of the cluster values that has error. Yao et al. proposed Entropy based Fuzzy Clustering (EFC) method is formed by means of threshold value [12]. Performance varies based on 2–3 level/multilevel heterogeneous WSNs. The scopes for further research in this concept still exist in the clustered network protocol. Enhanced SEP [13] is a clustering algorithm in a three tier node scenarios, that prolongs the network effective life time for homogenous and heterogeneous environment in terms of network lifetime and resource sharing. Deterministic SEP (DSEP) [14] is a two levels, three level and multi-level heterogeneous hierarchical WSN shows improvement in terms of energy consumption per round, data transmission and life time of sensor network. Liu et al. [15] proposed Energy-Efficient Prediction Clustering Algorithm (EEPCA). EEPCA determines node energy factor by comparing the energy of a node with the average energy of other nodes. Brahim et al. [16] proposed Stochastic and Balanced Development Distributed Energy Efficient Clustering (SBDEEC) protocol for CH selection to extend the network lifetime. The simulation result shows SBDEEC protocol performs better than SEP and DSEP in terms of network lifetime. However, some of the heterogeneous network algorithm such as SEP, DSEP, SBDEEC overcome the above issues.

3 Assumption and Properties

The assumption and properties of the protocol in a heterogeneous network is listed below:

- WSN model of the node in heterogeneous distribution carries initially same amount of energy.
- Sensor nodes are randomly distributed are stationary along with BS of the network field.
- Each sensor nodes know their own geographical position.
- Nodes are capable of adjusting their transmission power during data transmission phase.
- There is one sink station which is located at the centre of the sensing field.
- Cluster member transfer the data via cluster head (CH) to sink where nodes are randomly distributed.

4 Proposed Approach

The proposed work carried out in an autonomous distributed isotropic network where nodes are randomly placed. Here the mode of $N \times N$ square area is considered to measure the result matrix for transmitting data from Source to Sink. The proposed model is applied with Fuzzy Interference System for the process of clustered output consists of four modules: fuzzifier, fuzzy inference engine, fuzzy rules base, defuzzifier. In the proposed method, the Fuzzy Interference System used Mamdani Fuzzy Inference system [17, 18]. Generic algorithm is applied to the

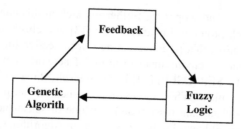

Fig. 1 Genetic algorithm system

output factors of Fuzzy Interference System as per the Fig. 1 for the optimization. The feedback is for improving the scaling factors.

The fundamental objective is to synthesize a clustered network control that is able to optimize the energy consumption analysis as close as possible. The density, energy consumption and energy estimation with respect to total energy and number of rounds mathematically formulate and applied the soft computing techniques. Here the Fuzzy approach is used to optimize the clustering.

The proposed approach is realized by considering the following attributes:

- The density of cluster from intra and inter clusters ρ
- Energy consumption in the CH for data transmission in terms of entropy is E_i
- The average energy estimation is given as $\overline{M}$
- The total initial energy of the heterogeneous network is given by E_{total}
- The total energy dissipated for given number of round (R) of the network is given by E_{round}
- Cluster head transmission energy is given as $CH_{Tx\ Energy}$
- Cluster head receiving energy is given as $CH_{Rx\ Energy}$
- The threshold value for optimum cluster head is $T(s)$
- The total energy of the cluster of N_i sensor node is given by $E_{Total\ Cluster}$.

4.1 Design of the Fuzzy Logic DFEC

The block diagram of DFEC using the Fuzzy logic is shown in Fig. 2. The block diagram consists of three inputs variables density ρ, Energy consumption E_i and Average energy (entropy) $\overline{M}$ are chosen to analyze the amount of energy consumption and the errors observed. Any estimated error is reduced based on the Fuzzy feedback inputs to membership scaling factor K_1, K_2 and K_3.

The inputs/output linguistic variables are defined as below:

$$\begin{cases} E_{total} = E_i\ \overline{M}\ \rho \\ E_{round} = \overline{M}\ E_i\ \rho \end{cases}$$

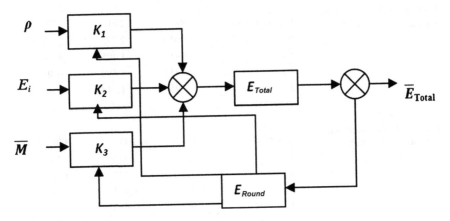

Fig. 2 Block diagram of DFEC

Table 1 DFEC rules' base

	ρ	E_i	$\overline{M}$
E_{Total}	E_i	$\overline{M}$	ρ
E_{Round}	$\overline{M}$	ρ	E_i

Triangular distribution in $[-1\ 1]$ interval were chosen as membership functions for the scaled inputs of ρ, E_i and $\overline{M}$.

The DFEC rule is based on Table 1. The DFEC algorithm considers these outputs of the Fuzzy inference system to determine the energy efficient sensor node eligibility factor. This measures the algorithm performance for a fixed amount of nodes alive and varies exponentially for all protocols. The estimation E_{total} is not the optimized so further optimization is done based on Genetic algorithm.

4.2 Design of the GDFEC Algorithm

The DFEC is tuned by genetic algorithm in order to reduce computational time and error control in the energy level measurement. Based on genetic fitness function the convergence criterion is used to measure the DFEC inputs scaling factor.

The fitness function J is given as:

$$J = \frac{1}{2} \int \left(\left(\frac{\rho}{\rho(1)} \right)^2 + \left(\frac{E_i}{E_i(1)} \right)^2 + \left(\frac{\overline{M}}{\overline{M}(1)} \right) \right) dt \tag{1}$$

where, $\rho(1)$, $E_i(1)$, and $\overline{M}(1)$ are the matrix error difference.

Table 2 GA's results with population size S = 300

Population size	50	200	350
K_1	2.014	2.175	2.223
K_2	1.412	1.405	1.464
K_3	0.411	0.399	0.405
J	1.954	1.948	1.949

Table 3 GA's results with generations G = 350

Generation	50	200	350
K_1	2.451	2.216	2.284
K_2	1.549	1.125	1.715
K_3	0.452	0.415	0.441
J	1.951	1.950	1.952

The genetic population and fitness function of genetic represents the genetic algorithm. The fitness function is representing as per populations.

The genetic algorithm steps are as follows:

Step 1: Initial generate random population of fixed size according to the variation range of scaling factor K_1, K_2 and K_3.

Step 2: Evaluate the fitness J of each individual population.

Step 3: Select the fittest of the reproduce population.

Step 4: Generate new population operator through reproduction, crossover and mutation.

Tables 2 and 3 show the genetic algorithm results for different size of population and generations' number.

The DFEC using Fuzzy when optimized by the execution of Genetic algorithm GDFEC for producing best population generates the three scaling factors K_1, K_2 and K_3. The GDFEC produce most significant population's size and generations' number are the since they have direct influence on the convergence of the GA to the optimal solution.

5 Simulation and Analysis

In this section we have simulated out distributed clustered wireless sensor networks in a sensing field of 100 m × 100 m size with 100 nodes that is autonomous distributed randomly placed using MATLAB 7.2. Simulation parameter used is listed in Table 4.

The Fig. 3a shows the number of nodes alive versus rounds for the lifetime of the network.

Table 4 Simulation parameter

Parameters	Value
Network sensing area	100 m, 100 m
Number of sensor nodes	100
Initial node energy	1 J
Data packet size	2,000 bits
E_{elec}	45 nJ/bit
E_{fs}	10 pJ/bit/m^2
E_{amp}	0.0008 pJ/bit/m^4
E_{DA}	4 nJ/bit/packet
Base station location	100 m
Aggregation	10 %

In this there is a significant improvement in number of nodes the lifetime of the network increases. The lifetime of GDFEC is longer and stable as compared to DFEC, LEACH-C, EEPCA, and SBDEFC. The algorithm of GDFEC is better than DFEC by 13.4 %, by 13.8 % to SBDEFC, to EEPCA by 14.3 % and by 14.8 % to LEACH-C. In EEPCA the nodes dies around 6,500 rounds while for GDFEC the node life attains till 8,930 rounds.

The Fig. 3b shows the comparison in terms of number of node alive for data packets delivery versus number of rounds to receive at the sink station.

The results show that GDFEC protocols linearly rise for around 4,070 rounds when compared to DFEC, LEACH-C, EEPCA, and SBDEFC. The GDFEC shows improvement to DFEC by 13.8 %, by 14.4 % to SBDEFC, to EEPCA by 15.8 % and by 16.4 % to LEACH-C. The throughput of GDFEC has improved as more numbers of data packets received at sink station when compared to other protocol.

In the Fig. 3c shows amount of total residual (remaining) energy over time for a given number of rounds.

The test carried out with different protocol shows total initial energy is 1 J that decreases linearly up to around 2,000–3,000 rounds except for GDFEC that is around 2,600. Energy per round is more in GDFEC as compared to DFEC, EEPCA, LEACH-C and SBDEEC up to around 6,100 rounds then other graph dies earlier. The GDFEC protocol retains most of the energy is consumed in the first 8,400 rounds.

In the Fig. 3d shows the stability of network energy on heterogeneous networks for fixed percentage of node life changes.

The network sends data to 17 % nodes of high quality and reliability. GDFEC shows more stable period throughout the network. It is observed that GDFEC shows better performance as compared to DFEC, EEPCA, LEACH-C and SBDEEC protocol. GDFEC obtain 42.3 % more stable period based on empirical study in heterogeneous network nodes. Therefore, with greater proportion of heterogeneous nodes, a more stable period is obtained in GDFEC algorithm protocol.

Fig. 3 **a** Number of alive
nodes. **b** Amount of packet
delivery. **c** Total remaining
energy over number of
rounds. **d** Stability of network
energy on heterogeneous
nodes changes

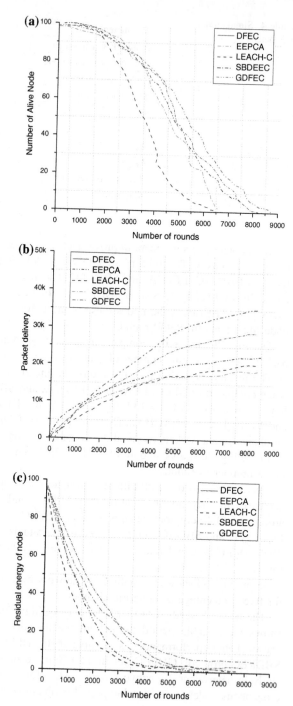

6 Conclusion

The proposed approach aims to design a protocol based on fuzzy logic techniques and improved by a genetic algorithm. The principal role of Fuzzy techniques is to determine the control objectives of DFEC. In this paper the GDFEC protocol is designed for improving the energy analysis of clustering algorithm for multilevel heterogeneous networks. The protocol overcomes the issues of black hole coverage as well as unequal energy distribution of nodes in different segment of network. The DFEC algorithm elects CH based on entropy condition using Fuzzy concept improves network black hole coverage and simultaneously minimize the energy wastage. The provided simulation results show that the proposed approach acts successfully with good performances in term of delivery and network life improvement. The use of a genetic algorithm GDFEC to tune the inputs' scaling parameters of the DFEC reduces considerably the energy wastage. The future work of paper will further extend to simulated annealing, Tabu search or other mathematical methods in terms of quality and time-complexity. In addition, this network optimization problem can also be extended to a mobile heterogeneous network for the improvement of lifetime and stability of network.

References

1. Akyildiz, I.F., Su, W., Sankarasubramaniam, Y., Cayirci, E.: Wireless sensor networks: a survey. Comput. Netw. **38**, 393–422 (2002)
2. Heinzelman, W.R., Chandrakasan, A., Balakrishnan, H.: Energy-efficient routing protocols for wireless microsensor networks. In: Proceeding 33rd Hawaii International Conference on System Sciences, vol. 8, pp. 8020–8030. (2000)
3. Smaragdakis, G., Matta, I., Bestavros, A.: A stable election protocol for clustered heterogeneous wireless sensor networks. Technical report, Boston University Computer Science Department. (2004)
4. Akkaya, K., Younis, M.: A survey on routing protocols for wireless sensor networks'. Ad Hoc Netw. **3**, 325–349 (2005)
5. Calvo, R., Figueiredo, M.: Reinforcement learning for hierarchical and modular network in autonomous robot navigation. In: Proceedings of the International Joint Conference on Neural Networks, vol. 1-4, pp. 1340–1345 (2003)
6. Ma, X., Liu, W., Li, Y., Song, R.: LVQ neural network based target differentiation method for mobile robot. In: Proceedings of the International Conference on Advanced Robotics, vol. 2005, pp. 680–685 (2005)
7. Li, T.S., Chang, S.J., Tong, W.: Fuzzy target tracking control of autonomous mobile robots by using infrared sensors. IEEE Trans. Fuzzy Syst. **12**(4), 491–501 (2004)
8. Ming, L., Zailin, G., Shuzi, Y.: Mobile robot fuzzy control optimization using genetic algorithm. Artif. Intell. Eng. **10**(4), 293–298 (1996)
9. Moreno, L., Armingol, J.M., Garrido, S., Escalera, A.D.L., Salishs, M.A.: Genetic algorithm for mobile robot localization using ultrasonic sensors. J. Intell. Rob. Syst. **34**(2), 135–154 (2002)

10. Heinzelman, W., Chandrakasan, A., Balakrishnan, H.: Energy-efficient communication protocol for wireless micro-sensor networks. In: Proceedings of the 33rd Hawaii International Conference on System Sciences (HICSS '00) (2000)
11. Fu, L., Medico, E.: FLAME: a novel fuzzy clustering method for the analysis of dna microarray data. BMC Bioinform. **8**, 3 (2007). doi:10.1186/1471-2105-8-3
12. Yao, J., Dash, M., Tan, S.T., Liu, H.: Entropy-based fuzzy clustering and fuzzy modeling. Fuzzy Sets Syst. **113**, 381–388 (2000)
13. Aderohunmu, F.A., Deng, J.D., Wu, X.H., Wang, S.: Performance comparison of LEACH and LEACH-C protocols by NS2. In: Proceedings of 9th International Symposium on Distributed Computing and Applications to Business, Engineering and Science, pp. 254–258. Hong Kong (2010)
14. Bala, M., Awasthi, L.: On proficiency of HEED protocol with heterogeneity for wireless sensor networks with BS and nodes mobility. Int. J. Intell. Syst. Appl. **4**(7), 58 (2012)
15. Liu, T., Peng, J., Yang, J., Wang, C.: Energy efficient prediction clustering algorithm for multilevel heterogeneous wireless sensor networks. Cite as: arXiv: 1105.6237 [cs.NI] (2011)
16. Brahim, E., Rachid, S., Zamora, A.P., Aboutajdine, D.: Stochastic and balanced distributed energy-efficient clustering (SBDEEC) for heterogeneous wireless sensor networks. INFOCOMP J. Comput. Sci. **8**(3), 11–20 (2009)
17. Zadeh, L.A.: Fuzzy logic, neural networks, and soft computing. Commun. ACM **37**, 77–84 (1994)
18. Kumar, S.S., Kumar, M.N., Sheeba, V.S., Kashwan, K.R.: Cluster based routing algorithm using dual staged fuzzy logic in wireless sensor networks. J. Inform. Comput. Sci. **9**(5), 1281–1297. ISSN: 1548–7741 (2012)

An ANN Model to Classify Multinomial Datasets with Optimized Target Using Particle Swarm Optimization Technique

Nilamadhab Dash, Rojalina Priyadarshini and Rachita Misra

Abstract In this paper we propose a particle swarm based back propagation neural network model which uses an optimized target to maximize the classification accuracy of the classifier. By using Particle swarm optimization technique an optimized target for each class was determined and there after the artificial neural network is used to classify the data using these targets. For this, some of the bench mark classification datasets are used, which are taken from UCI learning repository. An extensive experimental study has been carried out to compare the proposed method and existing method on the same datasets and a comparative analysis is done by taking several parameters like percentage of accuracy, time of response and complexity of the algorithm. During this study we have examined the performance improvement of the proposed PSO and BPN combined approach over the conventional BPN approach to generate classification inferences from the training and testing results.

Keywords Multinomial classification · Back propagation neural network · Normalization · Particle swarm optimization

1 Introduction

Data Classification and prediction are two of the prime tasks in Data mining. They continue to play a vital role in the area of data processing, financial analysis, stock market predictions, weather forecasting, disease predictions, pattern recognition, bioinformatics, image processing... etc. [1]. Clustering and classifications in Data

N. Dash (✉) · R. Priyadarshini · R. Misra
Department of IT, C.V. Raman College of Engineering, Bhubaneswar, Odisha, India
e-mail: nilamadhab04@gmail.com

R. Priyadarshini
e-mail: priyadarshini.rojalina@gmail.com

R. Misra
e-mail: rachita03@yahoo.com

© Springer India 2015
L.C. Jain et al. (eds.), *Computational Intelligence in Data Mining - Volume 1*,
Smart Innovation, Systems and Technologies 31, DOI 10.1007/978-81-322-2205-7_34

Mining are applied in various domains to give a meaning to the available data and also give some useful prediction results which can be applied to many of the crucial problem areas of the real world. Classification is the task of dividing different data from their known features to a particular group. Depending on the number of classes the classifier is a binary or multinomial classifier. Binary classifiers are applied to the data having two classes. In machine learning, multiclass or multinomial classification is the problem of classifying the given instances into more than two classes. While some classification algorithms naturally permit the use of more than two classes, others are by nature binary algorithms; these can, however, be turned into multinomial classifiers by a variety of strategies. Multiclass classification should not be confused with multi-label classification, where multiple classes are to be predicted for each problem instance. In this paper we have used artificial neural network as a classifier but tried to improvise the model by using particle swarm optimization technique with it. For measuring the performance the classification accuracy is taken as the prime criteria. The architectural complexity is taken care of by optimizing the number of nodes in the hidden layer. In this work we have taken some classification datasets from the UCI learning repository. They are: Iris, seeds and wine datasets.

The rest of the paper is sequentially arranged in the following order. Section 2 comprises the details of the dataset and pre-processing of the dataset. In Sect. 3, Particle Swarm optimization is briefly described, which is used to model the target output of the classifier. Section 4 describes the results, which contains the evaluation of the proposed model on the basis of different criteria. Finally Sect. 5 details the conclusion and future work.

2 Datasets and Preprocessing

In this work we have used 3 bench mark datasets, taken from UCI learning repository for verification and validation of the proposed model [2]. A brief of the datasets are as follows.

2.1 Iris Flower

The dataset consists of 150 samples which consists of a set of Iris flowers, where the goal is to predict three classes, setosa, versicolor and virginica. Based on sepal (green covering) length and width, and petal (the flower part) length and width.

2.2 Seed

This dataset consist of experimental high quality visualization of the internal kernel structure which is detected using a soft X-ray technique. The images were recorded on 13 × 18 cm X-ray KODAK plates. Studies were conducted using combine harvested wheat grain originating from experimental fields, explored at the Institute of Agro physics of the Polish academy of science Lubin. The data set can be used for the tasks of classification and cluster analysis. The dataset contains the following feature attributes, (1) area A, (2) perimeter P, (3) compactness C pi * A/P ^ 2, (4) length of kernel, (5) width of kernel, (6) asymmetry coefficient (7) length of kernel groove. All of these parameters were real-valued continuous.

2.3 Wine

These data are the results of a chemical analysis of wines grown in the same region in Italy but derived from three different cultivars. The analysis determined the quantities of 13 constituents found in each of the three types of wines. The data consists of (1) Alcohol (2) Malic acid (3) Ash (4) Alcalinity of ash (5) Magnesium, (6) Total phenols, (7) Flavanoids (8) Nonflavanoid phenols, (9) Proanthocyanins, (10) Color intensity, (11) Hue (12) OD280/OD315 of diluted wines, (13) Proline. All attributes are continuous.

2.4 Normalization

Normalization of input data is used for ranging the values to fall within an acceptable scope, and range [1]. These procedures help in obtaining faster and efficient training. If the neurons have nonlinear transfer functions (whose output range is from −1 to 1 or 0 to 1), the data is normalized for efficiency. As our outputs are falling within these ranges, each feature in each dataset is normalized using column normalization. The normalized data are used as the inputs to the machines.

3 Basic Principles of PSO

Particle swarm optimization (PSO) is a population based stochastic search and optimization technique, which was introduced by Kennedy and Eberhart in 1995 [3]. It is a multi-objective optimization method to find optimal solution to the problems having multiple objectives [4]. It is a technique, loosely modeled on the collective behavior of groups, such as flocks of birds and schools of fish. It mainly shows a natural behavior of a group of objects, search for some target (e.g., food). It is computer simulation of the coordinated behavior of a swarm of particles moving to achieve a common goal [5]. The goal is to reach to the global optimum of some multidimensional and possibly nonlinear function or system [6]. PSO's working principle is quiet similar to any other evolutionary algorithms like genetic algorithm. Initially particles are randomly distributed over the search space. So each particle gets a virtual position that represents a possible potential solution to optimization problem. It is an iterative process where in each iteration every particle moves to a new position, navigating through the entire search space. Each particle keeps track of its position in the search space and its best solution so far achieved. The personal best value is called as pBest and the ultimate goal is to find the global best called as gBest [7].

The standard PSO algorithm broadly consists of the following computational steps

(i)   Initialize particles with random positions and velocities;
     For each value of k [where 'k' represents the number of particles]
     Do   a) Evaluate fitness of each particle's position (p)

                 b) If fitness(p) better than fitness(pbest) then pbest = p

                 c) Set best of pBests as gBest

                 d) Update particles velocity and position

     End of for

(ii)   Stop: giving gBest, optimal solution.

     Here, a particle refers to a potential solution to a problem in d-dimensional design space with k particles. Each particle is characterized by Position vector..... $x_i(t)$ and Velocity vector......$v_i(t)$

Each particle has Individual knowledge pbest, its own best-so-far position, Social knowledge gbest, pbest of its best neighbour. The equations for velocity and position updates are given below

$$v(t + 1) = (w * v(t)) + (c1 * r1 \; * (p(t) - x(t)) + (c2 * r2 * (g(t) - x(t)) \qquad (1)$$

$$x(t + 1) = x(t) + v(t + 1) \qquad (2)$$

The first equation updates a particle's velocity. The term $v(t + 1)$ is the velocity at time $t + 1$. The new velocity depends on three terms. The first term is $w * v(t)$. The w factor is called the inertia weight and is just a constant between 0 and 1. Here the value of w is taken as 0.73, and $v(t)$ is the current velocity at time t. The second term is c1 * r1 * (p(t) − x(t)). The c1 factor is a constant called the cognitive (or personal or local) weight. The r1 factor is a random variable in the range [0, 1)—that is, greater than or equal to 0 and strictly less than 1. The p(t) vector value is the particle's best position found so far. The x(t) vector value is the particle's current position. The third term in the velocity update equation is (c2 * r2 * (g(t) − x(t)). The c2 factor is a constant called the social, or global, weight. The r2 factor is a random variable in the range [0, 1). The g (t) vector value is the best known position found by any particle in the swarm so far. Once the new velocity, v(t + 1) has been determined, it is used to compute the new particle position x(t + 1).

In case of a neural network, a particle's position represents the values for the network's weights and biases. Here, the goal is to find a position/weights so that the network generates computed outputs that match the outputs of the training data.

3.1 PSO for Target Optimization

In case of supervised learning, during the training process of neural network, three parameters are mostly required, the input, weight and target [8]. In practical approach the target of a neural network is either randomly chosen or depends on the feature of input dataset. e.g., if the input dataset has three different classes, three target outputs are generated. In case of unsupervised learning, the target does not exist at all, rather it is needed to be explored by the network itself. If any of the previously said problem occurs, the efficiency of the neural network reduces drastically due to the following reasons

1. The neural network may take longer time to get trained as randomly chosen target outputs may not be the optimal one.
2. Extensive computations are required as additional mapping is required in order to match up with the input values to the randomly chosen target values.
3. Usually for classification tasks, the number of output neurons in a neural network depends on the number of classes present in the dataset [9], whereas the number of input neurons depends on the number of features per input data. This particular architecture fails to maintain the relevance between the input and the randomly chosen target. (Feature to number of class mapping is done instead of feature to feature mapping which is more accurate). These problems are addressed by using the proposed PSO based technique, which generates a nearly optimized target by analyzing the input dataset of the neural network. The steps for designing the classifier is diagrammatically shown in Fig. 1.

Fig. 1 Proposed model of PSO based classifier

3.2 Training by BPNN and Simulation

Training in an artificial neural network is necessary to make the classifier learn by itself. It is the ability to approximate the behavior adaptively from the training data while generalization is the ability to predict the training data [9]. A BPN network learns by example. It gives us the desired output for a particular input if, provided with the known input by changing the network's weights so that, when training is finished, it will give the correct output. The change in weight takes place according to the error produced. The data sets are divided into two parts. i.e., training set (known data) and testing set (Unknown data) which is not used in the training process, and is used to verify the machine and then we have simulated our results with these datasets [10].

The training process of this work follows the following steps

1. On the input data, particle swarm optimization is applied to find out the optimized target. In other words the optimized inputs for the classifier are found out with the help of PSO technique.
2. The training data are prepared by normalizing the input and the optimized target that range from 0 to 1.
3. The artificial neural network is trained by previously got inputs till it gets converged.

The convergence criterion for the network is taken as minimum error condition.

Figures 2 and 3 show the convergence of different networks used to train the said datasets.

Fig. 2 Convergence graph of iris

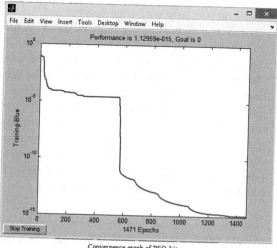

Convergence graph of PSO-Iris

Fig. 3 Convergence graph of seed

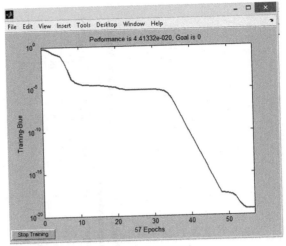

Convergence graph of PSO-Seed

In the proposed work the architecture of the network varies according to the dataset. This is due to the fact that, the output of the neural network depends on the number of features present in the dataset. The output of the neural network depends on the number of features present in the dataset. For example, a [4 * 5 * 4] network having 4 input data is used to train the iris dataset containing four features per data.

4 Results and Observations

The process of classification was carried out on the previously mentioned datasets by taking the conventional ANN with back propagation learning algorithm as well as the proposed approach. In both the cases the classification accuracy was taken as the most vital factor for performance evaluation. Number of misclassification is calculated by measuring the Euclidean distance between the target and actual output. Percentage of misclassification is ratio of incorrectly predicted class and total number of data present in the testing samples multiplied by 100. Table 1 shows the overall comparison between the PSO-BPNN and BPNN approach. It is clear from the result that the proposed approach shows significant improvement over BPNN. Tables 2, 3 and 4 give the simulation accuracy of the proposed work for different datasets up to 4 decimal places.

Table 1 Performance comparison of back propagation neural network and the proposed approach

Dataset	Number of classes	Number of samples	Number of features	Number of training samples	Number of testing samples	Number of misclassifications		Classification accuracy	
						BPNN	PSO-BPNN	BPNN (%)	PSO-BPNN (%)
Iris	3	150	4	120	30	11	1	78	98
Seed	3	210	7	150	60	17	9	72	85
Wine	3	178	13	150	28	6	1	79	98

Table 2 Simulation accuracy of wine dataset

No. of classes	Number of simulation result accurate up to				Total no. of test samples per class
	4 decimal places	3 decimal places	2 decimal places	1 decimal places	
Class 1	09	09	09	09	09
Class 2	10	10	11	11	11
Class 3	08	08	08	08	08
Percentage of accuracy	97 %	97 %	100 %	100 %	Total = 28

Table 3 Simulation accuracy of seed dataset

No. of classes	Number of simulation result accurate up to				Total no. of test samples
	4 decimal places	3 decimal places	2 decimal places	1 decimal places	
Class 1	16	16	17	20	20
Class 2	15	15	16	20	20
Class 3	17	18	18	20	20
Percentage of accuracy	80 %	83 %	85 %	100 %	Total = 60

Table 4 Simulation accuracy of iris dataset

No. of classes	Number of simulation result accurate up to				Total no. of test samples
	4 decimal places	3 decimal places	2 decimal places	1 decimal places	
Class 1	10	10	10	10	10
Class 2	10	10	10	10	10
Class 3	09	09	09	09	10
Accuracy %	98 %	98 %	98 %	98 %	Total = 30

5 Conclusion and Future Work

Though the role of back propagation neural network is inevitable in the field of classification, there is a certain need to assess its efficiency in terms of learning time, simulation accuracy and flexibility. In the proposed work we have tried to improve the classification accuracy, and got some promising results which verify that the proposed method shows a remarkable improvement over back propagation machine classifier alone. The particle swarm optimization technique also played a vital role to provide the optimized target that made the learning process easier and efficient. Considering the inspiring results obtained from the proposed work the future objectives are (1) to apply the proposed method on some real life problems with some benchmark datasets mainly in the area of computational biology and bioinformatics (2) to reduce the architectural and computational complexity of the network in terms of number of hidden neurons along with training and simulation time.

References

1. Dehuri, S., Cho, S.B.: A Comprehensive survey on functional link neural networks and adaptive PSO-BP learning for CFLNN. Neural Comput. Appl. (2009)
2. Bache, K., Lichman, M.: UCI Machine Learning Repository [http://archive.ics.uci.edu/ml]. University of California, School of Information and Computer Science, Irvine (2013)
3. Ince, T., Kiranyaz, S., Pulkkinen, J., Gabbouj, M.: Evaluation of global and local training techniques over feed-forward neural network architecture spaces for computer-aided medical diagnosis. Expert Syst. Appl. **13**, 8450–8461 (2010)
4. Dehuri, S., Chob, B.: Multi-criterion Pareto based particle swarm optimized polynomial neural network for classification: a review and state-of-the-art. Comput. Sci. Rev. **3**, 19–40 (2009)
5. Dehuri, S., Royb, R., Cho, S.B., Ghosh, A.: An improved swarm optimized functional link artificial neural network (ISO-FLANN) for classification. J. Syst. Softw. **85**, 1333–1345 (2012)
6. Kennedy, J., Eberhart, R.: Particle swarm optimization. IEEE Trans. 1942–1948 (1995)
7. Devi, S., Jagadev, A.K., Dehuri, S., Mall, R.: Knowledge discovery from bio-medical data using a hybrid PSO/Bayesian classifier. Tech. Int. J. Comput. Sci. Commun. Technol. **2**(1), 364–371 (2009)
8. Zhang, G.P.: Neural networks for classification: a survey. IEEE Trans. Syst. Man Cybern. C Appl. Rev. **30**(4), 451–463 (2013)
9. Haykin, S.: Neural networks—a comprehensive foundation. Prentice Hall, Englewood Cliffs (1999)
10. Priyadarshini, R., Dash, N., Swarnkar, T., Misra, R.: Functional analysis of artificial neural network for dataset classification. Int. J. Comput. Commun. Technol. **1**(2–4), 49–54 (2010)
11. Zhang, G.P.: Avoiding pitfalls in neural network research. IEEE Trans. Syst. Man Cybern. C Appl. Rev. **37**(1), 3–163 (2007)
12. van der Merwe, D.W., Engelbret, A.P.: Data clustering using particle swarm optimization. **1**, 215–220 (2003) (IEEE)

Novel Approach: Deduplication for Backup Systems Using Data Block Size

K.J. Latesh Kumar and R. Lawrance

Abstract The core confronts of today's Information Technology remains to be the funding and optimized management of storage infrastructure. Data maintenance and most importantly securing the data, since data is omitted by non IT infrastructure edging higher and hence storage appliances turning huge and breeding infrastructure capital investment, hence technology front is pointing at new research method that would cut and reduce the capital investments on storage front. Deduplication is one of the key components of storage efficiency technologies that enable customers to store the maximum amount of data for the lowest possible cost. This technology is implemented on storage to achieve efficient storage savings. Unlike any other storage technology deduplication is also crond to run at suitable clock across data centre. This research article aims to lower the storage cost and in achieving the higher deduplication rate.

Keywords RAID · NFS · WAN

1 Introduction

Today's IT world challenge is managing the rising e-data and digital data across data disaster and backup sites, since backup is a process where identical data will be dumped redundantly for recovery purposes. Deduplication [1] is enormously devised in these instances to reduce the cost of storage. The need for the deployment

K.J. Latesh Kumar (✉)
Computer Science and Engineering, Siddaganga Institute of Technology,
Karnataka 572103 Tumkur, India
e-mail: latesh.kj@hotmail.com

R. Lawrance
Department of MCA, Ayya Nadar Janaki Ammal College,
Sivakasi, Tamil Nadu, India
e-mail: lawrancer@yahoo.com

© Springer India 2015
L.C. Jain et al. (eds.), *Computational Intelligence in Data Mining - Volume 1*,
Smart Innovation, Systems and Technologies 31, DOI 10.1007/978-81-322-2205-7_35

of data deduplication technology [2] is most essential in today's IT infrastructure maintenance, business specific data deduplication essentially aimed for backups which is the most and critical process at the customer business. Deduplication [3] process, which uniquely identifies data fragments, cross verifies with the existing data and writes this data blocks to storage grid if found data blocks are new otherwise, if an incoming data block is a redundant only reference is created to it and the slice will not be repeatedly stored. For example, in a production environment every week backup usually holds the redundant data, there the deduplication process will store data if is unique and discard if it is redundant. This induces beneficial reduction in storage constraints; enterprise data consuming will highly help retention policies. This means that sites can store terabytes of backup data on terabyte of physical volume capacity, which has huge economic benefits.

In this approach the foremost advantage of using Data Block Size Deduplication is that it increases the performance, allows to size based computation that can be employed on big data sanitization and reliability on the storage system [4] to store maximum amount of data efficiently by making use of optimized storage space [3], that helps to reduce spending on the storage to inside data centre to accommodate to more data housing to run his business smoothly and efficiently. The guaranteed storage management is quite easy by employing this technology on the storage volume. The data block size deduplication has unique intelligence of deduplicating the data, based on the data block size and unique data hashing method. The crux of deduplication is odd data block and even data block data and unique data block hashing database. All these three components of the workflow intelligence will effort the deduplication process to achieve higher benefit and reduce the IO process by cutting down the stress across the storage grid to avoid redundant reads and writes. The unique data block hashing database enforces unique hashing ID's for every data block in the storage. In this approach the primary focus is on optimized quicker deduplication that is made possible onto storage appliance.

2 Our Approach

In this novel approach the data from clients before be stored onto storage pool is communicated onto three workflow that are data block size computation, data block category and finally unique data hashing method to complete the deduplication process as shown in the Fig. 1, two scenarios of deduplication implementation are analyzed. When started on production, "how deduplication works" [5] when the whole system is new production for deduplication [4] to start on, and the other is where business has existing data in on the storage, or may be migrating from other deduplication methods as well. Considering environments like these, the attempt in this research to achieve better data deduplication rate. In a environment of new setup being established and a client trying to write data onto storage volume where data duplication [data block size based] is enabled, the data block first passes through the data size computation which finds the data block size, passing this test

Fig. 1 Data block size deduplication workflow

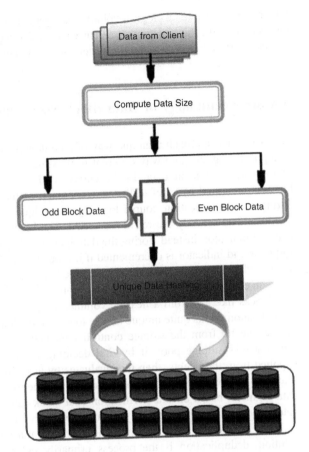

of producing the actual volume of the data block, categorization of the data block is performed to filter out entities odd data block and even data block, each block data is written to disk only after passing through this scan whether the file/object/ blocks are identical to the data block categories, after this process of recording the metadata about the data block, it does subset check of the data block size whether found in either of this category of data blocks, if found that the data block is identical to size and to its bit/byte level content, then each unit data length is compared, if found unique then it forwarded to the IO of the storage controller to store it onto disk and its metadata is also recorded to the categories of relevant data blocks, otherwise the data block set is disqualified and space on the disk is reclaimed and just a reference is created for that block instead of storing the data block, thus storage space is saved.

In this approach is induced on environments to shape the peer competitive results. The disk malfunction in the data centre is the primary challenge in devising large data protection environments. The disk aggregates are pool of thousands and more disks that are prone to have a disk failures inside when they operate continuously for long hours, to handle this logical volume management and RAID [6] intelligence are plugged to handle this situations to enhance the data storage capacity smoothly. Hence the tests are conducted on a flat file system design and environment.

3 Core Technology of Data Block Size Deduplication

In this novelty approach a unique way of data deduplication is proposed, this data deduplication goes quite way different in its structure to categorize the data set in filtering the redundant data. In the current deduplication mechanism Each data block is identified by a ref id that is component of the storage volume metadata, during the deduplication course locating the unique data set and discarding the duplicate data set only the ref id are varied. The block that residue on disk arrays with the indicator, instead storing the data blocks only the pointer count is updated and the ref id indicator is decremented if it finds duplicate data set. Otherwise there are ref id that is not matching any of the data segment will be automatically freed. In comparison to NetApp deduplication technology duplicating big chunk [7] size blocks like 4 k from a flexible volume, the proposed method data block size deduplication works quite unique in way to reduce the load on the storage controller so that the IO from the storage controller go better speed to perform reads and writes across storage pool, if by first deciding the block size, type of the data supplied from the clients, then this evaluates by applying the hashing method that scans the odd or even data blocks in the flex volume to locate if unique or repeated data segment, based on this verification data blocks are wrote disk pool or aggregates along with metadata if they are unique otherwise only reference is altered.

The Fig. 2 depicts quicker and smoother data filtering data reduction process solution, deduplication is the process primarily used at backup levels, since a backup system contains at least one copy of all data inside a vault. Since the data is getting updated on the vault it is a great challenge to manage and organize the efficiency [8] of storage usage. Henceforth effective Deduplication process must be involved to have these backup's created in best way so that managing storage shouldn't be an overhead for any organization and the data at the same time. In this approach prior data blocks are pushed to disk array/aggregates of the storage pool it is managed using the data block hashing method, which enables easy way of managing the open data segments and on production data activity. In the Fig. 3 the new scheme of data scanning is illustrated and described how data block size Deduplication is simple and effective when compared native and other Deduplication methods being used.

The core technology behind data block size deduplication is that it holds the size filter of the blocks like odd and even, which holds a virtual partition inside the data container of unique hashing database, this method induces much speedier scanning of the data blocks that would help the deduplication process to make the decision to whether data block is existing or it is non redundant block, since the intelligence can easily lookup to the size of the file and start probing for the one inside either in

Fig. 2 Data block size
deduplication process

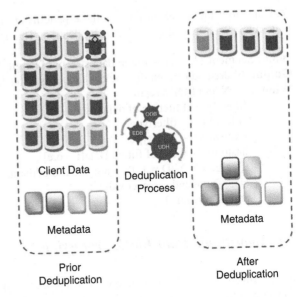

Fig. 3 Block level data
processing

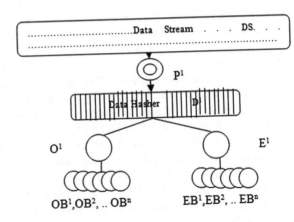

the odd sized data block or the even sized data block, this helps in achieving significantly quicker performance rate in either doing IO over onto storage volumes or probing the data blocks whether they are redundant or otherwise a unique data block so that it can pushed to storage volumes.

4 Implementation

In our approach the addressable data segments are not of fixed size and can vary to obtained length from the logical file data. Therefore the each data chunk [9] is first filtered by a data hash method that would identify to odd data/even data block segment from the offset, In process these patterns are later rolled to relevant data pattern and then it is plugged for deduplication. This is portrayed in Algorithm.

Input: Segment {1 – N} A file or data stream of length N
Output: filtered odd / even data blocks that falls to bucket odd and even [size]
While 1 ← N and EOF stream
      filtertype←Filtering of database
      Oddvalue ← filtertype
      Evenvalue← filtertype
      hashing database of filters Data[i – o,e] ;
      if value $\oplus$ ODD = 1 then append i to oddvalue ;
      else $\oplus$ append i to evenvalue ;
      end Algorithm 1:

4.1 Algorithmic Data Block Processing

As illustrated in Fig. 3, the data hasher scans for the two different data filters to update into the metadata framework, as the data hasher locates the odd/even block data segments it records that to a hashing table and then loads it odd/even data block area, thereby it is scanned for each bit/byte of data inside the blocks whether it exist in the data pool or not, if not then it is stored otherwise just only the reference is added the ref ID is copied to data hasher for further future data scanning for avoiding the redundant data set. This algorithm is implemented to enhance the deduplication rate by identifying the duplicate data using data block size algorithm and the algorithm as explained in article on the following infrastructure. Hardware 4 GB RAM, 2 dual-core 1.8 GHz DELL Optilex running on RHEL 3 (2.4.16-32smp) with 14 7200 RPM SATA [10] drives. *Operation*: Data stream 'DS' loads the data block onto root node (data processor) P^1 ← DS^1, P^1 processes the data from initial buffer and find block size the length of the $L(P^1)$ ← D^1 are now processed in P1 then moves it O^1 or E^1 accordingly till the last byte of the data stream. The prime consideration is data deduplication workarounds were at block size of 2 KB.

5 Results

The utilization of deduplication technology is growing very ubiquitously for distributed and grid computing environments where are parallel processing [11] and high speed wildfire servers are operating and storing fat quantity of duplicate data segments. Data block size deduplication initiates a deduplication technology that applies to file, image and blob data backups. This research method aims in saving

Fig. 4 Data block size deduplication—space save results

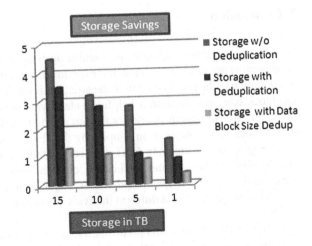

the storage appliances and capital investment from low scale [1] to large scale business functionalities. The Fig. 4 depicts the NFS based wire traffic of small data centre where in test cases are generated and tested for deduplication process, the data blocks originated from Windows work station and unix nodes that are sharing across samba server and NFS [12] shares. Data transmitted in kilobytes are ranging from small chunks to bigger chunks/second from both of the environments. Savings space is calculated based on the original aggregate volume and deduplicated volumes lying on storage via both unix and windows.

6 Benefits

The bulk data transfer from primary [13] backup/disaster site to secondary/remote data centre for every seconds inside a metro cluster design is a greater challenger since the bandwidth [14] of the network and latency issues that reduces the performance because the data is bulk since the backup usually contains the redundant data segments, data deduplication is highly beneficial in this arena by not storing the redundant data across the backup site that reduces the volume to be transferred from primary to remote. Technology that helps and induces quicker data transfer is highly appreciated and accepted. Transporting huge data set (backup) across WAN [15] is always a challenge, since longer transmission consumes higher the time, this can be avoided and faster data rate is achieved by empowering the data deduplication. This also eliminates the legacy storage appliances like tape and other disk drives that are expensive in maintenance like power and cooling across data centers.

7 Conclusion

To recap, the primary contribution is in setting up of the data block size deduplication by integrating novel design method. It is experimented the execution across vivid data sets as shown in the results section of the article. It is observed and demonstrated that data block size deduplication is helpful in identifying the different category of data blocks that are contiguous or non contiguous during modification. One shortcoming of data block size deduplication is updating data blocks, since all computation and processing from data stream to data hasher is repeated, this is highly beneficial where vault storing appliances and application environments, where system does not interested in making frequent updates. A slight memory compromising is also required since processing at different intervals of workflow is bit consuming of CPU resources.

To further this research and proposition to article, considerations can be given to the implication of extending this to all kinds of block level and file level data onto storage. The researcher must also consider the benefits and implications of applying this data protection method in and with various other different file systems and storage of a data center. Securing the data communication across data center by implementing the secured deduplication is look worth. The researcher must also consider the benefits and implications of applying this new data deduplication method in and with various other different file systems and storage of a data center.

References

1. Ateniese, R., Pietro, D., Mancini, L.V., Sudik, G.T.: Scalable and efficient provable data possession. In: Proceedings of the 4th International Conference on Security and Privacy in Communication Networks (SecureComm '08), pp. 9:1–9:10. ACM, New York (2008)
2. Russell, D.: Data deduplication will be even bigger in 2010. Gartner **8** (2010)
3. Douceur, J.R., Adya, A., Bolosky, W.J., Simon, D., Theimer, M.: Reclaiming space from duplicate files in a serverless distributed file system. In: Proceedings of the 22nd International Conference on Distributed Computing Systems (ICDCS '02), Washington, DC (2002)
4. Hamilton, J., Olsen, E.W.: Design and implementation of a storage repository using commonality factoring. In: Proceedings of the 20th IEEE/11th NASA Goddard Conference on Mass Storage Systems and Technologies (2003)
5. Software, how data deduplication works. http://www.falconstor.com/library/documents/white-papers/falconstorvtl-white-papers (2011)
6. Sahai, A.K.: Performance aspects of RAID architectures. In: IEEE International Performance, Computing, and Communications Conference, IPCCC 1997, Apple Comput. Inc., Cupertino, 0-7803-3873-1 (1997)
7. Bo, Z., Li, F., Can, W.: Research on chunking algorithm of data deduplication. Am. J. Eng. Technol. Res. **11**(9), 1353–1358 (2011)
8. McClure, T., Garrett, B.: EMC centera, optimizing archive efficiency. White paper (2009)
9. Bhagwat, D., Eshghi, K.; Long, D.D.E., Lillibridge, M.: Modeling, analysis and simulation of computer and telecommunication systems. In: IEEE International Symposium on, Extreme Binning: Scalable, Parallel Deduplication for Chunk-Based File Backup, MASCOTS '09. IEEE, pp. 21–23. University of California, Santa Cruz, Sept 2009

10. Chun-Ho. N.,Mingcao, M.: Conference: Middleware 2011—ACM/IFIP/USENIX 12th International Middleware Conference, Lisbon, Portugal, Dec 12–16, 2011. Proceedings Source: DBLPLive Deduplication Storage of Virtual Machine Images in an Open-Source Cloud. doi: 10.1007/978-3-642-25821-3_5
11. Cui, Z., Liang, Y. Rupnow, K.: An accurate GPU performance model for effective control flow divergence optimization. In: Proceedings of 26th IEEE International Parallel and Distributed Processing Symposium (2012)
12. Muthiacharoen, A., Chen, B., Mazieres, D.: A low-bandwidth network file system. In: Proceedings of the 18th ACM Symposium on Operating Systems Principles (2001)
13. George crump. Lab report: deduplication of primary storage. http://www.ocarinanetworks.com/products/products-overview (2009)
14. Xia, W., Jiang, H., Feng, D., Hua, Y., Lo, S.: A similarity-locality based near-exact deduplication scheme with low ram overhead and high throughput. In: Proceedings of the 2011 USENIX Conference on USENIX Annual Technical Conference (2011)
15. Zhang, Y., Ansari, N., Wu, M., Yu, H.: On wide area network optimization. Advanced Networking Lab., Department of Electrical and Computer Engineering, New Jersey Institute of Technology, Newark, 13 Oct 2011

Segmentation Google Earth Imagery Using K-Means Clustering and Normalized RGB Color Space

Nesdi Evrilyan Rozanda, M. Ismail and Inggih Permana

Abstract Image segmentation is defined as: "the search for homogenous regions in an image and later the classification of these regions". In this research, a remote sensing image, Pekanbaru city of Riau Province-Indonesia is provided for the green land segmentation. It is obtained by observing the surface of the earth using the Google Earth Imagery. To segment the green land of the given image, two different methods are used in this research, K-Means Clustering and Normalized RGB Color Space methods. This research is expected to have two clusters output: the spreading of green fields and not green fields. The result shows that the given Google Earth imagery can be segmented about 40.50 and 47.01 % pixels from all image pixels by K-Means Clustering and Normalized RGB Color Space respectively.

Keywords Google earth imagery · Image segmentation · K-Means clustering · Normalized RGB color space

1 Introduction

Remote sensing involves the use of airborne and space-imaging systems to inventory and monitor the Earth resources. Broadly defined, remote sensing is a methodology employed to study the characteristics of objects from a distance. Using the various remote sensing devices, the collected data can be analyzed to obtain information about the objects, areas, or phenomena of interest [1]. One way

N.E. Rozanda (✉) · M. Ismail · I. Permana
Information System Department, State Islamic University of Sultan
Syarif Kasim Riau, Pekanbaru, Indonesia
e-mail: nesdi.rozanda@uin-suska.ac.id

M. Ismail
e-mail: ismail.mz@uin-suska.ac.id

I. Permana
e-mail: inggihpermana@uin-suska.ac.id

© Springer India 2015
L.C. Jain et al. (eds.), *Computational Intelligence in Data Mining - Volume 1,*
Smart Innovation, Systems and Technologies 31, DOI 10.1007/978-81-322-2205-7_36

to acquire the remote sensing imagery is by utilizing the Google Earth Imagery. Furthermore, the rectification of the high resolution digital aerial images or satellite imagery for the large scale city mapping is a modern technology that needs well as a distributed and an accurate defined control points [2]. It can be obtained by using widely known software Google Earth and be applied for an accurate city maps construction.

As a data research, Google Earth Imagery has been widely used by many prior researches [3–9] captured by Google Earth. For different background interest, Google Earth is used since it is possible to zoom into the objects with a high resolution aside from the simplicity given in, such as an analyzing and an automatic of image segmentation, classification region for rainfall estimation, supporting the cloud computing application, sidewalk distance measurements, mapping gullies, tropical forest biomass assessments, oil palm age determination, and so forth.

In this research, the green land segmentation captured by Google Earth Imagery is investigated to show the converted green land for the residential development areas in Pekanbaru city-Riau Province, Indonesia. The converted Pekanbaru land had been increased approximately 60.11 % in 2004 which was mostly done for the residential. Spatial plan for settlement in 2000 amounted to 14,172 acres, while in 2004 the number increased to 35,531 hectares. Areas for the residential development occurred because the population was growing rapidly, whether locals or immigrants who took part in economic activities in Pekanbaru. The General Spatial Plan City of Pekanbaru (*Rencana Umum Tata Ruang Kota*/RUTRK) estimated that population in Pekanbaru in 2006 reached 704,220 people; while in 2002 it was only about 615,195 people. It means that there was about 12.64 % population increment during 2000–2006 [10]. The population increment will have an impact on land-use changing for residential, green area, allotment, or other conditions, and these are expected to continue to grow. The changing in dwindling condition of the green land gives its own interest for this study with the goal to test the power of Google Earth to show the distribution of the green land in Pekanbaru city of Riau province by using image processing techniques. Here, two different segmentation methods, K-Means Clustering and Normalized RGB Color Space are used to classify the expected output: green land fields and not green land fields; and then the comparing results of both implemented methods will be provided to show the performance of the two segmentation methods.

This paper is organized as follows. In Sect. 2, the proposed algorithms will be presented. In Sect. 3, we evaluate the proposed algorithms with several test images. Finally, we summarize and conclude the papers in Sect. 4.

2 The Proposed Method

Figure 1 shows a plot of this research method. Basically, in order to acquire a segmented imagery there are several steps that must be done. These stages can be divided into three general sections, namely input, process and output. For this

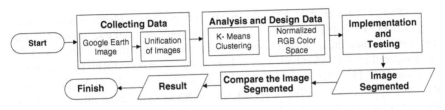

Fig. 1 Flowchart research methods

study, the required input is the Google Earth imagery obtained from the collecting data, and then it is processed including the finding and the counting process of the weight clustering Euclidean Distance. Furthermore, the weight clustering is used to obtain the output data explaining which areas are included in the cluster of green land and not green land.

2.1 Collecting Data

The meaning of the collecting data in this research is how to get the main data to be analyzed in this work from Google Earth Application. Figure 2 is the Pekanbaru

Fig. 2 Pekanbaru area from google earth

area captured from Google Earth obtained free by using Google Satellite Maps Downloader. This application will need the longitude and latitude values (coordinate values), then the application will download the image in accordance with the value entered into it. Pekanbaru area in Fig. 2 was downloaded with the coordinate values in 101° 18'–101° 36' of East Longitude and 0° 25'–0° 45' of North Latitude. By utilizing Google Satellite Maps Downloader, Google Earth imagery will be divided automatically to be 20 pieces image with the static large pixel resolution 256×256, the level zoom is 12, and the bit-depth image is 8 bit. Therefore, the entire piece image needs to be collected called by unification of image stage to show the whole main image of the Pekanbaru area. Actually, when the analysis and design data were staged, the image used to be analyzed is the real piece image downloaded by the Google Satellite Maps Downloader application. It is performed to avoid inaccurate results that may be caused by the unification process.

2.2 Analysis and Design Data

This step is used to describe the K-Means Clustering and the Normalized RGB Color Space method processes. Data utilized for this step is the real images of Pekanbaru area before the unification process is done. Figure 3 shows the process of K-Means Clustering to cluster the images and Fig. 4 is the block diagram of segmentation imaging using K-Means; while Fig. 5 is a figure regarding the Normalized RGB Color Space process.

2.2.1 K-Means Clustering Method

K-Means Clustering Method is one of unsupervised learning algorithm. This method is one of the simplest algorithms that solve the well-known clustering [11]. In our implementation as shown in Fig. 3, K-means works with a simple and an easy way to cluster any data set through the certain number of clusters fixed a priori (here, we expect to have two clusters, A and B clusters [12]. Cluster A is the green land cluster and cluster B is the not-green land cluster). It is started by input data-n (x, y). Data-n refers to x rows and y columns of image pixels. The goal of this method is to define A and B centroids. These centroids should be placed in a cunning way since the different location may cause different result. Thus, the better choice is to place them as much as possible far away from each other. It means that the K-means method performance depends on the initial positions of the cluster centers. Here, K-means shows that it is an inherently iterative algorithm. Besides, there is no guarantee about the convergence towards an optimum solution. The convergence centroids vary with the different initial points. It is also sensitive to noise and outliers. The K-Means method can be described as follows:

Fig. 3 Flowchart K-Means clustering

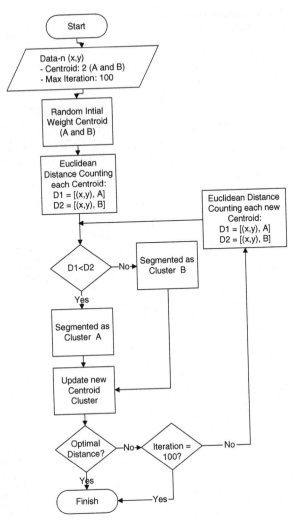

1. Initialize the number of clusters
2. Allocate data into the random cluster
3. Calculate the centroid/average of data in each cluster using the Euclidean distance
4. Allocate each data to the centroid/the nearest average
5. Go back to step 3, if there is still data to move or if the cluster centroid value changes.

Fig. 4 Segmentation imaging
block diagram

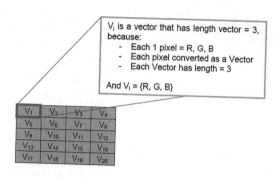

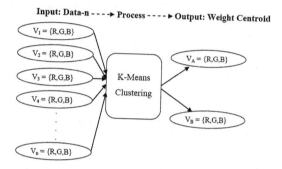

Fig. 5 RGB color space

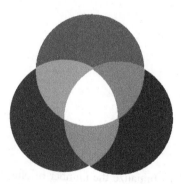

Furthermore, Fig. 4 explains that each input will be converted to a vector imagery to obtain the pixel rows. Each pixel represents the three color components: Red, Green, and Blue (RGB). It means that a single pixel is equal to the vector data which has members RGB color. Thus, each vector will be used as a data input for the K-means to obtain the weights in the cluster and will be categorized into two clusters with the same vector length as its input. The two cluster distributions will be derived after achieving the Euclidean distance results. In our implementation, the Euclidean distance is achieved by using formula in Eq. (1) [13].

2.2.2 Normalized RGB Color Space Method

Color model is an abstract mathematical model describing the way colors that can be represented in a collection of numbers/numeric. The color models usually have three or four color components. When the model linked to each other by a precise approach, it is formed to set a specific color called by color space or color spectral [13]. Thus, to form the colors of the RGB color set (one red, one green, and one blue) or called the Normalized RGB Color Space, each color must be superimposed on each other (overlay), as shown in Fig. 5, and each of the three colors overlay results will form blocks of color intensity mixed results respective RGB color components

$$d_{Euc}(P, Q) = \sum_{j=1}^{n} |P_j - Q_j|^2 \tag{1}$$

$$g = \frac{G}{\sqrt{R^2 + G^2 + B^2}} \tag{2}$$

In color with 24-bit depth, each of the three components R, G, and B for 8-bit can be represented in an integer ranging from 0 to 255 which is called the degree of color. A value of 0 means the components of the color black (no intensity), while 255 means full intensity so white. To reduce the sensitivity to changes in light distribution, the Normalized RGB Color Space or also called Normalized RGB Chromaticity Diagram is used to normalize the three components of RGB and RGB color space. And for normalized RGB can be done by performing the division of each value of the degree of the R, G, or B to the value of the accumulated squares of the three components of the color and squared roots. In the case of this study, normalization of color degree will only be made on the degree of greenish color image segmentation to obtain a green color which is green land distribution [14] as shown as in Eq. (2).

The process of normalization of the degree of greenness is used to ensure that the pixels are still at the threshold point can be identified from the greenish color of green, so it takes the value of the threshold point (threshold) is appropriate (optimum) for the identification and segmentation of color. And to position themselves threshold can be set and determined using the Normalized RGB Chromaticity Diagram, as shown in Fig. 6. From Fig. 6 it can be seen that to obtain the basic color (normal color) green, it is necessary to test the threshold point (threshold) for the green color, which ranges between 0.3 and 2.0. If the green degree threshold point (greenness degrees) is found, then the next step is to adjust the value of the threshold point is the condition in Eq. (3) below to obtain the image segmentation:

Fig. 6 Normalized RGB
chromaticity diagram

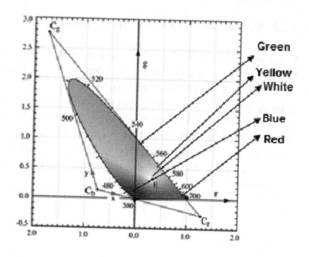

$$\textit{If}(\text{Greenness degrees} \; > = \text{Threshold})$$
$$\textit{then} \; \text{pixel is green} \qquad (3)$$

If the condition is met, it will automatically obtain the pixels (area) which is segmented the distribution of green and if not met then the result is a segmentation of the area which is not a distribution of green land.

3 Experimental Results

Referring of the 20 images Google Earth, 10 images are chosen randomly as the training data to obtain the best cluster weight and yield the best cluster using K-Means Clustering.

While Normalized RGB Color Space, it is only using one data training to get the best threshold value. It is chosen randomly but it is trained as many as 10 times. After training for 10 training images, the best cluster weights obtained on the fourth training data (K-Means Clustering) with the number of iterations = 100 iterations and the sixth training (Normalized RGB Color Space) could show the best threshold value. These two results can be seen in Table 1. Then, the results tested on 20 images in Google Earth (test data) and the results can be seen in Table 2. In Table 2 will be displayed only 6 of 20 result testing images.

The results segmentation testing of K-Means Cluster using weight vector of green land cluster R = 148.69789, G = 135.29292, B = 12.495605 and weight vector of not green land cluster R = 46.177826, G = 57.85212, B = 36.74251 and Normalized RGB Color Space Method using the best threshold value 0.6 of greenness degrees can be seen in Table 2. All images have the same large pixel resolution 256 × 256 pixels and bit-depth 8 bit. And the table can show that both methods have different results to segment the same image. If all images tested,

Table 1 Final results of training data of K-Means clustering and normalized RGB color space that chosen randomly from 10 images

No.	Original data training	K-Means result	Normalized RGB result
1			—
	It is the foth data training of ten of K-Means clustering training. We get the best weight vector in the first cluster (green land cluster) R = 148.69789, G = 135.29292, B = 12.495605. And the weight of second vector (not green land cluster) R = 46.177826, G = 57.85212, B = 36.74251 with 100 iteration		
2		—	
	It is the best result to get threshold greenes degrees using normalized RGB color space method. This is the sixth result after training the same original image with different threshold values. The best threshold is 0.6. It is chosen as it can segment the data training with the best segmentation result		

Table 2 Results testing segmentation of K-Means cluster and normalized RGB color space

No.	Original image	K-Means clustering results	Normalized RGB result
1		 Green cluster: 6.4 % Not green cluster: 93.6 %	 Green cluster: 16.5 % Not green cluster: 83.5 %
2		 Green cluster: 35.6 % Not green cluster: 64.4 %	 Green cluster: 42.9 % Not green cluster: 57.1 %
3		 Green cluster: 61.2 % Not green cluster: 38.8 %	 Green cluster: 65.7 % Not green cluster: 34.3 %
4		 Green cluster: 53.5 % Not green cluster: 46.5 %	 Green cluster: 57.5 % Not green cluster: 42.5 %
5		 Green cluster: 9.1 % Not green cluster: 90.9 %	 Green cluster: 34.0 % Not green cluster: 66.0 %

K-Means could show about 40.50 % pixels and Normalized RGB Color Space about 47.01 % pixels segmented as green land from all image pixels. However, K-means clustering shows the better accuracy to cluster images using color feature of image against of normalize RGB space method.

4 Conclusion

After comparing between K-Means Clustering and Normalized RGB Color Space Method, this research can be concluded that: Google Earth Imagery can be used as data research and proved in this research that by using K-Means Clustering and Normalized RGB Color Space method, it can be segmented well with the different results in each method. If 20 images are tested, K-Means could show about 40.50 % pixels and Normalized RGB Color Space about 47.01 % pixels segmented as the green land from all image pixels with the large pixels are 1.310.720 pixels. In this study, K-means clustering shows better accuracy to cluster the two expected clusters compared with normalized RGB space method. However, a clustering with the color feature of image usually considers the color depth of the pixels information. The better resolution and bit-depth of data input that will be clustered are given, the better accuracy cluster will be achieved.

References

1. Chipman, J., Kiefer, R.W., Lillesand, T.: Remote Sensing and Image Interpretation, 6th edn. Willey, Inc., USA (2007)
2. Ruzgiene, B., Xiang, Q.Y., Gecyte, S.: Rectification of satellite imagery from google earth: application for large scale city mapping. In: The 8th International Conference on Environmental Engineering, vol. 3, pp. 1451–1454. Vilnius, Lithuania (2011)
3. Li, Z., Liu, Z., Shi, W.: Semiautomatic airport runway extraction using a line-finder-aided level set evolution. IEEE J. Sel. Topics Appl. Earth Obs. 99, 1–13 (2014)
4. Aher, M., Pradhan, S., Dangdawate, Y.: Rainfall estimation over roof-top using land-cover classification of google earth images. In: International Conference on Electronic Systems, Signal Processing and Computing Technologies (ICESC), pp. 111–116. Nagpur, India (2014)
5. Coralini, A., Gidazzoli, A., Malfetti, P.: Cloud computing and virtual heritage: the social media-oriented paradigm experienced at cineca. In: IEEE 3rd Symposium on Network Cloud Computing and Applications (NCCA), pp. 121–124, Rome, Italy (2014)
6. Sahin, K., Ulusoy, I.: Automatic Multi-scale segmentation of high spatial resolution satellite images using waterhseds. In: IEEE International Geoscience and Remote Sensing Symposium (IGARSS), pp. 2505–2508, Melbourne (2013)
7. Almeer, M.H.: Vegetation extraction from free google earth images of deserts using a robust BPNN approach in HSV Space. Int. J. Adv. Res. Comput. Commun. Eng. 2(5), 1–8 (2012)
8. Janssen, I., Rosu, A.: Measuring sidewalk distances using google earth. J. BMC Med. Res. Methodol. 12(1), 1–10 (2012)
9. Ploton, P., et al.: Assessing aboveground tropical forest biomass using google earth canopy images. In: Ecol. Soc. Am. Ecolo. Appl. 22(3), 993–1003 (2012)

10. Stepanus, R.: Analisis kebutuhan ruang terbuka hijau di kota Pekanbaru. Master Thesis Institute of Pertanian Bogor, Bogor, Indonesia (2006)

11. Chinki, C., Soni. C., Khurshid, A.A.: An approach to image segmentation using K-Means clustering algorithm. Int. J. Info. Technol. (IJIT) 1(1), 11–17 (2012)

12. Narayana, M.: Comparison between euclidean distance metric and SVM for CBIR using level set features. Int. J. Eng. Sci. Technol. (IJEST) 4(1), 56–66 (2012)

13. Ravindra, S., Sangolli, H., Rajeshwari, K.: Segmentation of google map images based on color features. In: Proceedings of International Conference on Communication, Computation, Management and Nanotechnology (ICN), pp 77–80 (2011)

14. Moreno, F., Andrade-Cetto, J., Sanfeliu, A.: Localization of human faces fusing color segmentation and depth from stereo. In: Proceedings of the 8th IEEE International Conference on Emerging Technologies and Factory Automation, vol. 2, pp. 527–535. Antibes, France (2001)

Texture Based Associative Classifier—An Application of Data Mining for Mammogram Classification

Deepa S. Deshpande, Archana M. Rajurkar
and Ramchandra R. Manthalkar

Abstract The incidence of breast cancer is rapidly becoming the number one cancer in females. It is the serious health problem and leading cause of death for middle aged women. Mammography is one of the most reliable methods for early detection of breast cancer. But mammograms are the most difficult images for interpretation and may lead to false diagnosis. Therefore there is a significant need of automatic extraction of the actionable information from the mammogram data in order to ensure improvement in diagnosis. To address this issue, we have proposed an automatic classification system for breast cancer using Texture Based Associative Classifier (TBAC). Here we wish to automatically classify the breast mammograms into three basic categories i.e. normal, benign and malignant based on their texture associations. Our experimental results on MIAS dataset demonstrate that the proposed classifier TBAC is superior to existing associative classifiers for mammogram classification.

Keywords Association rules · Breast cancer · Mammogram classification · Associative classifier

D.S. Deshpande (✉)
Department of Computer Science and Engineering, MGM's Jawaharlal Nehru Engineering College, Aurangabad, Maharashtra, India
e-mail: deepadsd@yahoo.com

A.M. Rajurkar
Department of Computer Science and Engineering, MGM's College of Engineering College, Nanded, Maharashtra, India
e-mail: archana_rajurkar@yahoo.com

R.R. Manthalkar
Department of Electronics and Telecommunication Engineering, SGGS Institute of Engineering and Technology, Nanded, Maharashtra, India
e-mail: rmanthalkar@yahoo.com

© Springer India 2015
L.C. Jain et al. (eds.), *Computational Intelligence in Data Mining - Volume 1*, Smart Innovation, Systems and Technologies 31, DOI 10.1007/978-81-322-2205-7_37

1 Introduction

Breast cancer is the primary and most common disease found among women. The world health organization's International Agency for Research on cancer (IARC) estimates that more than 400,000 women expire each year due to this disease. The occurrence of breast cancer is increasing globally and disease remains a major public health problem. According to American college of Radiology (ACR) statistics one out of nine women will develop breast cancer during their life span. Statistics from breast cancer India website tells that the breast cancer accounts for 25–31 % of all cancers in women and the average age of developing the breast cancer has been shifted from 50–70 years to 30–50 years. Globocan (2008) data shows that there is a rapid growth in the death of women suffering from this disease i.e. one death for every two cases detected. The earlier the cancer detected the better treatment can be provided. Mammography is considered the most reliable radiological screening technique in early detection of breast cancer. Mammograms are relatively cheap to produce but have many disadvantages. Correct classification of anomalous areas in the mammograms through visual examination is a challenging task even for experts. It has been also found that radiologists misdiagnose 10–30 % of the malignant cases because of high volume of mammograms to be read by physicians and lack of useful analysis tools or computer aided diagnosis systems to realize hidden relationships and trends in digital mammograms for accurate diagnosis. Therefore there is a significant need of Computer Aided Diagnosis (CAD) system [1, 2] to assist medical staff. Such type of system can help to reduce the problems of screening mammograms. It can be used as a second reader and can improve the detection performance of a radiologist. Much research has been done during last few years in the field of mammogram classification [3–8], still there is no widely used method to classify mammograms due to the fact that medical domain requires high accuracy and adequate classification to handle the large amount of data made available by advancements in imaging.

Mammogram classification concerns about the automatic classification of mammogram images to one or multiple classes based on their content. Many different classification approaches were developed to classify mammograms, these approaches can be evaluated mainly by accuracy and the knowledge they produce. Some of them produce high accurate classifiers and others low accurate ones. However, one fundamental measurement criteria is the understandability of the end-user of the resulting classifiers. A new classification data mining technique called Associative Classification (AC) is developed which combines high accuracy and understandability output together based on association rule. AC is a high efficient method that builds more predictive and accurate classification systems than traditional classification methods such as probabilistic and the k-nearest neighbor algorithm according to many research experimental studies. AC produces rule's based classifiers that are easy to understand and manipulate by end-user. This research is devoted to develop a new model based on AC for mammogram classification problem. Mainly, we focused on three main steps in the mammogram classification problem and these are: (1) Developing efficient and fast method of association rule mining which can overcome the major drawbacks of traditional a priori algorithm [9] and adopting this method for classification of unstructured data like image database. (2) Proposing a novel rule scoring procedure that looks for probability estimate to indicate how likely the rules belong to interesting rules for classification during building the classifier. This novel method significantly handle imbalance class distribution of data when compared with existing AC models like CBA, CMAR, ARC-AC, ARC-BC etc. (3) Improving the accuracy by considering multiple association rules rather than class constrained rules where class is kept at the right hand side of the rules, in the classification step. Experimental results using mini mammography database made freely available by Mammography Image Analysis Society (MIAS) [10] indicated that the proposed model outperforms other existing classification techniques either traditional techniques or AC techniques. Thus the potential of association rule mining for mammogram classification is demonstrated in this paper using TBAC associative classifier.

2 Background Review

The classification of mammograms is a difficult and computationally enormous task. Much research in the field of mammogram classification has been done recently because of significant need of automated classification system to assist medical staff specially physicians and radiologists. Most of the research in the mammogram analysis has focused on content based image retrieval [11, 12]. For content based analysis mammograms are processed to extract a set of features that represents textures and shapes in the mammograms. Euclidian distance or Minkowshi distance methods are generally used for similarity matching among extracted features for mammogram classification. However such methods have not been found suitable to model perceptual similarity adequately. More effective methods for content based image analysis have been proposed to classify mammograms such as SVM based technique [13], wavelets [14, 15], fractal theory [16], statistical methods [17], fuzzy set theory [18, 19], markov model [20], neural network [21]. All of them use features extracted using image processing techniques. These methods work reasonably well, but still they require higher classification accuracy. Hence among numerous supervised classification methods, associative classification has been adopted by different researchers for mammogram classification. Liu et al. [22] proposed first classifier building algorithm (CBA) based on standard a priori algorithm using association rules to classify mammograms into different classes. CMAR [23] provides classification based on multiple class association rules using FP-Growth algorithm. Antonie et al. [5, 24, 25] introduced two techniques ARC-AC (association rule-based classification with all categories) and ARC-BC (association rule-based classification by categories) for association rule based classifier (ARC). ARC-BC is an improvement over ARC-AC where association rules are formed for each class separately rather than for entire dataset and it provides accuracy up to 80 %. All these classifiers generate class constrained rules where class is kept at the right hand side of the rules. Two major considerable limitations are that they consider only the presence of the feature but not its importance and do not able to handle unbalanced data. Ribeiro et al. [7] use texture features and association rules to classify mammogram images. But the major problem was segmentation of images and constraint of keeping class labels on the right side of the rule. Tseng et al. [26] applied multilevel association rules to cluster objects from various images and perform object based segmentation. But it is not widely applicable for medical images because they usually contain few objects. Yun et al. [27] used a combination of association rules with rough set theory to develop Joining Associative Classifier (JAC) for mammogram classification, which provides accuracy up to 77 %. Weighted Association Rule Based Classifier (WAR-BC) [28] having accuracy of 89 % was introduced by Dua et al. [29]. It uses intra-class weight and inter-class weight of each association rule for classification.

Although most of the researchers have been developed different classification techniques for mammogram classification, correct classification of mammograms with higher accuracy is still challenging task. Hence in this research work we considered the problem of mammogram classification and proposed Texture Based Associative Classifier. Here, we wish to automatically classify the mammogram either into normal, benign or malignant class based on associations among texture feature representation as the image content.

3 Data Collection

In the development of TBAC to access real mammograms for experimentation, we used breast cancer mammograms from the mini mammography database made freely available by Mammography Image Analysis Society (MIAS) [10]. It consists of 322 images which belong to three basic categories: normal, benign and malignant. There are 208 normal images, 63 benign and 51 malignant, which are considered abnormal. From this set of images we used 150 images (50 images per basic category) as training dataset and 25 images from each basic category are used for testing. The illumination conditions at the time mammograms were taken and noise introduced during digitization make mammogram interpretation difficult. Therefore mammogram images present in the image database need to preprocess in order to improve the quality of images. In image processing [30, 31] histogram equalization is a method of contrast adjustment using the image's histogram. Through this adjustment, the intensities can be better distributed on the histogram. Therefore we applied histogram equalization technique from the spatial domain in the image processing to make contrast adjustment so that the abnormalities of the mammogram images will be better visible. This helps to improve the efficiency of mining task.

4 Proposed Model of TBAC Classifier and Results

This section details the phases we considered in the development of TBAC. Feature extraction and subset selection is key for formation of transaction database in developing the proposed model. Next, we look at mining transaction database using association rule mining. Finally, we used output of mining task for construction of classification model.

4.1 Formation of Transaction Database

To create the transaction database to be mined for classification, feature extraction [28, 32] and subset selection represent the maximum relevant information that the image has to offer. The extracted features are then organized in a database, which is the input for the mining task of the classification system. In order to accurately classify the mammograms, total 18 texture features [33] like Autocorrelation, Contrast, Correlation, Cluster Prominence, Cluster Shade, Dissimilarity, Energy, Entropy, Homogeneity, Maximum Probability, Sum of Square Variance, Sum Average, Sum Variance, Sum Entropy, Difference Entropy, Information Measure of Correlation, Information Measure of Correlation2, Inverse Difference Moment Normalized are extracted using GLCM statistical method [34, 35]. Texture features are preferred as they are able to distinguish between normal and abnormal patterns. These features are then mapped into conventional database format where columns are texture features representing an image and rows have values corresponding to those texture features. Most of these features are irrelevant and redundant. Also use of all these 18 features may degrade the mining performance due to computational complexity. Therefore these features are then passed through feature subset selection process to limit the number of input features. The best five features that present high dissimilarity with the other features are selected using forward selection method. The selected features are Sum of squares variance (F_1), Auto correlation (F_2), Cluster shade (F_3), Sum variance (F_4) and Cluster prominence (F_5). These selected real value texture features are normalized to binary form by standard min-max data normalization technique [36] and organized in a database in the form of transactions, which in turn constitute the input for deriving association rules. The transactions are of the form [Image ID, F_1; F_2; ::::; F_5] where F_1:: F_5 are relevant 5 features for a given image. Thus transaction database is formed for the three basic categories of mammograms (normal, benign and malignant) using 50 images per category from the MIAS database and used as a training dataset for classification.

4.2 Association Rule Mining

Association rules [36] show attributes value conditions that occur frequently together in a given dataset. Association rule mining [9, 37] finds interesting associations and/or correlation relationships among large set of data items. An association rule is represented in the form $X \rightarrow Y$ where X, Y belongs to I and $X \cap Y = \emptyset$. Confidence c of the rule says that c % of transactions in T that support X also support Y. The support s of the rule tells that s % of the transactions in T contains $X \cup Y$. Traditional association rule mining algorithm [9] discovers the rules those have support and confidence greater than or equal to the user specified minimum support and minimum confidence. In our prior work [38], we have proposed novel method of association rule mining shown in Fig. 1. Thus the training dataset is mined to obtain frequent feature set using this algorithm and association rules are derived for all three predefined classes of mammograms. For an example frequent feature set for malignant class is shown here and sample of derived association rules for feature set {Sum of sq variance, Autocorrelation, Cluster prominence} is listed in Table 1.

Frequent feature set for malignant class = {{Sum of sq variance}, {Auto correlation}, {Cluster Shade}, {Sum Variance}, {Cluster prominence}, {Sum of sq variance, Auto correlation}, {Sum of sq variance, Cluster prominence}, {Auto correlation, Cluster prominence}, {Sum variance, Cluster prominence}, {Sum of sq variance, Autocorrelation, Cluster prominence}}.

Algorithm: Proposed method of association rule mining
Input: D – Transaction database, p – Set of items, min-supp – Minimum support threshold
Output: L- Large item set, R- Association rules
Begin
Step 1: k =1; C_k = Candidate item sets with length k.
Step 2: For i= 1 to p
    Scan Database to calculate Supp-Yes and Supp-No of i
    Add item i to C_k
    End for
Step 3: For each item set in C_k
    if Supp-Yes (item set) > min-supp then
    Add that item set to L_k
Step 4: for k > 1
    If L_{k-1} is not null then
    Generate C_k from L_{k-1}
    For each item set in C_k
    If Supp-Yes (each item from item set) > Max (Supp-No of each items in item set) then
    Add that item set to L_k
Step 5: $L = \cup L_k$
Step 6: k = k + 1
Step 7: Repeat steps 4 to 6 until no larger item set is found
End

Fig. 1 Algorithm for association rule mining

Table 1 Sample association rules for malignant class

Rule No.	Association rules
1	Sum of square variance, auto correlation $\Rightarrow$ cluster prominence
2	Sum of square variance, cluster prominence $\Rightarrow$ auto correlation
3	Auto correlation, cluster prominence => sum of square variance
4	Sum of square variance $\Rightarrow$ auto correlation, cluster prominence
5	Auto correlation $\Rightarrow$ sum of square variance, cluster prominence
6	Cluster prominence $\Rightarrow$ sum of square variance, auto correlation

4.3 Classification

Association rule mining aims to find all rules in data. Different rules may give different information. Many of them may give conflicting information which makes it difficult to know how likely the rules belong to interesting rules for classification. To address this issue we presented score function that makes use of all rules because every rule contains a certain amount of information. Support and confidence are two basic accuracy measures [39] to evaluate rules. But it finds many misleading rules which may not be useful for classification. Hence a different approach to assess association rules is required. In this paper, we have proposed a method for scoring of association rules in order to avoid significance of misleading rules for classification. Simply confidence can be used for scoring of the rule. The score is simply the confidence value of the rule in the respective class. Highest confidence indicates the likelihood that the rule belongs to respective class. However many rules that are put into specific class with highest confidence do not actually belong there due to imbalance class distribution. This may give poor results for classification. Hence we presented more reliable score function using weight which will push the rules towards two ends i.e. likely to be interesting or not interesting. Once the association rules have been discovered for three basic categories of mammograms, score is calculated for each rule of respective categories i.e. we assign probability estimate to indicate how likely the rule belongs to interesting rules. Score is formulated based on accuracy measures listed in Table 2. Since we have used four desirable effects of accuracy measure, weight should reflect their needs to define score of the rule. Therefore weight is calculated as

$$\text{Weight} = \text{Supp} * \text{Conf} * \text{CF} * \text{CP} \quad \text{if Lift} > 1. \tag{1}$$

$$\text{Weight} = (\text{Supp} * \text{Conf} * \text{CF} * \text{CP})/4 \quad \text{if} \quad \text{Lift} < 1. \tag{2}$$

Rules are interesting when they have high accuracy, high completeness, and high certainty factor but here interestingness is evaluated by score function given below and algorithm for rule scoring is presented in Fig. 2.

Table 2 Accuracy measures to access association rules

Accuracy measure	Formula	Description
Support (supp)	Supp(A $\Rightarrow$ C) = supp(A U C)	The probability that the feature set appears in a transaction DB. Support is the percentage of transactions where the rule holds
Confidence (confi)	Conf(A $\Rightarrow$ C) = supp(A U C)/supp(A) = Supp(A $\Rightarrow$ C)/supp(A)	Confidence is the conditional probability of C with respect to A or, in other words, the relative cardinality of C with respect to A
Lift	Lift(A $\Rightarrow$ C) =P(AÙC)/[P(A)* P(C)]	If Lift < 1 then occurrence of A is negatively correlated with occurrence of C If Lift > 1 then A and C are positively correlated If Lift = 1 then A and C are independent and there is no correlation between A and C
Certainty factor (CF)	if Conf(A $\Rightarrow$ C) > supp(C) CF(A $\Rightarrow$ C) = [Conf(A $\Rightarrow$ C)- supp(C)]/ [1-supp(C)] Else If Conf(A $\Rightarrow$ C) < supp(C) CF(A $\Rightarrow$ C) = [Conf(A $\Rightarrow$ C)- supp(C)]/ [supp(C)] Else If Conf(A $\Rightarrow$ C) = supp(C) CF = 0	The certainty factor is interpreted as a measure of variation of the probability that C is in a transaction when we consider only those transactions where A is
Completeness (CP)	CP(A $\Rightarrow$ C) =Supp(A $\Rightarrow$ C)/Supp(C)	How much of the target class a rule covers

```
Algorithm: Rule scoring is used for every rule present in the training database
Input   :  C: Number of classes, n: Number of rules in each class C_j
Output:    Scores of rules
Begin
For every class j ∀ j ≤ C
    For every rule R_iC_j ∀ i ≤ Cj, j ≤ C
        Compute Supp(R_iC_j) ,Conf(R_iC_j) ,CF(R_iC_j) ,CP(R_iC_j), Lift(R_iC_j)
        If Lift (R_iC_j) > 1 then Weight (R_iC_j) = Supp(R_iC_j) *Conf(R_iC_j) *CF(R_iC_j) *CP(R_iC_j)
        Else Weight (R_iC_j) = [Supp(R_iC_j) *Conf(R_iC_j) *CF(R_iC_j) *CP(R_iC_j)] /4
    End For
End For
For every class j ∀ j ≤ C
    For every rule R_iC_j ∀ i ≤ Cj, j ≤ C
```

$$\text{Score } (R_iC_j) = \sum_{i=1}^{n} [W(R_iC_j) * \text{Conf}(R_iC_j)] \ / \ \text{Conf}(R_iC_j)$$

```
    End For
End For
End
```

Fig. 2 Algorithm for rule scoring

$$Score = \sum_{i=1}^{N} [W_i * Conf_i]/Conf_i. \qquad (3)$$

Scores are computed for all these derived association rules based on their accuracy measures listed in Table 2. Association rules listed in Table 1 are presented along with their scores in Table 3. Rules 1–6 are expected rules as their confidence is close to one and value of completeness is also close to one. This shows that derived rules are with high accuracy and avoids the discovery of misleading rules. Such type of association rules derived from all three categories/classes of mammograms along with their scores are considered as global rule set and represent the classification model for TBAC. This model is used in the testing phase to classify previously unseen mammograms. For appropriate classification the algorithm given in Fig. 3 is applied on the query mammogram. Each association rule derived from query mammogram is processed to find match with the global rule set. Scores of matching rules are added on class by class basis and finally classification of the query mammogram is done to the class having highest cumulative sum.

Result for classification of digital mammograms obtained by proposed classifier TBAC is given in Table 9. The confusion matrix has been obtained from the testing part. 25 images from each basic category are used for testing. In this case out of 25

Table 3 Association rules with scores for malignant class

Rule No.	Confidence	Certainty factor	Completeness	Lift	Weight	Score
1	0.95	0.75	1	1.19	0.57	4.01
2	1	1	0.95	1.19	0.76	3.81
3	0.97	0.81	0.95	1.17	0.59	3.92
4	0.95	0.72	0.98	1.17	0.53	4.01
5	0.95	0.75	1	1.19	0.72	4.01
6	1	1	0.95	1.19	0.76	3.81

```
Algorithm:  Class prediction of query image
Input:      D: Transaction database of mammograms, T: set of all m test instances
            C₁: sets of scored association rules mined from normal class of D
            C₂: sets of scored association rules mined from benign class of D
            C₃: sets of scored association rules mined from malignant class of D
Output:         Predicted Class
Begin
For each t ∈ T do
        Find association rules for  t
        Match each association rule with global classification rules from C1, C₂ & C₃
        Scores of matching rules are added on class by class basis.
Predict class for t as the class with highest cumulative sum
End
```

Fig. 3 Algorithm for class prediction of query image

actual malignant images 4 images were classified as normal. In case of benign and normal all images are correctly classified. The confusion matrix is given in Table 4.

Given m classes, $CM_{i,j}$, an entry in a confusion matrix, indicates # of tuples in class i that are labeled by the classifier as class j.

The terms used to express accuracy measures are given in the contingency (Table 5). Where TP stands for True Positive i.e. images which are normal and labeled as normal by classifier. FP stands for False Positive i.e. images are abnormal but labeled as normal by classifier. FN stands for False Negative i.e. images which are normal but labeled as abnormal by classifier and TN stands for True Negatives i.e. images which are abnormal and labeled as abnormal by classifier. Resultant contingency tables for different categories (Tables 6, 7 and 8) are derived from confusion matrix and presented here.

In general, accuracy of classifier C, acc(C) is the percentage of test set tuples that are correctly classified by the model M and Error rate (misclassification rate) of C is calculated as 1—acc (C). Classification accuracy and error rate of different classes using TBAC is shown in Table 9. Also it is very much clear from Table 8 that out of 75 testing mammograms (25 normal + 50 abnormal) 71 testing mammograms (25 normal + 46 abnormal) are correctly classified. Therefore accuracy of TBAC classifier is 94.66 %.

Table 4 Confusion matrix

Actual class	Predicted class		
	Benign	Malignant	Normal
Benign	25	0	0
Malignant	0	21	4
Normal	0	0	25

Table 5 Contingency table

Category		Predicted class	
		Normal	Abnormal
Actual class	Normal	True positive (TP)	False negative (FN)
	Abnormal	False positive (FP)	True negative (TN)

Table 6 Contingency table (normal-benign category)

Category		Predicted class		Total	Recognition (%)
		Normal	Benign		
Actual class	Normal	25	00	25	100
	Benign	00	25	25	100
	Total	25	25	50	100

Table 7 Contingency table (normal-malignant category)

Category		Predicted class		Total	Recognition
		Normal	Malignant		
Actual class	Normal	25	00	25	100
	Malignant	04	21	25	84
	Total	29	21	50	92

Table 8 Contingency table (normal-abnormal category)

Category		Predicted class		Total	Recognition (%)
		Normal	Abnormal		
Actual class	Normal	25	00	25	100
	Abnormal	04	46	50	92
	Total	29	46	75	94.66

Table 9 Classification accuracy and error rate of the proposed model

Class	Accuracy (%)	Error rate (%)
Normal	100	0
Benign	100	0
Malignant	84	16
Abnormal	92	8

We have compared our proposed classifier with the other published classifiers. Table 10 presents the result in terms of accuracy for MIAS dataset using BPNN, ARC-AC, ARC-BC, JAC, WAR-BC and TBAC classifiers. Result shows that our proposed classifier TBAC performs well reaching over 94.66 % in accuracy as compared with the other classification techniques.

The merits of our proposed TBAC classifier are

1. Derived association rules represent actual associations among extracted texture features rather than class constraint rules where class is always kept at the right hand side of the rule.
2. The proposed classifier has been found to be performing well compared to the existing classifiers incurring accuracy as high as 94.66 % for MIAS image database.
3. This classifier can easily handle multiple classes with unbalanced data. This ability to handle unbalanced data is significant improvement to existing classifiers for mammogram classification.
4. It can be used as a second reader to improve the detection performance of radiologists at early stage of breast cancer. Also it can reduce the computation cost of mammogram image analysis.
5. It can be applied to other image analysis applications also.

Table 10 Comparison with different classification techniques

Sr. No.	Classification technique	Accuracy (%)	References
1	Back propagation neural network (BPNN)	81	[5]
2	Association rule-based classification with all categories (ARC-AC)	69	[24, 25]
3	Association rule-based classification by categories (ARC-BC)	80	[24, 25]
4	Joining associative classifier (JAC)	77	[27]
5	Weighted association rule based classifier (WAR-BC)	89	[29]
6	Texture based associative classifier (TBAC)	94.66	Proposed classifier

5 Conclusion

Mammography is one of the best methods helpful for early detection of breast cancer and increases the chance of successful treatment. But mammography tests are not perfect. Interpretation of mammograms is also very difficult and hence correct classification of mammograms through visual examination is a challenging task even for experts. To address this issue, a novel classifier TBAC is proposed in this paper to assist medical staff for correct classification of mammograms. The evaluation of the classifier is carried out on MIAS dataset. Mammograms are preprocessed to enhance the quality of images. Texture features are extracted from the mammograms and discretized for rule discovery. Association rules are derived by judging the importance of extracted texture features. These rules are then modeled for classification. Derived association rules represent actual associations among extracted texture features rather than class constraint rules where class is always kept at the right hand side of the rule. The proposed classifier TBAC has been found to be performing well compared to the existing classifiers incurring accuracy as high as 94.66 % for such dataset. This classifier can easily handle multiple classes with unbalanced data. This ability to handle unbalanced data is significant improvement to existing classifiers for mammogram classification. It also illustrates the use and effectiveness of association rule mining for mammogram classification in the area of medical image analysis. Moreover it can be used as a second reader to improve the detection performance of radiologists at early stage of breast cancer. Also it can reduce the computation cost of mammogram image analysis and can be applied to other image analysis applications.

References

1. Sampat, Mehul P., Mia, Markey, K., Bovik, A.C.: Computer-Aided Detection and Diagnosis in Mammography. Hand Book of Image and Video Processing (2nd edn) (2005), pp. 1195–1217
2. Rangayyan, R.M., Ayres, F.J., Desautels, J.E.L.: A review of computer-aided diagnosis of breast cancer: toward the detection of subtle signs. J. Franklin Inst. **344**, 312–348 (2007)
3. Verma, B., McLeod, P., Klevansky, A.: Classification of benign and malign patterns in digital mammograms for the diagnosis of breast cancer. Exp. Syst. Appl. **37**, 3344–3351 (2010)
4. Ferreira, C.B.R., Borges, D.L.: Automated mammogram classification using a multi resolution pattern recognition approach. In: Proceedings of XIV Brazilian Symposium on Computer Graphics and Image Processing, IEEE; Florianopolis, Brazil. pp. 76–83, (2001)
5. Antonie, M.L., Zaiane, O.R., Coman, A.: Application of data mining techniques for medical image classification. Proceeding of Second International Workshop On multimedia Data Mining 9MDM/KDD'2001) in Conjunction with ACM SIGKDD Conference, San Francisco, USA, pp. 94–101 (Aug 2001)
6. Nithya, R., Santhi, B.: Classification of normal and abnormal patterns in digital mammograms for the diagnosis of breast cancer. Int. J. Comput. Appl. **28**(6), (2011)
7. Ribeiro, M.X., Traina, A.J.M., Balan, A.G.R., Traina, C., Marques, PMA: SuGAR: A framework to support mammogram diagnosis, IEEE CBMS 2007; Maribor, Slovenia, pp. 47–52, (2007)
8. Surendiran, B., Vadivel, A.: An automated classification of mammogram masses using statistical measures. In Proceedings of 4th Indian International Conference on Artificial Intelligence (IICAI-09), pp. 1473–1485 (2009)
9. Srikant., R. :Fast algorithms for mining association rules in large databases. Proceedings of the 20th International Conference on Very Large Data Bases, Santiago, Chile, August 29–Sept 1 (1994)
10. MIAS Database. http://peipa.essex.ac.uk/info/mias.htm
11. Wei, L., Yang, Y., Nishikawa, R.M.: Micro calcification classification assisted by content-based image retrieval for breast cancer diagnosis. Pattern Recogn. **42**, 1126–1132 (2009)
12. Muller, H., Michoux, N., Bandon, D., Geissbuhler, A.: A review of content-based image retrieval systems in medical applications-clinical benefits and future directions. Int. J. Med. Informatics **73**, 1–23 (2003)
13. Goh, K.,Chang, F., Chang, T. :SVM binary classifier ensembles for image classification. ACM International Conference on Information and Knowledge Management, (Nov 2001)
14. Chen, C., Lee, G.: Image segmentation using multi-resolution wavelet analysis and expectation maximization (em) algorithm for digital mammography. Int. J. Imaging Syst. Technol. **8**(5), 491–504 (1997)
15. Wang, T., Karayiannis, N.: Detection of micro calcification in digital mammograms using wavelets. IEEE Trans. Med. Imaging **17**(4), 498–509 (1998)
16. Li, H., et al.: Fractal modeling and segmentation for the enhancement of micro calcifications in digital mammograms. IEEE Trans. Med. Imaging **16**(6), 785–798 (1997)
17. Chan, H., et al.: Computerized analysis of mammographic micro calcifications in morphological and feature spaces. Med. Phys. **25**(10), 2007–2019 (1998)
18. Brazokovic, D., Neskovic, M.: Mammogram screening using multi resolution-based image segmentation. Int. J. Pattern Recognit Artif Intell. **7**(6), 1437–1460 (1993)
19. Wang, S., Zhou, M., Geng, G.: Application of fuzzy cluster analysis for medical image data mining. Proceeding of the IEEE International Conference on Mecahtronics and Automation Niagara Falls, Canada, pp. 36–41 July (2005)
20. Li, H., et al.: Marcov random field for tumor detection in digital mammography. IEEE Trans. Med. Imaging **14**(3), 565–576 (1995)
21. Christoyianni, I et al.: Fast detection of masses in computer-aided mammography. IEEE Signal Process. Mag. pp. 54–64 (2000)

22. Liu, B., Hsu, W.,Ma, Y.: Integrating classification and association rule mining. Proceedings of ACM SIGKDD International Conference on Knowledge Discovery and Data Mining(KDD-98), pp. 80–86 (1998)
23. Li, W., Han, J.,Pei, J.: CMAR: Accurate and efficient classification based on multiple class-association rules. IEEE International Conference on Data Mining, (2001)
24. Antonie, M.L., Zaiane, O.R., Coman, A.: Associative classifiers for medical images, LNICS, vol. 2797, pp. 68–83. Springer, Berlin (2003)
25. Antonie, M.L., Zaiane, O.R., Coman, A.: Mammography classification by an association rule-based classifier. Proceedings of the Third International Workshop on Multimedia Data Mining, pp. 62–69 (2002)
26. Tseng, S.V.,Wang, M.H., Su, J.H.: A new method for image classification by using multilevel association rules. Presented at ICDE 05; Tokyo. pp. 1180–1187 (2005)
27. Yun, J., Zhanhuai, L., Yong, W., Longbo, Z.: Joining associative classifier for medical images, HIS (2005)
28. Panda, R.N., Panigrahi, B.K., Patro, M.R.: Feature extraction for classification of micro calcifications and mass lesions in mammograms. Int. J. Comput. Sci. Netw. Secur, 9(5), (2009)
29. Dua, S., Singh, H.,Thompson, H.W.: Associative classification of mammograms using weighted rules. Expert Syst Appl. 36(5), 9250–9259 (2009)
30. WILEY, J.: Image Processing The Fundamentals Maria Petrou University of SurreN Guildford, UK Panagiota Bosdogianni Technical Universify of Crete, Chania, Greece
31. Gonzalez, R. C.: Digital Image processing using MATLAB Pearson publication, (2005)
32. Sheshadri, H.S., Kandaswamy, A.: Breast tissue classification using statistical feature extraction of mammograms. Med. Imaging Inf. Sci. 23(3), 105–107 (2006)
33. Dinstein, I.H., Haralick, R.M., Shanmugam, S.K., Goel, D.: Texture tone study. Classification experiment, interim technical report. no. 4, KANSAS UNIV/CENTER FOR RESEARCH INC LAWRENCE REMOTE SENSING LAB, (Dec 1972)
34. Khuzi, A. M., Besar, R., Zaki, W. M. D. Wan: Texture features selection for masses detection in digital mammogram. 4th Kuala Lumpur International Conference on Biomedical Engineering 2008 IFMBE Proceedings, vol. 21, Part 3, Part 8, 629–632 (2008). doi: 10.1007/978-3-540-69139-6_157
35. Ke, L., Mu, N., Kang, Y.: Mass computer-aided diagnosis method in mammogram based on texture features, Biomedical Engineering and Informatics (BMEI). 3rd International Conference, IEEE Explore, November (2010) pp. 146—149 (2010) doi: 10.1109/BMEI.2010.5639662
36. Han, J., Kamber, M., Pei, J.: Data Mining: Concepts and Techniques, 3rd edn, The Morgan Kaufmann Series in Data Management Systems, (2011)
37. Agrawal, R., Imielinksi, T., Swami, A.: Vu: Database Mining a performance perspective. IEEE Trans. Knowl. Data Eng. 3(6), 914–925 (1993)
38. Deshpande, D.S., Rajurkar, Archana M., Manthalkar, Ramchandra R.: Mammogram classification using association rule mining. In: Proceedings of 10th International Conference on Data Mining (DMIN'14), Las Vegas Nevada, USA, pp. 95–101 (2014)
39. Frawley, P.S.G., Piatetsky-Shapiro, W.G.: Discovery, analysis, and presentation of strong rules. Knowl Discov. Databases. AAAI/MIT Press, pp. 229–238 (1991)

Particle Swarm Optimization Based Higher Order Neural Network for Classification

Janmenjoy Nayak, Bighnaraj Naik, H.S. Behera and Ajith Abraham

Abstract The maturity in the use of both the feed forward neural network and Multilayer perception brought the limitations of neural network like linear threshold unit and multi-layering in various applications. Hence, a higher order network can be useful to perform nonlinear mapping using the single layer of input units for overcoming the drawbacks of the above-mentioned neural networks. In this paper, a higher order neural network called Pi-Sigma neural network with standard back propagation Gradient descent learning and Particle Swarm Optimization algorithms has been coupled to develop an efficient robust hybrid training algorithm with the local and global searching capabilities for classification task. To demonstrate the capacity of the proposed PSO-PSNN model, the performance has been tested with various benchmark datasets from UCI machine learning repository and compared with the resulting performance of PSNN, GA-PSNN. Comparison result shows that the proposed model obtains a promising performance for classification problems.

Keywords Higher order neural network · Classification · PSO · Pi-sigma neural network · GA

J. Nayak (✉) · B. Naik · H.S. Behera
Department of Computer Science Engineering and Information Technology,
Veer Surendra Sai University of Technology, Burla, Sambalpur 768018,
Odisha, India
e-mail: jnayak@gmail.com

B. Naik
e-mail: bnaik@gmail.com

H.S. Behera
e-mail: hsbehera@gmail.com

A. Abraham
Machine Intelligence Research Labs (MIR Labs), Washington, USA
e-mail: ajith.abraham@ieee.org

A. Abraham
IT4 Innovations—Center of Excellence, VSB—Technical University of Ostrava,
Ostrava, Czech Republic

401

© Springer India 2015
L.C. Jain et al. (eds.), *Computational Intelligence in Data Mining - Volume 1,*
Smart Innovation, Systems and Technologies 31, DOI 10.1007/978-81-322-2205-7_38

1 Introduction

Over the last two decades, Particle Swarm Optimization (PSO) has been introduced for solving various real world problems. But, due to the non-linearity, high dimensionality, and multi-modality nature, finding the optimal solution for such problems is a quite challenging task. Many researchers have shown tremendous efforts by using the stochastic algorithms in solving various global optimization problems. Among the stochastic approaches, population-based evolutionary algorithms (EAs) [1] and swarm intelligence (SI) methods [2] offer a number of advantages [3] which make them more popular: simple implementation, inherent parallelism, vigorous and reliable performance, global search capability, zero necessity of particular information about the problem to solve, good insensitivity to noise.

Sun et al. [4] have proposed a new fitness estimation strategy for particle swarm optimization to reduce the number of fitness evaluations, thereby reducing the computational cost. They examined the performance of the proposed algorithm is on eight benchmark problems which show the proposed algorithm is easy to implement, effective and highly competitive. Imran et al. [5] introduced different variants of PSO with respect to initialization, inertia weight and mutation operators. Pan et al. [6] defined the state sequence of a single particle and the swarm state based on the proposed PSO based Markov Chain model, which calculates the one-step transition probability of particles by the random characteristics of PSO's acceleration factor. Babaei [7] presented a general form of PSO to approximately solve a great variety of linear and nonlinear ordinary differential equations (ODEs) independent of their form, order, and given conditions. An extensive review of literature available on concept, development and modification of Particle swarm optimization has been made by Khare and Rangnekar [8]. Authors claimed that due to simplicity and easy of implementation, PSO can be used in a wide variety range of problems. A comprehensive experimental study on Diversity enhanced particle swarm optimization is being conducted by Wang et al. [9] on a set of benchmark functions, including rotated multimodal and shifted high-dimensional problems, which obtains a promising results on various test problems. Neri et al. [10] addressed a compact particle swarm optimization, which employs a probabilistic representation of the swarm's behavior and allows a modest memory usage for the entire algorithmic functioning. A new hybrid approach for optimization combining Particle Swarm Optimization (PSO) and Genetic Algorithms (GAs) using Fuzzy Logic for modular neural network has been proposed by Valdez et al. [11].

Higher Order Neural Network (HONN) is a ongoing topic of research interest among the researchers. In this paper, the performance of a higher order neural network called Pi-Sigma neural network has been analyzed which were introduced by Shin and Ghosh [12]. The product units in HONNS are able to increase the capacity of information in neural networks and due to the presence of higher order terms they possesses the fast learning capabilities than some of the normal feed forward networks. A number of investigations have been carried out on the performance of HONN specially Pi-Sigma Neural Network (PSNN) in the applications like Online Algorithms [13–15], Cryptography [16], Classification [17], Control [18, 19] etc.

This paper proposes a particle swarm optimized higher order Pi-sigma neural network for data classification. The computational experiments of the proposed algorithm have been implemented in MATLAB and the accuracy measures have been tested by using the ANOVA statistical tool. Experimental results entail that the proposed method is robust, steady, reliable and provides better classification accuracy than other models. The remainder of the article is organized in the following way: Sect. 2 describes the preliminaries like Pi-Sigma neural network, Genetic Algorithm and PSO. Section 3 presents the proposed method. Experimental setup and Result Analysis have been presented in Sect. 4. Section 5 is devoted to Statistical analysis and Sect. 6 concludes the work with some future directions.

2 Preliminaries

2.1 Pi-Sigma Neural Network (PSNN)

The PSNN has the same structure like feed forward neural network with the set of input, hidden and output layers. PSNNs employ fewer weights and processing units than HONNs, but they are able to incorporate several of their capabilities and strengths in an indirect manner. PSNNs have efficiently addressed numerous complex tasks, where conventional Feed forward Neural Networks (FNNs) are having difficulties, such as zeroing polynomials [20] and polynomial factorization [21]. The structure of PSNN with n inputs and one output is shown in Fig. 1. The network has a regular structure with the summing of input layer which transfers the output as the product of summation unit to the output layer and uses less no. of weight unit processing nodes which makes it more efficient and accurate than the other neural networks. The reduction of number of weights allows the network for faster training.

Let the input $X = (X_0, X_1, \ldots, X_j \ldots X_n)^T$ be the $(n + 1)$ dimensional input vectors where x_j denotes the jth component of X. The $(n + 1)k$ dimensional weight vectors such that $W_{ij} = (Wij_0, Wij_1, Wij_2 \ldots Wij_n)^T, i = 1, 2 \ldots k$ are summed at a layer of k summing units, where k is the corresponding order of the network and Bj is the bias unit. The output at the hidden layer hj in Fig. 1 can be computed by Eq. 1.

$$h_j = B_j + \sum w_{ji} x_i \tag{1}$$

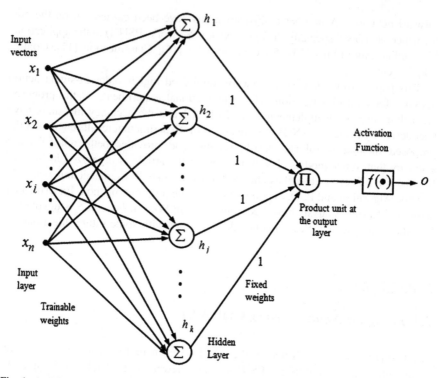

Fig. 1 Architecture of pi-sigma network

where w_{ij} represents the weight from the input to summing unit. As the weight in the hidden layer to output layer is fixed to 1, so the output O can be computed by Eq. 2.

$$O = f\left(\prod_{j=1}^{k} h_j\right)$$ (2)

where f(.) is an suitable activation function.

2.2 Genetic Algorithm (GA)

Holland [22] and Goldberg [23] initiated the work on Genetic Algorithms. The essence of the GA in both theoretical and practical domains has been well demonstrated [24]. The perception of applying a GA to resolve complex problems is practicable and sound. The algorithm starts with a population of chromosomes through random generation or from a set of known specimens. The population of chromosomes will be cycled through three steps, namely evaluation, selection, and reproduction [25]. Each chromosome represents one plausible solution to the problem or a rule in a classification problem. The simplest steps of a genetic algorithm can be expressed using the following steps:

Begin

Generate the initial population of chromosomes

Do

{

Evaluation of fitness of Chromosomes in the population;
Selection of Chromosome based on Suitable Selection Operator;
Reproduction of Chromosome through Crossover ;
number_generations ++;
} While (number _ generations <= maximum _ numbers _ generations);
Show results;

End

2.3 Particle Swarm Optimization (PSO)

PSO [26] is a population-based stochastic algorithm [9] that starts with an initial population of randomly generated particles. Since its inception, PSO has become an admired optimizer and has widely been applied in various practical problems. A set of particles move inside a decision space by adjusting their position and exchanging information about the current position in search space according to its own earlier experience and that of its neighbors [27] in order to detect their promising areas. While travelling in a group for either food or shelter [28], not only the behavior of various types of swarms indicates a unique indication towards the non-colliding nature between themselves, but also they adjust both their position and velocity. In this mechanism, the swarm members modify their positions as well as the velocities after communicating their group information according to the best position appeared in the current movement of the swarm [29]. The swarm particles would gradually get closer to the specified position and finally reach the optimal position with the help of interactive cooperation [30]. Each particle has to maintain their local best positions lbest and the global best position gbest among all of them.

$$V_i^{t+1} = V_i^{(t)} + c_1 * rand(1) * \left(l_{best_i}^{(t)} - X_i^{(t)} \right) + c_2 * rand(1) * \left(g_{best}^{(t)} - X_i^{(t)} \right) \quad (3)$$

$$X_i^{(t+1)} = X_i^{(t)} + V_i^{(t+1)} \quad (4)$$

Equation 3 controls both cognition and social behavior of particles and next position of the particles are updated using Eq. 4, $Vi^{(t)}$ and $Vi^{(t+1)}$ are the velocity of ith particle at time t and t + 1 in the population respectively, c_1 and c_2 are acceleration coefficient normally set between 0 and 2(may be same), $Xi^{(t)}$ is the position of ith particle and $l_{besti}^{(t)}$ and $g_{best}^{(t)}$ denotes local best particle of ith particle and global best particle among local bests at time t, rand(1) generates random value between 0 and 1.

3 Proposed Method

The feature of population distribution changes over the generations throughout the PSO search process. At a primary stage of the search, the particles may be spread over the entire search space. Consequently, the population allocation is dispersive. The proposed PSO-PSNN algorithm (Algorithm-1) uses a standard back propagation gradient learning algorithm proposed by Rumelhart et al. [31]. The algorithm starts with the initialization of weights to a small randomly generated value. The initial population of the algorithm will be the local best (lbest) at starting iteration. The fitness of all individuals will be evaluated by the "Fitness from training algorithm" (Algorithm-2) by using the root mean square error (RMSE) given in Eq. 5.

$$RMSE = \sqrt{\frac{\sum_{i=1}^n O_i - \hat{O}_i}{n}} \tag{5}$$

The network is trained with the production of errors and the overall estimated error function E can be calculated as in Eq. 6.

$$E_j(t) = d_j(t) - O_j(t) \tag{6}$$

where $dj^{(t)}$ indicates the final desired output at time $(t - 1)$. At each time $(t - 1)$, the output of each $Oj^{(t)}$ is calculated. The PSO-PSNN algorithm for classification is illustrated in algorithm 1 which produces the output of the network with some optimized weight set. The network is trained with the standard back propagation gradient descent learning (BP-GDL) during the fitness calculation. The weight change and the updating of weight value can be computed by Eqs. 7 and 8 respectively.

$$\Delta w_j = \eta \left(\prod_{j \neq 1}^m h_{ji} \right) x_k \tag{7}$$

where h_{ji} the output of summing is layer and η is the rate of learning.

$$w_i = w_i + \Delta w_i \qquad (8)$$

For accelerating the error convergence the momentum term α is added and the weight connection value can be computed as in Eq. 9.

$$w_i = w_i + \alpha \Delta w_i \qquad (9)$$

Algorithm 1. PSO- PSNN for Classification

INPUT: Dataset with target vector 't' , initial population of weight-sets 'P', Bias B.
OUTPUT: PSNN with optimized weight-set 'w'.
PSO-GA-PSNN -CLASSIFICATION (x, P, t, B)
Iter = 0;
    WHILE (1)
        Compute local best from population P based of fitness of individual weight-set of the population P.
        Iter = Iter + 1;
        IF (iter $==$ 1)
            lbest = P;
        ELSE
            Calculate fitness of all weight-sets by using algorithm-2 in the population. Select local best weight-sets
            'lbest' is generated by comparing fitness of current and previous weight-sets
    END
        Evaluate fitness of each weight-set (individuals) in population P based on RMSE.
        Select global best weight set from population based on fitness of all individuals (weight-sets) by using fitness vector F.
        Compute new velocity Vnew of all weight-set (individuals) by using eq. 3.
        Update positions of all weight-set (individuals) by using eq. 4.
        P = P + Vnew ;
        IF population is having 95 % similar weight sets (individuals) or if maximum number of iteration is reached,
            THEN stop the iteration.
    END
END

Algorithm 2. Fitness From Training Procedure

1. **FUNCTION** F= **Fitness From Training** (x, w, t, B)
2. **FOR** i = 1 to n, n is the length of the dataset
3.   Compute the output at the hidden layer by using (1)
4.   Compute the output of the network by using (2).
5.   Calculate the error term by using eq. (6) and compute fitness F(i)=1/RMSE.
6. **END FOR**
7. Compute root mean square error (RMSE) by using eq. 5 from target value and output
8. The weight changes by using the BP-GDL algorithm can be computed by using (7).
9. Update the weight by using eq. 8.
10. The weight value can be calculated after adding the momentum term by using (9).
11. **IF** the stopping criteria like training error or maximum no. of epochs are satisfied, then Stop.
    ELSE repeat the step from 2 to 11.
12. **END**

Table 1 Data set information

Dataset	Number of pattern	Number of attributes	Number of classes	Number of pattern in class-1	Number of pattern in class-2	Number of pattern in class-3
WBC (P)	194	32	02	148	46	–
PARKINSON	196	23	02	48	148	–
HEPATITIS	155	19	02	32	123	–
WBC (D)	569	30	02	357	212	–
IRIS	150	05	03	50	50	50

4 Experimental Setup and Result Analysis

Several tests of the PSO-PSNN method were performed with an implementation of the method by using MATLAB 9.0 on a system with an Intel Core Duo CPU T5800, 2 GHz processor, 2 GB RAM and Microsoft Windows-2007 OS. The neural network used in this paper is PSNN and the challenge is to find the optimal architecture of this type of higher order neural network, which means finding out the optimal number of layers and nodes of the neural network. Datasets (Table 1) for classification is prepared using fivefolds out of which fourfolds are used for training and onefold is used for testing. The classification accuracy is calculated from confusion matrix by using Eq. 10. If cm is the confusion matrix then accuracy of classification is computed as:

$$\text{Accuracy} = \frac{\sum_{i=1}^{n} \sum_{\substack{j=1, \\ i==j}}^{m} cm_{i,j}}{\sum_{i=1}^{n} \sum_{j=1}^{m} cm_{i,j}} * 100 \tag{10}$$

The normalization technique used is the Min-Max normalization, which maps the values of dataset v to v′ in the range [new_max$_A$ to new_min$_A$] of an attribute A using Eq. 11.

$$v' = \frac{v - \min_A}{\max_A - \min_A}(\text{new\_max}_A - \text{new\_min}_A) + \text{new\_min}_A \tag{11}$$

The proposed method has been designed for performing the classification task on various benchmark datasets like WBC (P), PARKINSON, HEPATITIS, WBC (D), and IRIS from the University of California at Irvine (UCI) machine learning repository [32]. For each of the datasets, runs are performed for every single fold. The data (Table 2) in our proposed method are prepared using the fivefold cross validation (processed by KEEL Data-Mining Software Tool [33]. Table 3 represents the result of average classification accuracy of the five datasets based on 10 runs. The results of average value for each dataset after performing the fivefold cross

Table 2 Fivefold cross validated iris dataset

Dataset	Data files	Number of patterns	Task	Number of pattern in class-1	Number of pattern in class-2	Number of pattern in class-3
IRIS	Iris-5-1trn.Dat	120	Training	40	40	40
	Iris-5-1tst.Dat	30	Testing	10	10	10
	Iris-5-2trn.Dat	120	Training	40	40	40
	Iris-5-2tst.Dat	30	Testing	10	10	10
	Iris-5-3trn.Dat	120	Training	40	40	40
	Iris-5-3tst.Dat	30	Testing	10	10	10
	Iris-5-4trn.Dat	120	Training	40	40	40
	Iris-5-4tst.Dat	30	Testing	10	10	10
	Iris-5-5trn.Dat	120	Training	40	40	40
	Iris-5-5tst.Dat	30	Testing	10	10	10

Table 3 Comparison of average performance of PSO-PSNN, GA-PSNN and PSNN

Dataset	Accuracy of classification in average					
	PSO-PSNN		GA-PSNN		PSNN	
	AHPT	AHPS	AHPT	AHPS	AHPT	AHPS
WBC (P)	86.352	85.127	82.007	81.832	76.651	75.390
PARKINSON	94.362	94.824	91.769	92.383	88.521	88.502
HEPATITIS	82.021	82.018	80.071	79.058	74.072	74.388
WBC(D)	94.031	94.038	90.042	91.348	88.304	87.103
IRIS	95.309	95.102	93.033	93.398	89.029	89.109

AHPT Average hit percentage in training, *AHPS* Average hits percentage in testing

validation of the proposed PSO-PSNN model clearly indicates that the classification accuracy of the proposed model is quite better than the others in both training and testing. The performances of all the models by considering 100 epochs and the RMSE is shown in Figs. 2, 3 and 4 for the three datasets.

5 Statistical Analysis

In this Section, the one-way ANOVA test [34] using the SPSS (version16.0) statistical tool has been used to prove the results are statistically significant. ANOVA can divide the sum variability [35] into the variability among the classifiers and the data sets as well as the residual (error) variability. The null hypothesis can be rejected

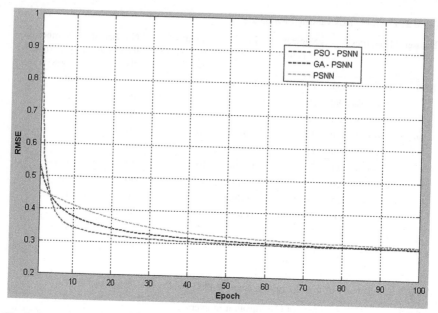

Fig. 2 Performance of three models on WBC (P) data

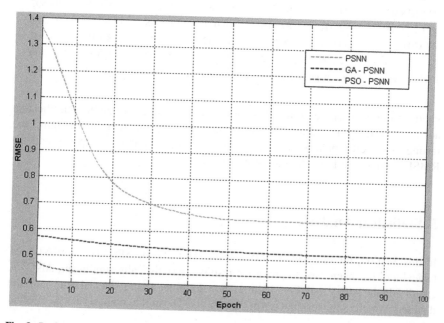

Fig. 3 Performance of three models on hepatitis data

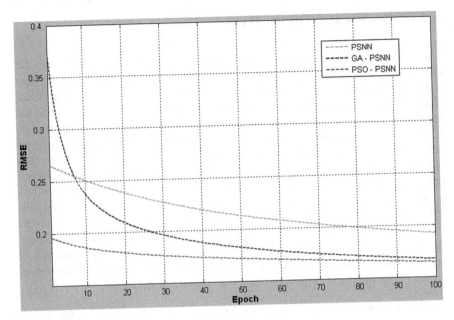

Fig. 4 Performance of three models on WBC (D) data

(a)

Descriptives

Sample	N	Mean	Std. Deviation	Std. Error	95% Confidence Inteval for Mean		Minimum	Maximum
					Lower Bound	Upper Bound		
1	10	90.3019	5.68609	1.79810	86.2343	94.36.95	82.02	95.31
2	10	87.5123	5.95293	1.88248	83.2538	91.7708	79.06	93.40
3	10	83.1380	6.95229	2.19851	78.1646	88.1114	74.07	89.11
Total	30	86.9841	6.71006	1.22508	84.4785	89.4896	74.07	95.31

(b)

ANOVA

Sample			Sum of Squares	df	Mean Square	F	Siq.
Between Groups	(Combined)		260.793	2	130.396	3.369	.049
	Linear Term	Contrast	256.607	1	256.607	6.630	.016
		Deviation	4.185	1	4.185	.108	.745
Within Groups			1044.930	27	38.701		
Total			1305.723	29			

Fig. 5 a, b ANOVA statistical results

to get some difference among the classifiers, based on some marginal better variability of the between classifier in compared to error variability. The test has been carried out using one-way ANOVA in Duncan multiple test range with 95 % confidence interval (Fig. 5), 0.05 significant levels and linear polynomial contrast.

6 Conclusions and Future Work

This paper presented a Particle swarm optimized HONN, which is analyzed for classification problems. The performance of the proposed PSO-PSNN is better than all the other models in terms of both the classification accuracy and computational efficiency. In the proposed method, a random generation of population selection has been done and the best m individuals were stored in an archive pool. Furthermore, as a possible justification of the results, it is important to notice that PSO performs better for the considered benchmark datasets and is a more robust technique than GA. So, the assimilation of global optimization methods like Evolutionary Algorithms instead of the use of classical local optimization methods is strongly recommended for classification problems. However, in case of some large data sets, PSO suffers from premature convergence because of rapid trailing of diversity. Further work will investigate the development of any nature inspired optimization algorithm, which may lead to better classification accuracy. Another direction of research may lead to the development of some other higher order neural network to reduce the architectural complexity in a less number of iterations.

Acknowledgments This work is supported by Department of Science and Technology (DST), Ministry of Science and Technology, New Delhi, Govt. of India, under grants No. DST/INSPIRE Fellowship/2013/585.

References

1. Eiben, A.E., Smith, J.E.: Introduction to Evolutionary Computing. Springer, Berlin (2003)
2. Engelbrecht, A.P.: Computational Intelligence An Introduction, 2nd edn. Wiley, London (2007)
3. Ugolottia, R., Nasheda, Y.S.G., Mesejoa, P., Ivekovi, S., Mussia, L., Cagnonia, S.: Particle swarm optimization and differential evolution for model-based object detection. Appl. Soft. Comput. **13**, 3092–3105 (2013)
4. Sun, C., Zeng, J., Pan, J., Xue, S., Jin, Y.: A new fitness estimation strategy for particle swarm optimization. Inf. Sci. **221**, 355–370 (2013)
5. Imran, M., Hashima, R., Khalid, N.E.A.: An overview of particle swarm optimization variants. Procedia Eng. **53**, 491–496 (2013)
6. Pan, F., Li, X.T., Zhou, Q., Li, W.X., Gao, Q.: Analysis of standard particle swarm optimization algorithm based on Markov chain. Acta Automatica Sinica **39**(4), 381–389 (2013)
7. Babaei, M.: A general approach to approximate solutions of nonlinear differential equations using particle swarm optimization. Appl. Soft Comput. **13**, 3354–3365 (2013)
8. Khare, A., Rangnekar, S.: A review of particle swarm optimization and its applications in solar photovoltaic system. Appl. Soft Comput. **13**, 2997–3006 (2013)
9. Wang, H., Sun, H., Li, C., Rahnamayan, S., Pan, J.S.: Diversity enhanced particle swarm optimization with neighborhood search. Inf. Sci. **223**, 119–135 (2013)
10. Neri, F., Mininno, E., Iacca, G.: Compact particle swarm optimization. Inf. Sci. **239**, 96–121 (2013)

11. Valdez, F., Melin, P., Castillo, O.: Modular neural networks architecture optimization with a new nature inspired method using a fuzzy combination of particle swarm optimization and genetic algorithms. Inf. Sci. **270**, 143–153 (2014)
12. Shin, Y., Ghosh, J.: The pi-sigma networks : an efficient higher order neural network for pattern classification and function approximation. In: Proceedings of International Joint Conference on Neural Networks, vol. 1, pp. 13–18. Seattle, Washington, (July 1991)
13. Liu,Y., Zhang, H., Yang, J., Wei, W.: Convergence of online gradient methods for pi-sigma neural network with a penalty term. In: IEEE International Conference on Anthology, pp. 1–4, (2013). doi:10.1109/ANTHOLOGY.2013.6784769
14. Yu, X., Tanga, L., Chena, Q., Xub, C.: Monotonicity and convergence of asynchronous update gradient method for ridge polynomial neural network. Neurocomputing. **129**, 437–444 (2014)
15. Yu, X., Deng, F.: Convergence of gradient method for training ridge polynomial neural network. Neural Comput. Appl. **22**(1), 333–339 (2013)
16. Deng, Y.Q., Song, G.: A verifiable visual cryptography scheme using neural networks. Adv. Mater. Res. **756–759**, 1361–1365 (2013)
17. Morissette, L., Chartier, S.: FEBAMSOM-BAM: Neural network model of human categorization of the N-bits parity problem. The 2013 International Joint Conference on Neural Networks (IJCNN), pp. 1–5 (2013)
18. Lin, Qing, Cai, ZhiHao, Wang, Ying, Xun,Yang, JinPeng,Chen, LiFang: Adaptive Flight Control Design for Quadrotor UAV Based on Dynamic Inversion and Neural Networks. Third IEEE International Conference on Instrumentation, Measurement, Computer, Communication and Control (IMCCC), pp. 1461–1466 (2013)
19. Lee, B.Y., Lee, H.I., Tahk, M.J.: Analysis of adaptive control using on-line neural networks for a quadrotor UAV. 13th IEEE International Conference on Control. Automation and Systems (ICCAS), pp. 1840–1844 (2013)
20. Huang, D.S., Ip, H.H.S., Law, K.C.K.: Chi: Zeroing polynomials using modified constrained neural network approach. IEEE Trans. Neural Netw. **16**(3), 721–732 (2005)
21. Perantonis, S., Ampazis, N., Varoufakis, S., Antoniou, G.: Constrained learning in neural networks: application to stable factorization of 2nd polynomials. Neural Process. Lett. **7**(1), 5–14 (1998)
22. Holland, J.H.: Adaption in Natural and Artificial Systems. MIT Press, Cambridge (1975)
23. Goldberg, D.E.: Genetic Algorithms In Search. Optimization and machine learning. Kluwer Academic Publishers, Boston (1989)
24. Man, K.F., Tang, K.S., Kwong, S.: Genetic Algorithms: Concepts and Designs. Springer, Berlin (1999)
25. Chen, C.H., Khoo, L.P., Chong, Y.T., Yin, X.F.: Knowledge discovery using genetic algorithm for maritime situational awareness. Expert Syst. Appl. **41**, 2742–2753 (2014)
26. Kennedy, J., Eberhart, R.: Particle swarm optimization. In: Proceedings of the 1995 IEEE International Conference on Neural Networks. vol. **4**, 1942–1948 (1995)
27. Wei, J., Guangbin, L., Dong, L.: Elite particle swarm optimizaion with mutation. IEEE Asia Simulation Conference—7th Intl Computing Conference on System. Simulation and Scientific, pp. 800–803 (2008)
28. Khare, A., Rangnekar, S.: A review of particle swarm optimization and its applications in solar photovoltaic system. Appl. Soft Comput. **13**, 2997–3006 (2013)
29. Babaei, M.: A general approach to approximate solutions of nonlinear differential equations using particle swarm optimization. Appl. Soft Comput. **13**, 3354–3365 (2013)
30. Sivanandam, S.N., Deepa, S.N.: Introduction to Genetic Algorithms. Springer, Heidelberg (2008)
31. Rumelhart, D.E., Hinton, G.E., Williams, R.J.: Learning representations by back-propagating errors. Nature **323**(9), 533–536 (1986)
32. Bache, K., Lichman, M.: UCI machine learning repository. http://archive.ics.uci.edu/ml. Irvine, CA: University of California, School of Information and Computer Science (2013)

33. Alcalá-Fdez, J., Fernandez, A., Luengo, J., Derrac, J., García, S., Sánchez, L., Herrera, F.: KEEL data-mining software tool: data set repository, integration of algorithms and experimental analysis framework. J Multiple-Valued Logic Soft Computing. **17**(2–3), 255–287 (2011)

34. Fisher, R.A.: Statistical methods and scientific inference, 2nd edn. Hafner Publishing Co., New York (1959)

35. Demsar, J.: Statistical comparisons of classifiers over multiple data sets. J. Maching Learn. Res. **7**, 1–30 (2006)

K-Strange Points Clustering Algorithm

Terence Johnson and Santosh Kumar Singh

Abstract The classical K-Means clustering algorithm yields means which can be called the final unchanging or fixed means around which all other points in the dataset get clustered. This is so because the K-Means clustering terminates when either the clusters repeat in the next iteration or when the means repeat in the next iteration. This reveals that if one is able to somehow calculate and find apriori the final unchanging means using the dataset, then the task of clustering reduces to only assigning the remaining points in the dataset into clusters, which are closest to these final fixed or unchanging means based on standard distance measures. Taking a cue from the result of the classical K-Means method, the K-Strange points clustering algorithm presented in this paper locates K points from the dataset equaling the number of required clusters which are farthest from each other and are hence called K-Strange points based on the Euclidean distance measure. The remaining points in the dataset are assigned to clusters formed by these K-Strange points.

Keywords K-Strange points clustering · Farthest points · Euclidean distance measure

1 Introduction

Data Mining is the process of detecting patterns from extremely huge quantities of data collection [1]. Data Mining explores large quantities of data in order to discover hidden rules and potentially meaningful patterns [2]. Data Mining can be

T. Johnson (✉)
AMET University, Chennai, India
e-mail: ykterence@rediffmail.com

S.K. Singh
Department of Information Technology, Thakur College of Science
and Commerce, Mumbai, India
e-mail: Singhsksingh14@gmail.com

© Springer India 2015
L.C. Jain et al. (eds.), *Computational Intelligence in Data Mining - Volume 1*,
Smart Innovation, Systems and Technologies 31, DOI 10.1007/978-81-322-2205-7_39

performed on various types of database and information repositories, but the kind of patterns to be found are specified by various data mining functionalities [3]. Grouping or bunching of data into a set of categories or clusters is one of the essential methods in manipulating and finding patterns from data [4]. Clustering is the most common data mining process which aims at dividing datasets into subsets or clusters in such a way that the objects in one subset are similar to each other with respect to a given similarity measure while objects in different subsets are dissimilar [5]. Clustering is a task that attempts to detect similar categories or groups of objects based on the implementation of their feature dimensions [6]. One can detect the predominant distribution patterns and interesting correlations that exist among data attributes by clustering which can determine dense and sparse areas [7]. Clustering organizes and partitions objects into groups whose members are alike in some way [8]. A cluster is a collection of data objects that are similar to one another within the same cluster and are dissimilar to the objects in other clusters [9]. A good clustering algorithm will produce high quality of clusters with high intra cluster similarity and low inter cluster similarity [10]. The purpose of clustering is to detect groups or clusters of similar objects where an object is represented as a vector of measurements or points in multidimensional space. The distance measure determines the dissimilarity between objects in various dimensions in the dataset [11]. Cluster analysis is an important technique to find the similar and dissimilar groups in data mining [12]. Clustering is commonly and heavily used in a variety of applications such as in market segmentation, medical science, environmental science, astronomy, geology, business intelligence and so on. It also helps users in understanding natural groupings in a data set or structure of the data set [13].

1.1 Motivation

The classical instantiation of the K-Means algorithm begins by randomly picking K prototype cluster centers called K-Means, assigning each point to the cluster whose mean is closest in a Euclidean sense, then computing the mean vectors of the points assigned to each cluster and using these as new centers in an iterative approach until the termination criteria is reached [14]. The complexity of the K-Means method is O(nktd) where n represents the number of data points, k represents the number of required clusters, t represents the number of iterations the algorithm should undergo if the cluster centers (means) do not repeat in the next iteration or if the clusters do not repeat in the next iteration and d represents the number of attributes or dimensions [15]. Clustering, using the classical K-Means method results in obtaining final fixed points which we call the final unchanging means around which all other points in the dataset get clustered. This suggests that if we are able to somehow calculate and find apriori the final unchanging means using the dataset, then the task of clustering reduces to only assigning the remaining points in the dataset into clusters, which are closest to these final fixed or unchanging means based on standard distance measures. Taking a cue from the result of the K-Means method the algorithm presented in this paper locates K points from the dataset equaling the number of required clusters which are farthest to each other based on the Euclidean distance measure. The remaining points in the dataset are assigned to clusters formed by these K-Strange points.

2 Proposed Work

This paper presents an algorithm for clustering by finding K points in a dataset equaling the number of required clusters which are most dissimilar to each other. The K points are referred to as K-Strange points because these K points are located farthest from each other or are the most dissimilar points to each other in the dataset. The Algorithm initially randomly chooses a point from the dataset representing the first of the K-Strange points (Fig. 1). It then locates a point which lies farthest from the first initially chosen point (Fig. 2). Then it finds a third point in the dataset which is farthest from the two strange (maximally separated) points located in the previous steps (Fig. 3). For k = 5 clusters, it finds the fourth point which is maximally separated from the previous 3 farthest points (Fig. 4). And eventually the fifth strange point from the four maximally separated farthest points is found thus forming five points which are strangers to each other or in simple words five points which are at maximum distance from each other (Fig. 5). If the required number of clusters is K = 5, then the five clusters can be formed by assigning the remaining points in the dataset into clusters formed by these 5 strange points (Fig. 6). If the clustering requirement is of K = T clusters then continue the procedure of finding

Fig. 1 First randomly chosen strange point

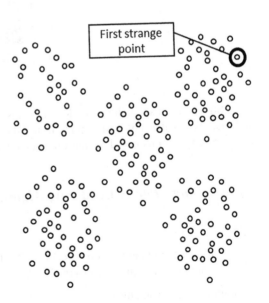

Fig. 2 Calculated second
strange point

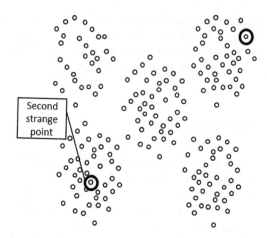

Fig. 3 Calculated third
strange point

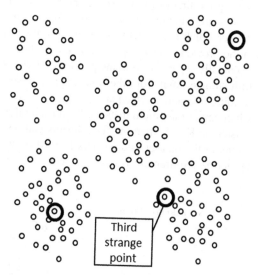

the T Strange points which are the K = T points which are farthest from each other
and then assign the remaining points in the dataset into clusters formed by these
K = T Strange points based on the Euclidean distance measure.

2.1 K-Strange Points Clustering Algorithm

Input:

(i) A database containing n objects. D = {D$_1$, D$_2$, D$_3$, D$_4$, ..., D$_n$}
(ii) The number of required clusters K = T

Output: A set of K clusters.

Fig. 4 Fourth strange point

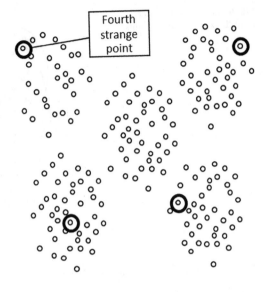

Fig. 5 Fifth strange point

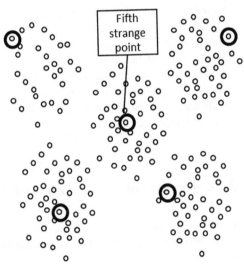

Step 1: Select two points D_k and D_w from the dataset which are at maximum distance from each other by finding distances between the all points in the dataset from each other using the Euclidean distance measure. The Euclidean distance between 2 points is defined as the square root of the sum of the squared differences [16]. The Euclidean distance between the points $i(w_1, x_1, y_1, z_1)$ and j (w_2, x_2, y_2, z_2) is given by:

Fig. 6 Five clusters formed
from the k = 5 strange points

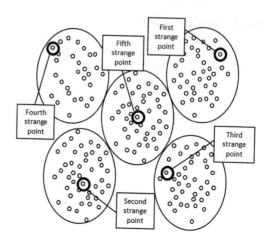

$$d(i,j) = \sqrt{(w_1 - w_2)^2 + (x_1 - x_2)^2 + (y_1 - y_2)^2 + (z_1 - z_2)^2}$$

Step 2: Locate a third point D_f which is farthest from D_k and Dw such that the sum of the distances between points D_f, D_k and D_w is larger than any other combination with D_k and D_w

Step 3: Repeat the above procedure until we locate K points equaling the number of required clusters mentioned in the problem

Step 4: Assign the remaining points in the dataset into clusters formed by these K-Strange (farthest) points using the Euclidean distance measure

Step 5: Output K clusters.

2.2 Implementation of the Proposed Algorithm

Consider a clustering requirement for 3 clusters of any dataset. The Euclidean distance between all points in the dataset can be found using the piece of code (1) given below.

```
for(int i=0;i<arrayRow;i++)
    for(int j=0;j<arrayRow;j++){
        double a = {d[i][0],d[i][1],d[i][2],d[i][3]};
        double b = {d[j][0],d[j][1],d[j][2],d[j][3]};
        double eucD = euclidDist(a,b);
        ed[k] = eucD;
```

On finding the Euclidean distance of all the data points from each other we find the two points which are at maximum distance from each other using the below piece of pseudo-code (2).

```
if(ed[k]>max){
        max = ed[k];
        Assign 1st strange point to f[][]
        Assign 2nd strange point to g[][]
        k++;
}
```

Then we locate a point which is farthest from these two points. If the nth data item D_{n-1} and the mth data item D_{m-1} are these two points then we locate a third point which is farthest from D_{n-1} and D_{m-1} such that the sum of the distances between the third point, D_{n-1} and D_{m-1} is larger than any other combination with D_{n-1} and D_{m-1}. This can be done as shown in the following pseudo-code (3).

```
double de = newMax + euclidDist(v1,v2,v3,v4,u1,u2,u3,u4) +
euclidDist(u1,u2,u3,u4,w1,w2,w3,w4);
dist[y]= de;
if(dist[y]>finalMax){
        finalMax = dist[y];
        Assign 3rd strange point to s[][]
        y++;
}
```

Once the third point is found using the above code, we stop finding any more farthest points since the clustering requirement is to group the points in the dataset into 3 clusters and as we have already found the K = 3 Strange points equaling the number of required clusters from the dataset we stop searching for any more farthest points. The next step is to assign the remaining points in the dataset into clusters formed by the K-Strange points. Finally, this is implemented as shown in the pseudo-code (4) below.

```
if((euclidDist(v,p)<=euclidDist(w,p))&(euclidDist(v,p)<=e
uclidDist(t,p)))          Assign p to Cluster 1
else
if((euclidDist(t,p)<=euclidDist(v,p))&(euclidDist(t,p)<=e
uclidDist(w,p)))          Assign p to Cluster 2
else
if((euclidDist(w,p)<=euclidDist(v,p))&(euclidDist(w,p)<=e
uclidDist(t,p)))          Assign p to Cluster 3
```

2.3 Experimental Results

The algorithm is tested with a 2D array dataset of 10,000 points each with 4 columns randomly generated by the following pseudo-code for finding 3 clusters.

```java
int arrayRow = 10000;
int arrayCol = 4;
int data[][] = new int[arrayRow][arrayCol];
for(int i=0; i<arrayRow; i++){
    for(int j=0; j<arrayCol; j++){
        data[i][j]= (int)(Math.random()*10 +10);
    }
    System.out.println();
}
```

Step 1: The Euclidean distance between all points in the dataset found using the pseudo-code (1)

Step 2: On finding the Euclidean distance of all the data points from each other we see that the two points which are at maximum distance from each other. Using pseudo-code (3) we locate the third farthest point D_k such that the sum of the distances between D_k, D_{n-1} and D_{m-1} is larger than any other combination with D_{n-1} and D_{m-1}. The snapshot of the 3 strange points can be seen in Fig. 7

Step 3: Here the remaining points in the dataset are assigned into clusters formed by the K-Strange points and this is done using the pseudo-code (4). On execution, the code gives the information on the formation of the 3 required clusters as follows:

```
Problems  @ Javadoc  Declaration  Console ⌖
<terminated> KStrange2DRandomDataset3Clusters [Java Application] C:\Program Files\Java\jdk1.8.0_05\bin\javaw.exe (Oct
K-Strange Points are:
f = 106.0 10.0 15.0 16.0
g = 10.0 101.0 109.0 107.0
s = 27.0 106.0 11.0 10.0
--------------------------------------------------------------
Number of points in Cluster1 = 3593
Number of points in Cluster2 = 2922
Number of points in Cluster3 = 3485
--------------------------------------------------------------
K=3 Strange Points Clustering for Random 2D Dataset of 10000 points took: = 1935 milliseconds
```

Fig. 7 K-Strange Clustering for a random array of size [1 0 0 0 0] [4] for 3 clusters

2.4 Comparison with K-Means and Inference for 3 Clusters

Table 1 shows the results of the K-Means and K-Strange points clustering algorithms for a random dataset of 1,000, 5,000, and 10,000 data points each with 4 dimensions for 3 clusters. Although the K-Strange clustering algorithm executes slower than the classical K-Means, the K-Means algorithm takes an exponential time to converge as the number of data points and dimensionality increases. Hence K-Means clustering algorithm uses t as the number of iterations to terminate the clustering process if it tends to go into an infinite loop. This will result in inaccurate clusters. Though the K-Strange Points Clustering algorithm takes a little more time for its execution than the K-Means algorithm in lower dimensions, it performs better than the K-Means in higher dimensions as seen from Fig. 8 and the Table 2 that follows.

We see that as the dimensions increase, the K-Strange Points Clustering algorithm gives us the results as seen in Fig. 8 but the K-Means algorithm doesn't converge.

	Algorithm	Data points	Execution time (ms)
Table 1 Comparison of K-means with K-strange for 4 dimensions	K-means	[1 0 0 0] [4]	16
	K-strange	[1 0 0 0] [4]	156
	K-means	[5 0 0 0] [4]	32
	K-strange	[5 0 0 0] [4]	546
	K-means	[1 0 0 0 0] [4]	202
	K-strange	[1 0 0 0 0] [4]	1,935

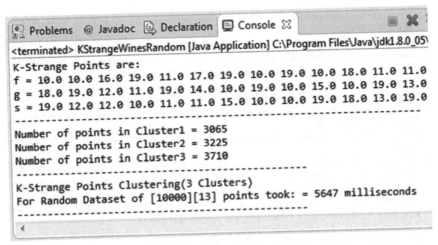

Fig. 8 K-Strange Clustering for a random array of size [1 0 0 0 0] [13] for 3 Clusters

Table 2 Comparison of K-means with K-strange for 13 dimensions

Algorithm	Data points	Execution time (ms)
K-means	[1 0 0 0] [13]	Not converging
K-strange	[1 0 0 0] [13]	203
K-means	[5 0 0 0] [13]	Not converging
K-strange	[5 0 0 0] [13]	1,467
K-means	[1 0 0 0 0] [13]	Not converging
K-strange	[1 0 0 0 0] [13]	5,647

3 Conclusion

The complexity of the K-Means Clustering method being O(nktd), there is a strong likely hood that the clusters so formed may not be accurate because according to the K-Means method, for clustering to yield accurate results, either the cluster centers (means) should repeat in the next iteration or the clusters should repeat in the next iteration. As dimensions increase, K-Means takes exponential time and so, abruptly terminating the clustering process after a certain number of specified iterations will not yield the desired accurate clusters. This issue is addressed by finding K points in any dataset equaling the number of required clusters which are at maximum distance from each other making them the most dissimilar or Strange points to each other and then assigning the remaining points in the dataset into clusters formed by these K = T strange points based on the Euclidean distance measure, and thereby eliminating the abrupt terminations associated with t, the number of iterations.

References

1. Abbas, O.: Comparisons between data clustering algorithms. Int. Arab J. Inf. Technol. **5**(3), 320–325 (2008)
2. Prabhu, P., Anbazhagan, N.: Improving the performance of k-means clustering for high dimensional dataset. Int. J. Comput. Sci. Eng. **3**(6), 2317–2322 (2011), ISSN: 0975-3397
3. Micheal, J.A.: Berry Gordon Linoff.: Mastering Data Mining. Wiley, Singapore (2001)
4. Bouveyrona, C., Girarda, S., Schmid, C.: High dimensional data clustering. J. Comput. Stat. Data Anal. **52**(1), 502–519 (2007)
5. Kaufman, L., Rousseeuw, P.J.: Finding Groups in Data: An Introduction to Cluster Analysis. Wiley, New York (1990)
6. Jain, A., Murty, M., Flynn, P.: Data clustering: a review. ACM Comput. Surv. **31**(3), 264–323 (1999)
7. Alijammaat, A., Khalilian, M., Mustapha, N.: A novel approach for high dimensional data clustering. In: Proceedings of the Third International Conference on Knowledge Discovery and Data Mining, Phuket, Iran, pp. 264–267 (2010)
8. Johnson, T.: Bisecting collinear clustering algorithm. Int. J. Comput. Sci. Eng. Inf. Technol. Res. **3**(5), 43–46 (2013), © TJPRC Pvt. Ltd., ISSN: 2249-6831
9. Johnson. T., Lobo, J.Z.: Collinear clustering algorithm in lower dimensions. IOSR J. Comput. Eng. **6**(5), 08–11 (2012), ISSN: 2278-0661, ISBN: 2278-8727

10. Singh, S.K., Johnson, T.: Improved collinear clustering algorithm in lower dimensions. In: Proceedings of Second International Conference on Emerging Research in Computing, Information, Communication and Applications (2014) (in press)

11. Nagi, S,. Bhattacharya, D.K., Kalita, J.K.: A preview on subspace clustering of high dimensional data. Int. J. Comput. Technol., 6(3), 441–448 (2013). ISSN: 22773061

12. Aravinder D.J., Naganathan, E.R.: Efficient centroids based clustering algorithm with data intelligence. J. Theor. Appl. Inf. Technol. 56(1), 126–130 (2013). ISSN: 1992-8645

13. Jahirabadkar, S., Kulkarni, P.: SCAF-An efficient approach to classify subspace clustering. Int. J. Data Mining Knowl. Manage. Process, 3(2) (2013)

14. Hand, D.J., Mannila, H., Smyth, P.: Principles of Data Mining, MIT Press, Cambridge, pp. 302–305 (2001)

15. Tan, P., Steinbach, M.K.: An Introduction to Data Mining. Wesley, London (2005)

16. A. Alfakih, A. Khandani, and H. Wolkowicz.: Solving Euclidean distance matrix completion problems via semide⁻nite programming. Comput. Optim. Appl. 12, 13–30 (1999)

Effective Detection of Kink in Helices from Amino Acid Sequence in Transmembrane Proteins Using Neural Network

Nivedita Mishra, Adikanda Khamari, Jayakishan Meher and Mukesh Kumar Raval

Abstract Transmembrane proteins play crucial roles in a wide variety of biochemical pathways which comprise around 20–30 % of a typical proteome and target for more than half of all available drugs. Knowledge of kinks or bends in helices plays an important role in its functions. Kink prediction from amino acid sequences is of great help in understanding the function of proteins and it is a computationally intensive task. In this paper we have developed Neural Network method based on radial basis function for prediction of kink in the helices with a prediction efficiency of 85 %. A feature vector generated using three physico-chemical properties such as alpha propensity, coil propensity, and EIIP constituted in kinked helices contains most of the necessary information in determining the kink location. The proposed method captures this information more effectively than existing methods.

Keywords Transmembrane proteins · Kink prediction · Radial basis function neural network · Physico-chemical properties · Amino acid sequence

N. Mishra (✉)
Department of Chemistry, Rajendra College,
Balangir 767002, Odisha, India
e-mail: nibedita1976@yahoo.com

A. Khamari
Department of Physics, Rajendra College,
Balangir 767002, Odisha, India
e-mail: akkhamari@gmail.com

J. Meher
Department of Computer Science and Engineering,
Vikash College of Engineering for Women,
Bargarh 768028, Odisha, India
e-mail: jk_meher@yahoo.co.in

M.K. Raval
Department of Chemistry, Gangadhar Meher College,
Sambalpur 768004, Odisha, India
e-mail: mraval@yahoo.com

427

© Springer India 2015
L.C. Jain et al. (eds.), *Computational Intelligence in Data Mining - Volume 1*,
Smart Innovation, Systems and Technologies 31, DOI 10.1007/978-81-322-2205-7_40

1 Introduction

Sequence based predictions on membrane spanning proteins are of great importance. Membrane proteins comprise about 25 % of all proteins encoded by most genomes. Transmembrane α-helix bundle is a common structural feature of membrane proteins except porins, which contains β-barrels. Membrane spanning α-helices differ from their globular counterpart by the presence of helix breakers, Pro and Gly, in the middle of helices. Pro is known to induce a kink in the helix [1, 2]. A hypothesis suggests that Pro is introduced by natural mutation to have a bend and later further mutated leaving the bend intact for required function during the course of evolution [3]. The role of Pro and kinks in transmembrane helices were extensively investigated both experimentally and theoretically to unravel the nature's architectural principles [2, 4]. Another observation suggests induction of kink at the juncture of α-helical and 3–10 helical structure in a transmembrane helix [2–6]. Mismatch of hydrophobicity of lipid bilayer and peptide may also result in distortion of α-helical structure [7]. Sequences of straight and kinked helices were further subjected to machine learning to develop a classifier for prediction of kink in a helix from amino acid sequences. Support vector machine (SVM) method [8] projects that helix breaking propensity of amino acid sequence determines kink in a helix. DWT has been applied on hydrophobicity signals in order to predict hydrophobic cores in proteins [9]. Protein sequence similarity has also been studied using DWT of a signal associated with the average energy states of all valence electrons of each amino acid [10]. Wavelet transform has been applied for transmembrane structure prediction [11]. Signal processing methods such as Fourier transform and wavelet transform can identify periodicies and variations in signals from a background noise. The presence of kink in amino acid sequence is determined effectively in transform domain analysis [12].

Kinked and straight helix of protein Type-4 Pilin and Chlorophyll a-b binding protein respectively are shown in Fig. 1.

A kink in a helix may be formed by helix-helix interaction. In such cases the intrinsic kink forming or helix breaking tendency may not be required. Even a helix forming tendency may be overridden. This possibility clamps a theoretical limit to predict a kink with high accuracy. Hence there is a need to develop advanced algorithm for faster and accurate prediction of kink in transmembrane helices. This motivates to develop novel approach based radial basis function neural network (RBFNN) to effectively predict kink in transmembrane α-helices.

2 Materials Preparation

2.1 Database

List of transmembrane proteins and their coordinate files were obtained from the Orientation of Proteins in Membranes (OPM) database at College of Pharmacy, University of Michigan (http://www.phar.umich.edu).

Fig. 1 Backbone
representation of **a** kinked
helix of type-4 pilin protein
(2pil), and **b** straight helix
(second helix) of chlorophyll
a-b binding protein (1rwt)

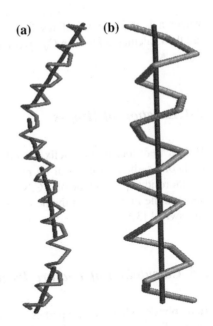

(a)   (b)

2.2 Determination of α-Helical Regions

Dihedral angles were computed using MAPMAK from coordinate files and listed
for each residue along with assignment of conformational status of the residue
namely right or left helical, β-strand. Molecular visual tools RasMol were used to
visually confirm the transmembrane α-helical regions.

2.3 Computation of Helix Axis

Helix axis was computed from the approximate local centroids $\theta'_i\,(x^0_i, y^0_i, z^0_i)$ of the
helix by taking a frame of tetrapeptide unit [13].

$$x^0_i = \frac{1}{4}\sum_i^{i+3} x_i, \quad y^0_i = \frac{1}{4}\sum_i^{i+3} y_i, \quad z^0_i = \frac{1}{4}\sum_i^{i+3} z_i \tag{1}$$

where x_i, y_i, and z_i are the coordinates of C_α atoms of the tetrapeptide frame. Unit
vector in the direction of resultant of vectors $\theta'_i\theta'_{i+1}$ yields direction cosines (l, m, n)
of axis of helix (A). The axis pass through the centroid of the helix $\theta^0 = (X^0, Y^0, Z^0)$.

$$X^0 = \frac{1}{n}\sum_{i=1}^n x_i, \quad Y^0 = \frac{1}{n}\sum_{i=1}^n y_i, \quad Z^0 = \frac{1}{n}\sum_{i=1}^n z_i \tag{2}$$

where n is the number of residues in a helix. Refined local centers θ_i of helix are then calculated for each C_α by computing the foot of perpendicular drawn from $C_{\alpha i}$ to A.

2.4 Location of Hinges

Hinges were located in a helix by a distance parameter $d(C_i N_{i+4})$, where C_i is the backbone carbonyl carbon of ith residue and N_{i+4} is backbone peptide nitrogen of i + 4th residue [13]. Value of $d(C_i N_{i+4})$ beyond the range $4.227 \pm 0.35\$$ Å reflects a hinge at the ith residue in the helix. Hinge was quantified by two parameters kink and swivel [3].

2.5 Calculation of Feature Parameters

Here physico-chemical properties of amino acids are used to draw the feature vector. These are alpha, coil and Electron ion pseudopotential interaction potential (EIIP) as shown in Table 1.

Table 1 Physicochemical parameters of amino acid residues used in algorithm for prediction of Ni-binding sites in proteins

Amino acid	Alpha	Coil	EIIP
A	1.372	0.824	0.0373
R	0.694	0.893	0.0959
N	0.473	1.167	0.0036
D	0.416	1.197	0.1263
C	1.021	0.953	0.0829
Q	0.765	0.947	0.0761
E	0.704	0.761	0.0058
G	0.913	1.251	0.0050
H	1.285	1.068	0.0242
L	1.471	0.810	0.0000
I	1.442	0.886	0.0000
K	0.681	0.897	0.0371
M	1.448	0.810	0.0823
F	1.459	0.797	0.0946
P	0.526	1.540	0.0198
S	0.903	1.130	0.0829
T	0.910	1.148	0.0941
W	1.393	0.941	0.0548
Y	0.907	1.109	0.0516
V	1.216	0.772	0.0057

3 Radial Basis Function Neural Network Classifier for Kink

In this paper we have introduced a low complexity radial basis function neural network (RBFNN) classifier to efficiently predict the sample class [14, 15]. The potential of the proposed approach is evaluated through an exhaustive study by many benchmark datasets.

The experimental results showed that the proposed method can be a useful approach for classification. A radial basis function network is an artificial neural network that uses radial basis functions as activation functions. It is a linear combination of radial basis functions. The radial basis function network (RBFNN) is suitable for function approximation and pattern classification problems because of their simple topological structure and their ability to learn in an explicit manner. In the classical RBF network, there is an input layer, a hidden layer consisting of nonlinear node function, an output layer and a set of weights to connect the hidden layer and output layer. Due to its simple structure it reduces the computational task as compared to conventional multi layer perception (MLP) network. The structure of a RBF network is shown in Fig. 2.

In the RBFNN based classifier, an input vector x is used as input to all radial basis functions, each with different parameters. The output of the network is a linear combination of the outputs from radial basis functions.

For an input feature vector x, the output y of the jth output node is given as.

$$y_j = \sum_{k=1}^{N} w_{kj}\varphi_k = \sum_{k=1}^{N} w_{kj}e^{-\frac{\|x(n)-C_k\|}{2\sigma_k^2}} \tag{3}$$

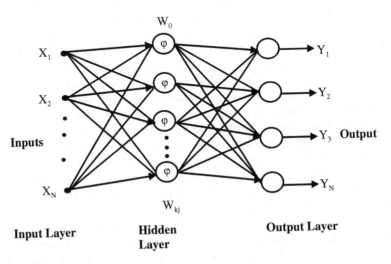

Fig. 2 The structure of a RBF network

The error occurs in the learning process is reduced by updating the three parameters, the positions of centers (C_k), the width of the Gaussian function (σ_k) and the connecting weights (w) of RBFNN by a stochastic gradient approach as defined below:

$$w(n+1) = w(n) - \mu_w \frac{\partial}{\partial w} J(n) \tag{4}$$

$$C_k(n+1) = C_k(n) - \mu_c \frac{\partial}{\partial C_k} J(n) \tag{5}$$

$$\sigma_k(n+1) = \sigma_k(n) - \mu_\sigma \frac{\partial}{\partial \sigma_k} J(n) \tag{6}$$

where, $J(n) = \frac{1}{2}|e(n)|^2$, $e(n) = d(n) - y(n)$ is the error, $d(n)$ is the target output and y (n) is the predicted output. μ_w, μ_C, and μ_σ are the learning parameters of the RBF network.

4 Simulation and Result Analysis

In order to compare the efficiency of the proposed method in predicting the class of the kink data we have used standard datasets. All the datasets categorized into two groups: binary class to assess the performance of the proposed method. The dataset consists of amino acid sequences of 9 characters. 400 sequences from kink dataset and 400 sequences from non-kink dataset are taken as training set. The feature selection process proposed in this paper includes alpha, coil and EIIP as shown in the Table 1. To implement the RBFNN classifier, we first read in the file of protein sequence which is represented with numerical values. The performance of the proposed feature extraction method is analyzed with the neural network classifiers: RBFNN. The leave one out cross validation (LOOCV) test is conducted by combining all the training and test samples for the classifiers with datasets [16]. LOOCV is a technique where the classifier is successively learned on n − 1 samples and tested on the remaining one. i.e., it removes one sample at a time for testing and takes other as training set. It involves leaving out all possible subsets so the entire process is run as many times as there are samples. This is repeated n times so that every sample was left out once. Repeating these procedure n times gives us n classifiers in the end. Our error score is the number of mispredictions. Out of 400 sequences from kink dataset all 400 samples are detected as true positive whereas out of all 400 sequences from non-kink dataset, all 400 samples are detected as true negative.

The prediction accuracy has been analyzed in terms of three measuring parameters such as accuracy (A), precision (P) and recall (R). These are defined in terms of four parameters true positive (t_p), false positive (f_p), true negative (t_n) and false negative (f_n). t_p denotes the number of kinks and are also predicted as kinks, f_p denotes the number of actually straight but are predicted to be kinks, t_n is the number of actually straight and also predicted to be straight, and fn is the number of actually kinks and predicted to be straights.

4.1 Accuracy

The accuracy of prediction of kinks in amino acid sequence is defined as the percentage of kinks correctly predicted of the total sequences present. It is computed as follows:

$$A = \frac{t_p + t_n}{t_p + f_p + t_n + f_n}. \tag{7}$$

4.2 Precision

Precision is defined as the percentage of kinks correctly predicted to be one class of the total kinks predicted to be of that class. It is computed as:

$$P = \frac{t_p}{t_p + f_p}. \tag{8}$$

4.3 Recall

Recall is defined as the percentage of the kinks that belong to a class that are predicted to be that class. Recall is computed as:

$$R = \frac{t_p}{t_p + f_n}. \tag{9}$$

A query sequence of 35 kink samples and 35 non-kink samples are tested for validation and the result obtained is shown in Table 2.

The accuracy, precision and recall are 0.85, 0.84, and 0.84 respectively. The accuracy of sequence based classifiers reported so far is about 85 %. Hence the present classifier appears to have high accuracy compared to existing sequence based classifiers.

Table 2 Measuring parameters for prediction accuracy

Actual/predicted	Kink	Non-kink
Kink	30 (t_p)	5 (f_p)
Non-kink	6 (f_n)	29 (t_n)

5 Conclusion

The proposed radial basis function neural network classifier approach plays a vital role in the prediction of kink in transmembrane α-helix. This method is not only fast but also has improved accuracy (85 %) as compared to SVM learning system (80 %) reported by us earlier [8]. However prediction of kink in a helix depends on the features of amino acid sequence. Feature vector with propensities of residues in helix and coil along with EIIP are only used for numerical representation in the present study. Although kink prediction has its own limitations, the present work is primary report in the area of helix kink prediction from amino acid sequence based on neural network algorithms. In future we plan to subject all kinked helices to molecular dynamics and filter out the helices having intrinsic tendency to form thus eliminating kink induced by external factor like hydrogen bonding and polar and nonpolar interaction thereby improving the accuracy.

Acknowledgments The authors wish to thank management members and the principal of the college for all kinds of supports to complete this work.

References

1. Ramachandran, G., Ramakrishnan, C., Sasisekharan, V.: Stereochemistry of polypeptide chain configuration. J. Mol. Biol. **7**, 95–97 (1963)
2. Sankararamakrishnan, R., Vishveshwara, S.: Conformational studies on peptides with proline in the right-handed α-helical region. Biopolymers **30**, 287–298 (1990)
3. Cordes, F., Bright, J., Sansom, M.P.: Proline induced distortions of transmembrane helices. J. Mol. Biol. **323**, 951–960 (2002)
4. Von Heijne, G.: Proline kinks in transmembrane α-helices. J. Mol. Biol. **218**, 499–503 (1991)
5. Yohannan, S., Faham, S., Whitelegge, J., Bowie, J.: The evolution of transmembrane helix kinks and the structural diversity of G-protein coupled receptors. Proc. Natl. Acad. Sci. U.S.A. **101**, 959–963 (2004)
6. Pal, L., Dasgupta, B., Chakrabarti, P.: 3(10)-Helix adjoining alpha-helix and beta-strand: sequence and structural features and their conservation. Bioploymers **78**, 147–162 (2005)
7. Daily, A., Greathouse, D., van der Wel, P., Koeppe, R.: Helical distortion in tryptophan-and lysine-anchored membrane-spanning alpha-helices as a function of hydrophobic mismatch: a solid-state deuterium NMR investigation using the geometric analysis of labeled alanines method. Biophys. J. **94**, 480–491 (2008)
8. Mishra, N., Khamari, A., Mohapatra, P.K., Meher, J.K., Raval, M.K.: Support vector machine method to predict kinks in transmembrane α-helices, pp. 399–404. Excel India Publishers, India (2010)
9. Hirakawa, H., Muta, S., Kuhara, S.: The hydrophobic cores of proteins predicted by wavelet analysis. Bioinformatics **15**, 141–148 (1999)
10. de Trad, C., Fang, Q., Cosic, I.: Protein sequence comparison based on the wavelet transform approach. Protein Eng. **15**, 193–203 (2002)
11. Murray, K.B., Gorse, D., Thornton, J.: Wavelet transforms for the characterization and detection of repeating motifs. J. Mol. Biol. **316**, 341–363 (2002)

12. Meher, J.K., Mishra, N., Mohapatra, P.K., Raval, M.K., Meher, P.K., Dash, G.N.: Signal processing approach for prediction kink in transmembrane α-helices. In: Proceeding of in the International Conference on Advances in Information Technology and Mobile Communication (AIM-2011), pp. 170–177. Springer CCIS, ISBN 978-3-642-20572-9 (2011)
13. Mohapatra, P.K., Khamari, A., Raval, M.K.: A method for structural analysis of α-helices of membrane proteins. J. Mol. Model. **10**, 393–398 (2004)
14. Chen, S., Cowan, C.F.N., Grant, P.M.: Orthogonal least squares learning algorithm for radial basis function networks. IEEE Trans. Neural Networks **2**, 302–309 (1991)
15. Powell, M.J.D.: Radial basis functions for multivariable interpolation: a review. In: IMA Conference on Algorithms for the Approximation of Functions and Data. RMCS, Shrivenham (1985)
16. Lachenbruch, P.A., Mickey, M.R.: Estimation of error rates in discriminant analysis. Technometrics **10**, 1–11 (1968)

A Novel Semantic Clustering Approach for Reasonable Diversity in News Recommendations

Punam Bedi, Shikha Agarwa, Archana Singhal, Ena Jain
and Gunjan Gupta

Abstract Experienced users expect the recommendations to be accurate as well as diverse. Unconditional diversity looses user's trust. Therefore a novel soft hierarchical semantic clustering approach is proposed to group users based on their semantic profiles, to bring reasonable diversity in news recommendations. To find ranked membership of user in a cluster, interest score along with rank of that category in profile is considered. Users are compared semantically using hierarchical structure of ontology, to bring positive serendipity along with reasonable diversity. New items are recommended with semantic ranking. Transparency helps user to understand and logically accept new unexpected recommendations. Clusters are formed by making variations in standard Jaccard similarity metric and are compared for homogeneity using Semi-Partial R-Squared (SPRS) metric. Result shows that formed clusters are better in terms of homogeneity and distribution of concepts per cluster. The approach is scalable, as number of users and items increases.

Keywords Collaborative filtering · Hierarchical semantic clustering · Soft clustering · Semantic user profile · Transparency

P. Bedi (✉) · S. Agarwa · A. Singhal · E. Jain · G. Gupta
Department of Computer Science, University of Delhi, Delhi, India
e-mail: punambedi@ieee.org

S. Agarwa
e-mail: shikha_8june@rediffmail.com

A. Singhal
e-mail: singhal_archana@yahoo.com

E. Jain
e-mail: ena.mcs.du.2012@gmail.com

G. Gupta
e-mail: gunjan.mcs.du.2012@gmail.com

© Springer India 2015
L.C. Jain et al. (eds.), *Computational Intelligence in Data Mining - Volume 1*,
Smart Innovation, Systems and Technologies 31, DOI 10.1007/978-81-322-2205-7_41

1 Introduction

Information overload and busy schedule of people has generated the demand of Personalized Recommendations. Different filtering approaches exist [1] for this purpose.

In a dynamic domain like news it is very productive to recommend items referred by similar minded users. For this purpose Collaborative Filtering (CF) [2] is used which is widely used for recommendations [3, 4]. Recent study shows that along with accuracy, diversity is also desirable [5, 6]. Diversity compromises accuracy. Therefore proposed approach semantically brings reasonable diversity, without much of the loss to accuracy.

In this paper a soft semantic hierarchical clustering approach is proposed with modified Jaccard similarity for making intersecting clusters. Making of implicit semantic profiles have already been proposed by authors [7], to capture dynamic and static likings of users, with temporal effects. Outliers in preferences were also identified and rectified by focused analysis, to bring accuracy in ranking of likings. Formed profiles are given as input to the proposed approach. Semi-Partial R-Squared (SPRS) metric [8] is used for comparing homogeneity of clusters. Result shows that approach makes homogeneous clusters along with well spread distribution of concepts in clusters.

In news domain user may belong to more than one cluster. Therefore soft clustering approach has been proposed. Each user belongs to one or more cluster with different membership score. Formed clusters are refined by eliminating weak members to reduce the computation cost. Similarity among users is also calculated and compared semantically to remove pairs which hold no merit for recommendation process. Recommendations are semantically ranked based on the average interest in profiles of other similar users. This approach brings reasonable diversity [9] because similarity among users is computed semantically at major concept levels as well as sub concepts level of Ontology. Positive serendipity [9] has also been achieved. User feedback shows that transparency in recommendations of unexpected items is desirable [10]. The proposed approach does not suffer with the scalability issue.

1.1 Background

Different filtering approaches exist in literature, each having its own benefits as well as limitations. Content based filtering approach focuses on accuracy in recommendations considering item specific features to identify user preferences. It faces the issue of over-specialization, sparsity and cold start problem for new item. CF identifies items preferred by likeminded persons. It faces the issue of scalability as the number of new items and number of new user increases. Recommendations based on CF is based on the assumption that users with similar past behaviors have similar preferences and recommends items that are preferred by other similar users. CF requires a clustering approach. Clustering is the task of grouping a set of objects in such a way that objects in the same group are more similar to each other than to those in other groups. Different types of clustering algorithms are: (1) K-means (2) fuzzy C-means (3) Hierarchical (4) Expectation-maximization (EM) is a Mixture of Gaussian algorithm. In first type data are grouped in an exclusive way. Second type uses fuzzy sets to cluster data, so that each point may belong to two or more clusters with difference between the two nearest clusters. In third type every datum is set as a cluster for beginning condition. After a few iterations it gives the final desired clusters. The fourth type uses a complete probabilistic approach.

2 Related Work

We have surveyed work done in this area in past few years and found given facts. In [11], authors have applied hierarchical clustering for fuzzy indexing of web documents to improve search engine efficiency. In [12], authors have used K-means clustering approach and cosine similarity to find similarity between users. In [13], CF is used to find new ad document related to user query. They had used click through rate of users and found matches in discrete categories: precise, approx, marginal and clear. Authors have used Pearson correlation to find relation between 2 queries. In [14], authors have compared Content Based, CF and hybrid approach for news recommendation and vouch for only CF as the domain is dynamic. Authors have classified news into probable 50 categories using K-means, making crisp clusters. In [1], authors have proposed a CF algorithm for news recommendation, which is a combination of memory based and model based CF algorithm. The approach just tries to achieve accuracy in recommendations. Moreover approach is not based on semantics. In [15], authors have used hierarchical clustering approach to group news items to be recommended to bring only diversity. Outliers are not handled in user preferences. In [16], User profiling is based on analysis of web log data without outlier analysis. In [17], recommendations are based just on binary values. In [18], author has stated that recommendation approaches can help to solve the problem of information overload. In [19], authors have focused deeply on the issues of CF. They emphasized that accuracy is not the only criteria for good RS and other evaluation criteria like diversity must be explored.

User profiling in proposed approach is based on preferences with temporal ranking. Outlier analysis of user preferences helps to improve the ranking of users' static preferences. News items to be recommended are classified into thousands of categories using multi label classification, and are arranged hierarchically in multi level news domain ontology. Unlike prior approaches preferences are semantically ranked based on interest scores. These user profiles are grouped semantically using hierarchical approach making intersecting clusters. Proposed modified Jaccard similarity metric considers user's ranked preferences in concepts instead of just the number of concepts. It makes homogenous clusters. Recommendations in proposed approach are semantically ranked. Earlier researchers have tried to bring just diversity using CF. In proposed approach similarity among users is compared at each level of the ontology which brings reasonable diversity as well as positive serendipity. It has also been shown that transparency in diverse recommendations improves understandability and acceptability. Moreover proposed approach is scalable also.

3 Proposed Approach

For bringing reasonable diversity Dual Hierarchical Soft Semantic Clustering Approach (DHSSC) has been proposed. The approach makes intersecting clusters of users considering ranked interest in their profiles. The approach has two phases: (1) Clustering of users using DHSSC, based on dual user preferences, in news concept and entity both. (2) Semantically ranked diverse and serendipitous recommendations with transparency.

Phase 1 This phase proposes a clustering algorithm DHSSC distributed in 4 steps:

Step 1 Assignment of users to clusters of individual major concepts (or entities), based on ranked preferences in different concepts (and entities). Ranked preferences are considered based on the fact that two users having same interest score for a category may have that category at different ranks in their profiles. It helps in comparison of rank of preferred concept in profiles of users and also to calculate their membership strength in cluster.

A user U_i has Interest Score 'S' in Concepts ($C1$–Cn) as $S(C1)...S(Cn)$. The scores are arranged in order such that $S(C1) > S(C2)...> S(Cn)$. Ranked Interest of a User [RI(U)] is calculated by giving highest rank to user's maximum interest score and decreasing the rank assigned to each subsequent interest score of user. Similar ranking method is used for all the users, to make their interest scores comparable. The formula is as follows (F1):

$$RI(Ui) = \frac{N}{N}(S(C1)),\ \frac{N-1}{N}(S(C2)),...,\ \frac{N-(n-1)}{N}(S(Cn)) \qquad (1)$$

Such that $n \leq N$. Where, N = total number of concepts (or entities) in the domain, n = total number of concepts (or entities) in the domain in which user has shown interest.

Unlike traditional hierarchical approach which has to consider all the users as individual data points, proposed approach assigns users to the limited number of news concepts or entities in which users have shown interest. It does not face the issue of scalability as the number of new user increases. We have individual news concepts (or entities) as initial data points, having all n number of users distributed among them, based on interests captures in profiles. This phase ensures making of intersecting clusters.

Step 2 Merging of individual clusters of Phase 1. To group similar users, similarity among the individual clusters is measured using modified Jaccard similarity metric. It considers ranked interest RI of user 'Ui', RI(Ui) in different formed clusters (calculated using formula F1), instead of just the number of preferred concepts. If a concept cluster Ca is preferred by a set {U1, U3, U4} of users and concept cluster Cb is preferred by {U3, U4} then similarity (sim) between Ca and Cb is calculated using formula (F2):

$$sim(Ca, Cb) = \frac{\sum_{i=1}^{z} \min(RI(Ui\ (Ca)), RI(Ui(Cb)))}{\sum_{i=1}^{z} \max(RI(Ui(Ca)), RI(Ui(Cb)))} \qquad (2)$$

where, z is no. of distinct users in both the clusters. Standard formula considers just the number of users. Jaccard similarity is used to ensure that similar concept clusters (based on number of intersecting users among clusters and their corresponding interest scores) are merged to form cluster of similar users. Evaluation result shows that homogeneity of formed clusters is improved using proposed approach.

Step 3 This step of refinement of users within merged concept clusters, (formed in step 2), removes the users which belong to a merged cluster with extremely low interest scores. It creates group of users with strong membership. To compute strength 'S' of a Ui within a merged concept cluster CLy, a formula has been given (F3):

$$S(Ui, CLy) = \frac{\sum_{j=1}^{m} RI(Ui(Cj))}{|CLy|} \qquad (3)$$

where m is the number of concepts in cluster CLy in which user Ui was interested and |CLy| is the total number of concepts in CLy. To reduce search space the users with strength below a threshold are not considered in next steps.

Step 4 Pair wise similarity among remaining users is computed to find highly related pair of users belonging to a particular cluster. This assists in recommending preferred concepts in profile of Ui which are new to Uj and vice versa. For computing pair wise similarity among users in a cluster, Jaccard similarity measure is given (F4):

$$Sim(Ui, Uj)CLy = \frac{\sum_{a=1}^{|CLy|} \min(UiCa, UjCa)}{\sum_{a=1}^{|CLy|} \max(UiCa, UjCa)} \qquad (4)$$

Phase 2 In this phase various conditions are taken into consideration for ranked recommendations with reasonable diversity and serendipity. Similar users are clustered into groups based on similarity in major concepts calculated in phase 1. It is observed that within a category, sub-categories are very diverse. Therefore it is required to semantically compare users within groups to check whether they are interested in a common sub category also. Approach considers similar users with high interest scores. However it has been observed that some similar users having low similarity score have good number of preferred common concepts. Therefore approach also considers these users for recommendations, which will improve their interest. Recommendations are arranged in ranks in descending order based on the average of score of other similar cluster member.

4 Experimental Study, Observations and Evaluations

A portal has been designed for testing of proposed approach. Portal is provided to under graduate and post graduate students of the University of Delhi. Captured user preferences are analyzed for homogeneity of clusters. None of the available benchmark datasets meet the required criteria as authors have used semantic approach. In traditional hierarchical clustering approach data points and the distances between them changes after each iteration. In proposed hierarchical clustering approach, maximum number of clusters in phase-I are fixed depending upon the total number of unique concepts having user interests. Interest score of users' is also fixed during analysis. Distances calculated among these concepts are computed offline and stored in a matrix.

In proposed approach individual concepts are initial data points, needed to be merged. A N × N similarity matrix is formed for all the pairs of major concepts. Clusters formed using standard and modified Jaccard metric, gives different results. Clusters are formed (Table 1) considering following three approaches:

Table 1 SPRS values of clusters for news concepts formed using three approaches

No.	SPRS value of Merged Clusters formed	No. of concepts
1	Merged Cluster-1 4.836988E-6 (0.011989694)	2
	Merged Cluster-2 8.0629225E-6 (0.01998598)	2
	Merged Cluster-3 5.100267E-5 (0.03436532)	2
	Merged Cluster-4 0.0	1
	Merged Cluster-5 0.0	1
	Merged Cluster-6 0.0	1
	Merged Cluster-7 0.0	1
	Merged Cluster-8 0.0	1
	Merged Cluster-9 0.0	1
2	Merged Cluster-1 0.011435431	3
	Merged Cluster-2 0.05623525	3
	Merged Cluster-3 0.037493944	3
	Merged Cluster-4 0.0	1
3	Merged Cluster-1 0.011435431	3
	Merged Cluster-2 0.014368579	3
	Merged Cluster-3 0.034721155	3
	Merged Cluster-4 0.0	1
	Merged Cluster-5 0.0	1
	Merged Cluster-6 0.0	1

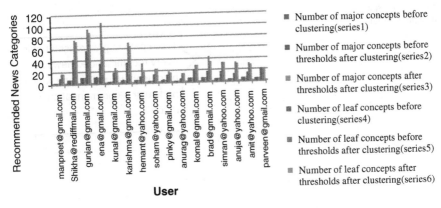

Fig. 1 Reasonable diversity in recommended major and leaf news concepts

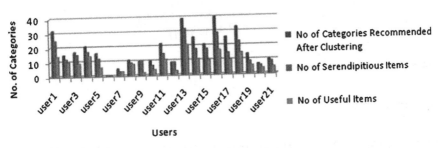

Fig. 2 Graph showing accepted serendipitous recommendations

(1) by computing Cluster centers, (2) by computing number of common users within concepts of clusters, (3) (Proposed) by computing ranked interest score of common users within concept of clusters. Formed clusters are compared using SPRS measure. SPRS gives results between 0 and 1. Values near 0 means homogeneity of merged clusters. We propose approach 3 which proves the best homogeneity along with well spread distribution of concepts.

In Fig. 1, series 3 shows reasonable diversity in major categories by compounding the span of diversity. Series 6 shows leaf categories recommended based on phase II of proposed approach to bring reasonable diversity. Figure 2 shows unexpected serendipitous recommendations and their usefulness. Series 1 shows recommended categories after clustering. Series 2 shows the serendipitous categories (which is not semantically related to user's individual preferences).

Series 2 is smaller than series 1 because all recommended categories in series 1 are not serendipitous. Series 3 shows the number of recommended serendipitous items, in which user has shown interest. Figure 3 shows transparency and Feedback for the question 'understand ability due to transparency' (Fig. 4) proves it is desirable.

Fig. 3 Transparency in designed portal

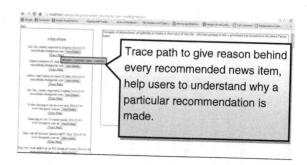

Trace path to give reason behind every recommended news item, help users to understand why a particular recommendation is made.

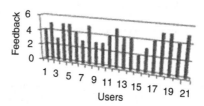

Fig. 4 User feedback on transparency

5 Conclusion

RS use similarity between users or items to generate recommendations which are novel and accurate but lack diversity. Soft semantic hierarchical clustering approach is proposed based on the preferences of other likeminded persons. In proposed approach Jaccard similarity metric has been modified to make homogeneous clusters. Semantic similarity among users brings reasonable diversity and serendipity. User feedback shows usefulness of transparency. Proposed semantics based approach is able to handle the scalability issue as the number of news items and number of new users increases.

References

1. Burke, R.: Knowledge-based recommender systems. In: Kent, A., Hall, C.M. (eds.) Encyclopedia of Library and Information Science, vol. 69, pp. 180–200. Marcel Dekker, New York (2000)
2. Su, X.Y., Khoshgoftaar, T.M.: A survey of collaborative filtering techniques. Adv. Artif. Intell. **2009**, p. 19 (2009). Article ID 421425
3. Adomavicius, G., Tuzhilin, A.: Toward the next generation of recommender systems: a survey of the state-of-the-art and possible extensions. IEEE Trans. Know. Data Eng **17**(6), 734–749 (2005)
4. Herlocker, J., Konstan, J., Terveen, L., Riedl, J.: Evaluating collaborative filtering recommender systems. ACM Trans. Inf. Syst. **22**(1), 5–53 (2004)

5. Vargas, S., Castells, P.: Rank and relevance in novelty and diversity metrics for recommender systems. In: Recsys 11. ACM 2011 987-1-4503-0683-6/11/10 (2011)
6. Belem, F., Martins, E., Almeida, J., et al.: Exploiting novelty and diversity in tag recommendation. In: ECIR 2013. LNCS, vol. 7814, pp. 380–391. Springer (2013)
7. Agarwal, S., Singhal, A.: Handling skewed results in news recommendations by focused analysis of semantic user profiles. In: International Conference on Reliability Optimization and Information Technology ICROIT 2014 IEEE Delhi Section (2014)
8. Halkidi, M., Batistakis, Y., Vazirgiannis, M.: Clustering validity checking methods: part II in ACM. SIGMOD Rec. **31**(3), 19–27 (2002)
9. McNee, S.M., Riedl, J., Konstan, J.A.: Being accurate is not enough: how accuracy metrics have hurt recommender systems. In: Conference on Human Factors in Computing Systems (CHI 2006), pp. 997–1001. ACM, Montreal (2006)
10. Johnson, J., Johnson, P.: Explanation facilities and interactive systems. Proc. Intell. User Interfaces **93**, 159–166 (1993)
11. Gupta, D., Bhatia, K.K., Sharma, A.K.: A novel indexing technique for web documents using hierarchical clustering. Int. J. Comput. Sci. Netw. Secur. **9**(9), 168–175 (2009)
12. Ju, C., Xu, C.: A new collaborative recommendation approach based on users clustering using artificial bee colony algorithm. Sci. World J. **2013**, 9 (2013). Article ID 869658
13. Anastasakos, T., Hillard, D., Kshetramade, S., Raghavan, H.: A collaborative filtering approach to sponsored search. In Yahoo! Labs, Technical Report No. YL-2009-006 (2009)
14. Garcin, F., Zhou, K., Faltings, B., Schickel, V.: Personalized news recommendation based on collaborative filtering. In: Proceedings of the 2012 IEEE/WIC/ACM International Joint Conferences on Web Intelligence and Intelligent Agent Technology, vol. 01, pp. 437–441 (2012)
15. Li, L., Wang, D., Li, T. et al.: SCENE: A scalable two-stage personalized news recommendation system. In: ACM 978-1-4503-0757-4/11/07, SIGIR'11. Beijing, China. Copyright 2011 (2011)
16. Castellano, G., Fanelli, A.M., Mencar, C. et al.: Similarity-based fuzzy clustering for user profiling. In: 2007 IEEE/WIC/ACM International Conferences on Web Intelligence and Intelligent Agent Technology—Workshops (2007)
17. Das, A., Datar, M., Garg, A.: Google news personalization: scalable online collaborative filtering. In: WWW 2007/Track: Industrial Practice and Experience, Alberta, Canada, 8–12 May 2007
18. Billsus, D., Pazzani, M.J.: A personal news agent that talks, learns and explains. In: Proceedings of the 3rd International Conference on Autonomous Agents, pp. 268–275. ACM Press, New York (1999)
19. Schafer, J.B. et al.: Collaborative filtering recommender systems. In: Brusilovsky, P., Kobsa, A., Nejdl, W. (eds.) The Adaptive Web. LNCS, vol. 4321, pp. 291–324 (2007)

Prediction of Blood Brain Barrier Permeability of Ligands Using Sequential Floating Forward Selection and Support Vector Machine

Pooja Gupta, Utkarsh Raj and Pritish K. Varadwaj

Abstract Prediction of Blood Brain Barrier (BBB) permeability index has been established as an important criterion for CNS active drug molecules. Various experimental and in silico approaches were being used for the prediction BBB permeability with accuracy level fall within 80 % on test dataset (r^2 = squared correlation coefficient; 0.65–0.91 derived from training set). In this study Sequential Floating Forward Selection (SFFS) feature selection method based Support Vector Machine (SVM) classification was carried out on a set of 453 chemically diverse compounds with known BBB permeability index. The prediction efficiency for the test set was found to be $r^2 = 0.95$ for 369 compounds (within the applicability domain after excluding four activity outliers). Classification accuracies for permeable (BBB +ve) and non-permeable (BBB −ve) were 96.84 and 98.21 % respectively.

Keywords Blood brain barrier · In silico · Support vector machine

1 Introduction

The blood-brain barrier (BBB) is a big paradox. On one side it protects the brain by being a constant systemic pouring of noxious substances. Similarly, on other side it prevents the delivery of most important drug or therapeutic agents to molecular receptors present in brain. BBB is an important interface between central nervous system (CNS) and peripheral blood circulation [1]. It is situated at brain capillary endothelial layer covering 12 m²/g of the brain parenchyma. The term "blood–brain

P. Gupta · U. Raj · P.K. Varadwaj (✉)
Indian Institute of Information Technology, Allahabad, India
e-mail: pritish.varadwaj@gmail.com

P. Gupta
e-mail: nanobiotech.pooja@gmail.com

U. Raj
e-mail: utkarsh01.raj@gmail.com

© Springer India 2015
L.C. Jain et al. (eds.), *Computational Intelligence in Data Mining - Volume 1*,
Smart Innovation, Systems and Technologies 31, DOI 10.1007/978-81-322-2205-7_42

barrier" was coined, by Lewandowsky, [2] when he along with Briedl and Kraus [3] found that neurotoxins affects the brain when injected directly, but not when injected intravenously.

The BBB is formed by epithelial-like high resistance tight junctions within the endothelium of capillaries perfusing the vertebrate brain. Because of the presence of the BBB, circulating molecules gain access to brain cells only via one of two processes: (i) lipid-mediated transport of small molecules through the BBB by free diffusion, or (ii) catalyzed transport. Drug delivery to the brain via endogenous transport systems within BBB requires drug reformation, such that it can access the blood brain barrier transport system and enters the brain. The brain drug discovery and brain drug targeting must be combined together to ensure the success of neuro-therapeutics development.

1.1 Models for BBB Prediction

BBB plays the rate limiting role in neuro-therapeutics development. The prediction of BBB permeability is very much needed for CNS-active drug candidates, as well as for ligands with non-CNS indications. However, most of the CNS-active drugs fail in clinical trials because of its poor BBB permeability. This has led to the development of different in vitro, in vivo, and in silico methods for the prediction of BBB drug permeability.

1.1.1 In Vitro Models

Isolated brain capillaries were being used as in vitro model of the BBB by isolating capillaries from brain tissues of various species. They can be used for binding and uptake assays and to study BBB transport systems for nutrients and peptides at the mRNA and protein level [4]. However, isolated brain capillaries are not metabolically viable [5, 6] and poses a high risk of contamination by other brain-derived cells [7].

1.1.2 In Vivo Models

A common method for the assessment of BBB permeability is the intravenous administration of a test substance into an animal. After a single bolus injection of a radiolabeled test compound, the animal is decapitated and the brain tissue is analyzed for radioactivity [8, 9]. Other methods include determination of brain uptake index (BUI) [10, 11], brain perfusion method [12] and micro-dialysis [13, 14]. The latest method used to study transport across the BBB in vivo is positron emission tomography (PET) [15].

Experimental determination of BBB permeability and non-permeability is very important but it is time taking and expensive process and suffers from variety of limitations. Thus, now-a-days in silico model of BBB permeability is gaining considerable importance.

1.1.3 In Silico Models

In silico models ought to predict BBB permeability from chemical structures. For the development of these computational models, experimental brain uptake data is correlated with molecular properties, such as lipo-philicity and molecular weight [16, 17] etc. So far, only passive permeability at the BBB has been modeled, because the knowledge about the relationship between molecular structure and active or facilitated transport is still limited [9]. Thus, **in silico** models may be a useful tool for the initial screening of lead compounds to predict passive BBB permeability.

A plethora of **in silico** models for BBB permeability prediction have been developed in recent past, by scientific community engaged in development of neurotherapeutics, based on Quantitative Structure Activity Relationship (QSAR) [18–21], probabilistic and statistical analysis [22–29] and artificial intelligence based machine learning classification [30–35]. All developed models are based on identification of molecular properties/features which directly or indirectly affects BBB permeability. Many such properties were identified by researchers like polarizability, hydrogen-bond acidity, hydrogen-bond basicity [18, 19], drug solvation free energy in water [20], polar surface area, variety of electro-topological indices [21, 36], octanol-water partition coefficient [16, 21, 27, 37], hydrogen-bond acceptors [37], molecular weight [16, 38], dipole, highest occupied molecular orbital (HOMO) energy [38], polarity, polarizability, and hydrogen bonding [16, 27] etc. These existing models have accuracy level well within 80 % on test dataset ($r^2 = 0.65$–0.91 derived from training set).

2 Materials and Methods

2.1 Dataset

We have selected BBB permeable and impermeable ligand data from literature study and ligand databases [30, 39, 40]. The selected dataset for this study comprises of chemically diverse 453 ligand molecules. Further we randomly split the whole dataset of 453 ligands, into training set of 80 ligands, comprised of 40 BBB permeable and 40 BBB non-permeable ligand data. Similarly 373 ligands were taken as test set which comprised of 249 BBB permeable molecules and 124 BBB non-permeable molecules. Training set data was used for training various support vector machine (SVM) classifiers, while the testing examples were not exposed to the system during learning, kernel selection and hyper-parameter selection phases.

2.2 Descriptor Selection (Feature Selection for BBB)

In most of the pattern recognition algorithm each input data to be classified is associated with large number of descriptors resulting in very large dimensions of feature space so makes classification a difficult task. Therefore dimensionality reduction is an essential step in addressing such problems. There exist two ways for feature selection viz filter method, which is based on intrinsic properties of data as the criterion for feature subset evaluation, and wrapper method, which takes the performance of the classifier into account. Recently reported least square (LS) bound measure [41] is a hybrid of filter and wrapper method which provides accurate classification with reasonable computational complexity. Unlike the classical wrapper method, it does not demand repeated trainings for cross validation. In present work, SFFS is used along with following measure:

$$M = \Sigma(\alpha_p^0 \left[(D_{min}^p)^2 + 2/\gamma \right] - 1) \tag{1}$$

where $(x)+ = \max(0, x)$. α^0 can be obtained from solving a set of linear equations in the LS-SVM. D_{min}^p is the distance between $\mathbf{x}_p$ and its nearest neighbor. $\gamma 1$ is a LS-SVM parameter. $\gamma 1$ plays very important role in the performance of LS bound method thus its appropriate value should be evaluated before testing. We employed our algorithm with a sequence of given values of $\gamma 1$ to select important genes on the entire dataset. Because the LS Bound measure indicates the generalization performance, the optimal value of $\gamma 1$ is chosen to be the one which gives the minimal LS Bound measure during the selection procedure.

2.3 Data Preprocessing and Descriptor Calculation

Different feature values for ligand dataset falls in different ranges hence to avoid the discrepancy we have further scaled down the numeric values between −1 and 1. Such scaling facilitates better representation of feature values in kernel function.

In electronic descriptor, Hydrogen Acceptor (H-A) and Hydrogen Donor (H-D) are taken into account. These are numeric representation of number of hydrogen bond donor and acceptors within the drug molecule; based upon Stein method [42]. In topological descriptors, we have used six important descriptors. viz. Randic Index-valence Connectivity Index chi-1 ($\chi 1v$) [43], Wiener Index (W) [44], Balban Distance Connectivity Index (J) [45], first Zagreb Index (ZM1), second Zagreb Index (ZM2) [46] and Polarity number (pol) [44, 47, 48]. These were calculated using empirical graph theory calculation.

Randic Index is inverse square root of the vertex degree. It measures sum of the bond connectivity over all bonds in the molecule and differentiate between drugs which have the same molecular weights and volume, but very different "branching" properties. It is the most widely used topological parameter. The first order Randic indices (Eq. 2) are given as:

$$\chi = \sum cij = \sum (\delta i \delta j) \qquad (2)$$

where, δi and δj are the free valences of the carbon of each bond vertex, i, with the adjacent, j, discounting hydrogen, because the molecule graph is, by definition, a hydrogen-free graph (Kenogram). cij represent relative bond accessibility areas (RBA). Whereas, wiener's index (Eq. 3) is defined as half the sum of the off-diagonal elements of the distance matrix of the relevant graph:

$$W = (1/2) \sum (\delta i,j) \qquad (3)$$

It counts the number of bonds between pair of atoms and sums the distance between all pairs by generating a distance matrix. It follows the shortest route. Balban Index is also known as 'averaged distance sum connectivity'. It is based on detour matrix and is denoted by J. It is based upon the Randic formula, the only difference is, it replaces vertex degrees by averaged distance sums.

For acyclic compounds:

$$Jacyclic = q \sum_{adji,j} (sisj)^{-0.5} \qquad (4)$$

For cyclic compounds:

$$Jcyclic = \frac{q}{\mu + 1} \sum_{adji,j} (sisj)^{-0.5} \qquad (5)$$

where, μ is the cyclomatic number, which may be defined as minimum number of edges that must be removed from the graph, to convert it in acyclic graph, q is the number of edges, si and sj are the distance sums. In molecular graph G, the first Zagreb index $M_1(G)$ (Eq. 6) and the second Zagreb index $M_2(G)$ (Eq. 7) are:

$$M_1(G) = \sum_{u \in v(G)} (d(u))^2 \qquad (6)$$

$$M_2(G) = \sum_{uv \in E(G)} d(u)d(v) \qquad (7)$$

where, $d(u)$ denotes the degree of the vertex u of G [49]. It represents molecular branching, complexity and other topological properties.

Like wiener, balaban and platt index, polarity of the molecule also depends upon the distance matrix. As the name suggests it gives the polarity of the molecule. It is sum of edge degree of a molecular graph. Pol was invented after the wiener but wiener gained all the popularity not only because it was first topological indices to be invented, but also it was simple to calculate and yet quite successful for many of the applications.

$$Pol = \frac{1}{2} \sum_{i=1}^{N} \sum_{j=1}^{N} (i+j) \quad \forall i, j \text{ where, } d_{ij} = 3 \tag{8}$$

Later, bulk properties are measured by molecular weight (MW) and mean atomic *van der waal* volume (MV)-scaled on carbon atom. LogP value was also used.

2.4 SVM Description

In this process of feature space mapping, training data of two types of ligand data (i) BBB permeable and (ii) BBB non-permeable were prepared with target labels +1 for *and* −1 respectively. So input vector for training as well as test set has been quantified as: $X^i = (X_1^i, X_2^i, ..., X_{11}^i)$ each labeled by corresponding $y^i = +1$ or $y^i = -1$ depending on whether it represents a BBB permeable or non-permeable ligand, respectively.

Training set data were used in SVM classifier. It fixes several hyper-parameters and their values are used in determining the function that SVM optimizes and therefore have a crucial effect on the performance of the trained classifier [50]. We have used several kernels: linear, polynomials and radial basis function (RBF) and found RBF as the suitable classifier function (as the number of features is not very large), for which training errors on BBB permeable ligand data (false negatives) outweigh errors on BBB non-permeable ligand data (false positives).

$$K(X_i, X_j) = \exp\left(-\gamma \|X_i - X_j\|^2\right), \quad \gamma > 0 \tag{9}$$

This kernel (Eq. 9) is basically suited best to deal with data that have a class-conditional probability distribution function approaching the Gaussian distribution. However, this kernel is difficult to design, in the sense that it is difficult to arrive at an optimum 'γ' and choose the corresponding C that works best for a given problem [51]. This has been taken care by running grid parameter search exploring all combinations of C and γ with each cross-validation routine, where γ ranged from 2^{-15} to 2^4 and C ranged from 2^{-5} to 2^{15} [52]. To identify an optimal hyper-parameter set we have performed a two step grid-search on C and γ using 10 folds cross-validation, by dividing training set into 10 subsets of equal size (8 ligands each). Iteratively each subset is tested using the classifier trained on the remaining 9 subsets. Pairs of (C; γ) have been tried and the one with the best cross-validation accuracy has been picked. The best cross-validation performance, for a value of $\gamma = 0.05$ and C = 325 was obtained by the RBF kernel, with parameter and cost factor (Fig. 1). The result obtained shows very good classification accuracy 99.25 % during the cross-validation.

3 Result and Discussion

SVM parameters γ and C, were optimized using 10-fold cross-validation on each of the training datasets bin, exploring various combinations of C (2^{-5} to 2^{15}) and γ (2^{-15} to 2^4). In 10-fold cross-validation, the training dataset (80 molecules, each of 11 feature vector length) was spilt into 10 subsets of 8 ligand entries (4 BBB permeable and 4 non-BBB permeable), where one of such subsets was used as the test dataset while the other subsets were used for training the classifier. The process was repeated 10 times using a different subset of corresponding test and training datasets, hence ensuring that all subsets are used for both training and testing. A two-fold grid optimization had been considered and result shown (Fig. 1) suggests the optimized C and γ were found to be 325 and 0.05 respectively.

The best combinations of γ and C obtained from the grid based optimization process were used for training the RBF kernel based SVM classifier using the entire training dataset of 80 ligand entries. The SVM classifier efficiency was further

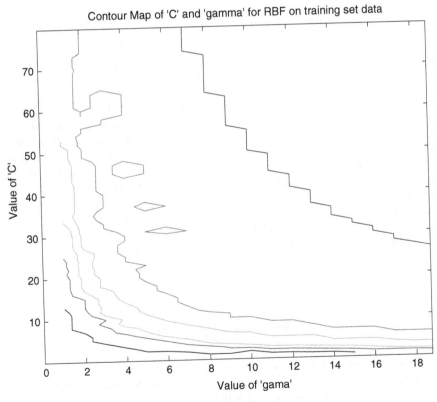

Fig. 1 Contour plot of grid search result showing optimum values of **hyper-parameter**

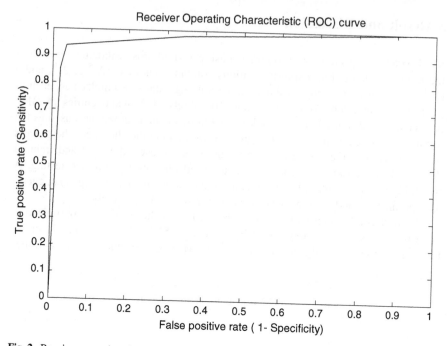

Fig. 2 Receiver operating characteristic (ROC) plot for classifier with optimized values of C and γ

evaluated by various quantitative variables: (a) *TN*, true negatives—the number of correctly classified BBB non-permeable ligand, (b) *FN*, false negatives—the number of incorrectly classified BBB permeable ligand, (c) *TP*, true positives—the number of correctly classified BBB permeable ligand, (d) *FP*, false positives—the number of incorrectly classified BBB non-permeable ligand. Using these variables several statistical metrics were calculated to measure the effectiveness of the proposed RBF-SVM classifier. *Sensitivity (Sn)* and *Specificity (Sp)* metrics, which indicates the ability of a prediction system to classify the BBB permeable and BBB non-permeable ligands, were calculated by Eqs. (10) and (11) and receiver operating characteristic curve (ROC) for the same has been plotted (Fig. 2).

$$Sn(\%) = \frac{TP}{TP + FN} \times 100 \tag{10}$$

$$Sp(\%) = \frac{TN}{TN + FP} \times 100 \tag{11}$$

To indicate an overall performance of the classifier system; (a) *Accuracy (Ac)*, for the percentage of correctly classified splice sites and the *Matthews Correlation Coefficient (MCC)* were computed as follows:

$$Ac(\%) = \frac{TP + TN}{TP + TN + FP + FN} \times 100 \qquad (12)$$

$$MCC = \frac{(TP \times TN) - (FP \times FN)}{\sqrt{(TN + FP)(TN + FN)(TP + FP)(TP + FN)}} \qquad (13)$$

Sensitivity (*Sn*) was found to be 98.2 % with false positive proportion (*FP*) 1.8 %, where as Specificity (*Sp*) was found to be 96.8 % with false negative (*FN*) proportion 3.2 %. Similarly *Youden's Index* (*Youden's Index = sensitivity + specificity* −1) was 0.9506 and *Matthews Correlation Coefficient* (*MCC*) *found to be 0.9501*. The overall accuracy (*Ac*) was calculated as 97.5 % and r^2 being 0.944 (r^2 is squared correlation coefficient between predicted and experimental result) which is significantly higher than existing methods. For existing methods the r^2 score falls between 0.65 and 0.91. Area under receiver operating curve (ROC) curve is found to be 0.9908 with standard error 0.007. We had chosen the RBF kernel with optimized parameters γ and C. Using 10-fold cross-validation, the parameters γ and C were optimized at 0.05 and 325 with an overall training datasets classification accuracy of 99.25 %, which is reasonably good. While the reported accuracy on the training datasets may indicate the effectiveness of a prediction method, it may not accurately portray how the method will perform on novel, hitherto undiscovered splice sites. Therefore, testing the SVM methodology on independent out-of-sample datasets, not used in the cross-validation is critical. Here, we applied the SVM classifiers, on the entire test datasets, the SVM method obtained an accuracy of 97.5 % using the RBF kernel with $\gamma = 0.05$ and $C = 325$. These findings suggested that the SVM-based prediction of BBB permeability might be helpful as a tool in drug discovery and development.

4 Conclusion

The BBB permeability prediction by this proposed SVM classifier on the test dataset was found to give 97.5 % accuracy with Sensitivity 98.2 % and Specificity 96.8 %. Further the MCC 0.950 and r^2 0.944 suggests the significance and superiority of this model over existing tools. Hence it may be suggested that while working on computer aided drug discovery and development, this SVM based BBB permeability prediction approach can be considered as an efficient tool.

References

1. Hawkins, B.T., Davis, P.: The blood-brain barrier/neurovascular unit in health and disease'. Pharmacol. Rev. **57**, 173–185 (2005)
2. Cuzner, M.L., Hayes, G.M., Newcombe, J., Woodroofe, M.N.: The nature of inflammatory components during demyelination in multiple sclerosis. J. Neuroimmunol. **20**, 203–209 (1988)

3. Esiri, M.M.: Immunoglobulin-containing cells in multiple-sclerosis plaques. Lancet **2**, 478–480 (1977)
4. Pardridge, W.M.: Isolated brain capillaries: an in vitro model of blood-brain barrier research. In: Pardridge, W.M. (ed.) Introduction to the Blood-Brain Barrier Methodology, Biology and Pathology. Cambridge University Press, Cambridge (1998)
5. Pardridge, W.M.: Blood-brain barrier biology and methodology. J. Neurovirol. **5**, 556–569 (1999)
6. Lasbennes, F., Sercombe, R., Seylaz, J.: Monoamine oxidase activity in brain microvessels determined using natural and artificial substrates: relevance to the blood-brain barrier. J. Cereb. Blood Flow Metab. **3**, 521–528 (1983)
7. Takakura, Y., Audus, K.L., Borchardt, T.: Blood-brain barrier: transport studies in isolated brain capillaries and in cultured brain endothelial cells. Adv. Pharmacol. **22**, 137–165 (1991)
8. Mater, S., Maickel, R.P., Brodie, B.B.: Kinetics of penetration of drugs and other foreign compounds into cerebrospinal fluid and brain. J. Pharmacol. Exp. Ther. **127**, 205–211 (1959)
9. Bickel, U.: How to measure drug transport across the blood-brain barrier. NeuroRx **2**, 15–26 (2005)
10. Oldendorf, W.H., Pardridge, W.M., Braun, L.D., Crane, P.D.: Measurement of cerebral glucose utilization using washout after carotid injection in the rat. J. Neurochem. **38**, 1413–1418 (1982)
11. Oldendorf, W.H.: Measurement of brain uptake of radiolabeled substances using a tritiated water internal standard. Brain Res. **24**, 372–376 (1970)
12. Takasato, Y., Rapoport, S.I., Smith, R.: An in situ brain perfusion technique to study cerebrovascular transport in the rat. Am. J. Cell Physiol. **247**, 484–493 (1984)
13. Aasmundstad, T.A., Morland, J., Paulsen, R.E.: Distribution of morphine 6-glucuronide and morphine across the blood-brain barrier in awake, freely moving rats investigated by in vivo microdialysis sampling. J. Pharmacol. Exp. Ther. **275**, 435–441 (1995)
14. Westergren, I., Nystrom, B., Hamberger, A., Johansson, B.B.: Intracerebral dialysis and the blood-brain barrier. J. Neurochem. **64**, 229–234 (1995)
15. Webb, S., Ott, R.J., Cherry, S.R., Quantization of blood-brain barrier permeability by positron emission tomography. Phys Med. Biol. **34**, 1767–171
16. Goodwin, J.T., Clark, E.: In silico predictions of blood-brain barrier penetration: considerations to "keep in mind". J. Pharmacol. Exp. Ther. **315**, 477–483 (2005)
17. Clark, D.E.: In-silico prediction of blood-brain barrier permeation. Drug Discov. Today **8**, 927–933 (2003)
18. Abraham, M.H., Chadha, H.S., Mitchell, R.C.: Hydrogen bonding. 33. Factors that influence the distribution of solutes between blood and brain. J. Pharm. Sci. **83**, 1257–1268 (1994)
19. Abraham, M.H., Chadha, H.C., Mitchell, R.C.: Hydrogen-bonding. Part 36. Determination of blood brain distribution using octanol-water partition coefficients. Drug Des. Discov. **13**, 123–131 (1995)
20. Lombardo, F., Blake, J.F., Curatolo, W.J.: Computation of brain–blood partitioning of organic solutes via free energy calculations. J. Med. Chem. **39**, 4750–4755 (2003)
21. Subramanian, G., Kitchen, D.B.: Computational models to predict blood–brain barrier permeation and CNS activity. J. Comput. Aided Mol. Des. **17**, 643–664 (2003)
22. Katritzky, A.R., Kuanar, M., Slavov, S., Dobchev, D.A., Fara, D.C., Karelson, M., Acree, W.E., Solov'ev, V.P., Varnek, A.: Correlation of blood–brain penetration using structural descriptors. Bioorg. Med. Chem. **14**, 4888–4917 (2006)
23. Hou, T.J., Xu, X.J.: ADME evaluation in drug discovery. 3. Modeling blood–brain barrier partitioning using simple molecular descriptors. J. Chem. Inf. Comput. Sci. **43**, 2137–2152 (2003)
24. Iyer, M., Mishru, R., Han, Y., Hopfinger, A.J.: Predicting blood–brain barrier partitioning of organic molecules using membrane-interaction QSAR analysis. Pharm. Res. **19**, 1611–1621 (2002)

25. Pan, D., Iyer, M., Liu, J., Li, Y., Hopfinger, A.J.: Constructing optimum blood brain barrier QSAR models using a combination of 4D-molecular similarity measures and cluster analysis. J. Chem. Inf. Comput. Sci. **44**, 2083–2098 (2004)

26. Ma, X.L., Chen, C., Yang, J.: Predictive model of blood–brain barrier penetration of organic compounds. Acta Pharmacol. Sin. **26**, 500–512 (2005)

27. Norinder, U., Haeberlein, H.: Computational approaches to the prediction of the blood–brain distribution. Adv. Drug Deliv. Rev. **54**, 291–313 (2002)

28. Platts, J.A., Abraham, M.H., Zhao, Y.H., Hersey, A., Ijaz, L., Butina, D.: Correlation and prediction of a large blood-brain distribution data set—an LFER study. Eur. J. Med. Chem. **36**, 719–730 (2001)

29. Hemmateenejad, B., Miri, R., Safarpour, M.A., Mehdipour, A.R.: Accurate prediction of the blood–brain partitioning of a large set of solutes using ab initio calculations and genetic neural network modeling. J. Comput. Chem. **27**, 1125–1135 (2006)

30. Zhang, L., Zhu, H., Oprea, T.I., Golbraikh, A., Tropsha, A.: QSAR modeling of the blood-brain barrier permeability for diverse organic compounds. Pharm. Res. **25**, 1902–1914 (2008)

31. Kortagere, S., Chekmarev, D., Welsh, W.J., Ekins, S.: New predictive models for blood-brain barrier permeability of drug-like molecules. Pharm. Res. **25**, 1836–1845 (2008)

32. Dureja, H., Madan, A.K.: Validation of topochemical models for the prediction of permeability through the blood-brain barrier. Acta Pharm. **57**, 451–467 (2007)

33. Doniger, S., Hofmann, T., Yeh, J.: Predicting CNS permeability of drug molecules: comparison of neural network and support vector machine algorithms. J. Comput. Biol. **9**, 849–864 (2002)

34. Guangli, M., Yiyu, C.: Predicting Caco-2 permeability using support vector machine and chemistry development kit. J. Pharm. Pharm. Sci. **9**, 210–221 (2006)

35. Yanga, S.Y., Huanga, Q., Lib, L.L., Maa, C.Y., Zhanga, H., Baia, R., Tenga, Q.Z., Xianga, M.L., Weia, Y.Q.: An integrated scheme for feature selection and parameter setting in the support vector machine modeling and its application to the prediction of pharmacokinetic properties of drugs. Artif. Intell. Med. **46**, 155–163 (2009)

36. Clark, D.E.: Rapid calculation of polar molecular surface area and its application to the prediction of transport phenomena. 2. Prediction of blood-brain barrier penetration. J. Pharm. Sci. **88**, 815–821 (1999)

37. Feher, M., Sourial, E., Schmidt, J.M.: A simple model for the prediction of blood–brain partitioning. Int. J. Pharm. **201**, 239–247 (2000)

38. Brewster, M.E., Pop, E., Huang, M.J., Bodor, N.: AM1-based model system for estimation of brain/blood concentration ratios. Int. J. Quantum Chem. **60**, 51–63 (1996)

39. Burns, J., Weaver, D.F.: A mathematical model for prediction of drug molecule diffusion across the blood-brain barrier. Can. J. Neurol. Sci. **31**, 520–527 (2004)

40. Li, H., Yap, C.W., Ung, C.Y., Xue, Y., Cao, Z.W., Chen, Y.Z.: Effect of selection of molecular descriptors on the prediction of blood-brain barrier penetrating and non-penetrating agents by statistical learning methods. J. Chem. Inf. Model. **45**, 1376–1384 (2005)

41. Zhou, X., Mao, K.Z.: LS bound based gene selection for DNA microarray data. Bioinformatics **21**, 1559–1564 (2005). Oxford university press

42. Stein, W.D.: The Movement of Molecules across Cell Membranes, p. 120. Academic Press, New York (1967)

43. Zhang, L.Z., Lu, M., Tian, F.: Maximum Randi´c index on trees with k-pendant vertices. J. Math. Chem. **41**, 161–171 (2007)

44. Wiener, H.: Structural determination of paraffin boiling points. J. Am. Chem. Soc. **69**, 17–20 (1947)

45. Balaban, A.T.: Distance connectivity index. Chem. Phys. Lett. **89**, 399–404 (1982)

46. Gutman, I., Ruscic, B., Trinajstic, N.S., Wilcox, C.F.: Graph theory and molecular orbitals XII acyclic polyenes. J. Chem. Phys. **62**, 3399–3405 (1975)

47. Wiener, H.: Correlation of heats of isomerization and differences in heats of vaporization of isomers among the paraffinic hydrocarbons. J. Am. Chem. Soc. **69**, 2636–2638 (1947)

48. Platt, J.R.: Prediction of isomeric differences in paraffin properties. J. Phys. Chem. **56**, 328–336 (1952)
49. Garg, P., Verma, J., Roy, N.: In Silico modeling for blood-brain barrier permeability predictions. Drug Absorption Stud. **8**, 289–297 (2008)
50. Cortes, C., Vapnik, V.: Support-vector networks. Mach. Learn. **20**, 273–297 (1995)
51. Varadwaj, V., Purohit, N., Arora, B.: Detection of splice sites using support vector machine. Contemp. Comput. **40**, 493–502 (2009)
52. Chang, C.C., Lin, C.J.: LIBSVM: a library for support vector machines. Software available at (2001) http://www.csie.ntu.edu.tw/~cjlin/libsvm

Feature Filtering of Amino Acid Sequences Using Rough Set Theory

Amit Paul, Jaya Sil and Chitrangada Das Mukhopadhyay

Abstract Numerous algorithms have been developed for extracting meaningful information from large dimensional biological data set. However, due to handling of large number of features and objects, the algorithms are often complex and procedures are lengthy. Feature selection procedure reduces complexity in analyzing high dimensional biological data and becoming essential step in bio-informatics research. The paper addresses feature selection problems by exploiting inter object feature distribution in protein sequence data where importance of amino acids are determined based on their appearance in protein. The proposed algorithm is compared with other well known feature selection methods revealing significant improvement in classification accuracy.

Keywords Protein · Importance factor · Oscillation factor · Classification

1 Introduction

In this drug discovery, characteristics of proteins are identified in order to separate active (binding) compounds from inactive (non-binding) ones, formed using twenty different amino acid sequences. Protein expression profiling [1–3] differences

A. Paul (✉)
Computer Science and Engineering, St. Thomas College of Engineering and Technology, Khidirpore, India
e-mail: amitpaul83@gmail.com

J. Sil
Computer Science and Technology, Bengal Engineering and Science University, Shibpur, India
e-mail: js@cs.becs.ac.in

C. Das Mukhopadhyay
Health Care Science and Technology, Bengal Engineering and Science University, Shibpur, India
e-mail: chitrangadadas@yahoo.com

© Springer India 2015
L.C. Jain et al. (eds.), *Computational Intelligence in Data Mining - Volume 1,*
Smart Innovation, Systems and Technologies 31, DOI 10.1007/978-81-322-2205-7_43

459

indicative of early cancer detection and provide promising results in improving diagnostics. Statistical analysis [4] of high dimensional proteomic data is challenging from different aspects such as dimension reduction, [5–7] feature subset selection and building of classification rules. An optimal feature subset selection (FSS) is an important step towards disease classification/diagnosis with biomarkers [6, 8–10].

Rough set theory (RST) [11–13] is used to reduce dimension of data by keeping only significant attributes that preserve indiscernibility relation between the objects of the information table. There are usually several such reduced attribute subsets, called reducts. Two main approaches are mentioned in the literature to finding reducts one is based on the degree of attribute dependency and another one is concerned with the discernibility matrix. It is not guaranteed to find a minimal subset of reduct [14–16] using the degree of dependency concept [17, 18] while the computational complexity is too high in case of discernibility matrix based methods. Finding all possible reducts [18] and then choosing one with minimal cardinality, is impractical and applicable for simple data set. Therefore, an approximate solution has been proposed in the paper to obtain a minimal reduct using protein sequence data set based on RST. In order to apply RST, data sets are discretized using linguistic variables which express correlation among the features.

2 Methods

In the paper protein sequence data set are taken for attribute selection based on the distribution of the attributes within the objects. Protein amino acid frequencies are obtained from highly parallel and quantitative two dimensional primary protein sequences. If a single amino acid is considered as a attribute then 20 such attributes are generated, similarly 400 attributes for two amino acid, 8,000 attributes for codon. However, the data are redundant or noisy, and the methods for exploring and extracting relationships within the data are still in its infancy. Classification or clustering methods are used for protein primary sequence data analysis, often infer contradictory decisions since a protein may fall in an incorrect class or may belong to several clusters. Fuzzy C-means clustering (FCM) algorithm [19, 20] can solve the problem by measuring degree of belongings of an amino acid in different clusters. However, FCM has its own limitations, which may create improper partitioning of protein sequence data. Correlation based feature selection (CFS) [21–23] algorithm invokes an appropriate correlation measure and a heuristic search strategy. CFS was evaluated using three different machine learning algorithms: a decision tree based learner, an instance based learner and Naive Bayes. Experiments on artificial data sets [24] show that CFS quickly identifies and screens irrelevant, redundant and noisy attributes, and selects important features. On natural domains, CFS typically eliminates well over half the features.

Recently application of RST for feature selection is widely used, but current methods are inadequate to finding computationally efficient and minimal reduct. The paper proposes an approach based on RST and applied on protein primary sequence data for dimension reduction by obtaining single reduct in one pass. Protein primary sequence data set is treated as decision table where different proteins and corresponding amino acids are represented as objects and condition attributes of the data sets along with their class labels, namely all-α, all-β, α/β, α + β. In the proposed reduct generation (RG) algorithm, importance factor of each condition attribute is evaluated and used as a priory information about the attribute. Importance factor of an attribute in a particular class is determined based on the maximum number of objects having similar attribute value belong to that class with respect to the total number of objects with different attribute values belong to that class. An attribute may have different importance in different classes. Importance factor represents influence of a feature to define a class. More importance factor value of an attribute in a particular class means the attribute is more essential to determine that class. In the paper we have used the term feature for representing corresponding attribute with its value.

2.1 Reduct Generation Algorithm (RG)

Reduct Generation Algorithm is divided into two parts

1. Normalization and Discretization.
2. Feature Selection.

Decision table (D) of a Protein sequence data set contains continuous attribute values which are discretized using fuzzy variables with proper semantics.

Discretized values are represented using Gaussian membership functions with varied mean and standard deviation. The membership value of an attribute represents the particular state of the respective object, which depends on the protein as defined by the *oscillation factor*.

Few terminologies are defined below for understanding the proposed RG algorithm.

1. *Oscillation factor*: Distribution of data in two-dimension space reveals that features of several objects oscillate more compare to others. *Oscillation factor*} determines the degree of oscillation and used to discretize the data set. *Oscillation factor* of *i*th object for *j*th attribute is defined in Eq. (1).

$$Oscillation\ factor_{(i,j)} = \$_{mean\_i}\} - D_{(i,j)}/\text{Standard deviation}_i \qquad (1)$$

The decision table is rebuilt with the discretized attribute values.

2. *Importance factor:* The feature selection procedure evaluates importance of each attribute in different classes. *Importance factor* of an attribute S_i in class C_j is defined by Eq. (2)

$$Importance\ factor_i = m/n \qquad (2)$$

where m = maximum number of objects having same value for attribute S_i in class C_j and n = total number of objects with class label C_j.

According to the definition given in Eq. (2), *importance factor* of an attribute lies between 0 and 1, inclusive of these two values where 1 means the single attribute is sufficient to define the class label of an object and 0 means the attribute has no necessity to defining the class. The *importance factor* between 0 and 1 measures the degree of significance of the attribute belonging to a particular class. Therefore, based on the value of *importance factor* most significant attributes (reduct) are generated using the proposed reduct generation algorithm.

In the proposed algorithm, the attributes are sorted in descending order depending on their *importance factor* with respect to a particular class. The attribute(s) with highest *importance factor* is considered and the objects, which contain a particular value appearing maximum number of times in the decision table, are marked. If there is some unmarked objects in that class then select next higher order attribute and repeat the procedure.

Algorithm 1: Normalization-Discretization Algorithm

Input : Decision matrix $D(n, m)$ having n number of object and m number of feature.
Output : $disc\_D(n, m)$ discretized decision matrix using linguistic term.
for $i \leftarrow 1$ **to** n **do**
  for $j \leftarrow 1$ **to** m **do**
    $mean\_object(i) \leftarrow mean(D(i, j))$
    $std\_object(i) \leftarrow$ standard deviation$(D(i,j))$
    $oscillation\ factor(i,j) \leftarrow (\frac{mean\_object(i)-D(i,j)}{std\_object(i)})$

for $i \leftarrow 1$ **to** n **do**
  for $j \leftarrow 1$ **to** m **do**
    if $oscillation\ factor(i, j) < -1$ **then** /* If normalized object's attribute value
    is below -1 then use linguistic term *Very Low(VL)* */
      | $disc\_D(i, j) \leftarrow$ very_low
    else if $-1 \leq oscillation\ factor(i, j) < 0$ **then** /* If normalized object's
    attribute value is between -1 and 0 then use linguistic term *Low(L)* */
      | $disc\_D(i, j) \leftarrow$ low
    else if $oscillation\ factor(i, j) = 0$ **then** /* If normalized object's attribute
    value is equals to 0 then use linguistic term *Medium(M)* */
      | $disc\_D(i, j) \leftarrow$ medium
    else if $0 < oscillation\ factor(i, j) < 1$ **then** /* If normalized object's attribute
    value is between 0 and 1 then use linguistic term *High(H)* */
      | $disc\_D(i, j) \leftarrow$ high
    else /* If normalized object's attribute value is above 1 then use linguistic
    term *Very High(VH)* */
      $\llcorner$ $disc\_D(i, j) \leftarrow$ very_high

Algorithm 2: Reduct Generation Algorithm

Input : Discretized Features
Output : Selected important attributes
for $i \leftarrow 1$ to *number of class* do
  for $k \leftarrow 1$ to m do
    for $j \leftarrow 1$ to *class index(i)* do
      if *disc_D(class(i,class index(i)),k)*= "low" then
        $lo \leftarrow lo+1$
        selected(1,object index(1)) $\leftarrow$ *class(i,class index(i))*
        object index(1) $\leftarrow$ *object index(1)*+1
      else if *disc_D(class(i,class index(i)),k)*= "medium" then
        $me \leftarrow me+1$
        selected(2,object index(2)) $\leftarrow$ *class(i,class index(i))*
        object index(2) $\leftarrow$ *object index(2)*+1
      else
        $ma \leftarrow ma+1$
        selected(3,object index(3)) $\leftarrow$ *class(i,class index(i))*
        object index(3) $\leftarrow$ *object index(3)*+1

    $count \leftarrow max(lo, me, ma)$
    if $count = lo$ then
      important(k).factor $\leftarrow \frac{lo}{class\ index(i)}$
      important(k).feature $\leftarrow k$, copy(*important(k).object,selected(1)*)
    else if $count = me$ then
      important(k).factor $\leftarrow \frac{me}{class\ index(i)}$
      important(k).feature $\leftarrow k$, copy(*important(k).object,selected(2)*)
    else
      important(k).factor $\leftarrow \frac{ma}{class\ index(i)}$
      important(k).feature $\leftarrow k$
      copy(*important(k).object,selected(3)*)

                          /* sort importance factor descending order */
  sort(*important*)
  mark object $\leftarrow \{ \}$
  while (*old mark object* $\neq$ *class index(i)*) do
    if $\|mark\ object\| \leq \|unique(mark\ object, important(next).object)\|$ then
      selected feature(next) $\leftarrow$ *important(next).feature*
      concatenate(*mark object, important(next).object*)
      mark object $\leftarrow$ unique(*mark object*)
    $next \leftarrow next+1$

2.2 Case Study

Reduct Generation (RG) algorithm has been applied on protein primary sequence data, described below.

Protein primary sequence data: The input is the protein primary single sequence derived from PDB [25–27] based on SCOP [28, 29] classification. The features are extracted for variable k = 1 to 4 and number of features are 20^1, 20^2, 20^3, and 20^4 respectively. For k = 1 the patterns are {A, C, D, E, F, G, H, I, K, L, M, N, P, Q, R, S, T, V, W, Y} (20 patterns/features); for k = 2, {AA, AC, ..., AY, CA, CC, ... CY, DA, DC, ... DY, EA, EC ... EY, ... YW, YY} (400 patterns/features) and so on.

A decision table (Table 1) for protein sequence data set is constructed with 8 instances consisting of 20 condition attributes {A, C, D, E, F, G, H, I, K, L, M, N, P, Q, R, S, T, V, W, Y} and four classes: {all-α, all-β, α/β, α + β}. The structural class all-α, all-β, α/β, α + β are assigned with values 1, 2, 3 and 4 respectively.

Table 1 is discretized using Algorithm (1) and *importance factor* of each attribute with respect to different classes are obtained.

Table 1 Protein primary sequence frequency data

Protein	A	C	D	E	F	...	W	Y	Class
P_1	12	7	16	13	7	...	9	10	1
P_2	11	12	7	5	4	...	4	11	1
P_3	30	14	30	9	11	...	8	8	2
P_4	23	4	18	23	7	...	9	14	2
P_5	25	0	15	20	6	...	5	9	3
P_6	12	4	7	13	7	...	3	3	3
P_7	39	2	34	32	36	...	5	22	4
P_8	42	3	28	39	32	...	5	20	4

3 Results and Discussion

The proposed reduct generation algorithm has been applied on primary protein sequence data sets for feature selection. Out of 400 features, 77 and 58 features are selected considering two protein databases *Swissport.fasta* and *igSeqProt.fasta* [30].

Principal component analysis (PCA) [31, 32] and Correlation Feature Selection (CFS) using genetic search algorithms [33, 34] are simple and widely applied methods for dimension reduction of data sets. These two procedures are compared with the proposed RG algorithm to measure its performance.

Assume, the PCA algorithm is applicable on matrix A of size $m \times n$. If $m \gg n$, the algorithm reduces m to k ($k < m$) and if $n \gg m$ then it reduces m to p where $p < n$. Otherwise, first few number of principal components are selected for reducing dimension of data set. With the increase of number of principal components the algorithm generates output, having more stable system. In the proposed RG algorithm, the number of attributes to be reduced is not supplied a priori. The algorithm only removes the attributes which are not necessary for prediction and therefore, the system becomes more stable.

The proposed method is applied on protein sequence data and compared with the three available methods, as described below.

1. Correlation Feature Selection (CFS) Subset evaluation
   (GeneticSearch) Evaluation mode: 10-fold cross-validation
2. CFS Subset evaluation (GeneticSearch)
   Evaluation mode: evaluate on all training data
3. Principal Components
   Evaluation mode: evaluate on all training data
4. Proposed method (RG)
   Evaluation mode: evaluate on all training data

Comparison results are given in Tables 2 and 3 using Bayes Net classifier.

Table 2 Comparison information of "igSeqProt" dataset using cross-validation

	CFS subset evaluation (1)	CFS subset evaluation (2)	Principal component (3)	Proposed method (4)
Correctly classified	2,846 76.075 %	2,785 74.44 %	2,708 72.38 %	2,893 77.33 %
Incorrectly classified	895 23.92 %	956 25.55 %	1,033 27.61 %	848 22.66 %
Kappa statistic	0.64	0.61	0.58	0.66
Mean absolute error	0.16	0.17	0.19	0.16

Table 3 Comparison information of "Swissport" data set using cross-validation

	CFS subset evaluation (1)	Proposed method (4)	Full data set
Attributes	96	77	400
Correctly classified	1,640 54.66 %	1,693 56.43 %	1,676 55.86 %
Incorrectly classified	1,360 45.33 %	1,307 43.56 %	1,324 44.13 %
Kappa statistic	0.32	0.35	0.33
Mean absolute error	0.30	0.29	0.29

4 Conclusion

The proposed reduct generation algorithm (RG) overcomes the limitation of PCA method applied on protein sequence data by selecting the exact number of amino acid(or codon). The RG algorithm achieves minimal reduct unlike the other RST based methods where feature selection procedure generates multiple reduct. Protein co-relation consists of redundant data which are more efficiently removed by the proposed method other than CFS using genetic search, thereby increasing prediction accuracy.

References

1. Donev, E.N., Tobias, Y.D., Donev, A.N., Tobias, R.D.: For drug discovery experiments (2010)
2. Kantardjieff, K., Rupp, B.: Structural bioinformatic approaches to the discovery of new antimyco bacterial drugs (2004)
3. Weston, J., Pérez-Cruz, F., Bousquet, O., Chapelle, O., Elisseeff, A., Schölkopf, B.: Feature selection and transduction for prediction of molecular bioactivity for drug design. Bioinformatics **19**(6), 764–771 (2003)
4. Semmes, O., Feng, Z., Adam, B., Banez, L., Bigbee, W., Campos, D., Cazares, L., Chan, D., Grizzle, W., Izbicka, E., Kagan, J., Malik, G., McLerran, D., Moul, J., Partin, A., Prasanna, P., Rosenzweig, J., Sokoll, L., Srivastava, S., Srivastava, S., Thompson, I., Welsh, M., White, N.,

Winget, M., Yasui, Y., Zhang, Z., Zhu, L.: Evaluation of serum protein profiling by surface-enhanced laser desorption/ionization time-of-flight mass spectrometry for the detection of prostate cancer: I. assessment of platform reproducibility. Clin. Chem. **51**(1), 102–112 (2005)

5. Arya, S., Mount, D.M., Netanyahu, N.S., Silverman, R., Wu, A.Y.: An optimal algorithm for approximate nearest neighbor searching in fixed dimensions. In: ACM-Siam Symposium on Discrete Algorithms, pp. 573–582. (1994)

6. Chang, Y.W.Z., Ying, Z., Zhu, L., Yang, Y.: A parsimonious threshold independent protein feature selection method through the area under receiver operating characteristic curve. Bioinformatics **23**(20), 2788–2794 (2007)

7. Roweis, S.T., Saul, L.K.: Nonlinear dimensionality reduction by locally linear embedding. Science **290**, 2323–2326 (2000)

8. John, G.H., Kohavi, R., Pfleger, K.: Irrelevant features and the subset selection problem. In: Machine learning: proceedings of the eleventh international. Morgan Kaufmann, Burlington, (1994) 121–129

9. Kohavi, R., John, G.H.: Wrappers for feature subset selection (1997)

10. Søndberg-madsen, N., Thomsen, C., Pea, J.M.: Unsupervised feature subset selection. In: In Proceedings of the Workshop on Probabilistic Graphical Models for Classification, pp. 71–82 (2003)

11. Lin, T.Y.: Rough set theory in very large databases. In: Proceedings of the IMACS Symposium on Modeling, Analysis and Simulation (CESA'96), pp. 936–941 (1996)

12. Pawlak, Z.: Rough sets: theoretical aspects of reasoning about data. Kluwer Academic Publishing, Dordrecht (1991)

13. Yao, Y.Y.: On generalizing rough set theory. In: Proceedings of 9th International Conference on Rough Sets, Fuzzy Sets, Data Mining, and Granular Computing, RSFDGrC03, pp. 44–51 (2003)

14. Lang, G., Li, Q., Guo, L.: Discernibility matrix simplification with new attribute dependency functions for incomplete information systems. Knowl. Inf. Syst. **37**(3), 611–638 (2012)

15. Yao, Y., Zhao, Y.: Discernibility matrix simplification for constructing attribute reducts. J. Am. Stat. Assoc. **179**(5), 867–882 (2009)

16. Zhao, Y., Yao, Y., Luo, F.: Data analysis based on discernibility and indiscernibility. Inf. Sci. **177**(4959–4976), 867–882 (2007)

17. Chouchoulas, A., Shen, Q.: Rough set-aided keyword reduction for text categorization. Appl. Artif. Intell. **15**(9), 843–873 (2001)

18. Jensen, R., Shen, Q.: Semantics-preserving dimensionality reduction: rough and fuzzy-rough based approaches. IEEE Trans. Knowl. Data Eng. **16**(12), 1457–1471 (2004)

19. Chiu, S.: Fuzzy model identification based on cluster estimation. J. Intell. Fuzzy Syst. **2**(3), 267–278 (1994)

20. Hore, P., Hall, L.O., Goldgof, D.B., Cheng, W.: Online fuzzy c means (2008)

21. Hall, M.A.: Correlation-based feature selection for machine learning. Technical report. University of Waikato, Hamilton (1998)

22. Hall, M.A.: Correlation-based feature selection for discrete and numeric class machine learning. In: ICML, pp. 359–366. Morgan Kaufmann, Burlington (2000)

23. Michalak, K., Kwaśnicka, H.: H.: Correlation-based feature selection strategy in classification problems. Int. J. Appl. Math. Comput. Sci. **16**, 503–511 (2006)

24. Zhang, H., Ling, C.X., Zhao, Z.: The learnability of naive bayes. In: Proceedings of Canadian Artificial Intelligence Conference, pp. 432–441. AAAI Press, California (2005)

25. Berman, H.M., Westbrook, J., Feng, Z., Gilliland, G., Bhat, T.N., Weissig, H., Shindyalov, I. N., Bourne, P.E.: The protein data bank. Nucleic Acids Res. **28**, 235–242 (2000)

26. Bhat, T.N., Bourne, P., Feng, Z., Gilliland, G., Jain, S., Ravichandran, V., Schneider, B., Schneider, K., Thanki, N., Weissig, H., Westbrook, J., Berman, H.: The pdb data uniformity project (2001)

27. Jonassen, I., Eidhammer, I.: Structure motif discovery and mining the pdb (2000)

28. Hubbard, T.J.P., Ailey, B., Brenner, S.E., Murzin, A.G., Chothia, C.: Scop, structural classification of proteins database: applications to evaluation of the effectiveness of sequence alignment methods and statistics of protein structural data (1998)

29. Watters, A.: The scop database (2000)
30. Bairoch, A., Apweiler, R.: The swiss-prot protein sequence database and its supplement tremble in 2000. Nucleic Acids Res. **27**, 49–54 (2000)
31. Jolliffe, I.: Principal component analysis. Springer Series in Statistics, New York (2002)
32. Sewell, M.: Principal component analysis (2007)
33. Frank, E., Hall, M.A., Holmes, G., Kirkby, R., Pfahringer, B.: Weka—a machine learning workbench for data mining. In: Maimon, O., Rokach, L., (eds.): The Data Mining and Knowledge Discovery Handbook, pp. 1305–1314. Springer, Berlin (2005)
34. Hall, M., Frank, E., Holmes, G., Pfahringer, B., Reutemann, P., Witten, I.H.: The weka data mining software: an update. SIGKDD Explorations **11**(1), 10–18 (2009)

Parameter Optimization in Genetic Algorithm and Its Impact on Scheduling Solutions

T. Amudha and B.L. Shivakumar

Abstract Parameter optimization is an ever fresh and less explored research area, which has ample scope for research investigation and to propose novel findings and interpretations. Identification of good parameter values is a highly challenging task which involves tedious and ad hoc course of actions with several heuristic choices. The complexity involved in parameter tweaking is primarily due to the unpredictable and heavily randomized nature of evolutionary algorithmic procedures. In this paper, an attempt was made to tweak the parameters and decision variables of Genetic Algorithm. GA with tweaked parameters was hybridized with Bacterial Foraging Algorithm, and applied to the Job shop and Permutation Flow Shop scheduling problem benchmarks. The results have proven that optimized parameter set tuning has obtained better scheduling performance.

Keywords Bacterial foraging · Genetic algorithm · Parameter optimization · Job shop scheduling

1 Introduction

Finding the right values of algorithm parameters for a given problem class is much significant, because the choice of exact parameters has a good scope to improve the performance of the chosen algorithm to a greater extent [1]. In evolutionary

T. Amudha (✉)
Department of Computer Applications, Bharathiar University, Coimbatore, India
e-mail: amudha.swamynathan@gmail.com

B.L. Shivakumar
Sri Ramakrishna Engineering College, Coimbatore, India
e-mail: blshiva@yahoo.com

© Springer India 2015
L.C. Jain et al. (eds.), *Computational Intelligence in Data Mining - Volume 1*,
Smart Innovation, Systems and Technologies 31, DOI 10.1007/978-81-322-2205-7_44

469

optimization techniques, the population size was found to have a control over representation of the search space; larger population size leads to better representation. In many cases, tweaking one particular parameter may indirectly have an influence on some other parameters. Tweaking the parameter values may be done either by simple hand tuning method in which, parameter values may be experimented on a trial basis based on past experiments or by tuning procedurally using some optimization algorithms. This concept was proven in several implementations of evolutionary algorithms that obtained maximum adaptability [2, 3]. To carry out parameter tuning in a systematic fashion, no standard procedure has been established so far to explore the entire range of combinatorial problems parameter setting [4]. Also, very few studies are available on design aspects such as confidence intervals for good parameter values and sensitivity analysis for parameter robustness.

In this paper, a hybrid optimization technique (GBSA) was proposed, which used Bacterial swarming and foraging algorithm for solution construction and Genetic Algorithm for solution improvement in solving complex benchmark instances of Job Shop and Flow Shop Scheduling. In any optimization metaheuristic, an optimal balance of diversification and intensification is required, and to maintain such a balance itself turns into an optimization procedure. Fine-tuning of parameters is very much required to improve the efficiency of the algorithms for solving a particular class of problems [2]. In GBSA, the parameters and decision variables of GA were fine-tuned and applied to the same instances of JSP and PFSSP. Experimentation of the proposed technique with optimized parameter set has shown a noticeable improvement in optimal solutions with scheduling problem classes.

2 Parameter Optimization in GA

The performance of Genetic Algorithm is greatly dependent on the selection of parameters [5]. Selection of individuals is the first and foremost phase of Genetic Algorithm. In this research, three types of selection mechanisms were used. They are Tournament Selection, Roulette Wheel Selection and Elitism. Several crossover operators have been proposed by GA researchers for the scheduling problems [6]. Four different crossover operators were used for parameter tuning and testing. They are Position Based Crossover, Partially Mapped Crossover, Linear Order Crossover and Order Based Crossover, detailed in the following sub section. Mutation helps for diversification which plays a key role to avoid fast convergence at local optima [4]. Three different mutation operators used for parameter tuning and testing are Bit Inversion Mutation, Boundary Mutation, Uniform Mutation and Non-Uniform Mutation.

Table 1 GA parameters and variants opted for tuning

Parameters	Variants
Population size	10–100
No. of generations	100–1,000
Crossover rate	0.2–0.9
Mutation rate	0.02–0.1
Selection strategy	(a) Tournament
	(b) Roulette wheel
	(c) Elitism
Crossover strategy	(a) Position based crossover—PBX
	(b) Partially mapped crossover—PMX
	(c) Linear order crossover—LOX
	(d) Order based crossover—OBX
Mutation strategy	(a) Single point mutation—SPM
	(b) Bit inversion mutation—BIM
	(c) Boundary mutation—BoM
	(d) Uniform mutation—UM

Seven parameters selected for tuning, with a possible set of variants were Population size, No. of Generations, Selection Strategy, Crossover rate, Crossover strategy, Mutation Rate and Mutation Strategy. The parameters of Genetic Algorithm considered for tuning and the variants used for implementation purpose are shown in Table 1.

3 Proposed Hybrid Technique—GBSA for Scheduling

Genetic algorithm has a wide range of practicality and it can handle any form of objective function and constraints, whether it is linear or non-linear, continuous or discrete. GA is a powerful method to quickly create high quality solutions to a problem. On the other hand, genetic algorithm has certain drawbacks such as premature convergence and insufficient local optimization ability and hence at all times, GA is not the most preferred choice for real time applications [7]. As long as the population remains distributed over the search space, GA adapts itself to changes in the objective function by reallocating the potential search efforts toward the highly favorable regions of the search space.

Bacterial swarming is a relatively new member of the metaheuristic family, primarily focused on optimization [8]. The foraging process of E. coli bacterium consists of a series moves towards food sources. The control system is in charge of evaluating changes from one state to the other states to provide reference information for E. coli bacterium's next state change. E. coli bacteria gradually approach their food sources under the influence of its control system. In general, if the bacteria are trapped into a region in deficiency of food, they might draw a conclusion based on past experience that other regions must be in abundance of food. Due to this conclusion, bacteria would change their states. Hence, each decision of state change is made under the physiological and environmental constraints with the final aim to maximize the obtained energy in unit time.

In our proposed GBSA, the bacterial swarming technique is used as the constructive heuristics for generation of the initial population and specific recombination methodology is devised as per the constraints of the scheduling problem. Then GA is employed as the improvement heuristic to be applied to the population and swarming surface built by the BSA. BSA identifies a potential swarming surface and identifies an initial locally optimal solution. After the creation of the initial population, the standard procedures of GA, selection, crossover and mutation take place. The significant aspect of GBSA is that it manages with the optimized solutions as inputs to the selection phase. It was of general opinion from the researchers that an improvement choice over the crossed over solutions will be effective to achieve competitive performance. The mutation is then performed on the locally optimized offspring. The solution obtained after mutation is again transformed into an optimized solution to keep the local optimality of the population. At this stage, mutation takes responsibility to minimize the possibility of return into previous local optima. The population replacement and swarming over a new and diverse region in the search space are specific to the bacterial strategy.

---

procedure GBSAMetaheuristic ( )

1 **for** Elimination-dispersal loop **do**
2 **for** Reproduction loop **do**
3 **for** Chemotaxis loop **do**
4 **for** Bacterium i **do**
5   Tumble: Generate a random vector $\varphi \in R^D$ in rand direction
6   Compute new move strategy and Move in the rand direction
7   Compute addl cost function for swarming
8   Initialize Swimlength
9   **while** Swimlength< No of steps**do**
10   perform a swim
11   **if** the new direction is promising **then**
12   **do** compute swarming positions & initialize optimized population
13   **for** No. of decision variables **do**
14   Evaluate the individuals
15   Perform Selection
16   Perform Crossover based on probability condition
17   Perform a swim or a tumble of bacteria over the
                                    resultant of crossover
18   Perform Mutation, if required based on probability condition
19   Update Swarming surface based on new search space
20   Compute the next bacteria move
21   **else**
22   Let swimlength = no. of steps
23   **end**
24   **end**
25   **end**
26 **end**
27   Sort bacteria as per ascending cost
28   Split the group, the bacteria with highest J value die &
                          other bacteria with the best value split
29   Update value of objective function and swarm nutrient cost accordingly.
30 **end**
31 Eliminate and disperse the bacteria to random locations with probability p_{ed}.
32 Update corresponding objective function and swarm nutrient locations
33 **End**

Bacterial swarming strategy guarantees a satisfactory degree of the diversity of the population, thereby trying to keep away from the drawbacks of GA such as premature convergence. The hybridization of BSA and GA has shown a remarkable improvement over the solution generation. This is mainly due to the fact that the population of local optima is maintained. Search space diversification, solution reconstruction and the periodic local improvement are the primary phases of this hybrid technique.

4 Experimental Setup and Results

This research was aimed at producing best possible solutions for scheduling problem benchmarks, by applying nature-inspired metaheuristic techniques with fine-tuned parameters of GA. The benchmark problem instances for Job Shop Scheduling Problems are taken from the OR-Library [9, 10]. The problems in OR-Library are highly challenging and are also considered as the most significant, classical and complex standard to test various algorithms and to prove the research efficiency. The benchmark problem instances for Permutation Flow Shop Scheduling Problems generated by Taillard [11] are used. Taillard has generated problem instances with 20 jobs, 50 jobs, 100 jobs, 200 jobs and 500 jobs. Problem instances ranging from 20 to 100 jobs are used in this work. A total of 63 JSSP benchmarks

Table 2 Fine-tuned GA parameter set for JSSP instances

JSSP instance size	No. of problems	Best parameter set identified				
		Crossover rate	Mutation rate	Selection strategy	Crossover strategy	Mutation strategy
10 × 5	5	0.2	0.04	Tournament	PMX	BIM
10 × 10	18	0.2	0.05			
15 × 5	5	0.2	0.05		LOX	
15 × 10	3	0.5	0.04	Elitism		
15 × 15	2	0.5	0.03			
20 × 5	6	0.4	0.04		PMX	
20 × 10	7	0.3	0.03			
20 × 15	8	0.3	0.03			
20 × 20	4	0.3	0.03			
50 × 10	5	0.3	0.04			

Table 3 Relative gap analysis of GBSA and fine-tuned GBSA for JSSP instances

JSSP instance size	GBSA gap (%)			Fine-tuned GBSA Gap (%)		
	Min	Max	Mean	Min	Max	Mean
10 × 5	0	0	0	0	0	0
10 × 10	0	9.7	0.3	0	4.2	0.1
15 × 5	8.5	15.1	2.5	5.3	10.8	2.5
15 × 10	10.3	12.4	3.8	5.9	7.23	2.8
15 × 15	10.3	11.7	5.5	6.1	5.4	1.4
20 × 5	6.4	18.1	1.8	3.3	10.6	2.8
20 × 10	7.4	10.9	1.3	4.3	4.7	2.3
20 × 15	3.1	10.9	1.0	1.0	4.6	0.2
20 × 20	12.0	25.1	4.5	7.2	10.8	0.5
50 × 10	3.9	7.1	1.1	1.1	3.2	0.7

and 90 PFSSP benchmarks are considered for testing the fine-tuned parameter set. The experiment was conducted with possible combinations of parameter variants.

Table 2 shows the fine-tuned parameter set obtained for solution improvement in JSSP benchmark. The relative gap analysis and comparison of GBSA and fine-tuned GBSA for JSSP instances are given in Table 3. The fine-tuned parameter set obtained for solution improvement in PFSSP benchmark is given in Table 4. The relative gap analysis and comparison of GBSA and fine-tuned GBSA for PFSSP instances are given in Table 5.

Tournament and Elitism were identified as the best strategy for Selection both in JSSP and PFSSP, as per our results for the tested problems. PMX and LOX were found to be the best Crossover strategy for JSSP whereas PMX performed well for all chosen PFSSP instances. In case of Mutation, BIM strategy gave the best performance in JSSP and PFSSP. The Population size and the No. of Generations were altered as per the size of the problem instance selected and hence, it was not

Table 4 Fine-tuned GA Parameter set for PFSSP instances

PFSSP instance size	No. of problems	Best parameter set identified				
		Crossover rate	Mutation rate	Selection strategy	Crossover strategy	Mutation strategy
20 × 5	10	0.3	0.03	Tournament	PMX	BIM
20 × 10	10	0.3	0.03			
20 × 20	10	0.3	0.03	Elitism		
50 × 5	10	0.2	0.04			
50 × 10	10	0.2	0.03			
50 × 20	10	0.2	0.04			
100 × 5	10	0.3	0.02			
100 × 10	10	0.2	0.02			
100 × 20	10	0.2	0.02			

Table 5 Relative gap analysis of GBSA and fine-tuned GBSA for PFSSP instances

PFSSP instance size	GBSA gap (%) (LB)			Fine-tuned GBSA gap (%) (LB)		
	Min	Max	Mean	Min	Max	Mean
20 × 5	0	1.0	0.2	0	0.4	0.1
20 × 10	0	1.2	0.3	0	0.3	0
20 × 20	0.1	9.4	3.2	0	3.2	0.6
50 × 5	0	0.6	0.2	0	0.2	0
50 × 10	0.3	3.8	1.9	0	1.2	0.3
50 × 20	4.9	10.9	7.8	2.3	3.5	0.7
100 × 5	0.6	1.2	1.0	0	0.4	0.1
100 × 10	0	0	0	0	0	0
100 × 20	2.3	6.4	3.7	0.3	2.1	0.3

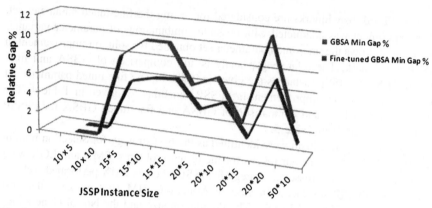

Fig. 1 Relative MinGap comparison of fine-tuned GBSA parameter Set—JSSP

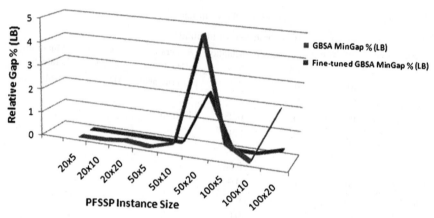

Fig. 2 Relative MinGap comparison of fine-tuned GBSA parameter Set—PFSSP

possible to exactly point out the values for these two parameters. Comparison of relative minimum gap values of JSSP is depicted in Fig. 1. Comparison of relative minimum gap values for Lower Bound of PFSSP is depicted in Fig. 2.

5 Conclusion

The efficiency of GA is highly motivated upon the selection of control parameters. In this research work, an attempt was made to tweak the parameters and decision variables of Genetic Algorithm, which was used in GBSA, and applied to the JSSP and PFSSP scheduling problem benchmarks. A total of 63 JSSP benchmarks and 90

PFSSP benchmarks are considered for testing the fine-tuned parameter set. The experiment was conducted with possible combinations of parameter variants. Results were consolidated exclusively for each different sized instance groups. Genetic algorithm has shown excellent performance in hybridization with tweaked parameter sets in case of Job shop as well as Permutation Flow shop problem instances. On an average, there was an improvement of 50–70 % in GBSA with fine-tuned parameter set when compared to the default parameter values applied earlier. It could be well inferred that parameter optimization in GA has a positive impact on scheduling effectiveness.

References

1. Nannen, V., Eiben, A.E.: Relevance estimation and value calibration of evolutionary algorithm parameters. In: International Joint Conference on Artificial Intelligence (IJCAI), pp. 975–980 (2007)
2. Eiben, E., Hinterding, R., Michalewicz, Z.: Parameter control in evolutionary algorithms. IEEE Trans. Evol. Comput. 3(2), 124–141 (1999)
3. Nannen, V., Eiben, A.E.: A method for parameter calibration and relevance estimation in evolutionary algorithms. In: Keijzer, M. et al. (eds.) GECCO 2006: Genetic and Evolutionary Computation Conference, pp. 183–190. ACM, New York (2006)
4. Reis, C., Paiva, L., Moutinho, J., Marques, V.M.: Genetic algorithms and sensitivity analysis in production planning optimization. New aspects of applied informatics, biomedical electronics and informatics and communications (2010)
5. Grefenstette, J.: Optimization of control parameters for genetic algorithms. IEEE Trans. Syst. Man Cybern. 16(1), 122–128 (1986)
6. Kahraman, C., Engin, O., Kaya, I., Yilmaz, M.K.: An application of effective genetic algorithms for solving hybrid flow shop scheduling problems. Int. J. Comput. Intell. Syst. 1(2), 134–147 (2008)
7. Kim, D.H., Abraham, A., Cho, J.H.: A hybrid genetic algorithm and bacterial foraging approach for global optimization. Inf. Sci. 177(18), 3918–3937 (2007)
8. Passino, K.M.: Biomimicry of bacterial foraging for distributed optimization and control. IEEE Control Syst. Mag. 22(3), 52–67 (2002)
9. Beasley, J.E.: Online document cited at http://people.brunel.ac.uk/∼mastjjb/jeb/info.html (2010)
10. Beasley, J.E., Online document cited at http://vlsicad.eecs.umich.edu/BK/Slots/cache/mscmga.ms.ic.ac.uk/jeb/or/contents.html (2010)
11. Taillard, E.: Benchmarks for basic scheduling problems (1989)
12. Gen, M., Cheng, R.: Genetic Algorithms and Engineering Optimization. Wiley, New York (2000)

Comparison of Performance of Different Functions in Functional Link Artificial Neural Network: A Case Study on Stock Index Forecasting

S.C. Nayak, B.B. Misra and H.S. Behera

Abstract The rapid growth of world economy and globalization has been attracting researchers to develop intelligent forecasting models for stock market prediction. In order to forecasting the stock market trend efficiently, this paper developed four single layer low complex forecasting models known as functional link artificial neural network (FLANN). Different basis functions such as Trigonometric, Chebysheb, Legendre and Lagurre polynomials are used for functional expansion of input signals to achieve higher dimensionality. The models are termed as TFLANN, CFLANN, LeFLANN and LFLANN respectively. The weight and bias vectors are optimized by genetic algorithm (GA). The number of functional expansion for each models are optimized by GA during the training process instead of fixing it earlier, which is the novelty of this research work. The models are employed to forecast the one-day-ahead prediction of three fast growing global stock markets. Different types of FLANN are considered and their comparative performance is investigated.

Keywords Stock market forecasting · Functional link artificial neural network · Genetic algorithm · Chebysheb polynomial · Legendre polynomial

S.C. Nayak (✉) · H.S. Behera
Veer Surendra Sai University of Technology, Burla, India
e-mail: sarat_silicon@yahoo.co.in

H.S. Behera
e-mail: hsbehera_india@yahoo.com

B.B. Misra
Silicon Institute of Technology, Bhubaneswar, India
e-mail: misrabijan@gmail.com

© Springer India 2015
L.C. Jain et al. (eds.), *Computational Intelligence in Data Mining - Volume 1*,
Smart Innovation, Systems and Technologies 31, DOI 10.1007/978-81-322-2205-7_45

1 Introduction

The stock market is very complex and dynamic by nature, and has been a subject of study for modeling its characteristics by various researchers. Stock index forecasting is challenging due to the nonlinear and non-stationary characteristics of the stock market. Hence, an accurate forecasting is both necessary and beneficial for all investors in the market including investment institutions as well as small individual investors. Hence there is a need to developing an automated forecasting model which can accurately estimate the risk level and the profit gained in return.

Traditionally statistical models can be applied on stationary data sets and can't be automated easily. At every stage it requires expert interpretation and development. They cannot be employed to mapping the nonlinearity and chaotic behavior of stock market. The most used statistical method is autoregressive moving average (ARMA) and autoregressive integrated moving average (ARIMA).

During last two decades there are tremendous development in the area of soft computing which includes artificial neural network (ANN), evolutionary algorithms, and fuzzy systems. This improvement in computational intelligence capabilities has enhances the modeling of complex, dynamic and multivariate nonlinear systems. These soft computing methodologies has been applied successfully to the area data classification, financial forecasting, credit scoring, portfolio management, risk level evaluation etc. and found to be producing better performance. The advantage of ANN applied to the area of stock market forecasting is that it incorporates prior knowledge in ANN to improve the prediction accuracy. It also allows the adaptive adjustment to the model and nonlinear description of the problems. ANNs are found to be good universal approximator which can approximate any continuous function to desired accuracy.

It has been found in most of the research work in financial forecasting area used ANN, particularly multilayer perceptron (MLP). The MLP contains one or more hidden layers, and each layer can contain more than one neuron. The input pattern is applied to the input layer of the network and its effect propagates through the network layer by layer. During the forward phase, the synaptic weights of the networks are fixed. In the backward phase, the weights are adjusted in accordance with the error correction rule. This algorithm is a popular one and called as Back Propagation learning algorithm. Suffering from slow convergence, sticking to local minima are the two well known lacuna of a MLP. In order to overcome the local minima, more number of nodes added to the hidden layers. Multiple hidden layers and more number of neurons in each layer also add more computational complexity to the network.

The Neuro-Genetic hybrid networks gain wide application in nonlinear forecasting due to its broad adaptive and learning ability [1]. At present, the most widely used neural network is back propagation neural network (BPNN), but it has many shortcomings such as the slow learning rate, large computation time, gets stuck to local minimum. RBF neural networks is also a very popular method to predict stock market, this network has better calculation and spreading abilities, it also has stronger nonlinear mapped ability [2]. The new hybrid iterative evolutionary learning algorithm is more effective than the conventional algorithm in terms of learning accuracy and prediction accuracy [3]. Many researchers have adopted a neural network model, which is trained by GA [4, 5]. The Neural Network-GA model to forecasting the index value has got wide acceptance.

FLANN was originally proposed by Pao [6]. It is a class of Higher Order Neural Networks (HONN) that utilizes higher combination of inputs [6, 7]. The properties of expanding the input space into a higher dimensional space huge number of high-order neural network architecture without hidden units were introduced. The performance of FLANN models have been experimented during several research works and found to be an effective approach both computationally as well as performance wise [8, 9]. The key factors which motivate us for research in this area are:

- Developing a neural forecasting model with less complex structural architecture.
- Improving the prediction accuracy by training the model with evolutionary optimization algorithm such as GA.
- Employing different basis functional expansions as compared to other research work, where using a single basis function for expansion of input signals.
- Using optimal number of functional expansions of input signals.

In this paper we have employed various FLANN models to forecast the next day's closing price in the stock market. Different function expansion has been considered for the expansion of input closing price. The variations of FLANN models are trained with genetic algorithm.

The rest of the paper is organized as follows. The stock market prediction problem and related research works has been explained in Sect. 1. Different functional expansion methods and development of FLANN based forecasting models have been discussed in Sect. 2. Section 3 explains about experimental settings and analyses the results obtained. The concluding remark has been given by Sect. 4 followed by a list of references.

2 Model Architecture

The FLANN is basically a single layer network, in which the need of hidden layers has been removed by incorporating functional expansion of the input pattern. In FLANN, each input to the network (the daily closing prices value in this experiment) undergoes functional expansion through a set of basis functions. The functional link acts on an element by generating a set of linearly independent functions. The input, expanded by a set of linearly independent functions in the functional expansion block causes an increase in the input vector dimensionality, which enables the FLANN to solve complex classification problems by generating nonlinear decision boundaries. The functional expansion effectively increases the dimensionality of the input vector, and hence the hyper planes generated by the FLANN provide greater discrimination capability in the input pattern space [7].

There have been several applications of FLANN including pattern classification and recognition, system identification and control, functional approximation, and digital communications channel equalization. In the research work [10], the author have been successively applied FLANN for the channel equalization problem. The FLANN also has been applied to the financial forecasting and proved to be computationally efficient [11].

In the functional expansion process each input closing price to the forecasting model is nonlinearly expanded to generate several input values. The functional expansion component will generate a set of linearly independent functions with each closing value in the input vector taken as argument of the function. This nonlinearity to the input elements reduces the number of layers and hence reduces the computational complexity.

2.1 Trigonometric Based FLANN (TFLANN)

The simple trigonometric basis functions of sine and cosine are used here to expand the original input value into higher dimensions. An input value x_i expanded to several terms by using the trigonometric expansion functions as in Eq. 1.

$$c_1(x_i) = (x_i), \quad c_2(x_i) = \sin(x_i), \quad c_3(x_i) = \cos(x_i), \quad c_4(x_i) = \sin(\pi x_i),$$
$$c_5(x_i) = \cos(\pi x_i), \quad c_6(x_i) = \sin(2\pi x_i), \quad c_7(x_i) = \cos(2\pi x_i). \tag{1}$$

2.2 Chebysheb Polynomial Based FLANN (CFLANN)

The basis function used here is named as Chebysheb polynomial function hence the model. Here an input value i.e. closing price x_i is expanded as in Eq. 2.

$$c_1(x_i) = (x_i), \quad c_2(x_i) = 2x_i^2 - 1, \quad c_3(x_i) = 4x_i^3 - 3x_i,$$
$$c_4(x_i) = 8x_i^4 - 8x_i^2 + 8x_i + 1 \tag{2}$$

2.3 Lagurre Polynomial Based FLANN (LFLANN)

The Lagurre polynomial basis function is used here to enhance the input pattern and the neural architecture. An input closing price value x_i to the model is expanded into several terms as in Eq. 3.

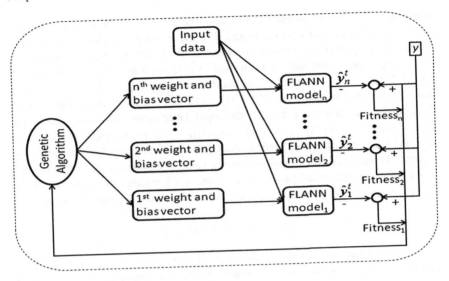

Fig. 1 FLANN model trained with genetic algorithm

$$c_1(x_1) = x_i, \quad c_2(x_i) = -x + 1, \quad c_3(x_i) = \frac{x^2}{2} - x + 1 \tag{3}$$

2.4 Legendre Polynomial Based FLANN (LeFLANN)

The basis function used here is Legendre polynomial function. An input value x_i is expanded as in Eq. 4.

$$c_1(x_i) = (x_i), \quad c_2(x_i) = (3x^2 - 1)/2, \quad c_3(x_i) = (5x^3 - 3x)/2,$$
$$c_4 = (35x^4 - 30x^2 + 3)/8 \tag{4}$$

The GA based FLANN model developed for this experimental has the general architecture as shown in Fig. 1.

3 Experimental Setting and Result Analysis

This section presents the forecasting results obtained by employing the above models. The closing prices are collected for each transaction day of the stock exchange for the year 2012. In order to normalize these data, several data normalization methods has been considered and out of these, the sigmoid normalization method has found to be superior [12]. The advantage of this normalization method is that it does not depend on the distribution of data, which may be unknown at the time of training the model. The normalized values are now considered as the input vector to the model. For the training purpose the normalized values are arranged in a two dimensional matrix of breadth d, termed as bed length. The sliding window method has been used to feed the input pattern to the model. Let $X(t) = (x_1, x_2, \ldots, x_d)$ be the normalized closing prices. This is first goes through functional expansion and let $X'(t) = f(x_1, x_2, \ldots, x_d)$ be the expand values. These expanded values are feed to the input layer neuron as actual input. Let $W(t) = [w_{11}(t), \ldots, w_{1j}(t), \ldots, w_{1N}(t), \ldots w_{d1}(t), \ldots w_{dN}(t)]$ represent the elements of a weight vector. Each expanded input pattern $X'(t)$ is applied to the model sequentially and the desired financial closing value is supplied at the output neuron. Given the input, the model produces an output $y'(t)$, which acts as an estimate to the desired value. The output of the linear part of the model is computed as in Eq. 5.

$$y'(t) = X'(t)^T * W(t) + b \tag{5}$$

where b represents the weighted bias input. This output is then passed through a nonlinear function, here sigmoid activation to produce the estimated output $\hat{y}$ by Eq. 6.

$$\hat{y}(t) = \frac{1}{1 + e^{-\lambda y'(t)}} \tag{6}$$

The error signal $e(t)$ is calculated as the difference between the desired response and the estimated output of the model as described in Eq. 7.

$$e(t) = y(t) - \hat{y}(t) \tag{7}$$

The error signal $e(t)$ and the input vector are employed to the weight update algorithm to compute the optimal weight vector.

To overcome the demerits of back propagation, we employed the GA which is a popular global search optimization. We adopted the binary encoding for GA. Each weight and bias value constitute of 17 binary bits. For calculation of weighted sum at output neuron, the decimal equivalent of the binary chromosome is considered. A randomly initialized population with 70 genotypes is considered. GA was run for maximum 200 generations with the same population size. Parents are selected from the population by elitism method in which first 20 % of the mating pools are selected from the best parents and the rest are selected by binary tournament selection method. A new offspring is generated from these parents using uniform crossover followed by mutation operator. In this experiment the crossover probability is taken as 0.7 and mutation probability is taken as 0.003. In this way the new population generated replaces the current population and the process continues until convergence occurs.

The fitness of the best and average individuals in each generation increases towards a global optimum. The uniformity of the individuals increases gradually leading to convergence. The major steps of the GA based FLANN models can be summarized as follows.

GA Based Training Algorithm:

1. Setting training data, i.e. choosing number of closing prices as input vector for the network.
2. Mapping of input patterns.
   Map each pattern from the lower dimension to higher dimension, i.e. expand each feature value according to the polynomial basis functions as presented above.
3. Random initialization of search spaces, i.e. populations.

   Initialize each search space, i.e. chromosome with values from the domain [0, 1].

4. While (termination criteria not met)
   For each chromosome in the search space
   Calculate the weighted sum and feed as an input to the node of output layer.
   Present the desired output, calculate the error signal and accumulate it.
   Fitness of the chromosome is equal to the accumulated error signal.
   End
   Apply crossover operator.
   Apply mutation operator.
   Select better fit solutions.

   End

5. Present the testing input vector, immediate to the training vectors.
   Calculated the weighted sum and calculate the error value.
6. Repeat the steps 1-5 for all training and testing patterns, calculate the total error signals.
7. Calculate the average percentage of errors (APE) for the whole financial time series.

APE has been considered for performance metric in order to have a comparable measure across experiments with different stocks. The formulas are represented by Eq. 8.

$$APE = \frac{1}{N}\sum_{i=1}^{N} \frac{|x_i - \hat{x}_i|}{xi} \times 100\ \% \tag{8}$$

The prediction performance of the evolutionary FLANN models experimented on BSE and DJIA closing prices of 2012 are shown in Table 1. It can be observed that for BSE, LeFLANN model outperforms other generating minimal APE of 0.0322. For DJIA, TFLANN gives superior result of APE value 0.0188 followed by LeFLANN which gives 0.0344 APE in this case. Again for NASDAQ the LeFL-ANN model outperforms other by generating APE value of 0.0083. The average performance of LeFLANN model over three data sets is 0.0249 followed by the TFLANN having 0.0672. Though all the four models are efficiently forecasting the

Table 1 Error signals generated by the models

FLANN model	BSE	DJIA	NASDAQ
TFLANN	0.0762	0.0188	0.1066
CFLANN	0.1275	0.1088	0.1216
LFLANN	0.0355	0.0675	0.3108
LeFLANN	0.0322	0.0344	0.0083

Table 2 Ranking of the models based on performance

FLANN model	BSE	DJIA	NASDAQ	Total rank
TFLANN	3	1	2	6
CFLANN	4	4	3	11
LFLANN	2	3	4	9
LeFLANN	1	2	1	4

trend of the stock data used, the trigonometric and Legendre polynomial based FLANN models are found to be superior and prominent models over others.

The models are ranked based on their performance and are presented by Table 2. For a particular data set, a rank is assigned to a particular model on the basis of APE obtained by that model. For example, in case of BSE, LeFLANN rank is 1, LFLANN obtain rank 2 followed by TFLANN and CFLANN. The total rank is the sum of ranks assigned to a model. It can be observed that the LeFLANN model performs better (i.e. total rank = 4). The second better performance has been obtained by TFLANN followed by LFLANN and CFLANN. However the performances of all these models are found to be satisfactory.

4 Conclusion

This paper developed and evaluated the effectiveness of variations of FLANN models trained with an evolutionary algorithm, GA. Four different polynomial basis functions such as trigonometric, Chebyshev, Lagurre and Legendre polynomial have been considered for expansion of input closing prices signals. The optimal parameters of these models as well as the number of functional expansions for each input signals are selected by GA. These models have been proved their effectiveness over a set of real stock markets. Particularly, the LeFLANN has a good average performance over all data sets. These FLANN models have less complex structural architecture and can be recommended as efficient forecasting model. However, the performance of these models varies significantly on choosing the expansion functions carefully.

References

1. Kwon, Y.K., Moon, B.R.: A hybrid neuro-genetic approach for stock forecasting. IEEE Trans. Neural Networks **18**(3), 851–864 (2007)
2. Guangxu, Z.: RBF based time-series forecasting. J. Comput. Appl. **9**, 2179–2183 (2005)
3. Yu, L., Zhang, Y. Q.: Evolutionary fuzzy neural networks for hybrid financial prediction. IEEE Trans. Syst. Man Cybern.—Part C Appl. Rev. **35**(2), 244–249 (2005)
4. Nayak, S.C., Misra, B.B., Behera, H.S.: Index prediction using neuro-genetic hybrid networks: a comparative analysis of performance. International Conference on Computing Communication and Application, pp. 22–24. IEEE (2012). doi: 10.1109/ICCCA.2012.6179215
5. Nayak, S.C., Misra, B.B., Behera, H.S.: Stock index prediction with neuro-genetic hybrid techniques. Int. J. Comput. Sci. Inform. **2**, 27–34 (2012)
6. Pao, Y.H.: Adaptive pattern recognition and neural networks. Addison-Wesley, Boston (1989)
7. Pao, Y.H., Takefuji, Y.: Functional-link net computing: theory, system architecture, and functionalities. Computer **25**, 76–79 (1992)
8. Patra, J.C., Bos, A.V.D.: Modeling of an intelligent pressure sensor using functional link artificial neural networks. ISA Trans. Elsevier **39**, 15–27 (2000)
9. Majhi, R., Panda, G., Sahoo, G.: Development and performance evaluation of FLANN based model for forecasting of stock markets. Expert Syst. Appl. **36**, 6800–6808 (2009)
10. Patra, J.C., Pal, R.N., Baliarsingh, R., Panda, G.: Nonlinear channel equilization for QAM signal constellation using artificial neural network. IEEE Trans. Syst. Man Cybern. Part B **29** (2), 262–271 (1999)
11. Patra, J.C., Kim, W., Meher, P.K., Ang, E.L.: Financial Prediction of Major Indices Using Computational Efficient Artificial Neural Networks, pp. 2114–2120. IJCNN, Vancouver (2006)
12. Nayak, S.C., Misra, B.B., Behera, H.S.: Impact of data normalization on stock index forecasting. Int. J. Comp. Inf. Syst. Ind. Manag. Appl. **6**, 357–369 (2014)

Application of RREQ Packet in Modified AODV(m-AODV) in the Contest of VANET

Soumen Saha, Utpal Roy and Devadutta Sinha

Abstract In Vehicular ad hoc network (VANET) the implementation and testing of the performance of routing protocols (www.ietf.org/rfc/rfc3561.txt, www.ietf.org/rfc/rfc4728.txt [1], Boppana et al. in An adaptive distance vector routing algorithm for mobile, ad hoc networks. INFOCOM 2001. IEEE Xplore, vol. 3, pp. 1753–1762, (2001) [2]) are very costly and difficult in real life situation. With the help of Network Simulators the Real-life situation can only be mimicked. Out of many simulators as available at present the NTCUns-6 (Wang et al. in Comput Netw 42(2):175–197, 2003 [3]) is much more versatile due to its inherent features. Most of the well known routing protocols are available for simulation procedure. But, it is found that, AODV is implemented in NCTUns in such a manner that it broadcast the route request within the same network, but multicast the route request between two different networks. Hence it seems a unnecessary wastage of Route Request (RREQ) (www.ietf.org/rfc/rfc3561.txt, www.ietf.org/rfc/rfc4728.txt [1]) packet for its neighbour networks (close network). To cope the problem the ADOV routing protocol has been modified accordingly within the simulator keeping other protocols unaltered. In suitable simulation scenario AODV routing protocol broadcast RREQ packets among the nodes of different networks functioning in the different lanes of the road and we got an enchantment in throughput for our modified AODV (m-AODV).

Keywords AODV · RREQ · Broadcast · Nctuns · VANET

S. Saha (✉)
Department of CST, ICVP, Jhargram, West Bengal, India
e-mail: soumen11@rediffmail.com

U. Roy
Department of Computer and System Sciences, Siksha-Bhavana,
Visva-Bharati, Santiniketan, India
e-mail: roy.utpal@gmail.com

D. Sinha
Department of CSE, University of Calcutta, Kolkata, India
e-mail: devadatta.sinha@gmail.com

489

© Springer India 2015
L.C. Jain et al. (eds.), *Computational Intelligence in Data Mining - Volume 1,*
Smart Innovation, Systems and Technologies 31, DOI 10.1007/978-81-322-2205-7_46

1 Introduction

For VANET the routing protocols work on ad hoc basis and infrastructure basis in network (Fig. 1). Where the ad hoc network is highly unstable due to vehicle's speed and lane change factors.

1.1 Ad hoc Routing Protocols

Ad hoc routing protocol setup the path, exchange information and take decision of runtime paths [4].

The topology based routing is classified (Fig. 2) in three way

1. Proactive (table-driven) routing protocols
2. Reactive (on-demand) routing protocols
3. Hybrid routing protocols (for both type).

1.1.1 Proactive Routing

Proactive routing protocols are based on shortest path first algorithms [4]. It maintains and update routing information's on routing in between all nodes of a supplied network at all times, even if the paths are not currently being used. Even if some paths are never used but updates for those paths are constantly broadcasted among nodes [5]. Route updates are periodically performed regardless of network payload, bandwidth constraints.

1.1.2 Reactive Routing

On demand or reactive routing protocols were planned to overcome the overhead problem, which was created by proactive routing protocols. Maintaining only those routes that are currently live and active [5]. These protocols implement route

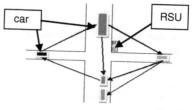

Fig. 1 Ad hoc VANET [12]

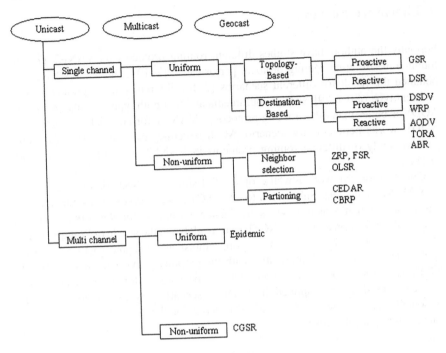

Fig. 2 Classification of Ad-hoc routing protocols [4]

determination on a demand basis or need basis and maintain only the routes that are currently in use. Therefore it reducing the burden and overhead on the network when only a subset of available routes is in active at any point of time [6].

AODV maintains and uses an efficient method of routing, which reduces network burden by broadcasting route discovery packet mechanism and by runtime updating routing information at each adjacent node. Route discovery in AODV can be perform by sending Route Request (RREQ) from a node when it needs a route to send the data to a particular destination. After sending RREQ, a node waits for the Route Reply (RREP) and if it does not receive any RREP within time threshold.

The node members of contracted ad hoc network when out of the range of the existing ad hoc network, it may fails to progress. Hence, we need some other helping equipments (road side equipment) to help those node (Vehicle) to progress. But, irrespective of that, if we taken the existing neighbor Ad hoc network, that can help to restart the communication with that isolate node (Vehicle), which is more economic, as we do not need any extra equipment or extra data communication.

2 Literature Survey

Formally the authors have studies different routing protocols on VANET [7] and found some different routing protocol performance [7–10]. The authors have also obtained the result for different scenarios [7, 8], different routing protocols [10], different type of data [9], such as multimedia and text both types of data. We found AODV is better than other routing protocol in VANET for non real time data. And AODV works better in city scenario (Mesh structure road shape) [8]. The comparative study of different routing protocols in VANET is available in the work of [11].

Out of various Network as well as of VANET simulators NS3, SUMO, OMneT++, MiXiM, NCTUns, Dia, Subversion (SVN), NCTUns-6 [12, 13] is a simulator very much popular mainly as it provides graphical environment for Network simulation. Details of application and use of NCTUns is available in the article of Wang et al. [3]. On the other hand the NCTUns supports almost all available routing protocols applicable in VANETs. Importantly, with the graphical environment supported in NCTUns various types of city scenarios can be realised for the simulation. The NCTUns simulator has applied in VANET scenarios by the authors [14] and reasonably accurate results have been observed for the Network parameters like packet drop, packet collision.

3 Proposed Work

The present is an attempt to propose an efficient algorithm, that the close neighbor network node (Vehicle) can send Route request (RREQ) and Route Reply (RREP) packet. It may be for some time more useful compared to multicast (2.0.255.255) of Internet Protocol (IP) scheme. In Fig. 3 we found left side scenario is working with only multicast running with AODV routing protocol. So, packets is only within the home network (2.0.0.0/16) i.e. left side scenario. Now let us assume there is a situation, where a node (vehicle) is out of range of a communication node of left

Fig. 3 m-AODV broadcast it's RREQ packet (simulation secnario)

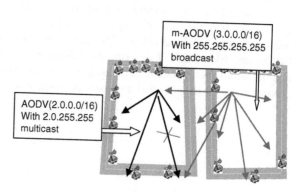

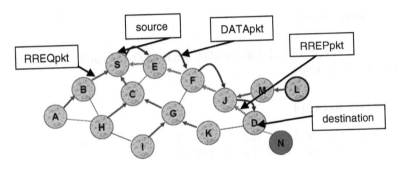

Fig. 4 AODV works [1, 15]

side (3.0.0.0/16) network. In this situation the out of range node will fail to communicate and propagate. But, according to our algorithm, right side (3.0.0.0/16) network running with modified AODV routing protocol (m-AODV) will help in this situation (Fig. 2) for the node within the range of this network, as it is designed for broadcast (255.255.255.255) IP. Hence if the node has link breakage of right side network, will also able to communicate with left side ad hoc network with the help of right side ad hoc network.

We have taken AODV (Fig. 4) protocol for ad hoc routing protocol as we found in our previous paper [7], the AODV is best performance compared to other routing (DSR, TORA, ADV) protocols. As we found in our previous paper [8], ADV and AODV is almost identical in performance, but ADV is better in séance of multimedia data only. Therefore, we try to focus the enhancement on AODV in this paper.

3.1 Proposed Algorithm

Algorithm m-AODV: [S = source node, D = destination node]
   Begin

1. Route discovery

   a. For node n! = destination or RERR packet received
      /* n is number of nodes [n = 0, 1, ... N − 1] */
   b. Send RREQ packet in broadcast to all networks with broadcast IP 255.255.255.255 starting from S node.
   c. If packet is not available from own network, receive packet from nearest neighbour.
      If packet received by destination at D, starts unicast.
      Else go to step a.
   d. On error Request Error (RERR) packet is broadcasted.

End.

3.2 Program Code

The part of modified code of Hello_packet, sendRREQ, forwardRREQ, bcastRERR packets are given below.

```
Hello_pack(){
.
p->rt_setgw(inet_addr("255.255.255.255"));
     p->pkt_setflow(PF_SEND);
.
}
int AODV::sendRREQ(u_longdst, constu_charttl) {
.
     strcpy(mypkt->pro_type, "AODV_RREQ");
     mypkt->dst_ip = inet_addr("255.255.255.255");
 // broadcast
.
}
int AODV::forwardRREQ(RREQ_msg * my_rreq,
constu_charcur_ttl) {
.
     strcpy(mypkt->pro_type,"AODV_RREQ");
     mypkt->dst_ip = inet_addr("255.255.255.255");
.
}
int AODV::bcastRERR(Unreach_list *p_ulist) {
     .
     strcpy(mypkt->pro_type,"AODV_RERR");
     mypkt->dst_ip = inet_addr("255.255.255.255");
     .
}
```

3.3 Performance Metrics

Different performance metrics are used to check the performance of routing protocols in various ad hoc network scenario. In this study we have opted throughput and broadcast of packets to check the performance of VANET routing protocols against each other. We select those performance metrics is to check the performance of routing protocols in highly mobile environment of VANET. Further, those performance metrics are used to check the effectiveness of VANET routing protocols i.e. how well the protocol deliver and receive packets and how efficiently the proposed algorithm for a routing protocol performs in order to discover the route towards destination. The selected metrics for modified AODV (m-AODV) protocols evaluation are as follows [11, 15].

3.3.1 Throughput

Throughput is the average number of successfully delivered data packets on a communication network or network node. In other words throughput describes as the total number of received packets at the destination out of total transmitted packets [11]. Throughput is calculated in bytes/sec or data packets per second. The simulation result for throughput in NCTUns6.0 shows the total received packets at destination in KB/Sec, mathematically throughput is shown as follows:

$$\text{Througput(bytes/sec)} = \frac{\text{Total number of received packets at destination} \times \text{packet size}}{\text{Total simulation time}}$$

Throughput of incoming packet is In throughput and incoming versus outgoing is IN-Out throughput. If it increases, the probability of link breakage is less.

3.3.2 Broadcast of Packets

Number of packet broadcast from the source and number of packet receive by a node. Number of packets out going versus incoming is called In-Out broadcast [14]. Lower packet drop rate shows higher protocol performance. If it incise, then probability of link breakage is less.

4 Result and Analysis

4.1 Simulator

There are different types of simulator, such as, NS2, NS3, SUMO, NCTUns, OMneT++, MiXiM, Dia, Subversion (SVN). But we opted chosen NCTUns-6.0 for simulation of our present work. It helps us to test before real-time.

The major characteristics [12] of NCTUns-6.0 are given below (Fig. 5)

- It directly uses the real-life Linux TCP/IP protocol stack to generate high-fidelity simulation results.
- It can run up any real-life UNIX-based application program on a simulated node without any modification.
- It can use any real-life UNIX network monitoring tools.
- Its setup and usage of a simulated network and application programs are exactly the same as those used in real-life IP networks.
- It simulates many important networks.
- It simulates many important protocols.
- It finishes a network simulation case quickly.
- It generates reliable and repeatable simulation results.

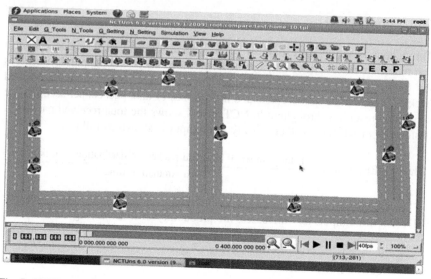

Fig. 5 NCTUns6.0 simulator

- It provides a highly-integrated and professional GUI environment.
- It adopts a module-based architecture.
- It can be easily used as an emulator.
- It supports seamless integration of emulation and simulation.

4.2 Testing Scenario Conditions of VANET

- Network is taken ad hoc and the path is absolutely dynamic in nature.
- Lane Width is taken 30 m.
- Initial average distance is ∼500 m in between different car.
- Simulation time is taken 400 s on average.
- RTS threshold is 3,000 bytes.
- The car profile is taken five.
- (20 %—speed is 20 km/h, 20 %—speed is 40 km/h, 20 %—speed is 60 km/h, 20 %—speed is 70 km/h, 20 %—speed is 90 km/h)
- Number of lane is taken 2.
- Number Network is taken 2 (one of 2.0.0.0/16 net id another is 3.0.0.0/16 network id).

The other simulation environment parameter is given in the Table 1.

The above testing scenario (Fig. 3) is planned with left side road structure with 2.0.0.0/16 sub net id and right side rode structure with 3.0.0.0/16 subnet id. And AODV run with Multicast in same network (2.0.255.255). But, m-AODV run with broadcasting (255.255.255.255) to left and right side both.

Table 1 Simulations environment parameter for VANET

Parameter	Value
Frequency (MHz)	2,400
Fading Var	10.0
Ricean K	10.0
Tx antenna height (m)	1.5
System loss	1.0
Trans power (dbm)	3.0
Average building height (m)	10
Street width (m)	30
Average building distance (m)	90
Path loss exponent	2.0
Shadowing standard deviation	4.0
Close In distance (m)	1.0
Rx antenna height (m)	1.5
Number of cars is taken for each network	5,10,15

4.3 Result

4.3.1 Car 5 of Each Network

1. In-Out broadcast of packets

   The In-Out packet broadcast output number is drastically incises in the m-AODV approach (Fig. 6). As the traffic is less and the left side (2.0.0.0/16) nodes absorbs the some right side (3.0.0.0/16) packet and this network packets is not confined within the network.

   Therefore we found the In packet for node is incises compared to original AODV.

Fig. 6 AODV versus m-AODV IP broadcast In-Out graph

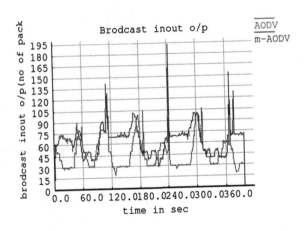

2. In throughput of Packets to nodes

   In throughput of packet also much improved in m-AODV scheme (Fig. 7), as it
   absorbs more compared to multicast.

3. In-Out packet throughput to nodes

   Naturally the below Fig. 8 indicates the in-out throughput is much more for right
   side network (3.0.0.0/16) as it receive the more numbers of packet from it's own
   network. The neighbour network (2.0.0.0/16) will not able to send any packet to
   the right side network as it is different network.

Therefore our proposal (m-AODV) is better in terms of In-out throughput of
packet.

Next we try to simulate further on more dance traffic situations.

Fig. 7 In throughput of
number of packets to nodes
AODV versus m-AODV

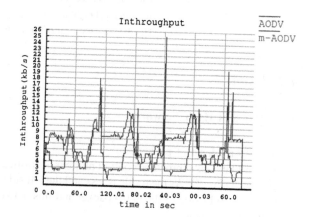

Fig. 8 In-Out throughput of
AODV versus m-AODV

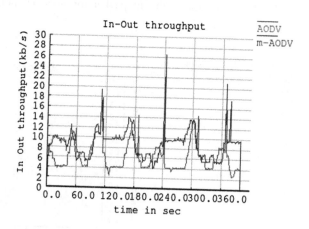

4.3.2 Car 10 of Each Network

1. In-Out broadcast of packets
2. In throughput of Packets to nodes
3. In-Out packet throughput to nodes

Hear we found the number of packet broadcast is again same for both (Fig. 9). But the In throughput (Fig. 10) and In-out throughput (Fig. 11) incising as the traffic incises for AODV compared to m-AODV. As, traffic incises, the LHS side traffic need less reuse of neighbour (RHS) side packet for maintaining Ad hoc network.

4.3.3 Car 15 of Each Network

1. In-Out broadcast of packets
2. In throughput of Packets to nodes
3. In-Out packet throughput to nodes

Fig. 9 AODV versus m-AODV IP broadcast In-Out graph

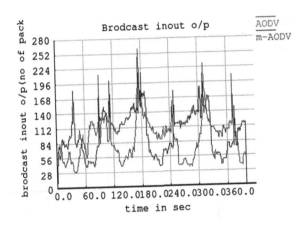

Fig. 10 In throughput of number of packets to nodes AODV versus m-AODV

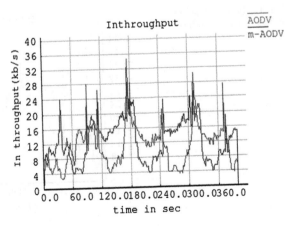

On third testing result set (Figs. 12, 13 and 14) indicates more clear enhancements of result compared to second test (10 car each side scenario) and first test set (5 car each side scenario).

Fig. 11 In-Out throughput of AODV versus m-AODV

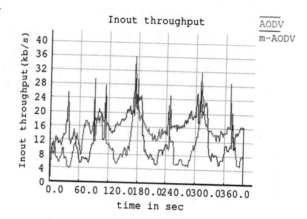

Fig. 12 AODV versus m-AODV IP broadcast In-Out graph

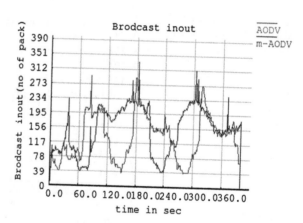

Fig. 13 In throughput of AODV versus m-AODV

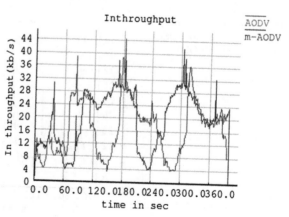

Fig. 14 In-Out throughput of
AODV versus m-AODV

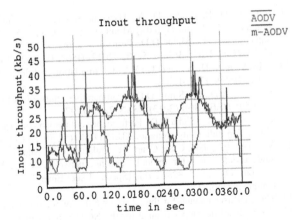

Finally we have found some very prospective result in that our modified AODV (m-AODV) working principal. It is far better in the link breakage situation for high density traffic situation, especially in city scenario (mesh structure). The performance with respect to Out throughput and In-out throughput situation is far better than conventional AODV working principal in NCTUns simulator.

5 Conclusion and Feature Work

We have found a very fruitful result through Figs. 6, 7, 8, 9, 10, 11, 12, 13 and 14 that, our modified AODV (m-AODV) working principal is far better in the link breakage and high traffic density situation. The performance with respect to broadcast, In-throughput and In-Out throughput, all the situation is better compared to conventional AODV working principal in NCTUns simulator.

The only drawback of this proposal is, it may increase collision, if traffic density is more. But, the number of packet as increase (Figs. 7, 8, 10, 11, 13 and 14), the performance is progressive in our proposal.

References

1. www.ietf.org/rfc/rfc3561.txt; www.ietf.org/rfc/rfc4728.txt
2. Boppana, R.V. et al.: An adaptive distance vector routing algorithm for mobile, ad hoc networks. INFOCOM 2001. IEEE Xplore, vol. 3, 1753–1762 (2001)
3. Wang, S.Y. et al.: The design and implementation of the NCTUns 1.0 network simulator. Comput. Netw. 42(2), 175–197 (2003)
4. Kuosmanen, P.: Classification of ad hoc routing protocols. Finnish Defence Forces, Naval Academy, Helsinki
5. Larsson, T., Hedman, N.: Routing protocols in wireless Ad Hoc networks—a simulation study. Department of Computer Science and Electrical Engineering, Luleå University of Technology, Stockholm (1998)

6. Johnson, D. B., Maltz, D. A.: Dynamic source routing in ad hoc wireless networks. In: Imielinski, T., Korth, H. (eds.) Mobile computing, pp. 153–181. Kluwer, Norwell (1996)
7. Saha, S., Roy, U., Sinha, D.D.: Performance analysis of VANET scenario in Ad-hoc network by NCTUns. IJICT 3(7), 575–581 (2013) (ISSN 0974–2239)
8. Saha, S., Roy, U., Sinha D.D.: VANET simulation in different Indian city scenario. IJEEE 3(9) (2013) (ISSN 2231–1297)
9. Saha, S., Roy, U., Sinha D.D.: Comparative study of Ad-Hoc Protocols in MANET and VANET. IJEEE 3(9) (2013) (ISSN 2231–1297)
10. Saha, S., Roy, U., Sinha D.D.: Performance comparison of various Ad-Hoc routing protocols of VANET in Indian city scenario. AIJRSTEM 1(5), 49–54 (2014) ([ISSN (Online):2328–3580])
11. Martinez, F.J., Toh, C.K., Cano, J.C, Calafate, C.T., Manzoni, P.: A survey and comparative study of simulators for vehicular and hoc networks (VANETs). Wairless Commun. Mobile Comput. 11(7), 813–828 (2011) (Special Issue: Emerging Techniques for Wireless Vehicular Communications)
12. The GUI user manual for the NCTUns 6.0 network simulator and emulator
13. The protocol developer manual for the NCTUns6.0 network simulator and emulator
14. Saha, S., Roy, U., Sinha D.D.: AODV routing protocol modification with broadcasting RREQ packet in VANET. IJETAE 4(8), 439–444 (2014)
15. Kawashima, H.: Japanese perspective of driver information systems. Transportation 17(3), 263–284 (1990)

Characterization of Transaction-Safe Cluster Allocation Strategies of TexFAT File System for Embedded Storage Devices

Keshava Munegowda, G.T. Raju, Veera Manikdandan Raju
and T.N. Manjunath

Abstract The Extended File Allocation Table (ExFAT) file system is optimized to use with SSD (Solid State Drives). The ExFAT file system supports higher storage size compared to conventional File Allocation Table (FAT) file system. The Transaction safe Extended FAT file system (TexFAT) is a variant of ExFAT file system with power fail safe feature. The TexFAT file system is available in Windows CE (Compact Embedded) version 6.0 and higher version Operating Systems (OS). This paper adopts the reverse engineering methodology to explore cluster allocation algorithms of TexFAT file system by conducting various combinations of file system operations in Windows CE 6.0 OS. This paper also records the performance benchmarking TexFAT and ExFAT file systems.

Keywords Cluster · ExFAT · FAT · Flash memory · FUSE · Heap · MMC · SD · SSD · TexFAT · TFAT · USB · Windows CE

1 Introduction

The ExFAT [1, 2] file system is the successor of FAT [3] file system. The ExFAT file system is optimized for flash memories and it supports the higher storage size. The SD card Specification [4] classifies the FAT and ExFAT as the standard file

K. Munegowda (✉)
EMC Corporation, Bagamane World Technology Center-SEZ,
Mahadevapura, K R Puram, Bengaluru 560048, India
e-mail: keshava.gowda@gmail.com; keshava.munegowda@emc.com

G.T. Raju
R N S Institute of Technology, Bengaluru, India

V.M. Raju
Texas Instruments, Bangalore, India

T.N. Manjunath
Acharya Institute of Technology, Bengaluru, India

© Springer India 2015
L.C. Jain et al. (eds.), *Computational Intelligence in Data Mining - Volume 1*,
Smart Innovation, Systems and Technologies 31, DOI 10.1007/978-81-322-2205-7_47

systems to be used with SD cards. The ExFAT file system is Multi OS supported and hence it is preferred file system to be used with embedded storage devices such as Multimedia card (MMC) [5], SD card and Universal Serial Bus (USB) flash drives. In ExFAT file system, a file/directory is a linked list of the data clusters. A cluster is a basic allocation unit for file/directory. A group of sectors is called as a "cluster". The default size of sector is 512 bytes. The ExFAT file system by default has only one FAT. The File Allocation Table contains the linked list of data clusters of Files/directories. The Cluster Heap [2] is a Bitmap of all data clusters of File system. The cluster heap is used to optimize the free cluster search while allocating to a file/directory and also to optimize write and delete operations of files with clusters allocated in contiguous numbers. In FAT and ExFAT file systems, if there is a sudden power loss during file system update operations such as file/directory creation, write, update, and deletion operations then user loses the existing file/directory data and further it may lead to a state in which existing data of file system is unreliable. Hence, both FAT and ExFAT file systems are not power fail safe. The TFAT [6, 7] is the extension of FAT file system and TexFAT [8] file system the extension of ExFAT file system. The TexFAT file system provides the power fail safe feature. The TexFAT file system is available in Microsoft's Windows CE 6.0 and later versions windows embedded operating systems. This file system is also supported in Windows Mobile 6.5 OS. But, it is not supported in Windows XP (eXPerience), windows 7/8 OSs. The TexFAT file system is backward compatible with conventional ExFAT file system. This means the file/directories created in the TexFAT file system are accessible for read and write operations in ExFAT file system implementations. As of today, August 2014, The Microsoft Corporation has not released any specification for TexFAT file system. This paper adopts the reverse engineering methodology by performing the various combinations of directory and file write operations in Windows CE platform and analyze the post file system write patterns to explore the transaction safe cluster allocation algorithms of TexFAT file system and hence determines the how the power fail safeness is achieved in these file systems.

2 Cluster Allocation Algorithms of TexFAT File System

A file system operation except the read operation involves following operations

(i) The Meta data update involves updating file directory entry [1], stream extension directory entry [1] and file name extension directory entry [1] of file/directory, Updating the File Allocation Table and updating the Cluster Heap.
(ii) The User data update involves the updating the data clusters of the file.

Note that, if there is sudden power-loss or abrupt storage device removal from computer system or system crash during user data update, then file/directory content is corrupted or unreliable. But, if there is sudden power-loss during Meta data update then it causes entire file system to be corrupted/unreliable and enforces user/

system administrator to format the file system and erase all valid data of file system. In TexFAT file system, power fail safe feature is provided by performing atomic the file systems update operations. The TexFAT file system implements the transaction safe file write/udpate/truncation operations and transaction safe directory update operation. The ExFAT file system uses only one instance of FAT where as the TexFAT file system uses two instances of FATs. The TexFAT uses two instances of cluster heap to optimize the transaction safe file write operation. If the contiguous clusters are available for allocation, then the ExFAT file system does not update the cluster status in FAT entries, instead "No FAT chain" [2] bit is set 1, to indicate the contiguous number of clusters are allocated. The allocated cluster status bit is indicated in cluster heap. The "No FAT chain" bit is a part of generic primary flags of stream extension directory entry of files/directory.

2.1 Transaction Safe File Write Operation

The file/folder creation, file write/update and file/folder removal operations causes the update in the data of the clusters of the parent directory in which file/folder exists/ created. The TexFAT file system provides the transaction safe update/write operation of directory. Following are the two scenarios of directory update operations

1. File/Folder creation operation
2. File/Folder update/deletion operation

2.1.1 Directory Update by File/Folder Creation

The creation of a file or sub directory/folder causes either update in the allocated clusters of the directory or allocation of new cluster to the directory. Whenever a file is created, following are the steps performed with automicity.

1. Search for sufficient number of free bytes available in all the clusters of directory to write the Meta data of the file to be created. In TexFAT, The Meta data of the file consists of 32 bytes file directory entry, 32 bytes stream extension entry and multiple 32 bytes file name extension entries.
2. If the sufficient number of free bytes are not available then allocate a new cluster to the tail of the cluster chain of the directory and create the file by writing 32 bytes file directory entry, 32 bytes stream extension entry and multiple 32 bytes file name extension entries.
3. If the sufficient number of free bytes are available in any allocated cluster of cluster chain of directory, then copy the content of this cluster a new cluster. Create the file by writing the Meta data of the file in the new cluster. This new cluster replaces the cluster in which is sufficient number of free bytes are found.
4. Copy the content of active FAT and cluster heap to backup FAT and cluster heap.

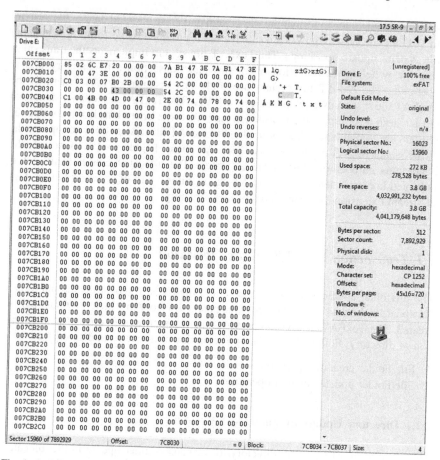

Fig. 1 Snapshot of the root directory containing the file KMG.txt; the cluster 0 × 43 is the first data cluster of the file KMG.txt

Figure 1 shows the WinHex [9] tool display snapshot of the root directory of the ExFAT file system of Win CE 6.0 OS on 4 GB (Giga Bytes) SD (Secure Digital) card. The root directory contains the file named KMG.txt with contiguous clusters 0 × 43 (hexadecimal number 43), 0 × 44 and 0 × 45. Note that, data cluster number 0 × 42 is allocated to root directory in which the file KMG.txt exists.

2.1.2 Directory Update by File/Folder Deletion/Update

Whenever the file write is performed the starting data cluster number, file size, last updated time changes and it causes write operation to the data cluster of the parent directory. Similarly, whenever the file/folder deleted, the 32 bytes file/folder entry, stream extension entry and file name extension entries are marked as deleted and it cause the write operation to the data cluster of the directory in which file/folder exists. During file write and file/folder deletions following steps are performed with atomicity.

(i)   Search for the data cluster in which file/folder need to be updated/deleted.
(ii)   Allocate the new data cluster in the active FAT and Cluster Heap.
(iii) Write the content of the data cluster, in which file/folder need to be updated to the new data cluster.
(iv)   Update the 32 bytes file entry, stream extension entry and file name entries in the new data clusters.
(v)   Replace the old cluster number by new cluster in cluster linked list (chain) in FAT.
(vi)   Mark the old cluster as free in cluster heap.
(vii) Copy the content of active FAT and cluster heap to backup FAT and cluster heap.

As an example, If the file KMG.txt, shown in Fig. 1, is deleted, the Meta data of the KMG.txt such as 32 bytes file entry, stream extension entry and file name extension entries are need to be marked as deleted. Hence, new cluster 0×46 is allocated to root directory, as shown in Fig. 2; the content of the cluster 0×42 is

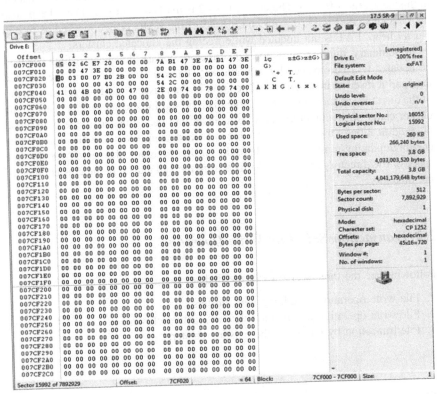

Fig. 2 Snapshot of cluster 0×46 of root directory in which file KMG.txt is marked as deleted

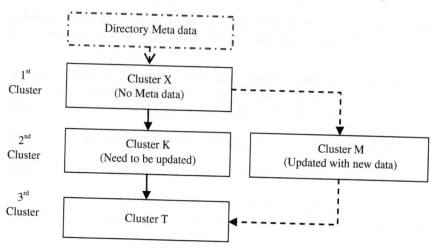

Fig. 3 Directory with clusters X, K and T; the cluster K is replaced with cluster M in the cluster chain

copied to this new cluster. The Meta data of the file KMG.txt are marked as deleted in this new cluster 0×46. The cluster 0×42 is marked as free in cluster heap, also the data clusters 0×43, 0×44 and 0×45 of file KMG.txt are marked as free in cluster heap. FAT is not updated while freeing the allocated clusters.

In Summary, during transaction safe directory update, the cluster which needs to be updated will be replaced by new cluster. The new cluster contains the updated data and it inserted in same logical position in the cluster linked list (cluster chain) of the directory as shown in Fig. 3. Note that, the starting cluster or 1st cluster of the cluster chain of the directory does not contain any data; this technique is to avoid the modification to the first cluster value in Meta data update of the directory. While creating/updating/removing the file/folder in the directory, the search for the free space or file/directory name is performed from the 2nd cluster of the cluster chain of the directory.

2.2 Transaction Safe File Write Operation

Following are the three scenarios of file update operations

1. File write operation by adding new clusters.
2. File write operation by updating data in existing clusters.
3. File truncation by removing the existing clusters.

2.2.1 File Write Operation by Adding New Clusters

Whenever a file is extended by adding new clusters, following are the steps performed with atomicity.

(i)   Search for new set of data clusters.
(ii) If the available free clusters are contiguous then these clusters marked as allocated in the active cluster heap and FAT is not updated. If the available free clusters are not contiguous then these clusters marked as allocated in the active cluster heap and FAT.
(iii) The data of existing clusters are copied to the new clusters.
(iv) The new data is appended to the old data copied in step (iii).
(v)   If the contiguous clusters are allocated then content of active Cluster Heap are copied to backup Cluster Heap.
(vi) If the non-contiguous clusters are allocated then content of active Cluster heap and active FAT are copied to backup clusters heap and backup FAT.
(vii) The Meta data of the file such as 32 bytes file/folder directory entry and the 32 bytes stream extension directory entry of the file is updated with new starting data cluster, updated time and new file size. The updating of the Meta data of the file is the write operation of data clusters of the parent directory in which the file exists. The data cluster updating of the directory is the performed as transaction safe operation as described in the Sect. 2.1.2.

Figure 1 shows that an example file named KMG.txt is created in the root directory. The cluster numbered 0×43, 0×44 and 0×45 are allocated to the file. If the file is extended by adding new data then new set of clusters 0×47, 0×49, $0 \times 4A$, $0 \times 4B$ are allocated to the file. The data of the clusters 0×43, 0×44 and 0×45 are copied to the clusters 0×47, 0×49, $0 \times 4A$ respectively. The new data to be appended is copied to the cluster $0 \times 4B$. Figure 4 shows the updated 32 byte file entry, stream extension and file name extension entries of the file KMG.txt with new starting data cluster 0×47. The new cluster $0 \times 4C$ is allocated to the root directory to replace the cluster 0×46.

2.2.2 File Write Operation by Updating Data in Existing Clusters

Whenever existing data of file is modified, following are the steps performed with atomicity.

(i)   Search for new set of data clusters.
(ii) If the available free clusters are contiguous then these clusters marked as allocated in the active cluster heap and FAT is not updated. If the available free clusters are not contiguous then these clusters marked as allocated in the active cluster heap and FAT.
(iii) The data of existing clusters is updated with new data.
(iv) This updated data is copied to the newly allocated clusters.

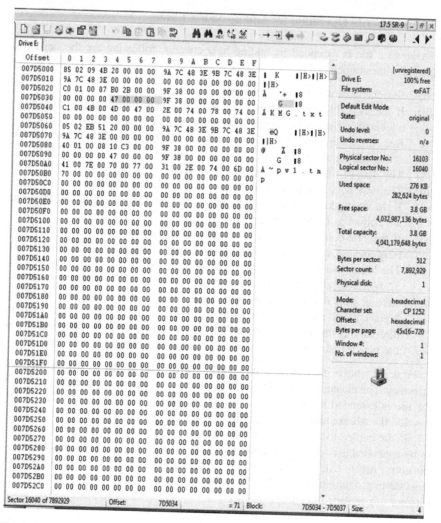

Fig. 4 Snapshot of the cluster 0 × 4C of root directory with updated Meta data of the file KMG.txt

(v)   If the contiguous clusters are allocated then content of active Cluster Heap are copied to backup Cluster Heap.

(vi) If the non-contiguous clusters are allocated then content of active Cluster heap and active FAT are copied to backup clusters heap and backup FAT.

(vii) The Meta data of the file such as 32 bytes file/folder directory entry and the 32 bytes stream extension directory entry of the file is updated with new starting data cluster, updated time and new file size. The updating of the metadata of the file is the write operation of data clusters of the parent directory in which the file exists. The data cluster updating of the directory is the performed as transaction safe operation as described in the Sect. 2.1.2.

For Example, if the data of the cluster 0 × 45 of file KMG.txt, as shown in Fig. 1 is modified then new set of clusters 0 × 47, 0 × 49 and 0 × 4A are allocated to the file KMG.txt as shown in Fig. 5. The clusters 0 × 47 and 0 × 49 contain the older data of clusters 0 × 43, 0 × 44 of the file but the cluster 0 × 4A contains the modified data of the cluster 0 × 45. The stream extension entry of the file KMG.TXT is updated with new first cluster number 0 × 47.

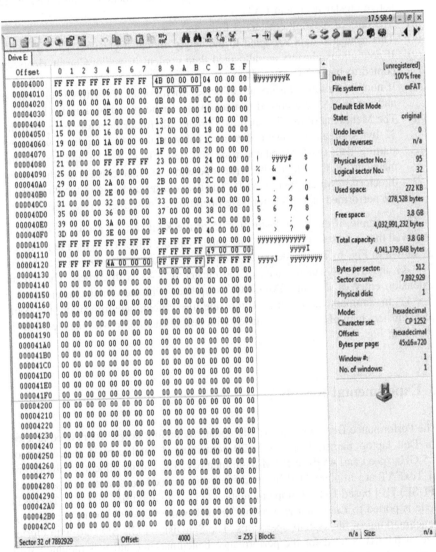

Fig. 5 Snapshot of the FAT with new cluster 0 × 4B allocated to root directory; new set of clusters 0 × 47, 0 × 49, and 0 × 4A are allocated to file KMG.txt

2.2.3 File Truncation by Removing Existing Clusters

Whenever a file is truncated, following are the steps performed with atomicity.

(i)   Search for new set of data clusters.
(ii) If the available free clusters are contiguous then these clusters marked as allocated in the active cluster heap and FAT is not updated. If the available free clusters are not contiguous then these clusters marked as allocated in the active cluster heap and FAT.
(iii) The data of existing clusters is truncated.
(iv) This truncated data is copied to the newly allocated clusters.
(v)   If contiguous clusters are allocated then content of active Cluster Heap are copied to backup Cluster Heap.
(vi) If non-contiguous clusters are allocated then content of active Cluster heap and active FAT are copied to backup cluster heap and backup FAT.
(vii) The Meta data of the file such as 32 bytes file/folder directory entry and the 32 bytes stream extension directory entry of the file is updated with new starting data cluster, updated time and new file size. The updating of the metadata of the file is the write operation of data clusters of the parent directory in which the file exists. The data cluster updating of the directory is the performed as transaction safe operation as described in the Sect. 2.1.2.

For Example, if the data cluster 0×45 of file KMG.txt, as shown in Fig. 1, is removed then new set of clusters 0×47 and 0×49 are allocated to the file KMG.txt as shown in Fig. 6. The clusters 0×47 and 0×49 contain the data of the clusters 0×43 and 0×44 but the data of the cluster 0×45 is discarded. The stream extension entry of the file KMG.TXT is updated with new first cluster number 0×47. In summary, during transaction safe file write/update/truncation, the complete cluster chain of the file is replaced with new cluster chain with updated data as shown in Fig. 7.

3 Experimental Results

The Performance Benchmarking of file update by adding new clusters is conducted on Dell laptop named Inspiron 5,520 containing 4 Intel core i5 processors of 2.5 GHz speed and 4 GB of RAM/main memory. The cluster allocation algorithms of TexFAT are implemented by modifying the Linux File system in User Space (FUSE) [10] based ExFAT implementation [11]. This modified ExFAT (TexFAT) code is ported to Linux kernel version 3.8 of Ubuntu 12.4 OS for the accuracy of benchmarking of file system operations. The SanDisk 64 GB SD card with cluster size of 64 KB is used for the performance benchmarking and it is mounted with buffer cache disabled in the Linux kernel. Each write operation is performed with the 4 MB or record size. The file update operation is performed on the existing test file of size 64 MB (Mega Bytes). The file update write operation by adding new

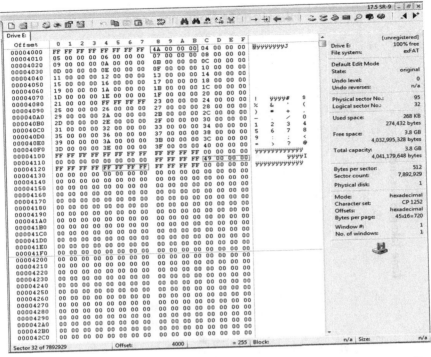

Fig. 6 Snapshot of the FAT with new set of clusters 0 × 47 and 0 × 49 are allocated to file KMG.txt

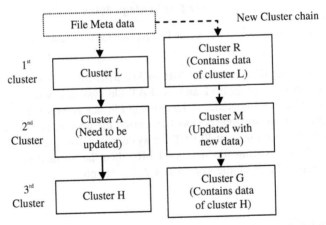

Fig. 7 File with clusters L, A and H; the updated file has new cluster chain with clusters R, M and G

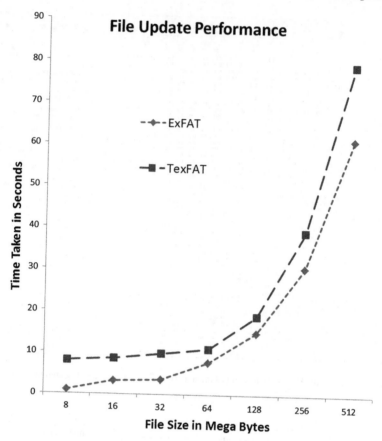

Fig. 8 File Update operation performance of TexFAT and ExFAT file systems

clusters involves the data sizes of 8 MB to 512 MB. Figure 8 shows file update operation performance of ExFAT and TexFAT file systems. Note that, file write performance of TexFAT degrades if the existing file size is larger. Thus, larger the file size higher the performance degradation during file update. The average update (re-write) performance range of ExFAT file system is 6.8–8.2 MBpS (MegaBytes per Second). If the existing file size is 64 MB, then the average file update performance range of TexFAT file system is 4.2–5.2 MBpS.

4 Conclusions

The TexFAT file system supports transaction safe file update and directory update operations. The file write operation in TexFAT file system secures the previous copy of the file data in case of abrupt power failures, but the performance is degraded. The TexFAT file system always ensures that both user data of the file and Meta data of the file system are safe across uncontrolled power loss or abrupt device removal or system/OS crashes. Similar to ExFAT file system, in TexFAT File System also, during file/directory deletion, the allocated data clusters are marked as free only cluster heap and FAT is not updated. The file truncation operation in ExFAT involves only the removal of data clusters from the cluster chain of the file. In TexFAT file system, the file truncation operation includes the file write operation along with removal of existing data clusters of the file.

References

1. Pudipeddi, R.V., Ghotge, V.V., Thind, R.S.: Quick file name look up using name hash. USPTO Patent 8321439, (2012)
2. Munegowda, K., Raju, G.T., Raju, V.M.: Cluster allocation strategies of the ExFAT and FAT file systems: a comparative study in embedded storage systems. Proceedings of International Conference on Advances in Computing (ICAdC), vol. 174. Springer, Berlin (2012)
3. Microsoft Corporation.: FAT32 File System Specification, FAT: General Overview of On-Disk Format. Version 1.03 (2000)
4. SD Specifications Part 1: Physical Layer Simplified Specification version 4.10. SD card Association (2013)
5. Multimedia card specification version 4.4. JEDEC (Joint Electron Devices Engineering Council) standard, (2008)
6. Malueg, M.D., Li, H., Gopalan, Y.N., Radko, R.O., Polivy, D.J., Drasnin, S., Farmer, J. R., Huang, D.: Transaction safe FAT file system. USPTO Patent 7174420 (2007)
7. Patel, S., Gopalan, Y.N.: Transaction safe FAT file system improvements. USPTO Patent 8024507 (2011)
8. Microsoft, TexFAT overview (windows Embedded CE 6.0). http://msdn.microsoft.com/en-us/library/ee490643(v=winembedded.60)aspx (2010)
9. WinHex, Universal Hex Editor. http://www.winhex.com/winhex/hex-editor.html, Version 7.5 (2014)
10. FUSE (File System in User Space). http://fuse.sourceforge.net/
11. Fuse based ExFAT implementation for Linux. http://code.google.com/p/exfat/. Version 1.0 (2014)

An Efficient PSO-GA Based Back Propagation Learning-MLP (PSO-GA-BP-MLP) for Classification

Chanda Prasad, S. Mohanty, Bighnaraj Naik, Janmenjoy Nayak and H.S. Behera

Abstract In last few decades, Evolutionary computation and Swarm intelligence are two hot favorites for almost all types of researchers. Moreover, many contributions have been made in two directions: Genetic Algorithm (GA) and Particle Swarm optimization (PSO). But, some limitations in both the algorithms (complicated operator like crossover and mutation in GA and early convergence in PSO), are the major restricted boundaries for solving complex problems. In this paper, a hybridization of Particle swarm optimization and Genetic algorithm has been proposed with the back propagation learning based Multilayer perceptron neural network. The effectiveness of the proposed algorithm is shown through a no. of simulation steps with the help of the benchmark datasets considered from UCI machine learning repository. The performance of the algorithm is compared with other standard algorithms to show the steadiness and efficiency as well as statically significant.

Keywords Particle swarm optimization · Genetic algorithm · MLP · Classification · Data mining

C. Prasad (✉) · S. Mohanty
School of Computer Science and Engineering, Kalinga Institute of Industrial Technology University, Bhubaneswar 751024, Odisha, India
e-mail: chanda.prasad49@gmail.com

S. Mohanty
e-mail: satarupamohanty2007@gmail.com

B. Naik · J. Nayak · H.S. Behera
Department of Comp. Science Engineering and Information Technology, Veer Surendra Sai University of Technology, Burla, Sambalpur 768018, Odisha, India
e-mail: mailtobnaik@gmail.com

J. Nayak
e-mail: mailforjnayak@gmail.com

H.S. Behera
e-mail: mailtohsbehera@gmail.com

© Springer India 2015
L.C. Jain et al. (eds.), *Computational Intelligence in Data Mining - Volume 1*,
Smart Innovation, Systems and Technologies 31, DOI 10.1007/978-81-322-2205-7_48

517

1 Introduction

Evolutionary computation and Swarm intelligence are diversified area of research which make them more favorites than other optimization algorithms. Evolutionary algorithms (like GA) [1], Differential Evolution (DE) [2] and Swarm Intelligence (like PSO) [3]; Group Search Optimization (GSO), [4]; Ant Colony algorithm (ACO) [5]; Bee Colony algorithm(BCO) [6, 7], Cat Swarm Optimization (CSO) [8, 9], Glowworm Swarm Optimization(GSO) [10], are widely used in various engineering applications and these methodologies have been combined with different neural network to perform various data mining tasks like classification, prediction and forecasting.

Shi et al. [11] have introduced an improved GA algorithm and a new hybrid algorithm using PSO & GA. They performed the simulations for optimization problems and found that the hybrid algorithm performs better than other techniques. A novel Chaotic Hybrid Algorithm (CHA) by using the hybridized PSO-GA method for circle detection has been developed by Dong et al. [12]. Liu et al. [13] have made an investigation on two Wavelet-MLP hybrid frameworks for wind speed prediction using GA and PSO optimization. The comparison performance between the two networks helps to prove that the Wavelet network is more statistically significant than MLP. Ludermir et al. [14] have used the PSO technique for the identification of factors related to Common Mental Disorders. A hybrid algorithm combining Regrouping Particle Swarm Optimization (RegPSO) with wavelet radial basis function neural network is presented by Nasir et al. [15] which is used to detect, identify and characterize the acoustic signals due to surface discharge activity. Sahoo et al. [16] have developed a hybrid method by considering both PSO and GA for solving mixed integer nonlinear reliability optimization problems in series, series–parallel and bridge systems. An efficient classification method based on PSO and GA based hybrid ANN has been proposed by Naik et al. [17] and it is found relatively better in performance as compared to other alternatives.

Most of the conventional optimization techniques are iterative methods in which the selection of initial solution is based on the nature of the problem [18] and are revised using deterministic update rules which usually depend upon the problem structure. However, the improvement of heuristic techniques, like genetic algorithm, particle swarm optimization, attracts the researcher's attention towards these methods, due to the efficiency of solving a complex iterative optimization problem within a rational time complexity. But, for the improvement in computational efficiency, hybridization between two or more algorithms are required. With this intention, in this paper a hybridized PSO-GA based multilayer perceptron has been proposed to perform classification task. The experiments for this purpose will comprise of the following performance comparisons: (1) GA-MLP, (2) PSO-MLP, (3) Hybrid PSO/GA-MLP. Remaining of this paper is organized as follows: Preliminaries, Proposed Model, Experimental Setup and Result Analysis, Conclusion and References.

2 Preliminaries

2.1 Particle Swarm Optimization

Particle swarm optimization (PSO) [3, 19] is a widely used stochastic based search algorithm and it is able to search global optimized solution. Like other population based optimization methods the particle swarm optimization starts with randomly initialized population for individuals and it works on the social behavior of particle to find out the global best solution by adjusting each individual's positions with respect to global best position of particle of the entire population (Society). Each individual is adjusting by altering the velocity according to its own experience and by observing the experience of the other particles in search space by use of Eqs. 1 and 2. Equation 1 is responsible for social and cognition behavior of particles respectively where c1 and c2 are the constants in between 0–2 and rand(1) is random function which generates random number in between 0–1.

$$V_i(t+1) = V_i(t+1) + c_1 * rand(1) * (lbest_i - X_i) + c_2 * rand(1) * (gbest_i - X_i) \tag{1}$$

$$X_i(t+1) = X_i(t) + V_i(t+1) \tag{2}$$

The basic steps of problem solving strategy of PSO are:

Initialize the position of particles $V_i(t)$ (population of particles) and velocity of each particle $X_i(t)$.
Do

> *Compute fitness of each particle in the population*
> *Generate local best particles (LBest) by comparing fitness of particles in previous population with new population*
> *Choose particle with higher fitness from local best population as global best particle (GBest).*
> *Compute new velocity $V_i(t+1)$ by using* Eq. (1).
> *Generate new position $X_i(t+1)$ of the particles by using* Eq. (2).

While *(iteration <= maximum iteration OR velocity exceeds predefined velocity range);*

2.2 Genetic Algorithm

Genetic algorithm (GA) [1] is a computational model of machine learning inspired by evolution. The development of GA has now reached a stage of maturity, due to the effort made in the last decade by academics and engineers all over the world. They are less vulnerable to getting 'stuck' at local optima than gradient search methods. The pioneering work is contributed by J.H. Holland for various scientific and engineering applications. GA is inspired by the mechanism of natural selection, a biological process in which stronger individuals are likely be the winners in a competing environment. Fatnesses (goodness) of the chromosomes are used for solving the problem and in each cycle of genetic operation (known as evolving process) a successive generation is created from the chromosomes with respect to the current population. To facilitate the GA evolution cycle, an appropriate selection procedure and two major fundamental operators called crossover and mutation are required to create a population of solutions in order to find the optimal solution (chromosome).

2.3 Multi Layer Perceptron

Multi Layer Perceptron (Fig. 1) is the simplest neural network model which is consists of neurons called (Rosenblatt 1958). From multiple real valued inputs, the perceptron compute a single output according to its weights and non-linear activation functions. Basically MLP network is consists of input layer, one or more hidden layer and output layer of computation perceptron.

MLP is a model for supervised learning which uses back propagation algorithm. This consists of two phases. In the 1st phase, error (Eq. 4) based on the predicted outputs (Eq. 3) corresponding to the given input is computed (forward phase) and in the 2nd phase, the resultant error is propagated back to the network based on that weight of the network are adjusted to minimize the error (Back Propagation phase).

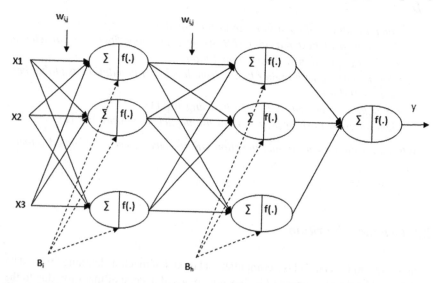

Fig. 1 MLP with input layer, single hidden layer and output layer, Where f(.) is non-linear activation function

$$y = f\left(\sum_{i=1}^{n} w_i x_i + b\right) \tag{3}$$

where w is the weight vector, x is the input vector, b is the bias and $f(.)$ is the non-linear activation function.

$$\delta_k = (t_k - y_k)f(y_{in_k}) \tag{4}$$

where t_k and y_k is the given target value and predicted output value of input kth pattern and δ_k is the error term for kth input pattern.

The popularity of MLP increases among the neural network research community due to its properties like nonlinearity, robustness, adaptability and ease of use. Also it has been applied successfully in many applications [20–28].

3 Proposed Model

In this section, we have proposed a PSO and GA based back propagation learning-MLP (PSO-GA-BP-MLP) for classification. Here basic concepts and problem solving strategy of PSO and GA evolutionary algorithm are used for better result of MLP classifier.

Algorithm. PSO-GA based Back Propagation Learning- MLP (PSO-GA-BP-MLP) for classification

---

% Initialization of population
        P = round(rand(n,(c-1)*(c-1)));
        Where n is the number of weight-sets (chromosomes) in the population 'P' and c is the number of attributes in dataset (excluding class label).
% Initialization of velocity
        v = rand(n,(c-1)*(c-1));

% Neural Setup for MLP
        Bh =(rand(c-1,1))';   **% Bh: Bias of hidden layer.**
        Bo = rand(1);   **% Bo: Bias of output layer.**

% PSO-GA Iteration
Iter=0;   **% iter: Iteration**
while(1)

        1. Selecting local best weight-sets (lbest) by comparing with weight-sets in previous population. If it is first iteration, then initial population(P) is considered to be local best(lbest). Otherwise, new 'lbest' population is formed by selecting best weight-sets from previous population (P) and current local best (lbest).
                Iter = Iter+1;
                If (Iter == 1)
                        lbest = P;
                else
                        [lbest] = lbestselection (lbest, P, tdata, t);
                end

2. Compute fitness of all weight-sets in local best 'lbest'. Each weight-sets are set individually in MLP and trained with training data 'tdata'. RMSE for each weight-sets are calculated with respect to target 't'. Based on RMSE, fitness of weight-sets are calculated by using 'fitfromtrain' procedure.

[F] = fitfromtrain (lbest, tdata, t);

3. Select a global best 'gbest' from local best 'lbest' based on their fitness(F) (calculated by using 'fitfromtrain' procedure) by using 'gbestselection' procedure.

[gbest] = gbestselection (P, F, rmse);

4. Create mating pool by replacing weak individuals (having minimum fitness value) with global best 'gbest' weight set.

5. Perform two point crossovers in mating pool mating pool and replace all the weigh set of population P with weight sets of mating pool.

6. Compute new velocity 'vnew' from population (P), velocity (v), local best 'lbest' and global best 'gbest' by using 'calcnewvelocity' procedure.

vnew = calcnewvelocity (P, v, lbest, gbest);

7. Update next position by using current population (P) and new velocity (vnew).

P = P + vnew;

8. If maximum iteration is reached or fatnesses are remain unchanged for a pre-defined number of iteration then stop.

If (Iter == maxIter)

break;

end

end

---

function [lbest] = lbestselection (lbest, P, tdata, t)

1. Compute fitness of all weight-sets in local best 'lbest' and previous population 'P'. Each weight-sets are set individually in MLP and trained with training data 'tdata' by using 'fitfromtrain'. RMSE for each weight-sets are calculated with respect to target 't'. Based on RMSE, fitness of weight-sets is calculated.

[F1] = fitfromtrain (lbest, tdata, t);
[F2] = fitfromtrain (P, tdata, t);

2. Compare fitness of weight-sets in lbest and P by comparing F1 and F2 , where F1 and F2 are fitness vector of lbest and P respectively. Based on this comparison, generate new lbest for next generation.

for i = 1:1:number of weight-sets in P or lbset.

If (F1(i,1) <= F2(i,1))

lbest(i,:) = P(i,:);

else

lbest(i,:) = lbest(i,:);

end

end

end

function [F] = fitfromtrain(data, tdata, t)

    Step-1: Repeat step 1 to 6 for all the weight-set 'w' in 'P'.

    Error=0.

    Here tdata is the training data without class label and it is considered to be tdata={$x_1, x_2, \ldots, x_L$}, L is the number of input patterns.

    Step-2: Repeat step- to step- for all the patterns $x_k = (x_{k1}, x_{k2}, \ldots, x_{kn})$ in given dataset tdata, where n is the number of attribute value in a single input pattern.

    Step-3: Feed forward stage.

$$Z_{in_j} = Bh_j + \sum_{i=1}^{n} x_{ki} w_{ij}$$

    Here Z_{in_j} represents j^{th} hidden unit and i=1,2..n, where n is the number of input which are connected to a single hidden unit.

    $y_{in_k} = B_{ok} + \sum_{j=1}^{m} Z_j W h_{jk}$, Where j=1,2..m, m is the number of hidden unit connected to one output unit.

    $y_k = f(y_{in_k})$, for k=1,2,..p, where p is the number of output unit in output layer and $f(.)$ is the sigmoid activation function and can defined as follows.

$$f(y_{in_k}) = \frac{1}{1 + e^{-y_{in_k}}}$$

    Step-4: Calculation of error terms in hidden layer.

    $\delta_k = (t_k - y_k)f(y_{in_k})$, for each output unit y_k, k=1,2...m. For each hidden unit Z_j, j=1,2...n, sums of delta input is computed as follows.

$$\delta_{in_j} = \sum_{i=1}^{m} \delta_k\, w_{jk}$$

    Error terms are calculated as: $\Delta_j = \delta_{in_j} f_1(z_{in_j})$, where $f_1(.)$ is the activation function which is defined as:

$$f_1(z_{in_j}) = f(z_{in_j})\left(1 - f(z_{in_j})\right)$$
$$= \frac{1}{1+e^{-z_{in_j}}}\left(1 - \frac{1}{1+e^{-z_{in_j}}}\right)$$

    Finally error of the network in calculated as: Error = Error + Δ_j.

    Step-5: if all input patterns in dataset are processed then goto step-6. Else goto step-1.

    Step-6: Compute RMSE based on total error of the network and compute fitness of the weight-set as F(i)=1/RMSE., where F(i) is the fitness of ith weight-set in P.

end

---

function vnew = calcnewvelocity (P, v, lbest, gbest)

    for i = 1:1:number of row in population 'P'.

        for j = 1:1: of column in population 'P'.

        vnew (i,j) = v (i,j) + r1 * c1 * (lbest(i,j) - P(i,j)) + r2 * c2 * (gbest(1,j) - P(i,j));

        end

    end

end

4 Experimental Setup and Result Analysis

The proposed method has been implemented using MATLAB 9.0 on a system with an Intel Core Duo CPU T2300, 1.66 GHz, 2 GB RAM and Microsoft Windows XP Professional 2002 OS. In this section, the comparative study on the efficiency of our proposed method has been presented. Benchmark datasets (Table 1) from UCI machine learning repository [29] and KEEL Data Set Repository [30] have been used for classification and the result of proposed PSO-GA-MLP model is compared with MLP, GA-MLP based on Genetic Algorithm and PSO-FLANN based on Particle Swarm Optimization. Datasets have been normalized and scaled in the interval −1 to +1 using Min-Max normalization before training and testing is made. Classification accuracy (Eq. 5) of models has been calculated in terms of number of classified patterns are listed in Table 2.

If cm is confusion matrix of order m × n then, accuracy of classification is computed as:

$$Classification\ Accuracy = \frac{\sum_{i=1}^{n} \sum_{\substack{j=1, \\ i==j}}^{m} cm_{i,j}}{\sum_{i=1}^{n} \sum_{j=1}^{m} cm_{i,j}} \times 100 \tag{5}$$

4.1 Parameter Setting

During simulation, c1 and c2 constants of PSO have been set to 2 and two point crossovers is used throughout the experiment. In MLP, one input layer, one hidden layer and one output layer for the neural network has been set during training and testing.

Table 1 Data set information

Dataset	Number of pattern	Number of features (excluding class label)	Number of classes	Number of pattern in class-1	Number of pattern in class-2	Number of pattern in class-3
Monk 2	256	06	02	121	135	–
Hayesroth	160	04	03	65	64	31
Heart	256	13	02	142	114	–
New thyroid	215	05	03	150	35	30
Iris	150	04	03	50	50	50
Pima	768	08	02	500	268	–
Wine	178	13	03	71	59	48
Bupa	345	06	02	145	200	–

Table 2 Performance comparison in terms of accuracy

Dataset	Accuracy of classification in average							
	MLP		GA-MLP		PSO-MLP		PSO-GA-MLP	
	Train	Test	Train	Test	Train	Test	Train	Test
Monk 2	86.94648	85.27453	87.23734	87.85732	90.19375	92.44732	93.18283	92.01343
Hayesroth	83.48576	82.38657	85.43675	81.04653	88.38271	81.97365	90.28576	82.25314
Heart	82.84653	74.77453	85.44937	75.03645	86.23755	75.84651	88.62356	78.81449
New Thyroid	92.03782	73.26876	92.74834	73.92756	93.02785	75.92784	93.18599	76.13558
Iris	90.87365	92.15368	92. 56873	95.93158	92.51736	93.36158	94.93756	96.20485
Pima	73.73645	73.88647	76.82642	77.38275	78.17464	78.28746	78.27583	78.54764
Wine	80.69731	80.87294	88.93645	77.37565	90.75389	91.77319	92.53647	93.07364
Bupa	68.66251	69.82637	70.97485	71.274459	70.27841	71.21465	72.05168	72.47364

5 Conclusion

The analysis of experimental of the combined optimization technique considered in this paper, PSO+GA, for training MLP network with back propagation learning is found to better as compare to MLP, GA-MLP and PSO-MLP for all most all the datasets. By using such hybrid approach, it is easier to optimize the weights of MLP network. Simulation results in Table 2 show that hybrid PSO-GA-MLP outperforms other alternatives. In future, the above work may be extended to other promising area of data mining by using it in higher order neural network.

References

1. Holland, J.H.: Adaption in Natural and Artificial Systems. MIT Press, Cambridge (1975)
2. Storn, R.: System design by constraint adaptation and differential evolution. IEEE Trans. Evol. Comput. 3(1), 22–34 (1999)
3. Kennedy, J., Eberhart, R.: Particle swarm optimization. In: Proceedings of the 1995 IEEE International Conference on Neural Networks, vol. 4, pp. 1942–1948 (1995)
4. He, S., Wu, Q.H., Saunders, J.R.: Group search optimizer: an optimization algorithm inspired by animal searching behavior. IEEE Trans. Evol. Comput. 13(5), 973–990 (2009)
5. Alatas, B., Akın, E.: FCACO: fuzzy classification rules mining algorithm with ant colony optimization, ICNC 2005. Lecture Notes in Computer Science, vol. 3612, pp. 787–797. Springer, Heidelberg (2005)
6. Alatas, B.: Chaotic bee colony algorithms for global numerical optimization. Expert Syst. Appl. 37(8), 5682–5687 (2010)
7. Karaboga, D., Basturk, B.: A powerful and efficient algorithm for numerical function optimization: Artificial bee colony (ABC) algorithm. J. Global Optim. 39(3), 459–471 (2007)
8. Chu, S., Tsai, P., Pan, J.: Cat swarm optimization. Lecture Notes in Computer Science, vol. 4099, pp. 854–858. Springer, Heidelberg (2006)
9. Pradhan, P.M., Panda, G.: Solving multiobjective problems using cat swarm optimization. Expert Syst. Appl. 39(3), 2956–2964 (2012)
10. Krishnanand, K.N., Ghose, D.: Glowworm swarm based optimization algorithm for multimodal functions with collective robotics applications. Multiagent Grid Syst. 2(3), 209–222 (2006)
11. Shi, X.H., Liang, Y.C., Lee, H.P., Lu, C., Wanga, L.M.: An improved GA and a novel PSO-GA-based hybrid algorithm. Inf. Process. Lett. 93, 255–261 (2005)
12. Dong, N., Wu, C.H., Ip, W.H., Chen, Z.Q., Chan, C.Y., Yung, K.L.: An opposition-based chaotic GA/PSO hybrid algorithm and its application in circle detection. Comp. Math. Appl. 64(6), 1886–1902 (2012)
13. Liu, Hui, Tian, Hong-qi, Chen, Chao, Li, Yan-fei: An experimental investigation of two Wavelet-MLP hybrid frameworks for wind speed prediction using GA and PSO optimization. Electr. Power Energy Syst. 52, 161–173 (2013)
14. Ludermir, Teresa B., de Oliveira, W.R.: Particle swarm optimization of MLP for the identification of factors related to common mental disorders. Expert Syst. Appl. 40, 4648–4652 (2013)
15. Al-geelania, N.A., Piah, M.A.M., Adzis, Z., Algeelani, M.A.: Hybrid regrouping PSO based wavelet neural networks for characterization of acoustic signals due to surface discharges on HV glass insulators. Appl. Soft Comput. 13, 4622–4632 (2013)

16. Sahoo, L. et al.: An efficient GA–PSO approach for solving mixed-integer nonlinear programming problem in reliability optimization. Swarm Evol. Comput. (2014). http://dx.doi.org/10.1016/j.swevo.2014.07.002i
17. Naik, B., Nayak, J., Behera, H.S.: A novel FLANN with a hybrid PSO and GA based gradient descent learning for classification. In: Proceedings of the 3rd International Conference on Frontiers of Intelligent Computing (FICTA). Advances in Intelligent Systems and Computing 327, vol. 1, pp. 745–754 (2014). doi: 10.1007/978-3-319-11933-5_84
18. Thakur, M.: A new genetic algorithm for global optimization of multimodal continuous functions. J. Comput. Sci. 5, 298–311 (2014)
19. Kennedy, J. Eberhart, R.: Swarm Intelligence Morgan Kaufmann, 3rd edn. Academic Press, New Delhi (2001)
20. Hamedi, M., et al.: Comparison of Multilayer Perceptron and Radial Basis Function Neural Networks for EMG-Based Facial Gesture Recognition. The 8th International Conference on Robotic, Vision, Signal Processing & Power Applications. Springer Singapore (2014)
21. Ndiaye, A., et al.: Development of a multilayer perceptron (MLP) based neural network controller for grid connected photovoltaic system. Int. J. Phys. Sci. 9(3), 41–47 (2014)
22. Roy, M., et al.: Ensemble of multilayer perceptrons for change detection in remotely sensed images. Geosci. Remote Sens. Lett. IEEE 11(1), 49–53 (2014)
23. Hassanien, A.E., et al.: MRI breast cancer diagnosis hybrid approach using adaptive ant-based segmentation and multilayer perceptron neural networks classifier. Appl. Soft Comput. 14, 62–71 (2014)
24. Aydin, K., Kisi, O.: Damage detection in Timoshenko beam structures by multilayer perceptron and radial basis function networks. Neural Comput. Appl. 24(3–4), 583–597 (2014)
25. Velo, R., et al.: Wind speed estimation using multilayer perceptron. Energy Convers. Manage. 81, 1–9 (2014)
26. Lee, S., Choeh, J.Y.: Predicting the helpfulness of online reviews using multilayer perceptron neural networks. Expert Syst. Appl. 41(6), 3041–3046 (2014)
27. Azim, S., Aggarwal, S.: Hybrid model for data imputation: using fuzzy c means and multi layer perceptron. IEEE International Advance Computing Conference (IACC) (2014)
28. Chaudhuri, S., et al.: Medium-range forecast of cyclogenesis over North Indian Ocean with multilayer perceptron model using satellite data. Nat. Hazards 70(1), 173–193 (2014)
29. Alcalá-Fdez, J., et al.: KEEL data-mining software tool: data set repository, integration of algorithms and experimental analysis framework. J. Multiple-Valued Logic Soft Comput. 17 (2–3), 255–287 (2011)
30. Bache, K., Lichman, M.: UCI Machine Learning Repository. School of Information and Computer Science, University of California, Irvine (2013). [http://archive.ics.uci.edu/ml]

Analysis of New Data Sources in Modern Teaching and Learning Processes in the Perspective of Personalized Recommendation

G.M. Shivanagowda, R.H. Goudar and U.P. Kulkarni

Abstract The increased variety of learning resources have substantially affected learning styles of students, like e-books with modern collaborative tools, video lectures of different teachers across the world, lively discussion boards etc. Having accepted such forms of learning materials, teaching and learning processes in conventional set up do not have a way to capture the data generated out of students' learning activities involving such resources and use them effectively. This paper analyses data generated by the student's activities in Compiler of Resources in Engineering and Technology to Aid Learning (CRETAL) restricted to video resources and asserts that they are indeed critically helpful data for teachers/tutoring systems in generating personalised recommendations which are possible only because of said data. CRETAL is the modern learning station, an intelligent system, being developed at author's institution to facilitate variety of learning resources created and adapted by the faculty and the teachers worldwide to students.

Keywords Learning data · Recommendation system · Personalised learning · Collaborative learning · Video learning resources

G.M. Shivanagowda (✉) · U.P. Kulkarni
Department of Computer Science and Engineering, SDMCET, Dharwad 580004, India
e-mail: shivana.gowda@gmail.com

U.P. Kulkarni
e-mail: upkulkarni@yahoo.com

R.H. Goudar
Department of Computer Network Engineering, Visvesvaraya Technological University,
Belgaum 590018, India
e-mail: rhgoudar@vtu.ac.in

529

L.C. Jain et al. (eds.), *Computational Intelligence in Data Mining - Volume 1*,
Smart Innovation, Systems and Technologies 31, DOI 10.1007/978-81-322-2205-7_49

1 Introduction

CRETAL offers personalized learning environment with intelligent web services which were used for author's courses at undergraduate level at his institution. Though CRETAL offers several other types of resources like any other hypermedia system; study and analysis in this paper is limited to the video learning resources and data related them. CRETAL is loaded with the video lectures produced by different people around the world like National Program on Technology Enhanced Learning (NPTEL), Khan Academy etc. and we refer them as Video Learning Resource (VLR). Purpose of CRETAL's services with respect to VLR is to enable students to use VLRs like a modern E books providing annotation tools. There are good numbers of annotation tools available in web as part of bigger system (like few at http://www.annotations.harvard.edu); most of them have wider purposes like movies, surveillance, medicine etc.... The multimodal system by Haojin Yang et al. and VideoJot by Michael Riegler are too complex for a focused experiment of our type which needs support only for the temporal text annotations [1, 2]. All of them are less sensitive to the data that our experiment wanted to capture. In this regard CRETAL have the features like, table of content in beginning of the VLR so that user can seek the video stream directly from the point of his interest, tagging VLRs at a particular instant, attach chit notes, marking, highlighting, preserving these tags and allowing the user to search them during quick references etc. CRETAL also lets users to share their tags and chit notes to their Buddy group (see (f) of Fig. 1) to accommodate the collaborative and active learning styles. Among these, table of content is the production time tagging done by the VLR producers or by the teacher in the resources compiler mode. CREATL also have features to construct the users

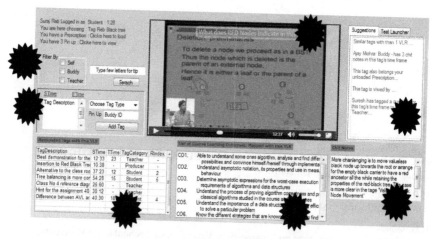

Fig. 1 CRETAL's snap shot in the student mode. **a** Optional pop up **b** context based scrolling suggestions. **c** Chit Notes if available. **d** Outcomes mapped by teacher in compiler mode. **e** Associated tags with current VLR. **f** Interface for tagging. **g** Interface for searching tags

learning trajectory by following their activity in a session and across the sessions and generate dynamic suggestions. A typical structure of tag is <Tag_ID, Tag_Description, Start_Time, End_Time, Tag_Type, User_ID, Relevance_Index, VLR_ID>. Among the components of a tag, tag_type is one which helps in distinguishing the tag as Chit Note, Marker, Connector, Launcher, and Question to Teacher or Buddy etc. All these tags become searching contents for any user who logs into CRETAL looking for VLR of his interests (see (g) of Fig. 1). Such tags can be suggested by the course teacher as references to his students and a set of VLR can be adapted as references to the courses like reference books. It is not the scope of this paper to either discuss the design of CRETAL or to evaluate its performance against other existing E-learning systems.

Rest of the paper is organized as follows. Section 2 describes data used by E learning systems for assessing and assisting learning, which are outcome of some previous research. Section 3 introduces the new data sources because of VLR by CRETAL and their estimation. Section 4 describes the outcomes of the CRETAL's experiment by author in his courses along with analysis of data. Finally Sect. 5 provides the concluding remark.

2 Nature of Data Required to Assist and Assess Learning: Survey

Various methods and systems developed based on the information communication have doubled in the recent past to improve the learning effectiveness of students at all levels of education. They can be sorted as web based systems from just course material presentation to personalised learning systems. We considered the recent most appreciated systems like DEPTHS (Design Patter Teaching Help System), ZOSMAT, Personalised Learning Course Planner (PLCP) and Personalised E-Learning System (PELS) to understand the kinds of data along with their sources which help these systems in achieving their objective of enhanced learning and its effectiveness. Most commonly, data in any such system will depend on the nature of the content of the learning material, its availability, method of measuring the learning outcomes of the targeted concepts etc.

2.1 Depths

E helping system developed by Jeremic et al. [3] is restricted to non VLR and for few topics of Software Patterns concepts. Three most important data that DEPTHS uses for the student modeling are Personal data which is not essential for the adaption of the teaching material, Performance data which is mixture of dynamic and static data with respect to cognitive and individual characteristics, Teaching history which is DEPTHS actions, time taken to solve test, time spent of reading slides/lessons and time taken for activities etc. All these data are used to calculate the degree of mastery on particular concepts of the course content.

2.2 PLCP

Choi et al. [4] have used E-Learning Decision Support System (EL-DSS) to achieve personalised course content organization; effort is to prove that if the course is planned as per the requirements of the students then their learning performance is improved. The data they record and use for this purpose are the user profile data which give due importance to the pre-knowledge of the students, which is analysed along with priorities of the course content set by the user to advise a learning trajectory using EL-DDS and decision matrix.

2.3 ZOSMAT

An Intelligent Tutoring System (ITS) for mathematics which follows students in every learning step is developed by Ayturk Keles et al. with flash animation also as one of learning material [5]. This needs good amount of data which is temporal in nature. Important components of the student's model in ZOSMAT are personal information, learning status with respect the learning objectives and series of test data which are attached to the topics in the content of the course.

2.4 PELS

This is also an E-Learning System much similar to the PLCP but it differentiates itself with self-regulated learning feature. This effort of Chen [6] is rich in data acquisition with respect to learning. Learning time, number of modules learnt, degree of concentration, values of learning ability etc. are used along with the data extracted from the self evaluation form to estimate the performance of learning using the Achievement index for the concentrated learning, Reading rate, Effort level of learning. These play a vital role in generating recommendation to students.

All the above mentioned systems and few other commercial systems in web, more or less have the reading type material as their resources without collaborative features and except ZOSMAT none make attempt to consider the offline data of the real class room or have characteristic to make themselves supplement the evaluation process of students of the conventional set up. Other than this, Catherine Mulwa et al. [7] in their survey of hypermedia system with respect to Technology Enhanced Learning; identify 14 different factors that have influence on learning. All learning systems fail or not able to consider one or the other data which is a contributor to performance assessment or recommendation generation [8]; the obvious one in conventional set up is the student's activities outside the class room or during the non contact hours toward learning. The data related to student's efforts in learning plays an important role in understanding the student's differences [9]. Rest of the paper presents new data sources created only because of the VLR which supplements to the regular class room activity of teaching-learning-assessing in the conventional set up and how these data becomes relevant in generating the personalised recommendations.

3 CRETAL's Data Sources

The course teacher is believed to have mastery over the content of course irrespective of the type of the resource material. Like every other user, teachers also have their own tags on different VLR belonging to CRETALs repository. Several collections of such tags bind to popup can be regarded as teacher's prescription to a student like <P_ID, Student_ID>, which is nothing but advising a student to follow predefined learning trajectory aided by VLR and short notes, they can be treated similar to reading assignment in conventional setup. A typical structure of a prescription is <P_ID, (Tag_ID1, Tag_ID2...), (PopUp_ID1, PopUp_ID2...)>. Popup are similar to ads in YouTube, but launch short text notes, questions, tasks or simply collect responses etc.... A typical structure of a popup is <PopUp_ID, Pop_Time, Popup_Type, Genus, Author_Response_Time, Tag_ID, P_ID>; where Type of a popup means, question which is to be answered, chit note which need to read, locator which asks to create a tag of marker type etc. and Genus is to indicate whether popup is optional or compulsory. There will be a predefined response time fixed by the author of the popup in Author_Response_time.

As user watch VLR, he might initiate and involve in lot of activities using CRETAL's services like tagging, appreciate, depreciate other's tag, following his teacher's prescription, pausing, repeating, seeking back and front, quitting, switching to dynamic suggestion by CRETAL etc. For better imagination a snap shot of the CRETAL users view in student mode is given in the Fig. 1.

All the said activities of the users are the new sources of data that we talked about in the beginning; these can contribute substantially in the computation of competence and performance indicators that many tutoring system use for the student modeling. They do also contribute when teachers create blended or hybrid learning scenarios by blending Technology-Enhanced Learning and traditional approaches to teaching and learning [7]. Few of the contributing data their sources are as follows;

3.1 Learning Time Invested

This is an important contributor in calculating performance index of an individual learner; the number of tags searched, their relevance index, total number of tags followed and time invested [see Eqs. (1) and (2)] by the user per session. These data can increase the correctness of the 'Achievement Index of learning Time' which has direct influence on the performance index of individual learners [6, 10, 11]. Even if it is not self regulated system, the data plays a vital role in estimating prescription's Achievement Index.

$$
\begin{aligned}
T_{Invest}(Students\_ID) = \sum_{Tag_i \in F, Tag_i.User\_ID \neq Students\_ID} length\_of(Tag_i) \\
+ T_{PopUp}(Tag_i.Tag\_ID) - T_{Idle}(Students\_ID)
\end{aligned}
\tag{1}
$$

$$
T_{PopUp}(Tag\_ID) = \sum_{PopUp_j.Tag\_ID = Tag\_ID} response\_time\_of(PopUp_j)
\tag{2}
$$

All T in above equations indicate the total time of the subscript in common unit in a session. Tag_i is the ith tag and F is the set of tags followed by the student in his free mode sessions. If T_{Invest} is to be calculated for a prescription P_k, it is be noted that, $\forall i \, \forall j \, Tag_i \, \& \, PopUp_j \in P_k$ which is given by $T_{Invest}(Students\_ID, P_k)$ across sessions. Response time of any $PopUp_j$ is simply the VLR paused time, but its intensity and interpretation depends on its type.

T_{Idle} is the Idle time; in learning environments it is usually based on the mouse and key board actions. It might be misleading if adapted to VLR without deliberations. Sometimes the VLR are too long; in CRETAL we advise to consider responses to the dynamic popup to estimate the idle time [see Eq. (3)]. For example in prescription mode, VLR has $PopUp_k$ popped at time t_0, its author's response time is r_k units and the user has not responded to the popup till $t_0 + 2r_k$ then total idle time till the moment is given by the equation below;

$$
\begin{aligned}
T_{Idle}(Students\_ID) = T'_{Idle}(Students\_ID) + start\_time\_of(PopUp_k) \\
- start\_time\_of(PopUp_k) \\
- author\_response\_time\_of(PopUp_{k-1}) - 2r_k
\end{aligned}
\tag{3}
$$

3.2 Learnability Index and Degree of Understanding

Popularly known as learning ability Index, used in the estimation of learning gain and as performance indicator to understand and quantify the attainment of learning outcomes [6, 11]; the popup and the dynamic suggestion box can be connected with Questioning System which supplies questions of required denomination, type, targeted concept, difficulty level as per the outcome mapper (see (d) of Fig. 1) similar to intelligent question bank agent of the web based self-assessment system developed by Antal and Koncz [12]. The response time of such tests can be used to estimate the learnability of the students according to the Baker's item response theory; like answering two questions belonging to the same concept with incremental difficulty where second question's answer is built over the first one. It could be this way in CRETAL, two popup $PopUp_p$ at t_1 and $PopUp_q$ at t_2 where both are of the Type question and belongs to same prescription or timeline of the tag; in the learning span of $(t_2 - t_1)$ if the user's average response time $r \leq (r_p + r_q)/2$ then it can be considered towards the higher ability.

For the degree of mastery DM of the concepts learned (learning gain) can be estimated from the popup having Popup_Type as question along with question's difficulty level [3, 12]. In CRETAL environment it depends on the objective of the prescription and is directly proportional to the level of mastery like Familiarity, Usage, and Assessment etc. Other than popup of question type, tags having the Tag_Type as chit notes and their relevance reward index R_{index} (see Eq. (4)) can contribute to the estimation of DM.

$$R_{index}(Students\_ID) = \alpha \sum_{Tag_i.User\_ID=Students\_ID, Tag_i.Type=Chit\_Notes} number\_of\_appreciations\_of(Tag_i)$$

$$(4)$$

where, α is the stabilization constant fixed by the teacher in his own method like cross verification of number of students participated in R_{index} or an offline discussion with students. R_{index} is calculated out of the accumulated appreciations for all student's tags by the community also contributes to the learning achievement. If the prescription of the teacher is to create chit notes by the students, then its offline evaluation index value along with Eq. (4) can be used for the estimation of students Attainment of Outcome of the concepts covered by the tags or VLR followed.

3.3 Learning Style Assessments

Input about the student's learning style and pre-knowledge is vital for personalized recommendation [10]; prescription by the teacher using CRETAL is an effort to help the students in similar line. The automatic student modeling approach proposed by Graf et al. [13] for detecting the learning style out of the four dimensions theory of Felder and Silverman; depends on student's activities in learning management system like Moodle. The data related to student's activity in the CRETAL either in prescription mode or in free learning mode are helpful in assessing the students learning style which in turn can support any recommendation system. Number of sessions, number of times the similar content accessed, number of seek backs for an VLR in a session, total pause time across the tags of marker type (this type of tags are used by the CRETAL users for marking diagrams, charts, table etc.

in an VLR), number of dynamic advises generated versus the number of advises taken etc. can increase the correctness of the estimation of Index of Learning Style.

3.4 Learning Rate and Concentration Index

These are the data in learning systems without VLR in knowing the attention level of the students during the session. In prescription mode of CRETAL's environment, responding to the optional popup can be used observe attention of the user (see (a) of Fig. 1).

4 CRETAL's Data Analysis

We have taught two different courses for the same set of undergraduate students at our institute during 2012–2014 in separate semesters with 135 student's registrations; we adapted the Open Assessment Methodology for the student's achievement assessments and grading [14]. We adapted CRETAL only for last course where most of the VLR were from MIT and NPTEL. To study the effectiveness and relevance of the data generated by CRETAL, we compared assessment data of both the courses, we found significant differences in the student's liveliness and their participation in the course conduction. We will have the context and content of the courses excluded to keep the further discussion a data centric; we will refer the Course2 as the course which adapted CRETAL and Course1 as the other. During the Course2, class room coverage is extensively mapped with the teacher's tags and prescriptions with help of Class Activity Sheet (CAS) [14]. The mapping involved more than 65 VLR from NPTEL, MIT and few from YouTube for extra examples and deep dives. With nearly 200 teachers tag, there were 1,000 tags altogether. Though CRETAL do not have analytics tool yet to make automatic analysis of Buddy shares in the community; the manual analysis of the raw data showed nearly 3,000 plus sharing. Figure 2 shows total number of offline consultations being uniformly distributed in the semester weeks for Courses2 compared to the Course1; the availability of the reference material (VLR) beyond contact hours with pin pointed tags have reduced number consultation with teacher. This is because the students have found a way to clear certain level of doubts by repeating the VLR, Chit notes of buddies and teachers. We confirm this from the Fig. 3 which shows the number of effective hours (excluded idle time with CRETAL) invested by the students in using CRETAL which otherwise might have been spent on non VLRs as non fruitful exercise and hence this effort of students would have gone unaccounted. As result of this enhanced investment of time in effective and innovative way has resulted nearly 20 % of grade shift; we can confirm this from Fig. 4.

Fig. 2 Variation plot of total number consultation by students with course teacher

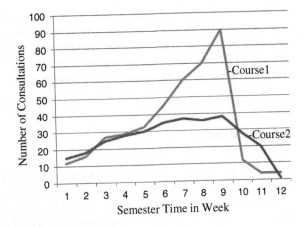

Fig. 3 Variation plot of total effective time that students put in using the VLR during Course2

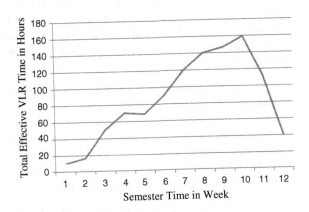

Fig. 4 Variation plot of grade distribution of students for the two courses

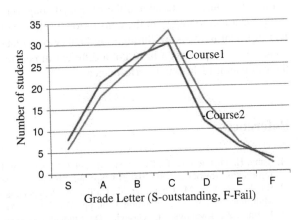

The main reasons for this grade shift are; (1) effective facilitation VLR as supplementary (not alternatives) course materials and (2) Enable the course teacher with new set of data about the students efforts in non contact hours.

Some of the positive changes that we noticed were quality of discussions and arguments by the students to defend the correctness of their assignments had significantly improved; we could not quantify this change in the students as it is difficult to interpret objectively, but we could confirm this to ourselves with grade distribution and the satisfaction levels of the students.

5 Conclusion

Producing the video material for learners without bothering individuals learning style is a one-fits-all kind of method; another is to effectively adapt such VLR along with VLRs which are byproduct of premium institution's regular teaching-learning process to facilitate students in conventional setup of other institutions. CRETAL has high potential to enable the students with different learning styles by facilitating variety of supplementary learning resources. CREATL can help teachers and the institutions to adapt VLR and generate meaningful data to all of these types of users. With CRETAL and its data, three most significant benefits that we would like confirm are;

1. Dynamic suggestions to a fresh learner based on the previously established trajectory by students of similar type.
2. Data on students activities and their effective time in CRETAL helps teacher to know about off-line (non contact hours) effort of a student in achieving the course outcomes, which can be used for grading.
3. Institutions adapting such system along with conventional set up helps them to move towards personalisation of education.

The data given by the CRETAL are certainly from the new source because of the nature of resource material and their blended adaption to the courses through open assessment methodology in the conventional setup. Not to forget the CRETAL has feel of social networking through the facility of collaborative learning, because of which the data produced have got a different dimension to them in contributing to the estimation of learning performance.

References

1. Yang, H., Gr¨unewald, F., Bauer, M., Meinel, C., Lecture Video Browsing Using Multimodal Information Resources, pp. 204–213. Springer, ICWL, Heidelberg (2013)
2. Riegler, M., Lux, M., Charvillat, V., Carlier, A., Vliegendhart, R., Larson, M.: VideoJot—A Multifunctional Video Annotation Tool, pp. 534–538. ACM, ICMR (2014)
3. Jeremic, Z., Jovanovic, J., Gaševic, D.: Student modeling and assessment in intelligent tutoring of software patterns. J. Expert Syst. Appl. **39**, 210–222 (2012) (Elsevier)
4. Choi, C.R., Song, Y.J., Jeong, H.Y.: Personalized learning course planner with E-learning DSS using user profile. J. Expert Syst. Appl. **39**, 2567–2577 (2012) (Elsevier)

5. ZOSMAT: Web-based intelligent tutoring system for teaching–learning process. J. Expert Syst. Appl. **36**, 1229–1239 (2009) (Elsevier)
6. Chen, C.M.: Personalized E-learning system with self-regulated learning assisted mechanisms for promoting learning performance. J. Expert Syst. Appl. **36**, 8816–8829 (2009)
7. Mulwa, C., Lawless, S., Sharp, M.: Inmaculada Arnedillo-Sanchez and Vincent Wade, Adaptive educational hypermedia systems in technology enhanced learning: A Literature Review, pp. 73–84. ACM, Information Technology Education, New York (2010)
8. Klamma, R.: Community Learning Analytics—Challenges and Opportunities, pp. 284–293. Springer, Heidelberg (2014)
9. Felder, R.M., Brent, R.: Understanding student differences. J. Eng. Edu. **94**, 57–72 (2005)
10. Wang, K.H., Wang, T.H., Wang, W.L., Huang, S.C.: Learning styles and formative assessment strategy: enhancing student achievement in web-based learning. J. Comp. Assist. Learn. **22**, 207–217 (2006)
11. Chrysafiadi, K., Virvou, M.: PeRSIVA: An empirical evaluation method of a student model of an intelligent e-learning environment for computer programming. J. Comp. Edu. **68**, 322–333 (2013) (Elsevier)
12. Antal, M., Koncz, S.: Student modeling for a web-based self assessment system. J. Expert Syst. Appl. **38**, 6492–6497 (2011) (Elsevier)
13. Graf, S., Viola, S.R.: Kinshuk: Automatic Student Modelling for Detecting Learning Style Preferences in Learning Management Systems, pp. 72–179. IADIS, Algarve (2007)
14. Shivanagowda, G.M., Goudar, R.H., Kulakrni, U.P.: Open assessment method for better understanding of student's learnabilty to create personalised recommendations. In: Proceedings of CTIEE, Springer, Heidelberg (2014) (in press)
15. Devedžić: A survey of modern knowledge modeling techniques. J. Expert Syst. Appl. **17**, 275–294 (1999) (Elsevier)
16. Costello, R., Mundy, D.P.: The Adaptive Intelligent Personalised Learning Environment, IEEE, Advanced Learning Technologies, pp. 606–610 (2009)
17. Wanga, S.L., Wub, C.Y.: Application of context-aware and personalized recommendation to implement an adaptive ubiquitous learning system. J. Expert Syst. Appl. **38** 10831–10838 (2011) (Elsevier)

Hybrid Single Electron Transistor Based Low Power Consuming Odd Parity Generator and Parity Checker Circuit in 22 nm Technology

Sudipta Mukherjee, Anindya Jana and Subir Kumar Sarkar

Abstract Co-fabrication between single electron transistor (SET) and CMOS technology has already proved to be feasible in production of future low power ultra dense circuitry. Mutual integration between this two can thus be efficient in computing applications. Here, an odd parity generator and parity checker circuit is build up with hybridization of SET-CMOS technology using Mahapatra-Ionescu-Banerjee (MIB model) and BSIM 4.6.1 model. Power consumption and PDP are also calculated and compared numerically and graphically with the conventional CMOS technology.

Keywords MIB · BSIM4.6.1 · Parity generator · Parity checker · Hybrid SET-CMOS

1 Introduction

In every processing system to accomplish data transfer, some parity bits are to be generated and checked whether transmission of data is successful or not. Those parity generator and checker circuits are hereby designed with hybrid SET-CMOS technology primarily to reduce power consumption. Single electron transistor (SET) [1] is capable of delivering ultra low power as well as ultra dense circuitry,

S. Mukherjee (✉) · A. Jana · S.K. Sarkar
Department of Electronics and Telecommunication Engineering, Jadavpur University,
Kolkata 700032, India
e-mail: sudipta.conference@gmail.com

A. Jana
e-mail: anindya.jana@rediffmail.com

S.K. Sarkar
e-mail: su_sircir@yahoo.co.in

© Springer India 2015
L.C. Jain et al. (eds.), *Computational Intelligence in Data Mining - Volume 1*,
Smart Innovation, Systems and Technologies 31, DOI 10.1007/978-81-322-2205-7_50

but speed of operation is not so high. Moreover, it shows some background charge effect and non-availability of room temperature operable technology. On the other hand, CMOS has high current drive and well established fabrication techniques found in many years. That is why hybridization [2] was needed between them.

2 Basic Parity Generator Circuit

For any instruction to direct computer in making an operation some extra binary bits are added to the original binary number to produce that number in even or odd parity. Say, 4 bits of given input are 0110; this has even parity. Therefore when applied to an Ex-or gate, it will generate low logic at output. If an inverter is connected before that ex-or output, final bit will be high logic (1). Thus conjunctively 5 bits are presented with odd parity (10110).With the help of EX-NOR gate, we can easily generate that parity bit. Figure 1 reflects this circuit with hybridization between SET and CMOS.

Fig. 1 Odd parity generator circuit with CMOS-SET hybridization

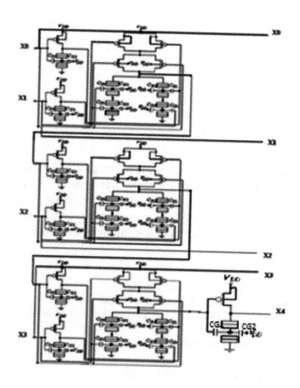

Fig. 2 Hybrid SET-CMOS
based parity checker circuit

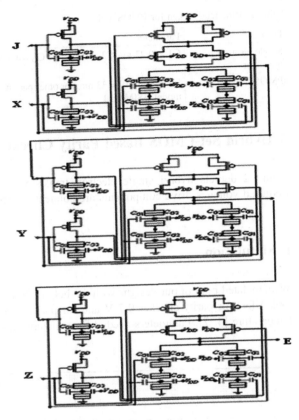

3 Concept of Parity Checker Circuit

Parity checking technique is mainly used in stored code groups. Ex-or gates are compatible for parity checking as they produce high level output when there are odd number of 1's present at input. For even number of 1's, it will surely generate low logic at output. To check parity of an n-bit number, $(n - 1)$ Ex-or gates are needed to be cascaded. This circuit in hybrid form is vividly depicted in Fig. 2.

4 Hybrid Set-CMOS Based Odd Parity Generator Circuit

Following Fig. 1 represents hybrid [3] odd parity generator where X0–X3 are initially given to the circuit as inputs. Now bit X4 is generated resulting from and along with them.

Table 1 Parameters taken for PMOS and SET

Device	Parameter
Single electron transistor	C1 = 0.27 aF, C2 = 0.12 5aF C_D = C_S = 0.1 aF, R_{TD} = R_{TS} = 1 MΩ
PMOS	L = 22 nm, W = 33 nm; for other parameters standard values taken from model BSIM4.6.1

5 Hybrid Set-CMOS Based Parity Checker Circuit

Figure 2 depicts hybrid single electron transistor based parity checker circuit. Required voltages drive components are connected accordingly.

6 Simulation

We simulated both of our designs with the help of MIB and BSIM 4.6.1 model and took VDD for parity checker 0.7 V and for parity generator to be 0.9 V at room temperature [4, 7]. Parameters needed for realization [4] are shown in Table 1.

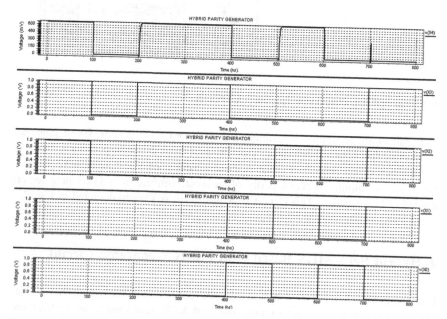

Fig. 3 Waveforms for hybrid SET-CMOS based parity generator

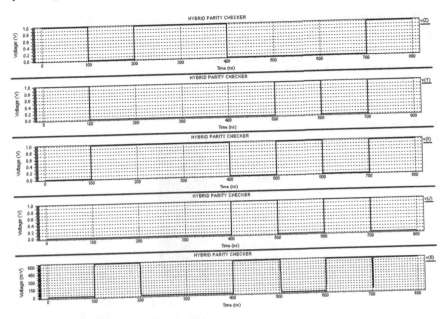

Fig. 4 Responses from hybrid SET-CMOS based parity checker

Figure 3 shows input output voltage waveforms found at simulating the hybrid odd parity generator circuit and Fig. 4 reflects the responses we got from hybrid parity checker circuit.

7 Power and PDP Analysis

An impressive reduction in power consumption is demonstrated both for parity generator as well as parity checker in comparison with so-called CMOS logic numerically over following Tables 2 and 3.

Above comparison is graphically shown in Fig. 5.

Above comparison is graphically shown in Fig. 6.

Power-delay product for hybrid SET-CMOS [5, 6] based circuit and conventional CMOS based encoder is shown in Tables 4 and 5.

The graphical comparison plot is reflected in Fig. 7.

The graphical comparison plot is reflected in Fig. 8.

Table 2 Comparison of average power consumption for odd parity generator circuit

Parameter	Conventional CMOS based odd parity generator	Hybrid SET-CMOS based odd parity generator
Power consumption (W)	4.59E–09	4.44E–11

Table 3 Comparison of average power consumption for parity checker circuit

Parameter	Conventional CMOS based parity checker	Hybrid SET-CMOS based parity checker
Power consumption (W)	4.59E−09	4.44E−11

Fig. 5 Comparison of average power consumption

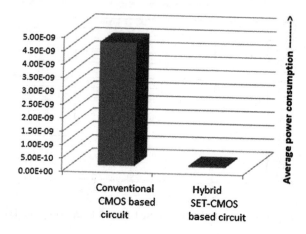

Fig. 6 Comparison of average power consumption

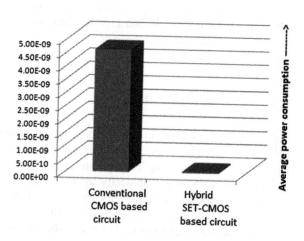

Table 4 Power-delay product comparison for odd parity generator circuit

Parameter	Conventional CMOS based odd parity generator	Hybrid SET-CMOS based odd parity generator
Power-delay product (W s)	1.61E−15	1.32E−17

Table 5 Power-delay product comparison for parity checker circuit

Parameter	Conventional CMOS based parity checker	Hybrid SET-CMOS based parity checker
Power-delay product (W s)	1.61E–15	1.33E–17

Fig. 7 Power-delay product comparison

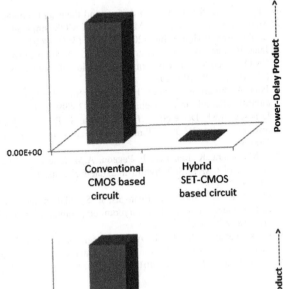

Fig. 8 Power-delay product comparison

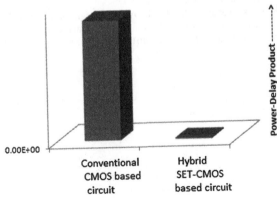

8 Conclusion

This work has proved itself to be implemented in future low power, ultra dense computing circuit as presented in Sect. 7. We have made the total power consumption reduced by providing supply voltage below 1 V as well as making all the island related capacitance values to lie within some atto-farad range. Therefore our designed odd parity generator and parity checker circuit is feasible to be operated at room temperature [7, 8] in latest nano-technological trend.

Acknowledgments Subir Kumar Sarkar thankfully acknowledges the financial support obtained in the form of fellowship from UGC UPE PHASE-II, "Nano Science and Technology".

References

1. Likharev, K.: Single-electron devices and their applications. Proc. IEEE **87**, 606–632 (1999)
2. Mahapatra, S., Ionescu, A.M.: Hybrid CMOS single-electron-transistor device and circuit design. Artech House, Inc. (2006). ISBN 1596930691
3. Jana, B., Jana, A., Sing, J.K., Sarkar, S.K.: A comparative performance study of hybrid SET-CMOS based logic circuits for the estimation of robustness. J. Nano Electron. Phys. **5**(3), 3057 (1)–3057(6), 96–100 (2013)
4. Jana, A., Singh, N.B., Sing, J.K., Sarkar, S.K.: Design and simulation of hybrid CMOS-SET circuits. Microelectronic. Reliab **53**(4), 592–599 (2013)
5. Lonescu, A.M., Declercq, M.J., Mahapatra, S., Banerjee, K., Gautier, J.: Few electron devices: towards hybrid CMOS-SET integrated circuits. In: IEEE Conference Publications, pp. 88–93 (2002)
6. S. Mahapatra, K. Banerjee, F. Pegeon, A.M. Ionescu : A CAD framework for co-design and analysis of CMOS-SET hybrid integrated circuits. In: Proceedings of ICCAD, pp. 497–502 (2003)
7. Parekh, R., Beaumont, A., Beauvais, J.: Simulation and design methodology for hybrid SET-CMOS integrated logic at 22-nm room-temperature operation. IEEE Trans. Electron Devices **59** (4), 918–923 (2012)
8. Venkataratnam, A., Goel, A.K.: Design and simulation of logic circuits with hybrid architectures of single-electron transistors and conventional MOS devices at room temperature. Microelectron. J. **39**(12), 1461–1468 (2008)

Comparative Analysis of Decision Tree Algorithms: ID3, C4.5 and Random Forest

Shiju Sathyadevan and Remya R. Nair

Abstract To analyze the raw data manually and find the correct information from it is a tough process. But Data mining technique automatically detect the relevant patterns or information from the raw data, using the data mining algorithms. In Data mining algorithms, Decision trees are the best and commonly used approach for representing the data. Using these Decision trees, data can be represented as a most visualizing form. Many different decision tree algorithms are used for the data mining technique. Each algorithm gives a unique decision tree from the input data. This paper focus on the comparison of different decision tree algorithms for data analysis.

Keywords Iterative dichotomiser 3 (ID3) · C4.5 · Randomforest

1 Introduction

We live in a world where vast amounts of data are collected daily. Analyzing such data for acquiring the meaningful information is an important need. Human analysts with no special tool might not yield in producing the right result sets. But data mining can meet this need by providing tools to discover knowledge from data, Classification and prediction are two forms of data analysis tools.

Classification is one of the fundamental and Useful technique in data mining. Using classification algorithm we can construct a model and used to predict the class label of the testing instances. It has been successfully applied to many real world application areas, such as medical diagnosis, weather prediction, credit approval etc. In classification, several approaches are adopted to classify the data.

S. Sathyadevan (✉) · R.R. Nair
Amrita Center for Cyber Security Systems and Networks, Amrita University, Kollam, India
e-mail: shiju.s@am.amrita.edu

R.R. Nair
e-mail: remyarnair@am.amrita.edu

549

© Springer India 2015
L.C. Jain et al. (eds.), *Computational Intelligence in Data Mining - Volume 1,*
Smart Innovation, Systems and Technologies 31, DOI 10.1007/978-81-322-2205-7_51

Decision trees are a very popular and commonly used approach for classification [1]. This Decision tree classifiers start with the Training set with their associated class labels. The root node is the main feature. Each internal node represents the test attributes, each branch represents the outcome of the test and each leaf node represents the class labels. To identify the class label for an unknown sample, the Decision tree classifier will trace path from root to the leaf node, which holds the class label for that sample.

This paper focus on the comparison of three different decision tree algorithms ID3, C4.5 and Random Forest [2]. These three decision tree algorithms are different in their features and hence in the accuracy of their result sets. ID3 and C4.5 build a single tree from the input data. But there are some differences in these two algorithms. ID3 only work with Discrete or nominal data, but C4.5 work with both Discrete and Continuous data. Random Forest is entirely different from ID3 and C4.5, it builds several trees from a single data set, and select the best decision among the forest of trees it generate. The rest of the paper is organized in four sections. In Sect. 2 the related work and literature of decision tree algorithms is presented. In Sect. 3 the three algorithms', ID3, C4.5 and Random Forest is explained with a brief discussion about the algorithm along with the pseudo code. Section 4 compares the three algorithms and explains the differences among them. In Sect. 5, an application is selected where the three algorithms are applied and compared with respect to their prediction accuracy.

2 Related Work

There are so many researches are done for Decision tree learning algorithms. ID3 and C4.5 are two most popular decision tree algorithms. Among these algorithms, ID3 is the basic algorithm. But it has many drawbacks. So To improve ID3, researchers have proposed many methods, such as, use weighting instead of information gain [3], user's interestingness [4] and attribute similarity to information gain as weight. Chun and Zeng [5] also have proposed improved ID3 based on weighted modified information Gain called ωID3.

Quinlan [6] has developed tree based classification algorithm known as C4.5, an extension to the ID3 algorithm. It uses the gain ratio for building the tree. The C4.5 deals with continuous attributes which was not supported by ID3. It divides the values of a continuous attribute in a two subsets. He also proposed method of pruning, which deals with the removal of unwanted branches which are generated by noise or too small size of training data [7].

In the case of Random Forest algorithm, it builds random trees from a single input dataset. This algorithm is efficient for predicting the accurate results. Because, we compare many trees and obtain the best from them. So the result obtained from Random Forest is more accurate than ID3 and C4.5. So our main goal is to compare the prediction accuracy of these 3 decision tree algorithms ID3, C4.5 and Random Forest and achieve high performance and accurate results.

3 Understanding Decision Tree Classifiers

3.1 ID3 Classifier

ID3 (Iterative Dichotomiser 3) is the basic algorithm for inducing decision trees. This algorithm builds a decision tree from the data which are discrete in nature. For each node, select the best attribute. And this best attribute is selected using the selection criteria—Information Gain [8]. It indicates how much informative a node is. And the attribute which has the highest Information Gain is selected as split node.

3.2 C4.5 Classifier

C4.5 Algorithm is developed based on the Decision tree Algorithm ID3 [9]. ID3 is also used to generate decision trees. But it does not guarantee an optimal solution to analyze continuous data. But C4.5 algorithm overcomes these short comings. They are:

```
Algorithm 1 pseudo code
```

```
Input: - Dataset S
Output: - Decision Tree
  Begin
1.Create a Root node for the tree.
2.If all Examples are +ve, Return single Root with label = +.
3.If all Examples are -ve, Return single Root with label = -.
4.If Attributes is empty, Return the single-node tree with
  label = most common value of Target attributes in Examples.
5.Otherwise Begin
  (a) A, the attribute that best Classifies the input data.
  (b) The decision attribute for Root <- A.
  (c) For each possible value, vᵢ , of A,
     (i)   Add a new tree branch below Root corresponding to
           the test A = vᵢ
     (ii)  Let Examples be the subset of Examples that have
           value vᵢ for A
     (iii) If Examples is empty

           (A) Then below this new branch add a leaf node with
               label = most common value of  Target attribute in
               Examples
           (B) Else below this new branch add the sub tree.
6.Building the decision tree nodes and branches recursively
  until a certain subset of the instances belonging to the same
  category.
  End
```
---

1. Attribute Selection Criteria

ID3 uses Information Gain as the Selection criteria.

$$Gain(S) = Entropy(S) - \sum_{c=1}^{n} \frac{|S_c|}{S} Entropy(S_v) \qquad (1)$$

If the input is continuous, this Gain doesn't give an optimum result. That's why ID3 is only applicable for discrete datas. But C4.5 overcomes this by introducing new method known as Gain Ratio: This method contains two concepts, Gain and Split Info.

$$Gain\, Ratio(S) = \frac{Gain(S)}{Split\, Info(S)} \qquad (2)$$

So If the attribute is continuous, this selection criteria gives the optimum result.

3.3 Random Forest Classifier

Random forest is another Decision tree technique that operates by constructing multiple decision trees [10]. This algorithm is based on bagging (Bootstrap aggregating) [11], i.e. after building multiple random training subsets, the algorithm construct one tree per random training subsets. This technique is called random split selection method and the trees known as random trees.

```
Algorithm 2 pseudo code
Input: -  A data set S
Output: - Random Number of Trees
   Begin
1.Choose T - number of trees to grow
2.For b = 1 to T
   a)Draw a Bootstrap sample Z  of size N from the training Data
   b)Grow a random-forest tree T b to the bootstrapped data, by
     re -cursively repeating until the minimum node size nmin is
     reached.
     1)Select m variables at random from the p variables.
     2)Pick the best variable/split-point among the m.
     3)Split the node into daughter nodes.
3.  Output the ensemble of trees fTBg
   End
```

4 Comparisons

Among the three algorithms: ID3, C4.5 and Random Forest. We know that, ID3 is the basic decision tree algorithm. The selection criteria used for ID3 is not suit for continuous datasets. So this algorithm is only applicable for discrete cases. To overcome this problem, Quinlan extend the ID3 to C4.5 by introducing a new selection criteria called Gain Ratio [12]. This gives optimum result with both discrete and continuous case data sets. It is not possible to assure that the tree build from ID3 and C4.5 is an accurate tree, because it only generates a single tree for a given set of input data. So if the new data set is applied to the model(tree) so generated, we get only one prediction result. This Prediction may or may not be correct. Hence the correctness and accuracy of these two algorithms cannot be assured. To overcome this accuracy problem, Random Forest was implemented. This algorithm generates several trees (many Random trees are generated) from a single data set. So the method is to apply the new data set to every tree in the model and list the output generated from it. Best prediction is determined by selecting the majority class value. That is, which ever class among the lot is predicted the most number of times that class is selected as the final prediction.

5 Experimental Results

Dataset used for this application is Credit Approval Dataset, obtained from the Machine Learning repository UCI [13].

In this Credit Approval dataset some of the attributes are linear and some are nominal. Table 1 shows the characteristics of selected Credit Dataset.

Now a day, banks play a crucial role in market economies. For markets and society to function, individuals and companies need access to credit [14]. To a bank, a good risk prediction model for loan allocation is necessary so that the bank can provide as much credits as possible without any risk. This report describes an approach for predicting the loan approving application using ID3, C4.5 and Random Forest [15]. Using these algorithms, we can determine whether the loan can be approved or not. But before approving a loan, bank will look at all the relevant financial history of the borrower. These features are represented as the attributes of the dataset are shown in Table 2.

As a first step, we check the class labels of all the instances in the dataset. If all tends to fall under the same class, it is possible to stop the tree with the leaf node as

Table 1 Training dataset matrix

Dataset	Credit approval
No of instances	25
No of attributes	21
No of classes	2

Table 2 Training dataset attribute details

Attribute name	Description
Checking account	Checking the account status
Credit history	No credits taken/all credits paid back duly
Purpose	Car, furniture, business, education etc.
Credit amount	Credit amount
Savings status	Current saving status of the borrower
Years employed	Unemployed/present employed since
Installment rate	Installment rate in % of disposable income
Personal status	Single, married, divorced, separated
Other debtors	Guarantors/co-applicant
Resident since	Present residence since
Property	Real estate, savings, car, no property
Age	Age of borrower in years
Other plans	Banks, stores, none
Noncredit at bank	Number of existing credits at this bank
Job	Check the borrower is employed or not
Dependents	Number of dependents including family
Telephone	None or yes
Foreign	None or yes
Housing	Rent/own/for free
Approve	Good or bad

the common class label. Otherwise, select the best attribute that partition the training set.

Figure 1 shows the decision tree built by the ID3 algorithm where it uses information gain as the splitting measure. This tree does not represent an optimum result because some of the attributes are continuous and some are discrete. In the case of continuous datasets, the highest gain is for the attribute that has more number of partitions. Here, out of 20 attributes, 7 attributes are continuous (Credit Amount, Duration, Installment rate, Resident Since, Age, Num Credits At Bank and Dependents). Information Gain is used as the selection method. Out of this 7 continuous attributes, highest gain is for the attribute "Credit Amount" (because it has 25 different values). So based on the Information Gain, Credit Amount is selected as the Root node. But this tree cannot be treated as an optimum solution for this application, because there are no other checking conditions (sub node) in this tree.

So if it is tested against a new data set, then it will predict only based on the attribute "Credit Amount". For example, Table 3 represents testing data consisting of 10 instances, where 30 % of the test data is the training data without class label

Fig. 1 Credit approval tree using ID3

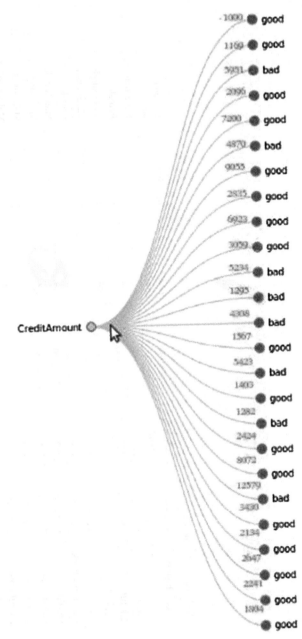

and 70 % of the test data is new test instances. Each instance in this test data will be predicted with the help of the ID3 tree in Fig. 2. So when this data is applied on ID3, each instance traverses the ID3 tree and reaches the class values (in this case either good or bad). This class values will be taken as the predicted result. ID3 tree

Table 3 Test data applied against the model

Checking account	Credit history	Purpose	Credit amount	Savings status	Years employed	Installment rate	Personal status	
<0	Ok	Car	7,200	<100	<7	3	Male-married	...
<200	Ok-till-now	Furniture	6,923	<100	<4	4	Female	...
None	Critical	Car	5,423	<500	<4	4	Male-single	...
None	Ok	Television	426	<100	>=7	4	Male-married	...
>=200	Ok-at-this-bank	Television	409	>=1000	<4	3	Female	...
<0	Ok-till-now	Furniture	7,882	<100	<7	2	Male-single	...
<200	Ok-till-now	Television	2,415	<100	<4	3	Male-single	...
<0	Past-delays	Car	4,870	<100	<4	3	Male-single	...
<0	Ok-till-now	Furniture	1,374	<100	<4	1	Male-married	...
<0	Past-delays	Business	6,836	<100	>=7	3	Male-divorced	...

#	Checking	Duration	CreditHistory	Purpose	CreditAmount	Savings	YearsEmplo	Installr	PersonalStatus	OtherDeb	Residents	Property	Age	Other	House	NumCh	Job	Depende	Telepho	Foreign	Approve
1	<0	0	ok	furniture	-1000000000	unknown	unemployed	-10000	female_single	none	0	car	-293	none	own	-1000	skilled	-185	yes	no	good
2	<0	0	ok	furniture	1169	unknown	>=7	4	male_single	none	4	real_estate	67	none	own	2	skilled	1	yes	yes	good
3	<0	6	critical	television	5951	<100	<4	2	female	none	2	real_estate	22	none	own	1	unskilled	1	no	yes	bad
4	<200	48	ok_til_now	television	2096	<100	<7	2	male_single	none	3	real_estate	49	none	own	1	unskilled	2	no	yes	good
5	none	12	critical	education	7882	<100	<7	2	male_single	guarantor	4	savings	45	none	free	1	skilled	2	no	yes	good
6	<0	42	ok_til_now	furniture	4870	<100	<4	3	male_single	none	4	unknown	53	none	free	2	skilled	2	yes	yes	bad
7	<0	24	past_delays	car_new	9055	unknown	<4	2	male_single	none	4	unknown	35	none	free	1	unskilled	2	yes	yes	good
8	none	36	ok_til_now	education	2835	<1000	>=7	3	male_single	none	4	savings	53	none	own	1	skilled	2	yes	yes	good
9	none	24	ok_til_now	furniture	6948	<100	<4	2	male_single	none	2	car	35	none	rent	1	management	1	yes	yes	good
10	<200	36	ok_til_now	car_used																	

Fig. 2 Training dataset for building the model

contains 1 root node "Credit Amount" and 25 leaf nodes. Prediction result are as listed in Table 4. So each instance in the above testing data will only check their Credit Amount value, based on which the results will be predicted.

So in actual practice, it is not possible to make a decision whether to approved or reject a loan based on just one attribute which in this case is "Credit Amount". So this cannot be considered as an optimum result. To overcome this, the same data set is pushed through C4.5.

If we apply this Credit dataset on C4.5, it builds a decision tree which is totally different from ID3's output.

In C4.5 we use Gain Ratio as the splitting criteria. Figure 3 shows the C4.5 Decision Tree. But the working of Gain Ratio is different from Information Gain as used in ID3. Gain Ratio doesn't depend on the attribute type, instead calculates Information Gain and Split Info for each attribute. It will select the attribute which has the highest Gain Ratio as the best split-point. So by using this Ratio format formula Eq. (2), it is now possible to overcome the problems faced while using ID3.

The same 3 data instances as shown in Table 3, when applied to C4.5, each instance traverse through the model and predict the output. Here the attribute "Personal status" is selected as the root node as it had the highest gain ratio. "Installment rate" is selected as the sub node to "Persona Status" as it has the highest Gain ratio within that tree. Going forward when new data sets are applied to this model, it will only check "Personal Status" and "Installment Rate", because in this tree these are the 2 nodes whose value will predict the result set. Prediction results are as listed in Table 5.

It is not possible to assure that the tree build from C4.5 is an accurate one because here also only a single tree is generated from the input data. So if a new test data is applied to the model (tree), there will only be one prediction result which may or may not be correct.

If it is possible to compare more than one prediction and pick the best prediction from the lot, then it can provide better accuracy. Random Forest Algorithm can build number of trees from the training data set and is represented in the form of rules [16]. For each tree, corresponding rules are generated. Testing data (as shown in Table 3) is applied on each rule generated from the Random Forest, and pick the one that occurs most frequent as the best decision.

Figure 4 shows the rules generated by applying credit dataset on Random Forest. When a new test data set is applied on to these rules, then whichever class value (Good or bad) occurs the most for an instance against each tree, that class value will be set as the best prediction. The same test data set show under Table 3 is applied against Random Forest algorithm.

Table 4 Prediction generated by ID3

Checking account	Credit history	Purpose	Credit amount	Savings status	Years employed	Installment rat		Approve
<0	Ok	Car	7,200	<100	<7	3	⋮	Good
<200	Ok-till-now	Furniture	6,923	<100	<4	4	⋮	Good
None	Critical	Car	5,423	<500	<4	4	⋮	Bad
None	Ok	Television	426	<100	>=7	4	⋮	Bad
>=200	Ok-at-this-bank	Television	409	>=1,000	<4	3	⋮	Bad
<0	Ok-till-now	Furniture	7,882	<100	<7	2	⋮	Good
<200	Ok-till-now	Television	2415	<100	<4	3	⋮	Bad
<0	Past-delays	Car	4870	<100	<4	3	⋮	Bad
<0	Ok-till-now	Furniture	1374	<100	<4	1	⋮	Bad
<0	Past-delays	Business	6836	<100	>=7	3	⋮	Good

Fig. 3 Credit approval tree using C4.5

Table 5 Prediction generated by C4.5

Checking account	Credit history	Purpose		Installment rate	Personal status		Approve
<0	Ok	Car	..	3	Male-married	..	Bad
<200	Ok-till-now	Furniture	..	4	Female	..	Good
None	Critical	Car	..	4	Male-single	..	Good
None	Ok	Television	..	2	Male-married	..	Bad
>=200	Ok-at-this-bank	Television	..	3	Female	...	Bad
<0	Ok-till-now	Furniture	..	2	Male-single	...	Good
<200	Ok-till-now	Television	..	3	Male-single	...	Good
<0	Past-delays	Car	..	3	Male-single	...	Good
<0	Ok-till-now	Furniture	..	1	Male-married	...	Bad
<0	Past-delays	Business	..	3	Male-single	...	Good

As shown in Table 6, in the 1st instance, Tree1 and Tree 3 predicted "good" where as Tree 2 predicted as bad. So class value with the maximum occurrence is "good". So this be selected as the final prediction, where as in the 2nd instances, class value "bad" have the maximum occurrence, and hence its final prediction was set as "bad". Similarly for the 3rd instance class value "Good" happen to repeat the most.

The output got from these three algorithms is different. In the case of ID3 and C4.5, only one prediction result was generated. Hence it was not easy to confirm that the prediction was correct as there were no way to compare the result sets. But in the case of Random Forest, three different decision trees were built, and hence three predictions were generated. From these three predictions, the class with the most occurrence was treated as the best prediction.

```
Tree3.model
IF {Property=car AND  Duration=<24.0} THEN Approve IS [good]
IF {Property=car AND  Duration=>24.0} THEN Approve IS [good]
IF {Property=unknown} THEN Approve IS [bad]
IF {Property=savings} THEN Approve IS [good]
IF {Property=real_estate AND  CheckingAccount=<200} THEN Approve IS [bad]
IF {Property=real_estate AND  CheckingAccount=none} THEN Approve IS [good]
IF {Property=real_estate AND  CheckingAccount=<0} THEN Approve IS [good]
Tree1.model
IF {SavingsAccount=<100 AND  Purpose=car_new AND  CreditAmount=<4870.0} THEN Approve IS [bad]
IF {SavingsAccount=<100 AND  Purpose=car_new AND  CreditAmount=>4870.0} THEN Approve IS [bad]
IF {Purpose=television AND  CheckingAccount=<200 AND  InstallmentRate=<2.0} THEN Approve IS [bad]
IF {Purpose=television AND  CheckingAccount=<200 AND  InstallmentRate=>2.0} THEN Approve IS [bad]
IF {Purpose=car_used} THEN Approve IS [bad]
IF {Purpose=education} THEN Approve IS [good]
IF {SavingsAccount=<1000} THEN Approve IS [good]
IF {SavingsAccount=<500} THEN Approve IS [bad]
IF {SavingsAccount=unknown} THEN Approve IS [good]
IF {SavingsAccount=>=1000} THEN Approve IS [good]
Tree2.model
IF {NumCreditsAtBank=<3.0} THEN Approve IS [good]
IF {NumCreditsAtBank=>3.0} THEN Approve IS [good]
IF {NumCreditsAtBank=>2.0 AND  InstallmentRate=<4.0 AND  CheckingAccount=none} THEN Approve IS [good]
IF {NumCreditsAtBank=>2.0 AND  InstallmentRate=<4.0 AND  CheckingAccount=<0} THEN Approve IS [bad]
IF {InstallmentRate=>4.0 AND  CreditAmount=<2134.0} THEN Approve IS [good]
IF {InstallmentRate=>4.0 AND  CreditAmount=>2134.0} THEN Approve IS [good]

--------------------
No of trees generated : 3
--------------------
```

Fig. 4 Credit approval rules—using random forest

Table 6 Prediction generated by random forest

Instances	Tree1	Tree2	Tree3	Final prediction
1st	Good	Bad	Good	Good
2nd	Bad	Bad	Good	Bad
3rd	Bad	Good	Good	Good
4th	Bad	Good	Good	Good
5th	Bad	Good	Bad	Bad
6th	Good	Good	Good	Good
7th	Bad	Good	Good	Good
8th	Good	Good	Good	Good
9th	Bad	Good	Bad	Bad
10th	Good	Good	Bad	Good

In the case of Prediction accuracy, Random Forest is better when compared to other two algorithms. The Prediction Accuracy of three algorithms is shown in Fig. 5.

This bar chart gives the overall idea about the Prediction Accuracy of the three Algorithms: ID3, C4.5 and Random For-est. For testing, 30 % of the training data without class and 70 % of the new data was selected. So from the above testing, ID3 gives prediction only based on 1 attribute (i.e. Root node-Credit Amount) only 30 % of the data is correctly classified. In the case of C4.5, it gives prediction based on 2 attributes (i.e., Personal-Status and Installment Rate), here 60 % of the testing data is correctly classified where as Random Forest compare multiple prediction

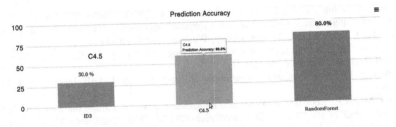

Fig. 5 Prediction accuracy—bar chart

results and select the best prediction from it based on the occurrence of the class values, here out of 10 instances, 8 instances are correctly classified, i.e. 80 % are correctly predicted.. So the Prediction Accuracy of Random Forest is better when compared to ID3 and C4.5. From the test results it is evident that from the three different forms of decision trees, Random Forest could generate better decision.

6 Conclusion

This paper provides a brief explanation of Decision tree learning method and classification techniques. It explains the main three Decision tree algorithms, ID3, C4.5 and Random Forest. Paper highlights the fact that these three decision tree algorithms are different in their prediction accuracy. Experimental Result section, compares these algorithms against a common testing dataset and the result set so generated are also compared. After analyzing and comparing the results, Random Forest gives the better prediction result. So Among these algorithms, Random Forest is best for accurate classification.

Acknowledgments We thank our college Amrita School of Engineering, Amritapuri and Amrita Center of Cyber Security, Amritapuri for giving us an opportunity to be a part of the internship program that leads to the development of this work. Many thanks to Shiju Sathyadevan for countless discussions and feedback that help me to complete the work successfully.

References

1. Rokach, L., Maimon, O.: Decision trees. In Maimon, O.Z., Rokach, L. (eds.) Data Mining and Knowledge Discovery Handbook vol. 6, pp. 165–192. Springer, Heidelberg (2005)
2. Peng, X., Guo, H., Pang, J.: Performance analysis between different decision trees for uncertain data. In: Proceedings of the 2012 International Conference on Computer Science and Service System, CSSS '12, pp. 574–577. Washington, DC (2012)
3. Jin, C., De-lin, L., Fen-xiang, M.: An improved ID3 decision tree algorithm. In: Proceedings of 2009 4th International Conference on Computer Science and Education, vol. 1 (2009)
4. Wang, M., Chai, R.: Improved classification attribute selection scheme for decision for decision tree. Comp. Eng. Appl. **3**, 127–129 (2010)

5. Guan, C., Zeng, X.: An improved ID3 based on weighted modified information gain. In: Seventh International Conference on Computational Intelligence and Security, vol. 1, pp. 1283–1285. IEEE Computer Society, Washington, DC (2011)
6. Quinlan, J.R.: C4.5 program for machine learning, vol. 16, pp. 21–30. Morgan Kaufmann Publishers Inc., San Francisco (1993)
7. Quinlan, J.R.: Improved use of continuous attributes inc 4.5. J. Artif. Intell. Res. 4, 77–90 (1996)
8. de Vries, A.P., Roelleke, T.: Relevance information: a loss of entropy but a gain for idf? In: Proceedings of the 28th Annual International ACM SIGIR Conference on Research and Development in Information Retrieval, pp. 282–289 (2005)
9. Ruggieri, S.: Efficient C4.5. IEEE Trans. Knowl. Data Eng. 2, 438–444 (2002)
10. Kotsiantis, S.: A hybrid decision tree classifier. J. Intell. Fuzzy Syst. 1, 327–336 (2014)
11. Fern, A., Givan, R.: Online ensemble learning: an empirical study. Mach. Learn 53, 279–286 (2000) (Morgan Kaufmann Publishers Inc., San Francisco)
12. Quinlan, J.R.: C4.5: Programs for machine learning, vol. 16, pp. 235–240. Morgan Kaufmann Publishers Inc., San Francisco, CA (1993)
13. Bache, K., Lichman, M.: UCI machine learning repository. Digit. Libr. 1 (2013)
14. Abele, J., Mantas, C.J.: Improving experimental studies about ensembles of classifiers for bankruptcy prediction and credit scoring. Expert Syst. Appl. Int. J. 41, 3825–3830. Pergamon Press Inc., USA (2014)
15. Olson, D.L., Delen, D., Meng, Y.: Comparative analysis of data mining methods for bankruptcy prediction. Decis. Support Syst. 2, 464–473 (2012)
16. Grzymala-Busse, J.W.: Rule induction. In: Maimon, O.Z., Rokach, L. (eds.) Data Mining and Knowledge Discovery Handbook, vol. 6, pp. 277–294. Springer, Heidelberg (2005)
17. Maimon, O., Rokach, L.: The Data Mining and Knowledge Discovery Handbook, vol. 2, pp. 1–17. Springer, Heidelerg (2005)
18. Podgorelec, V., Kokol, P., Stiglic, B., Rozman, I.: Decision trees: an overview and their use in medicine. J. Med. Syst. 5, 445–463 (2002)
19. Povalej, P., Kokol, P.: End user friendly data mining with decision trees: a reality or a wish? In: Proceedings of the 2007 Annual Conference on International Conference on Computer Engineering and Applications, vol. 3, pp. 35–40. Stevens Point, Wisconsin (2007)

Digital Forensic in Cloudsim

Jagruti Shah and L.G. Malik

Abstract Digital forensic is the method of providing digital evidence in order to prove the crime in digital world. In context of cloud environment the term digital forensics becomes more challenging due to dynamic and decentralized nature of cloud. The research proposes the digital forensic technique using Cloudsim. Cloud simulator is a powerful tool for modeling, simulating and carry out experimentation. This research presents digital forensic technique in context of cloud environment using cloudsim. The scenario of crime is created by hacking the client's sensitive data, which is stored in cloud. The tiled bitmap algorithm is used to detect tampering of database on cloud server which presents the potential evidence. Also the proposed method is used to store the file on cloud server using timestamp and encryption method to avoid hacking. The research focuses on one of the category of crime that is tampering of data on cloud server.

Keywords Cloudsim · Tiled bitmap

1 Introduction

Cloud computing is sharing of resources on a larger scale which is cost effective and location independent. Resources on the cloud can be deployed by the vendor, and used by the client. It also shares necessary software's and on-demand tools for various IT Industries.

When it comes to Security, cloud really suffers a lot. The vendor for Cloud must make sure that the customer does not face any problem such as loss of data or data theft. There is also a possibility where a malicious user can penetrate the cloud by

J. Shah (✉) · L.G. Malik
G.H. Raisoni College of Engineering, Nagpur, India
e-mail: shah.jagruti13@gmail.com

L.G. Malik
e-mail: latesh.malik@raisoni.net

© Springer India 2015
L.C. Jain et al. (eds.), *Computational Intelligence in Data Mining - Volume 1*,
Smart Innovation, Systems and Technologies 31, DOI 10.1007/978-81-322-2205-7_52

impersonating a legitimate user, there by infecting the entire cloud thus affecting many customers who are sharing the infected cloud. Data integrity, Data theft, Privacy issues, Infected application, Data loss, Data location, Security on Vendor level, Security on user level are various cloud crimes. In order to overcome these crimes a new branch of Digital forensic came into existence that is cloud forensic. A cloud forensics is a cross-discipline between cloud computing and digital forensics. Cloud forensics as a subset of network forensics (DFRWS 2001), as network forensics deals with forensic investigations in any kind of public or private networks, and cloud computing is based on broad network access, thus technically, cloud forensics should follow the main phases of network forensic process with extended or novel techniques tailored for cloud computing environment in each phase [1].

The digital forensics is unique in cloud environment due to multitenant architecture of cloud. The evidences are diversified and are very large. Some mining techniques are used to classify the logs into relevant and irrelevant data [2]. The cloud simulator can be used for simulating the cloud as well as crime related to cloud. Every activity in cloud has unique log at corresponding component of cloud layer. These logs would be basis for initiating cloud forensics process. In this work the cloudsim is used for creating cloud server for banking application. The deployment model the work uses is Infrastructure as a service. The digital forensic techniques are cumbersome and dependent step on cloud provider. The forensic investigator has to rely on cloud providers for evidences. This work implements the digital forensic technique using tiled bitmap algorithm for detecting server database tampering. In this work the trust on cloud provider is required.

Consider the cloud server that maintains the database of cloud customers for banking purpose. The customers logged in through authenticate logging mechanism and process the transaction. The hacker hacks the credential of customers, the customer now asks for invalid transaction that had happened. This new forensic technique gives details of individual customer's logging information through tiled bitmap algorithm.

The cloudsim is a tool for modeling datacenters, network topologies, message passing applications etc. [3]. It is developed by CLOUDLABS from Melbourne. The architecture of cloudsim includes components like network, cloud resources, cloud services, virtual machine services user interface structures scheduling policies etc. The VM provisioning, datacenter and virtual machines are the services provided by cloudsim. This work uses the cloudsim for simulating attack on cloud environment and enabling digital forensic in cloudsim.

The tiled bitmap algorithm [4] is used for generating hash value of each operation of server and clients. This algorithm is used for detecting tampering in database with the help of candidate set. The candidate set is created by ANDING of granules of target bit pattern.

The rest of the paper is organized as follows Sect. 2 illustrates the related work, Sect. 3 includes implementation details, results are discussed in Sects. 4 and 5 concludes the paper.

2 Related Work

Every digital forensic technique consists of identification, data extraction, preservation and collection, analysis and presentation Phase. Every phase has its own significance in context of cloud environment. Many authors had given different approaches for all this stages of cloud forensics because of problems faced due to multilayered, multitenant architecture of cloud. Identification is reporting of malicious activity in cloud such as illegal use of cloud for storing files, deleting files and so on. This phase arise in cloud by the complaint made by individual, by CSP authority reporting misuse of cloud or any other. Now there are many case studies related to this phase who had reported the crime related to cloud which have involvement of CSP and client as well. In [5] author discussed the digital forensic difficulties using two hypothetical case studies.

2.1 Data Extraction

Data extraction refers to physical acquisition of forensic data. The data collection phase in cloud environment is very critical due to abstract and magic box nature of cloud which guarantees 100 % availability of resources hiding architectural detail of cloud. Any digital forensic procedure consists of physically taking the custody of hard disk being investigated and then taking bitwise copy of same maintaining the integrity of data. But in case of cloud this is impossible. Physical seizure is difficult and dependent step in cloud. The investigator has to contact CSP for physical acquisition of data which is distributed among many data centers. According to [6] data collection phase of cloud forensics should also consider the storage capacity for collecting evidence. Data extraction in cloud varies with different deployment model. According to different deployment model user has varying control at different layer. According to NIST definition IAAS model provides more control, PAAS lesser and SAAS again lesser.

In [1] author suggests that data can be extracted at different level of cloud viz. virtual cloud instance, network layer and at client side. In [7] the virtual machine snapshot is taken in order to generate the periodic events of crime scene creation. The code module of VM is compared with threshold value which is already calculated with previously known attack scenarios. The data extraction in [8] is done with help of traditional forensic examination tool FTK and ENCASE. The author suggests that during data extraction phase for forensic investigation the trust is required at different layer of cloud. The trust that is mentioned in research violates the principle of digital forensic investigation process. In [9] the author extracts data from three different components. The data retrieval is done from firewall logs, Nova compute API logs and virtual disk image. The firewall logs tells us the investigators the attacker scanning target VM, Nova compute API logs gives information about unauthorized access to virtual machine and virtual disk of user may tell what an attacker done after getting access to virtual instance. The Openstack creates a directory for individual virtual instance which it could provide as an incident response. The authenticated logging service The Authenticated Logging Service uses Merkle trees as the data structure for storing API and firewall log data. When a new user joins the cloud service, a sub tree is created for the user under the root. The user's tree root is signed using the user's public key. All API logs associated with that user are stored in his or her sub tree. Data under the user's root are organized in five layers, corresponding to the machine instance year, month, and day of the respective client. The value of each node is calculated by concatenating the values of the children which would ensure integrity of retrieved data. When we talk about acquisition of API logs of particular instance the CSP provides the raw log messages and hash value of particular node in order to validate the integrity of digital evidence. The next levels of logs are firewall logs. The firewall logs in Openstack are stored in / var/log/syslog. The research adds special prefix that is instance id to each of log messages so that each of the log messages can be scanned for particular instance id. When the match found the corresponding logs are retrieved. Disk images are stored in the file system of the host operating system. The file path includes the name of the instance, which is used to identify the correct image to return to the user. In this way the strong data acquisition in order to prove the data as evidence is retrieved. In [10] the author presents some monitoring tools in order to collect logs. The research uses snort, syslog and log analyzer in order to collect data for forensic purpose. From the logs in /var/eucalyptus/jetty-request-05-09-xx file on Cloud Controller (CC) machine, it is possible to identify the attacking machine IP, browser type, and content requested. From these logs, it is also possible to determine the total number of VMs, controlled by single Eucalyptus user and VMs communication patterns.

2.2 Data Analysis

In [11] virtual snapshot technique is used for regenerating events. LVM2 that is Logical Volume Manager is used to take periodic snapshot of virtual network environment in Linux Based Operating System. To record the attack properly or accurately a fuzzy clustering based algorithm is used. The research assumes that enough documentation about attack is available. A partitioning clustering technique is used to divide attacks into divisions and the hierarchical clustering is used again to divide it into disjoints set of attack. Then the attacks have some degree of similarities in them so fuzzy clustering technique is used. The current code module that is running is continuously compared with known threshold value. The threshold value is calculated by observing memory usage, bandwidth usage, pro-cessing power requirement if the distance compared is less than the threshold value then the VM is unsafe and it is put in the unsafe zone. In [8] FTK and Encase is used to in order to acquire evidences. This Research gives us insight about the trust required at each level for data acquisition process. In [9] the API logs, the firewall logs and virtual disk image are retrieved. In order to retrieve API logs the personal identifier of client is used. The Openstack stores the NOVA API logs in /var/log/

nova/on. When the match found it retrieves the logs corresponding to UUID of that virtual machine. While retrieving firewall logs new rule is appended that logs all dropped packets to /var/log/syslog. For each instance, the research appends a special prefix to the log messages that labels the UUID of the machine. Doing so enables us to parse the log file and identify those lines that correspond to the particular virtual machine that the user requests. While retrieving snapshot of volume the research supports the retrieval of disk images from virtual machines that are powered off. In [12] logs are collected from various components of eucalyptus cloud and usage of processor in normal condition and attack condition is given in order to provide forensic evidences.

3 Implementation Details

The cloudsim is used for simulating cloud as well as crimes related with it. The research tries to initiate cloud forensic process by getting log of different activities happening in cloud environment. For this cloudsim is used initially so that experiments could be further extended to real time cloud as well.

3.1 Experimental Setup

Cloudsim is a free package for deploying Infrastructure as a Service instance. The cloudsim requires Net beans IDE. The simulator enables the simulation of cloud environment as well as cloud crimes. This work can be extended for getting evidences in real time scenario as well. The MySQL database is used for storing client information and server database. For each operation the hash value is generated and is stored in database. Whenever the single operation takes place for example login either at server end or client end the new hash value is generated and checked with previously generated hash value. The Fig. 1 shows the hash value generated during server login and corresponding logs of same.

```
HASH Value generate
HASH Value generate77
In server log Trying to login with username :admin and password :adminat :Hours: 5 Minutes:46Seconds:20has77
connecting to the database
valid user
HASH Value generate
HASH Value generate82
In server log loged in succes with username : admin and password  :admin at : Hours: 5 Minutes:46Seconds:20has82
BUILD SUCCESSFUL (total time: 24 seconds)
```

Fig. 1 Hash values of server login

In this way the hash value is generated for each operation and is stored in database for forensic examination. The server maintains the information about all the clients and related information about the same. The cloud crime takes one of the two forms [9] in one the crime is committed using cloud as an accessory and another is crime against cloud. The research takes into account all the possibilities.

The client uses authenticate login mechanism for new transaction. In this approach the hash value is generated for client login and is saved in database of corresponding client. The client uploads and downloads files in the cloud. During this operation a partial hash value is generated. This partial hash value is used for further checking.

The simulation of cloud crime where a suspicious person alters the legitimate users' files on cloud is done. In this mechanism as already stated the partial hash value is generated again and is matched with the previously generated hash values of clients stored in database. The Fig. 2 shows the hash value generated when hacking is done.

The hash value generated at hacker side does not match with client's hash value and can be proved as an evidence for forensic investigation process. The Fig. 3 shows the results of audit logs of proposed systems.

3.2 Methodology

The experimental setup gives us the hash values during various transactions occurring during various operations on cloud server. These hash values would be basis for further analysis.

The tiled bitmap algorithm [4] is implemented for forensic analysis. The tiled bitmap algorithm [4] is based on the notion of candidate set. The candidate set is a

```
in side try
HASH Value generate
HASH Value generate62
In user log Trying to update user information with account number :jjs at :Hours: 3 Minutes:33Seconds:30ha
HASH Value generate
HASH Value generate46
In user log Updated user information Successfully at :Hours: 3 Minutes:33Seconds:30has46
```

Fig. 2 Hash value generated hacker side

```
run:
user log44
user audit log30
server log55
server audit log54
BUILD SUCCESSFUL (total time: 27 seconds)
```

Fig. 3 Hash value generated at client side and hacker side do not match

set of possible corruption events. It is derived from single target. The target is binary number which is the result of partial hash chains of tampered tile. In this approach a validator is run every 16 h. When a validator returns false result, a partial hash chain of tampered tile is conducted and the result of each partial hash values is ANDED to form target binary number. This target binary number is input to form candidate set. The algorithm is used to generate target timestamps which is used for forensic analysis. The forensic analysis investigation process, with the help of tiled bitmap algorithm is able to detect when and what of data is altered. The process initiates the forensic examination process. The audit logs of server and clients are generated separately. The forensic analysis is able to decide whether the threat is from client side or server side. After determining the exact location of crime it would be easy for investigation team to prove the malicious user. To avoid this, the research also proposed a timestamp based secure login in cloud to avoid the tampering of database in cloud. The files are uploaded on cloud server. Each of the file on cloud server is given the unique security key and timestamp to access the file as shown in Fig. 7. The security key is valid for that timestamp only. Whenever the client logs in, the client is given the security key and timestamp to access the file on cloud server. The key get expires after accessing the file for that timestamp. This method would secure the files from unauthentic accessing of files on cloud server. Because only those user can get access to the files on the server who has security key from secure cloud server. The Fig. 4 shows how the timestamp is given to each of the file on cloud server. The security key is also given to each file to access that file. Users with security key can access that file.

The Fig. 5 shows the security key and timestamp given to client is used for downloading and accessing the files on cloud server.

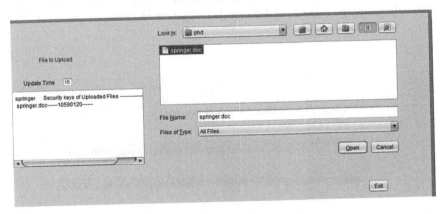

Fig. 4 Security key and timestamp assigned to files on server side

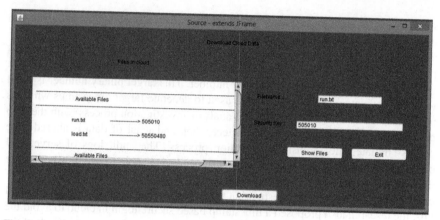

Fig. 5 Security key and timestamp used to access files on client side

4 Result and Discussion

This research tries to provide basis for preliminary cloud forensic analysis. Identification of crime in cloud is also another big issue. As cloud is very big infrastructure consisting of various and diversified components, getting evidences at each level and correlating those with each other to prove all evidences pointing towards same victim is also big issue. If this research provides the logs at different level that gives us the identification of crime that would give the investigator the starting point of evidence collection. The results shows us if during transaction process if any malicious activity takes place then the cloud forensic analysis process aids to detects the fault either at client or at server end as shown in Fig. 6. This definitely eases the process of forensic investigation in cloud environment. The

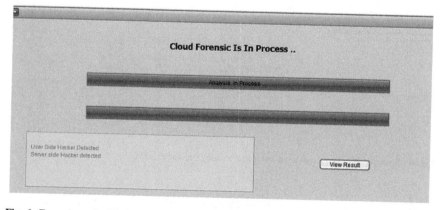

Fig. 6 Forensic analysis process

Fig. 7 Results of security
key generated

```
BB25
namerun.txt
name11------run.txt
sk--------run.txt
skey--------505010
Time Expired
Cname------------run.txt
afterpass20250110
successfull!!!!!!
```

encryption technique proposed with timestamp to each of the files on cloud server definitely secure files from unauthenticated accessing (Fig. 7).

The results shows cloud evidences which is be presented in order to prove any crime. The limitation of this research is every parameter is analyzed at simulator level and only one of the categories of cloud crime is covered. The cloud simulator is chosen for preliminary analysis of algorithm and results so that it can be validated in real time. In real time every aspect of cloud crime would be covered and combination of methodology to do forensic investigation in cloud environment.

5 Conclusion

With advent of cloud computing and crimes related with the same the cloud forensics has new dimension in this area. The cloud crime could be altering data files on cloud, DDOS attack, cross channel attack and many more. The research aims at providing forensic solution to avoid tampering of data in cloud environment. The data in cloud server is accessed through different clients at diversified area. From where the data is accessed is unknown to user due to missing architectural detail of cloud. The crime related to cloud is increasing day by day. The research focuses on malicious accessing of data on cloud. Also it tries to provide the security solution for the same. The security solution provides re-encryption technique to provide more security on cloud server. The timestamp is given to access the file for client whenever user logins to access file on cloud server. The user accesses the file within that timestamp. If the user fails to do so the time expires and user would not be able to access the file. In this way the security is provided for unauthorized access to data on cloud server. The research also focuses on providing digital evidences if a malicious person tries to tamper the database on cloud server. The Tiled Bitmap algorithm provides hash values as an evidence to prove something unauthorized is happened. This unmatched hash value initiates the process of investigation. The novel technique of re-encrypting the file and giving timestamp for accessing the files uploaded on cloud server provides more secure level to customers of cloud. Our future work would try to include different other forms of cloud crime and with capability of providing potential evidences as well.

References

1. Birk, D., Wegener, C.:Technical issues of forensic investigations in cloud computing environments. In: Proceedings of the 6th International Workshop on Systematic Approaches to Digital Forensic Engineering (SADFE) Oakland, CA, USA (2011)
2. Shah, J.J., Malik, L.G.: An approach towards digital forensic framework for cloud. In: Advance Computing Conference (IACC), pp. 798–801. 21–22 Feb 2014
3. Calheiros, R.N., Ranjan, R., De Rose, C.A.F., Buyya, R.: CloudSim: a novel framework for modeling and simulation of cloud computing infrastructures and services
4. Pavlou, K.E., Snodgrass, R.T.: The tiled bitmap forensic analysis algorithm. IEEE Trans. knowl. Data Eng. 22, 590–601 (2010)
5. Dykstra, J., Sherman, A.T.: Understanding issues in cloud forensics: two hypothetical case studies. J. Netw. Forensics 3(1), 19–31 (2011). (Autumn)
6. Grisspos, G., storer, T., Glisson, W.B.: Calm before the storm: the challenges of cloud computing in digital forensics. Int. J. Digit. Crime Forensics 4(2), 28–48 (2012)
7. Belorkar, A., Geethakumari, G.: Regeneration of events using system snapshots for cloud forensic analysis. In: India Conference (indicon), vol. 16–18, pp. 1–4 (2011)
8. Dykstra, J., Sherman, A.T.: Acquiring forensic evidence from infrastructure-as-a-service cloud computing: exploring and evaluating tools, trust, and techniques. Digit. Invest. 9, S90–S98 (2012)
9. Dykstra, J., Sherman, A.T.: Design and implementation of FROST: digital forensic tools for the OpenStack cloud computing platform. In: Digital Investigation. The Proceedings of the Thirteenth Annual DFRWS Conference 13th Annual Digital Forensics Research Conference, vol. 10 (Supplement), pp. S87–S95 (2013)
10. Zafarullah, Z., Anwar, F., Anwar, Z.: Digital forensics for eucalyptus. In: Frontiers of Information Technology (FIT), IEEE, pp. 110–116. 19–21 Dec (2011)
11. Geethakumari, G., Belorkar, A.: Regenerating cloud attack scenarios using LVM2 based system snapshots for forensic analysis. Int. J. Cloud Comput. Serv. Sci. (IJ-CLOSER) 1(3): 134–141 (2012). (ISSN: 2089-3337)

A Novel Genetic Based Framework for the Detection and Destabilization of Influencing Nodes in Terrorist Network

Saumil Maheshwari and Akhilesh Tiwari

Abstract The Social Network Analysis analyses social network on web. SNA has led the law enforcement agencies to study the behavior of terrorist networks for the identification of relationships that may exist between nodes. Recently, Terrorist Network Mining (special branch of SNA) has been in vogue in Data mining community because of its ability to identify key nodes present in the network. This paper proposes a new approach for Terrorist network mining. The proposed work is carried out in two phases. The first phase proposes Genetic based optimization mechanism. Proposed mechanism is suitable for effective optimization of large social network containing terrorist and non-terrorist nodes. During optimization process removal of non-terrorist nodes from the network has been performed and resultant represents the reduced graph containing only the set of potential nodes. The second phase proposes a weighted degree centrality measure (considers frequency of communication) for effectively neutralizing of the terrorist network.

Keywords Social network analysis · Terrorist network mining · Genetic algorithm

1 Introduction

Today with the improved lifestyle of individual, advancement in the technology and widespread use of internet, assuring the security of the mankind has become one of the major challenges. The modernization of technology has laid the criminals to use this cyber era for fulfilling their inhuman goals against the mankind. Thus raising concerns about the effects that attacks from cyberspace can cause to the overall society. Therefore it is necessary for law enforcement agencies to provide the security to

S. Maheshwari (✉) · A. Tiwari
Department of CSE and IT, Madhav Institute of Technology
and Science, Gwalior, Madhya Pradesh, India
e-mail: saumilmaheshwari@yahoo.co.in

A. Tiwari
e-mail: atiwari.mits@gmail.com

© Springer India 2015
L.C. Jain et al. (eds.), *Computational Intelligence in Data Mining - Volume 1*,
Smart Innovation, Systems and Technologies 31, DOI 10.1007/978-81-322-2205-7_53

civilians from these criminals. To do so these agencies need to ascertain the terrorist network knowledge effectively and efficiently. A key challenge faced by the law enforcement agencies is the large crime 'raw' data volumes and the lack of sophisticated network tools and techniques to utilize the data effectively and efficiently [1].

This explosively growing, widely available, and gigantic body of data makes our time truly the data age [2]. This bulk of data created and available in fast computing world has laid the data mining as an emerging tool for detecting and preventing terrorism [3]. Data Mining is seen as a most prominent technique because of its ability to mine useful information from large amount of data available. Data mining is useful for decision makers to make decision, discovering hidden relation and pattern, foreshowing possible behaviours which may occur in the future by analysing historical and current data [3].

Genetic algorithm is the most important technology in many mining technology, which can select information from large amounts of data in the data warehouse, find possible operating mode in market, and mining facts people hard to find out [4]. Genetic based optimization of social network is done in order to remove nonterrorist nodes from the network and the optimization resultant represents only the set of potential nodes. These potential nodes are the possible terrorist nodes that have a potential to cause a high impact on the society. Discovering or detecting these potential nodes prior to the event (such as terrorist attack) may destroy their plan for executing the event [4].

The use of web by these inhuman people has unintentionally allowed them for fulfilling their inhuman goals by exchanging messages, planning their strategy etc. Therefore, security concerns on the web too have increased. In this context, data mining has established itself as the demanding field for studying social networks on web by applying one of its feasible techniques named Social Network Analysis (SNA) [5]. Social network analysis studies networks for the identification of relationships and associations that may exist between nodes [6]. The SNA uses a centrality measures pointing out who is the central node(s) in the network. It is because of this that SNA is utmost utilized technique by the law enforcement agencies for studying trends of hidden terrorist networks [1].

When the SNA is applied for investigating of terrorist networks on web then it is acknowledged as Investigative Data Mining (IDM), also known as Terrorist Network Mining [7]. Defined centrality measures for SNA have been embodied successfully for destabilization of terrorist network [8]. The technique discovers the most promising node(s) (central node) within the network and the goal is to remove this node(s) from the network in order to neutralize the network activities [9].

For the analysis, the terrorist network is considered as a graph and users are considered as its nodes. Using the SNA centrality measures i.e. degree and eigenvector, various roles are estimated from the graph. Making use of these measures hierarchy is constructed in the form of a tree utilizing the algorithm defined for destabilization of terrorist network. The case study of some of the attacks is performed by Memon, Hicks, Hussain, Larsen [10–12].

The paper suggests the use of genetic based optimization mechanism in order to to remove non-terrorist nodes from the network and the optimization resultant represents only the set of potential nodes. Also the paper urges to use the weighted degree centrality measure. The use of weighted degree centrality measure for finding the central node(s) also considers the frequency (weights) of communication between the nodes. This would aid in an effective estimation of hierarchy followed by the terrorist networks.

2 Gaps Available in the Current Knowledge

Though the present centrality measures are quite capable of determining the hidden hierarchy of the terrorist network, lists are certain short comings of the currently used centrality measures:

(a) Currently used centrality measures for terrorist network mining does not consider the frequency of communication among the nodes while determining the central node in a network. Hence some new centrality measure is required which considers the frequency of communication among the nodes for finding central node in Terrorist network. The new centrality measure for terrorist network mining is proposed with the view for improving the efficiency of the algorithm.

(b) Also some optimization technique need to be executed before applying the centrality measure in order to obtained the desired results. The Genetic based optimization technique extracts possible set of potential (terrorist) nodes from social network of terrorist and non-terrorist nodes.

3 Proposed Methodology

The proposed methodology works in two phases. During the first phase genetic based optimization has been performed on the large social network. The fitness function in genetic based optimization predicts and minimizes the social network containing only the potential nodes.

3.1 Proposed Genetic Based Optimization Mechanism

GA works with the coding of solution set and not with the solution itself. Coding of a solution set is encoding of a solution into chromosome, which is just an abstract representation. Coding all the possible solutions into chromosome is the first part.

The structure of information associated with each node is shown in Table 1.

Five chromosomes are used to define each individual in a graph namely: age of individual; Gender, with 00 means female, 01 means male; Income level as low,

Table 1 Structure of information associated with each node

Age	Gender	Income level	Health condition	Crime record
56	F	Med.	Good	Ordinary
36	F	High	General	No record
20	M	High	Poor	Ordinary
30	M	Low	Good	Serious

medium or high expressed as 02, 01 or 00 respectively; Health condition as good, general or bad expressed as 02, 01 or 00 respectively; and Crime record as serious, ordinary or no record expressed as 02, 01 or 00 respectively.

Appropriate representation, fitness function and reproduction operators are the determining factors, as the behavior of the Genetic algorithm is extremely dependent on it. Brief explanation and algorithm as follows:

(a) **Coding Strategy**

The encoding depends mainly on solving the problem. For example, one can encode directly real or inter numbers. There are many parameters associated with each node hence multi-parameter coding technology can be used. Basic technique or idea is to encode every parameter associated with each node and then combining these encoded parameters into a complete chromosome. Thus for each node the associated parameters are converted into chromosome. For instance, 25|male|Low|General| Serious, are the parameters associated with a node. Each parameter is encoded and combined to get a complete chromosome. Hence for the parameters 25|male|Low|General| Serious, the encoded chromosome will be 25|01|02|01|02.

(b) **Fitness Function for the Optimization of Terrorist Network**

The fitness function is so selected that it optimizes the graph such that a clear distinction can be made between the possible terrorist and non-terrorist nodes. The terrorist node in a graph is assumed to be the one with parameters defined as male, income level as low, health condition to be good, crime record as serious crime and age between 18 and 60 based on the data set of 26/11 and other attacks. Fitness value is computed for each chromosome by the following fitness function:

$$f(x) = \frac{1}{1 + (100 - \grave{A})} + \frac{1}{1 + \sqrt[2]{\breve{g} + 1}} + \frac{1}{2 + \ln(\alpha + 1)} + \frac{1}{3 + (\beta)^2} + \frac{1}{1 + \exp{(\acute{C}r.)}}$$

where in the equation, $\grave{A}$, $\breve{g}$, α, β and $\acute{C}r.$ corresponds to the parameters associated with the nodes namely: Age, Gender, Health condition, Income level and Crime record respectively. Here in the function a high weightage is given for the parameter crime record by calculating the exponential of the value.

Based on the graph obtained as an output of first phase, during the second phase, tree hierarchy is generated by applying algorithm for destabilization of Terrorist Network. The inputs for the algorithm are optimized graph (output of first phase) and centralities value. The proposed new centrality measure i.e. degree centrality measure for weighted graph is used for destabilization of Terrorist Network which considers the frequency of communication between the nodes for calculating the central node of a graph.

3.2 Degree Centrality in Terrorist Network (Considering Frequency of Communication)

Centrality measurement is considered as a crucial parameter for understanding and analyzing the actors' roles in social networks [8] and is used to identify the strength of the information influence.

Till now the focus has been on the Degree centrality on non-weighted graph, i.e. the calculation of relationship number at each node. This study will focus on degree centrality on weighted graph.

Input: A Terrorist Network represented as a Graph G(V, E)

Output: Weighted Degree Centralities values of all nodes

Algorithm DCWeighted (V, E, Weight, C)

```
1.   {
2.   GP := graphallshortestpath (sparse(G));
3.   for i := 1 to n                          // n is length (GP)
4.       for j :=1 to n                       // n is length (GP)
5.       e(i , k) := GP(i , k) / Sum(GP(i , :)) ;
6.   end
7.   H(i) := entropy (e(i , : ));             // H(i) = -∑_{i=1}^{z} hi log₂ hi
8.   end
9.   CDCA := H(i)*2;
10.  C := CDCA;
11.  return C
12.  }
```

The comment on line 7 reads: $H(i) = -\sum_{i=1}^{z} hi\ log_2 hi$

Algorithm for Destabilization of Terrorist Network

The framework of the proposed algorithm is as follows:

1. Take any node n of graph G and find neighbors N
2. Take a node m such that m∈N
3. Compare DCweighted of each node to its neighbor node
     If (DCweighted of m > DCweighted of n
           Add node m to parent set of n
     Else if (DCweighted of m < DCweighted of n)
           Add node m to children set of n
     Else
           Ignore the link
4. After calculating the Parent and Children sets, find the hierarchy.
5. If a node has no parent, add root of tree T as its parent and mark n as children of root.
6. For Parent set with one value, node is the child of that set value node.
7. For node with Parent set with more than one value, maximum [N(P1)∩N(n)] is estimated and the parent node with maximum value is set as parent of a node. For N(P1) ∩N(n) = 0, node is overlooked.
8. Even then the Parent set has multiple values the node is attached to root.

4 Experimental Results

For testing the usefulness of developed optimization mechanism and centrality measure experimentation has been performed using matlab (version R2012a). For experimental purpose synthetic dataset (containing 60 nodes and connections between them) has been utilized. Weights on the edges indicate the number of communication between the nodes connected by edge. Our work is carried out in two phases:

(a) Firstly, optimization has been performed on the data set. Optimization resultant produces an optimized/reduced graph containing 15 nodes. Figure 1 shows the graph/Social network corresponding to considered dataset. In the graph each node is assigned with certain weights such as age, gender, health condition, income level and crime record.

    After applying the Genetic based optimization mechanism the social network represented by graph is reduced to a graph containing only 15 nodes. The fitness calculation uses the above discussed formula to reveal the fitness value for each node. Fitness function is applied to each chromosome and corresponding fitness values are generated with respect to each chromosome. Now to distinguish between the terrorist and non-terrorist nodes we define a threshold value so that the chromosomes with fitness value satisfying or less than the threshold value

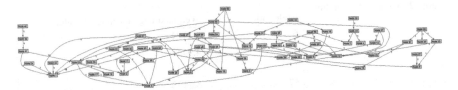

Fig. 1 Graph representing 60 nodes with connections between them

are the set of possible terrorist nodes. Possible set of terrorist nodes are those whose fitness value is smaller. Optimized graph contains 15 nodes (indexed with their original serial no.) as depicted in the Fig. 2.

(b) Secondly above defined algorithm for destabilization of Terrorist Network with new centrality measure is applied on the resultant of the proposed optimization and a tree hierarchy is generated. Table 2 consist the values of centrality calculated for the algorithm:

With the centrality value obtained the tree hierarchy is generated using the algorithm. With the hierarchy of tree generated the crucial node among the reduced graph can be identified. The output obtained after accessing the algorithm could be visualized through Fig. 3. Here the crucial nodes identified are

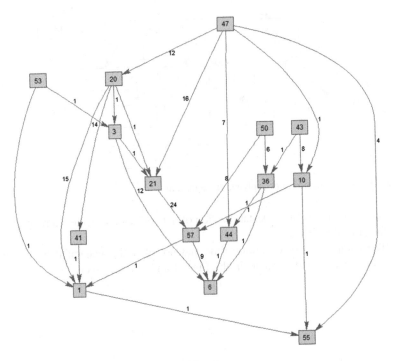

Fig. 2 Reduced/optimized graph containing 15 nodes

Table 2 Values of centrality

Node index	Weighted degree centrality values
6	6.1131
55	4.9458
1	4.7312
57	5.2125
44	6.1131
41	5.3785
36	5.9471
21	5.9118
3	5.4791
20	5.9118
10	5.3131
47	6.4804
53	4.9979
50	5.5798
43	6.1131

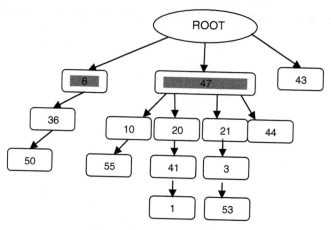

Fig. 3 Hierarchy obtained after applying the algorithm

shown with red color. Destabilization of network can be achieved by capturing the crucial nodes.

The comparison of execution time and no. of nodes generated by the algorithms is depicted by chart in the Figs. 4 and 5 respectively.

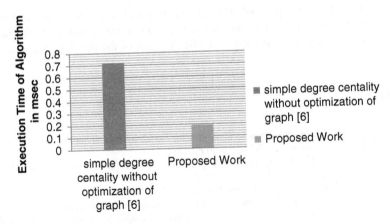

Fig. 4 Execution time of algorithms in milliseconds

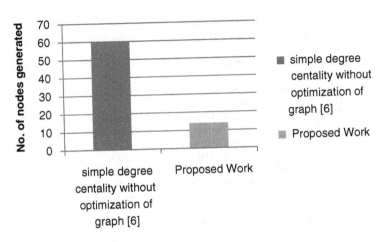

Fig. 5 Number of nodes generated

5 Conclusion

This paper proposes a novel Genetic Based Terrorist Network Mining algorithm. Proposed algorithm works effectively for the detection and destabilization of Terrorist Network/influencing nodes present in the network. Due to the incorporation of Genetic Based Optimization mechanism within developed algorithm, now it is very easy to focus only on the set of potential nodes for the identification and removal of actual terrorist nodes. Furthermore, considering the frequency of communication between the nodes while determining the crucial node in a network has led the effective destabilization of terrorist network. Experimentation has been performed with MATLAB tool. During experimental analysis synthetic dataset containing 60 nodes and connections between them has been considered. The result is as per the expectation and shows the effectiveness of the developed algorithm.

References

1. Chaurasia, N., Tiwari, A.: Efficient algorithm for destabilization of terrorist networks. Int. J. Inf. Technol. Comput. Sci. **5**(12), 21–30 (2013)
2. Han, J., Kamber, M., Pei, J.: Data Mining : Concepts and Techniques, 3rd edn. Morgan Kaufmann Publications, Massachusetts (2006)
3. Pujari, A.K.: Data Mining Techniques, Universities Press (2001)
4. Maheshwari, S., Tiwari, A.: Genetic Based optimization of social network with special reference to terrorist network mining. TECHNIA-Int. J. Comput. Sci. Commun. Technol. 7:1. [ISSN 0974-3375] (2014)
5. Chaurasia, N., Tiwari, A.: On the use of Brokerage approach to discover influencing nodes in terrorist networks. In: Social Networking, Intelligent Systems Reference Library, vol. 65, pp. 271–295. (doi:10.1007/978-3-319-05164-2_11) (2014)
6. Marin, A., Wellman, B.: Social network analysis: an introduction. The Sage Handbook of Social Network Analysis
7. Shaikh, M.A., Jiaxin, W.: Investigative data mining: identifying key nodes in terrorist networks (2006)
8. Wasserman, S., Faust, K.: Social Network Analysis, Methods and Applications. Cambridge University Press, Cambridge (1994)
9. Chaurasia, N., Dhakar, M., Tiwari, A., Gupta, R.K.: A survey on terrorist network mining: currenttrends and opportunities. Int. J. Comput. Sci. Eng. Surv. (IJCSES) **3**(4), 59–66 (2012)
10. Memon, N., Larsen, H.L.: Investigative data mining toolkit: a software prototype for visualizing, analyzing and destabilizing terrorist networks (2006)
11. Memon, N., Hicks, D.L., Hussain, D.M.K., Larsen, H.L.: Practical algorithms and mathematical models for destabilizing terrorist networks. In: Mehrotra, S., Zeng, D.D., Chen, H., Thuraisingham, B.M., Wang, F.-Y. (eds.) ISI 2006, LNCS 3975, pp. 389. Springer-Verlag, Berlin (2006)
12. Memon, N., Larsen, H.L., Hicks, D.L., Harkiolakis, N.: Detecting hidden hierarchy in terrorist networks: some case studies. In: Proceedings of ISI 2008 Workshops, LNCS 5075, pp. 477–489. Springer-Verlag, Berlin (2008)
13. Memon, N., Larsen, H.L.: Practical approaches for analysis, visualization and destabilizing terrorist networks. In: Proceedings of the First International Conference on Availability, Reliability and Security, ARES (2006)

Analyzing User's Comments to Peer Recommendations in Virtual Communities

Silvana Aciar and Gabriela Aciar

Abstract Social networks and virtual communities has become a popular communication tool among Internet users. Millions of users share publications about different aspects: educational, personal, cultural, etc. Therefore these social sites are rich sources of information about who can help us solve any problems. In this paper, we focus on using the written comments to recommend a person who can answer a request. An automatic analysis of information using text mining techniques was proposed to select the most suitable users. Experimental evaluations show that the proposed techniques are efficient and perform better than a standard search.

Keywords Social networks · Text mining · Recommender systems

1 Introduction

Social networks have become in a few years, a global phenomenon that is expanding every day. Twitter, Facebook, Google+ and LinkedIn have emerged with overwhelming force in society. Social technologies are radically changing the way people communicate and interact. People often consult the Internet to find answers to your questions. Perform queries on Google, which returns thousands of web pages with possible answers. A person can take a long time to find the right answer among all received pages. This leads to abandon the search or accept an answer that may not be the most appropriate. Due to overload of information on the Internet, people prefer to ask a friend. The way to communicate is using social media, today

S. Aciar (✉) · G. Aciar
Universidad Nacional de San Juan, San Juan, Argentina
e-mail: saciar@iinfo.unsj.edu.ar

G. Aciar
e-mail: gaby_aciar@yahoo.com.ar

© Springer India 2015
L.C. Jain et al. (eds.), *Computational Intelligence in Data Mining - Volume 1*,
Smart Innovation, Systems and Technologies 31, DOI 10.1007/978-81-322-2205-7_54

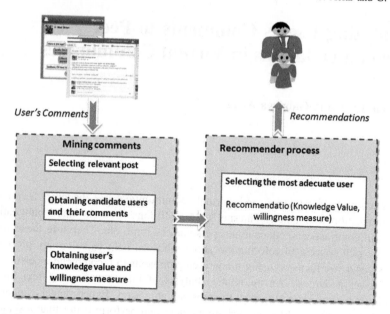

Fig. 1 Process to recommend user based on information from comments

a person is part of at least 3 social networks and has more than 100 contacts. In large networks, identify the right person who can answer a specific question is not an easy task for one person, sometimes the right person is not directly related to him. One way to facilitate interaction and solve the problems of information overload is through the use of recommender systems [1, 2]. The purpose recommender systems are to simplify the search process by providing to users information, products or services based on their needs, interests, and preferences. The recommender systems use techniques from several disciplines such as artificial intelligence, information retrieval, data mining and machine learning to identify items of interest for a user in a particular context of recommendations [2–4]. This paper proposes to use recommender systems to suggest suitable users to answer queries of people. In the literature there are several studies that recommend users to interact in a social network [5–7] none of which used the information posted by users. In this paper we consider necessary to analyze the content of user's post to get their knowledge about a particular topic, and then suggest users to answer questions based on that value. Despite the importance and value of the information introduced by users in a comment, there is no comprehensive mechanism that formalizes in an application: The process of selection and retrieval of user comments; Information processing to obtain a value representing the knowledge of users on a topic; Recommendation of a user based on that value. Part of the problem lies in the complexity of information extraction from textual and turns that information into recommendations. Figure 1 shows the structure of the proposed recommender system.

The structure of the paper is as follows. Section 2 presents related works about the recommendation of users using information from social networks. In Sect. 3 we describe in details the mining comments process, which uses text mining techniques. The set of measures used to calculate a recommender value of a user is also briefly explained in this section. Section 4 presents the experiments performed that demonstrates the effectiveness of proposed approach. The example is in a virtual community on the e-learning domain. Finally, the conclusions, the limitations of the work and the directions for future research are discussed in Sect. 5.

2 Related Works

While in the community of recommender systems has been much progress in the study of methods to make recommendations of products and services [2–4, 8], there are few work of recommending people. The methods used in recommending users take into account the similarity between the profiles, reputation within a community and network of contacts. In [9, 10] recommendations are made based on the similarity of user profile on a social network. The profile is constructed explicitly. The profile and reputation of individuals in a virtual community are considered to make recommendations in [11, 12]. Recommendations are currently being conducted using the network of contacts of people [13, 14]. The virtual community is modeled as a graph consisting of nodes representing individuals and links between those nodes that usually represents the distance between people. Once the network structure is obtained, recommendations are made based on the type, organization and network properties [5–7, 15, 16]. From the review, has not found work that uses information from user's comment to recommend people to interact. The content of the comment have to be analyzed in order to decide who can answer a question for those who cannot. The following sections detail the process for recommending users based on comments written in virtual communities.

3 Recommendation of Users Based on Comment Posted

The following tasks are required to select and retrieve information from the comments, process that information and use it to know if a user is suitable for recommendation:

- Implementation of text mining methods to obtain information from written comments by users.
- The definition of metrics that allow prioritized users by the degree of their knowledge on a specific topic.
- Developing recommender mechanisms to filter and present the most appropriate users.

The process will be done through an application that given a user request Q, it searches user comments containing Q obtaining set of relevant post. A set of metrics for obtaining recommending users from the set of relevant post have been defined. In next sections are explained the process to obtain such measures.

3.1 Selection of Relevant Post and Identification of Candidate Users

Given a user request Q of a user and Q is composed of keywords such that $Q = (q_1, q_2, ..., q_n)$, the first step is to find all the post containing Q. The search for relevant post is a simple search where those publications containing Q chains (q_i) are only selected. The selected post constitute a collection of relevant publications $Pr = \{p_1, p_2, p_3, ... p_n\}$ where each publication can have an associated set of comment and users who wrote them. Figure 2 shows the structure of the data obtained from relevant publications.

Users who made comments and those who made the publication constitute a set of candidate users $UC = \{u_1, u_2, u_3, ..., u_n\}$. For each candidate user, comments that were made and the times they interacted in the virtual community are obtained.

3.2 Analyzing User Comments

A user can answer a request Q if he has knowledge about Q and he is available to answer questions. We have defined a set of metrics to obtain such information. The

Fig. 2 Information obtained from relevant post

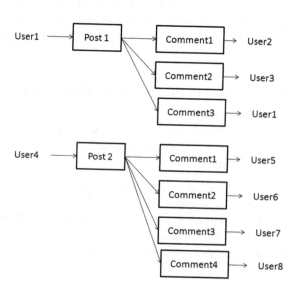

User Information about a Topic (UIT) measure is computed with information written on comments while the **User Interaction (UI)** and **Willingness to Respond (WR)** measures are computed using information from interactions in the virtual community. **User Information about a Topic (UIT)** is defined as the sum of weights (w_i) of the words q_i multiplied by the times C that these words appear together in user comments. It is calculated using Eq. 1.

$$UIT = \sum_{i=1}^{n} w_i * C \tag{1}$$

w_i is obtained by means of the measure tf-idf (Term Frequency-Inverse Document Frequency) widely used in Information Retrieval. In our proposal it means that the q_i terms that appear most frequently written in a user's comments, but less in the other reviews, is more probable that it be most used by the user. Equations 2–4 are employed to compute the measure.

$$w_i = f_i * idf_i \tag{2}$$

$$f_i = \frac{frec_i}{MaxFrec} \tag{3}$$

$$idf_i = \log \frac{N}{n_i} \tag{4}$$

f_j is the normalized term frequency of q_i in the comments made by a user. Idf_i represents the inverted term frequency of q_i. $frec_i$ is the frequency of term q_i in all comments. $MaxFrec$ represents the maximum frequency of all frequencies of words in the comments. N is the number of comments and n_i is the number of comments containing the term q_i. **Interactions User (IU)**: is defined as the times that a user has posted or has commented NCU over the total number of post and comments on the virtual community TNC, as indicated by Eq. 5.

$$IU = \frac{NCU}{TNC} \tag{5}$$

User's knowledge on the topic (UKT): it is obtained based on the user information on the topic UIT multiplied by the number of interactions of the user in the virtual network IU. Equation 6 is used to implement this measure.

$$UKT(u) = UIT * IU \tag{6}$$

Willingness to Respond (WR) is defined as the probability that a user responds to a request based on past behavior. It is probably that a person who has frequently answered questions in the past, now answer a question. Equation 7 is used to calculate the willingness of reply.

$$WR(u)\frac{RPOS + 1}{(RPOS + 1) + (RNEG + 1)} \tag{7}$$

where *RPOS* is the number of times that the user has responded request and RNEG is the total number of time that he has not responded.

3.3 Selecting the Most Suitable User

The most suitable and reliable user are chosen to make the recommendations. A selection algorithm has been defined to make the choice automatically. The algorithm is composed by 3 elements, a set (*U*) of candidate users, a selection function *Selection(UKT(u),WR(u))* to obtain the most relevant and reliable user which uses the values of user's knowledge about a topic *UKT(u)* and willingness to respond WDR(u) as parameters and a solution set (*F*) containing the users selected (*F U*). With every step the algorithm chooses a user of *U*, let us call it *u*. Next it checks if the *u F* can lead to a solution; if it cannot, it eliminates *u* from the set *U*, includes the user in *F* and goes back to choose another. If the users run out, it has finished; if not, it continues.

Algorithm to select relevant users

        Algorithm (U: Set of candidates users)
            F: =∅   ;
                while (U <>∅) **do**
                    if Selection(UKT(u),WR(u)) > threshold **then**
                        F := F∪ u;
                            Eliminate (U, u)
                end if
                end while
                return F;

The selection function has as parameters the information of the user about a topic UKT(u) and the availability to respond WR(u). This function is obtained through equation:

$$selection(UKT(u), WR(u)) = UKT(u) * WR(u) \tag{8}$$

The lists of relevant users are ordered from highest to lowest selection value and the top of the list are recommended.

4 Experimental Results

The domain of education was chosen to carry out the experiments and evaluate the feasibility of the proposal. A user recommender system has been implemented in the Moodle platform where post made in forums for teachers and students are analyzed. Moodle forums allow users to post information about a topic through comments. A post on a Moodle forum includes a section containing the user name of the writer who can be real or fictitious. A section describing the content of the post written by the user and a comment section where other users can comment the post. For the experiments have been analyzed data from the participation of 22 students and teachers in several forums of Moodle during 2013. Data collection for testing consists of: 22 users, 45 posts and 134 comments. The recommender system implemented can be seen in Fig. 3.

Two experiments were performed using the same data set.

Experiment 1 Recommendation based on UKT and WR: The recommender system used the proposed method in this paper to suggest users. 10 users requested by users using the recommender systems. They introduced request about several subjects. The recommender analyzed forums to obtain candidate users. For each candidate user, the system performances the metrics defined in Sect. 3.1 with information from their comments. The most suitable user is selected applying the proposed algorithm. When a person has been recommended, the system solicits to user the evaluation of people's behavior. Therefore, the system presents to user a form asking the user to answer yes if the person recommended answered your request. This information is used to compute the *WR* measure presented in Sect. 3.

Experiment 2 Recommendation based on keyword search: In this experiment the recommender system perform a basic search to recommend users. Others 10 users have been requested recommendation. They introduced request about several

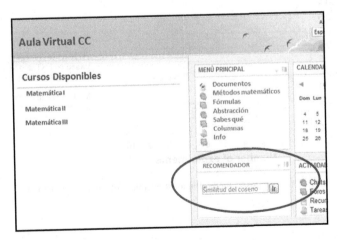

Fig. 3 User recommender system implemented on Moodle platform

subjects. The system searches for the keywords in the comments and it recommends users who mentioned more times the keywords in your comments. When a person has been recommended, the system solicits to user the evaluation of people's behavior. Therefore, the system presents to user a form asking the user to answer yes if the person recommended answered your request. This information is used to compute the WR measure presented in Sect. 3.

4.1 Performance Metric

In order to evaluate the result of recommendations made in both experiments the accuracy rate calculated by Eq. 9 is used. The evaluation is based on good recommendations over all recommendations, in which a good recommendation is defined as: given a request, a person who will already answer the question is recommended; this information is retrieval from the feedback given by the user after contact with the recommended person.

$$Accurracy\,Rate = \frac{NG}{N} \tag{9}$$

where NG is the number of the users who are correctly recommended (recommended person who was responded the question); N is the total number of the recommended person. Both experiments were performed 10 times with different users each time. Experimental results, as shown in Fig. 4 and Table 1, display that the proposed method has a higher accuracy.

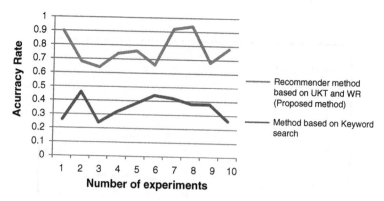

Fig. 4 Accuracy of our proposed method against accuracy of a baseline method

Table 1 Evaluation proposed recommendation method. Rate accuracy resulting of the experimentation

Times	Accuracy rate recommendations based on UKT and WR	Accuracy rate recommendations based keywords search
1	0.90	0.26
2	0.68	0.46
3	0.64	0.24
4	0.74	0.32
5	0.76	0.38
6	0.66	0.44
7	0.92	0.42
8	0.94	0.38
9	0.68	0.38
10	0.78	0.26

5 Conclusions

The large amount of information available nowadays makes the process of detecting user for interact more and more difficult. Recommender systems have been used to make this task easier, but the use of these systems does not guarantee that the recommended person be the most suitable. We propose a recommender system that uses information from user's comments to obtain the most suitable to interact with other user. We have defined a set of metrics that allow us know if the user has relevant information or not about a subject to answer a request. A user recommender system has been implemented in the Moodle platform where post made in forums for teachers and students are analyzed. Moodle forums allow users to post information about a topic through comments. Two experiments have been performed to test our approach. Accuracy of our proposed method against accuracy of a baseline method has been compared. Observing the results of the experiments carried out in this paper, we note that the recommendations are better if we take into account the knowledge and willingness of user to present users in order to response a question. In our future work, we intend to evaluate our approach with open social networks such as Facebook and Google+.

References

1. Konstan, J., Riedl, J.: Recommender systems: from algorithms to user experience. User Model. User-Adap. Inter. **22**, 101–123 (2012)
2. Deuk, H., Hyea, K., Young, C., Jae, K.: A literature review and classification of recommender systems research. Expert Syst. Appl. **39**, 10059–10072 (2012)
3. Bobadilla, J., Ortega, F., Hernando, A., Gutiérrez, A.: Recommender systems survey. Knowl.-Based Syst. **46**, 109–132 (2013)

4. Al-Shamri, M.Y.H.: Power coefficient as a similarity measure for memory-based collaborative recommender systems. Expert Syst. Appl. **41**, 5680–5688 (2014)
5. Konstas, I., Stathopoulos, V., Joemon M.: On social networks and collaborative recommendation. In: Proceedings of the 32nd International ACM SIGIR Conference on Research and Development in Information Retrieval (SIGIR '09), pp. 195–202 (2009)
6. Jamali, M., Ester, M.: A matrix factorization technique with trust propagation for recommendation in social networks. In: Proceedings of the Fourth ACM Conference on Recommender Systems (RecSys '10), pp. 135–142 (2010)
7. Yang, X., Steck, H., Liu, Y.: Circle-based recommendation in online social networks. In: Proceedings of the 18th ACM SIGKDD International Conference on Knowledge Discovery and Data Mining (KDD '12), pp. 1267–1275 (2012)
8. Briguez, C., Budán, M., Deagustini, C., Maguitman, A., Capobianco, M., Simari, G.: Argument-based mixed recommenders and their application to movie suggestion. Expert Syst. Appl. **41**, 6467–6482 (2014)
9. Kautz, H., Selman, B., Shah, M.: Referral web: combining social networks and collaborative filtering. Commun. ACM **40**, 63–65 (1997)
10. Zhao, K., Wang, X., Yu, M., Gao, B.: User recommendations in reciprocal and bipartite social networks–an online dating case study. IEEE Intell. Syst. **29**, 27–35 (2014)
11. Aciar, G., Aciar, S.: Recomendador de usuarios en una platafoma colaborativa. In: 14th Argentine Symposium on Artificial Intelligence ASAI, pp. 1–11 (2013)
12. Pujol, J., Sangüesa, R., Delgado, J.: Extracting reputation in multi agent systems by means of social network topology. In: Proceedings of the First International Joint Conference on Autonomous Agents And Multiagent Systems AAMAS '02, pp. 467–474 (2002)
13. Silva, N., Ing-Ren, T., Cavalcanti, G., Ing-Jyh, T.: A graph-based friend recommendation system using genetic algorithm. evolutionary computation (CEC), pp. 1–7 (2010)
14. Leskovec, J., Huttenlocher, D., Kleinberg, J.: Signed networks in social media. In: Proceedings of the SIGCHI Conference on Human Factors in Computing Systems (CHI '10), pp. 1361–1370 (2010)
15. Leskovec, J., Huttenlocher, D., Kleinberg, J.: Predicting positive and negative links in online social networks. In: Proceedings of the 19th International Conference on World wide web (WWW '10), pp. 641–650 (2010)
16. Wang, Z., Tan, Y., Zhang, M.: Graph-based recommendation on social networks. In: Proceeding of Web Conference (APWEB), pp. 116–122 (2010)

Fuzzy-Based Reliable Spectrum Tree Formation for Efficient Communication in Cognitive Radio Ad Hoc Network

Ashima Rout, Srinivas Sethi and P.K. Banerjee

Abstract In view of the scarce major radio resource vis-a-vis to resolve the bottleneck experienced in the present scenario of revolutionary technology, researches are under progress in the domain of cognitive radio ad hoc network (CRAHN). In this context, spectrum sensing followed by reliable as well as efficient spectrum detection and its effective utilization has been a main feature to achieve the quality at par with the intelligibility during communication in CRAHN. In this paper, performance of spectrum sensing has been discussed and tried to meet the challenges for a secured communication. A new approach has been formulated here using spectrum tree formation in cognitive radio ad hoc network environment at each intermediate node in between source and destination points. Analyzing this novel concept using fuzzy spectrum tree, the proposed model proves to perform better in cognitive radio ad hoc network environment in terms of throughput with minimum overheads.

Keywords CRAHN · Spectrum analysis · Spectrum sensing · Fuzzy based spectrum detection · Spectrum tree

A. Rout (✉) · P.K. Banerjee
Department of ETC Engineering, IGIT, Saranag, Dhenkanal, Odisha, India
e-mail: ashimarout@gmail.com

S. Sethi
Department of CSEA, IGIT, Saranag, Dhenkanal, Odisha, India
e-mail: igitsethi@gmail.com

P.K. Banerjee
Department of ETC Engineering, Jadavpur University, Kolkata, India
e-mail: pkbju65@gmail.com

© Springer India 2015
L.C. Jain et al. (eds.), *Computational Intelligence in Data Mining - Volume 1*,
Smart Innovation, Systems and Technologies 31, DOI 10.1007/978-81-322-2205-7_55

593

1 Introduction

Federal Communication Commission (FCC) [1] provides the information regarding the fixed spectrum assignment strategy which is inappropriate in the recent wireless communication era and most of the allotted licensed spectrum bands are under-utilized. The detection and deployment of the unused spectrum bands can be realized by sensing its radio environment which carry rigorous challenges and vital functionalities like spectrum sensing, sharing, management and mobility for insightful ideas on cognitive radio [2, 3]. The maximum spaces of licensed and under-utilized spectrum can be resolved by actual utilization of spectrum which implies the implementation of cognitive radio network (CRN). The CRN can be outlined from the aspects of Software Defined Radio and Intelligent Signal Processing. Further basing on the concept of software defined radio; the Cognitive Radio (CR) can sense its environment and track the changes for observational outcomes dynamically. The implementation of the CR is executed by intelligent signal processing to perform functions of resource management and access to communication media [2, 4].

Cognitive radio (CR) is a proficient technology that recognizes the Dynamic Spectrum Access [5] of licensed bands of the spectrum in the primary system environment. It can be performed by taking the benefits of spectrum utilization with an access to the cognitive/secondary users which are treated as unlicensed users. Hence, the theory of spectrum reuse by secondary user (SUs) has been evolved to capture the licensed spectrums of primary users (PUs). The SUs enter the unused licensed band of PUs opportunistically and quit their tenancy when PU is on the verge of possession into that band [6]. To maximize channel utilization based on the concepts of opportunistically assigning licensed channels it is desirable to make the whole activity a consistent and well-organized one that has been justified in this paper.

The rest of the paper is structured as follows. Related works have been discussed in Sect. 2 followed by fuzzy-based spectrum tree formation in Sect. 3. Section 4 discussed the performance evaluation with result analysis of proposed model. The conclusion has been discussed in Sect. 5.

2 Related Works

Distinguishing the spectrum availability and its re-configuration play vital role in dynamic network where the later is the process to identify the spectrum accessibility. The re-configuration may be completed according to the environment and different transmission parameters in order to make use of the available spectrum opportunistically [3, 7].

The secondary or cognitive radio network does not have a licensed spectrum and tries to utilize the unused spectrum in an opportunistic way but such a dynamic network may have its own base station [7]. Sensing the radio frequency and spectrum discovery are the foremost duties of cognitive radio in CRN. This may sense the spectrum holes which is needed at the receiver to articulate exposure and transmit power control. The management of the spectrum selects the communication power levels and frequency holes for communication through nodes. The exertion is passed in receiver and transmitter which involve some procedure of feedback between receiver and transmitter [8].

A base station schedules the data transmission in the centralized architecture of CR network which can be taken as one of the entities in the network. Another one is responsible for assigning the radio resource to spectrum broker or users in the dynamic network. In this situation, the users may be primary, secondary or both. The job of the spectrum broker may be thought of performing the roles of a primary or secondary base station or else accomplishing the sensing functionalities [4, 10, 11]. In the cooperative and non- cooperative network of the distributed CRN, there is no spectrum broker or a base station to coordinate spectrum access of secondary users. The users share the interference information in cooperative network and determine the spectrum allocation based on this. But, there is no communication for interference information in non-cooperative network. In this paper, the choice of improved spectrum in terms of reliability and spectrum tree formation have been proposed for enhanced communication in cognitive radio ad hoc network.

3 Fuzzy-Based Spectrum Tree Formation

In this section, fuzzy based reliable spectrum selection algorithm has been discussed. Same radio channels have been shared by the Primary or Secondary users to communicate with their base stations in the network. The SUs depend on the activities of primary users. The specified spectrum assigned to the primary network is divided into number of channels, where the primary network consists of number of primary users in the same area, a CR network is organized which consists of number of cognitive or secondary users. The secondary users are active when the primary user is inactive and vice versa.

In a dynamic environment like CRAHN, effective channel sensing and its management are significant issues to perform better result. The channel need not be dedicated so that it can be shared to transfer numerous packets from one or more nodes to a number of destination points. The intermediate nodes forward the same message signal to other neighborhood nodes in an ad hoc manner in cognitive radio network. In this paper, primary user activity has been assumed through the statistical information and channel capacity as in Fig. 1, to detect the reliable spectrum in CRAHN environment. Definition of primary user activities may be established as the percentage in average of the usage of dedicated channel by primary users as per the statistical information whereas; channel capacity is treated as the maximum data rate over an assigned channel. A novel concept of spectrum tree formation has been proposed using fuzzy logic considering two basic parameters as primary user activity and channel capacity [13, 15, 16].

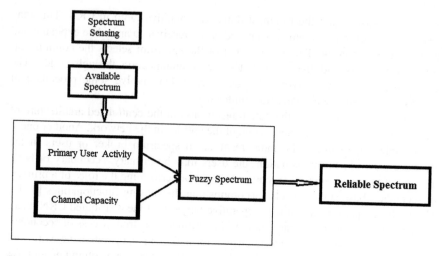

Fig. 1 Architecture of the reliable spectrum sensing mechanism in CRAHN

Here, estimation of reliable spectrum process has been justified by using fuzzy logic thereby taking into account primary user activity and channel capacity in CRAHN environment. During this approach, the fuzzifier converts the primary user activity and channel capacity to perform fuzzification process, which is used to find out the reliable spectrum as an output. It has two inputs as primary user activity and channel capacity during the evaluation process. It decides the grade to which they belong to each of the suitable fuzzy sets via membership functions and evaluates the reliable value of a spectrum through the defined inputs.

The output fuzzy sets for reliable spectrum sensing are evaluated by rule base method after creating input fuzzy sets, which consists of "if-then"-statements. The reliability of the resulted spectrum can be determined by means of the estimation process carried out through fuzzy inference sys (FIS), which applies a set of fuzzy rules. The fuzzy sets of inputs and outputs are reflected as in Figs. 2, 3 and 4.

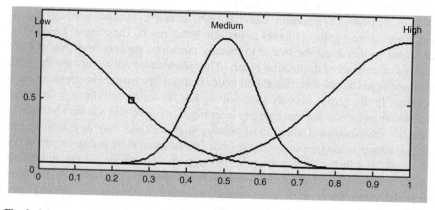

Fig. 2 Primary user activity as per statistical information

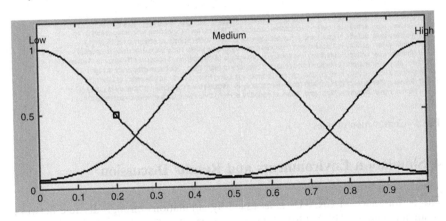

Fig. 3 Channel capacity

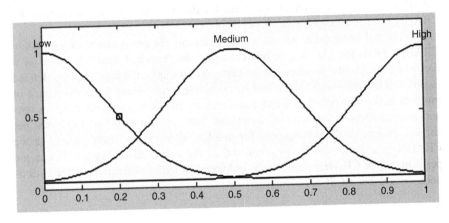

Fig. 4 Spectrum reliability

Rules are defined through a series of "IF-THEN" statement for the output with the help of the input conditions using the rule structure of fuzzy logic. There are sets of nine (3 × 3) possible logical output response decisions, as shown in Fig. 5.

The reliable spectrum sensing can be done by using fuzzy concept with the help of the primary user activity in terms of statistical information and channel capacity in CRAHN environment. A set of SUs form a tree in each available spectrum band, in the network called spectrum-tree [18].

A spectrum-tree can be formed by using cognitive/secondary users in cognitive radio ad hoc network through each available reliable spectrum band. Each spectrum-tree has only one root and it keeps the elementary information about each node in the tree in a dynamic topology, such as the routes for other nodes. Each node has its unique secondary user identification in one spectrum-tree. This is proactive communication in the network and may help to make a superior and reliable communication in the CRAHN which is main objective of this paper.

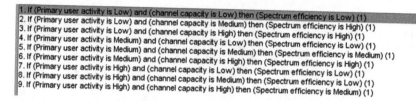

598

Fig. 5 Construction of rules

4 Simulation Environments and Results Discussion

The performance of CRAHN is assessed by simulations through discrete network simulator named NS-2 [20], integrated with the Cognitive Radio Cognitive Network simulator [21]. The source node and sink nodes have been considered in CRAHN to work in different spectrum bands. The Simulation parameters for CRAHN have been considered as per Table 1.

The control overhead is considered for analyzing the performance of protocol in CRAHN. From the Fig. 6 it is observed that the proposed model is better as it consumed less control overhead in the network. Similarly, the throughput is defined as an essential parameter which can be considered to evaluate the performance of protocol in the network. It is the rate at which the nodes in a network receives data and a good channel capacity of a network and rated in terms of bits/bytes per second. In Fig. 7, the throughputs for number of secondary users present in the CRAHN give better results as compared to the conventional model. Hence the performance of CRAHN seems to be enhanced in terms of throughput for various node dimensions of SUs.

Table 1 Simulation parameters for CRAHN

S. No	Parameters	Values
1	Area size	500 m × 500 m
2	Transmission range of PU	250 m
3	Transmission range of SU	100 m
4	Simulation time	1,000 s
5	Nodes speed	5 m/s
6	Pause times	10 s
7	Data rate	5 Kbps
8	Mobility model	Random any point
9	Interface	1
10	Number of channel	5
11	Numbers of SU Node	10, 20, 30, 40, 50, 60, 70, 80, 90, 100
12	Numbers of PU	5
13	No. of simulation	5

Fig. 6 Overhead versus number of secondary users (SU)

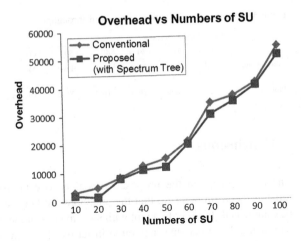

Fig. 7 System throughput versus number of secondary users (SU)

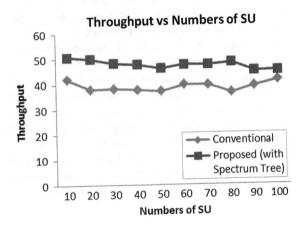

Further, keeping in view the comparison status between the existing and proposed algorithms in terms of low, high and moderate (mod.) the performance evaluation as regards to Overhead and Throughput have been mentioned in the Tables 2 and 3 respectively.

Table 2 Performance evaluation in terms of overhead

| | No. of secondary users | | | | | | | | | |
	10	20	30	40	50	60	70	80	90	100
Conventional	Mod.	High	Mod.	Mod.	High	Mod.	High	Mod.	Low	High
Proposed	Low	Low	Mod.	Low	Low	Mod.	Low	Low	Low	Low

Table 3 Performance evaluation in terms of throughput

| | No. of secondary users | | | | | | | | | |
	10	20	30	40	50	60	70	80	90	100
Conventional	Mod.	Too low	Low	Low	Low	Mod.	Mod.	Too low	Mod.	Mod.
Proposed	High	High	High	High	Mod.	High	High	High	Mod.	Mod.

5 Conclusions

The key concept in the proposed model is to establish a spectrum-tree in each reliable spectrum band, so that the collaboration between spectrum decisions could be established. In order to adapt with system environment, consideration of different parameters have become inevitable in terms of system throughput to measure the channel excellence, effectiveness and efficiency in the CRAHN for observation purpose. Correspondingly, significance of the system has also been studied under this scenario in the said network. Irrespective of number of secondary users a line of comparison has been established showing whether throughput and overhead are low, moderate or high as regards to conventional and proposed spectrum tree algorithm, thus resulting the proposed model to supersede the conventional one. However, the comparison table may attract alteration of present state in different cognitive radio environment. As other parameters also influence towards the increase and decrease of throughput and overhead, the performance of spectrum sensing can be evaluated by trading the number of secondary users with our assumed parameters like overhead and throughput. Inferences could be justified and rationalized as compared to the existing and conventional one towards achieving the quality and efficiency of spectrum for SUs node size in a CRAHN.

References

1. Federal Communications Commission: Facilitating opportunities forflexible, efficient, and reliable spectrum use employing cognitive radio technologies, FCC Report, ET Docket, (2003) 03–322
2. Mitola, J.: Cognitive radio for flexible multimedia communications. In: IEEE International Workshop on Mobile Multimedia Communications (MoMuC '99), pp. 3–10 (1999)
3. Lassila, P., Penttinen, A.: Survey on Performance Analysis of Cognitive Radio Networks. COMNET Department, Helsinki University of Technology, Finland (2009)
4. Haykin, S.: Cognitive radio: brain-empowered wireless communications. IEEE J. Sel. Areas Commun. 23(2), 201–220 (2005)
5. Li, C., Li, C.: Dynamic channel selection algorithm for cognitive radios. In: 4th IEEE International Conference on Circuits and Systems for Communications (ICCSC 2008), pp. 175–178 (2008)
6. Chen, T., Zhang, H., Maggio, G.M., Chlamtac, I.: CogMesh: a cluster-based cognitive radio network. In: Proceedings of IEEE DySPAN (2007)

7. Issariyakul, T., Pillutla, L.S., Krishnamurthy, V.: Tuning radio resource in an overlay cognitive radio network for TCP: greed isn't good. In: IEEE Communication Magazin, pp. 57–63 (2009)
8. VIT RESEARCH REPORT VIT-R-02219-08
9. Khalid, Q., Hasari, C., Muneer, M., Sabit, E.: Performance analysis of ad hoc dispersed spectrum cognitive radio networks over fading channels, URASIP J. Wireless Commun. Networking (2011)
10. Hampshire, F.: Cognitive radio technology: a study. QinetiQ Ltd Cody Technology Park, vol. 1(1.1) (2007)
11. Clancy, T., Walker, B.: Predictive dynamic spectrum access. In: SDR Forum Conference (2006)
12. Akyildiz, I.F., Lee, W., Vuran, M., Mohanty, S.: Next generation/dynamic spectrum access/ cognitive radio wireless networks a survey. Comput. Netw. **50**(13), 2127–2159 (2006)
13. Wong, Y.F., Wong, W.C.: A fuzzy-decision-based routing protocol for mobile ad hoc networks. In: 10th IEEE International Conference on Network, pp. 317–322 (2002)
14. Raju, G.V.S., Hernandez, G., Zou, Q.: Quality of service routing in ad hoc networks. IEEE Wireless Commun. Networking Conf. **1**, 263–265 (2000)
15. Pedrycz, W., Gomide, F.: An introduction to fuzzy sets: analysis and design (complex adaptive systems). MIT Press, Cambridge (1998)
16. Buckley, J.J., Eslami, E., Esfandiar, E.: An introduction to fuzzy logic and fuzzy sets (advances in soft computing). Physica Verlag (2002)
17. Rout, A., Sethi, S.: Throughput analysis of spectrum in cognitive radio ad hoc network. Int. J. Appl. Innov. Eng. Manage. (IJAIEM) **2**(8) (2013)
18. Zhu, G.M., Akyildiz, I.F., Kuo, G.S.: STOD-RP: A spectrum-tree based on-demand routing protocol for multi-hop cognitive radio networks. In: IEEE Global Telecommunications Conference. IEEE GLOBECOM, pp. 1–5 (2008)
19. Rout, A., Sethi, S., Banerjee, P.K.: Fuzzy-based reliable and efficient communication in cognitive radio ad hoc network. In: Proceeding of IEEE International Conference on Control, Instrumentation, Energy and Communication (CIEC-2014), Kolkota (2014)
20. Ns-2 Manual, Internet Draft: http://www.isi.edu/nsnam/ns/nsdocumentation.html (2009)
21. CRCN integration: http://stuweb.ee.mtu.edu/~ljialian/

Realization of Digital Down Convertor Using Xilinx System Generator

**Manoj Kumar Sahoo, Kaliprasanna Swain
and Amiya Kumar Rath**

Abstract In some secure communication system, to detect the frequently varied baseband signal, a digital down converter (DDC) with a variable digital filter is used. In this paper, a reconfiguration design process of DDC is discussed for a GSM application with the help of Xilinx system generator (XSG) on field programmable gate array (FPGA). The approach is based on hardware Co-simulation based on XSG platform which integrates itself with the Matlab based Simulink graphics environment and implemented on Virtex-II based xc2v200-4fg676 FPGA device. Optimal equiripple technique implements DDC which reduces the resource requirement. To solve the complexity, a novel type of polyphase decomposition structure is used. Keeping the view of the system performance, such as area and speed, this paper proposes a model which is implemented by using embedded multiplier, LUTs and BRAM of FPGA. It is seen that the proposed design consumes very less resources available on target devices.

Keywords DDC · FPGA · Matlab · Xilinx system generator · Co-HW simulation

1 Introduction

DSP algorithm is hidden inside a number of consumer electronic products such as mobile phones, satellite phones, PDAs and many other wireless products. The DSP algorithms are modified time to time in order to increase the speed, area and power

M.K. Sahoo (✉) · K. Swain
GITA, Badaraghunathpur, Madanpur, Bhubaneswar 752054, Odisha, India
e-mail: mksahoo@hotmail.com

K. Swain
e-mail: kaleep.swain@gmail.com

A.K. Rath
VSSUT, Burla 768018, Odisha, India
e-mail: amiyaamiya@rediffmail.com

603

© Springer India 2015
L.C. Jain et al. (eds.), *Computational Intelligence in Data Mining - Volume 1*,
Smart Innovation, Systems and Technologies 31, DOI 10.1007/978-81-322-2205-7_56

consumption. The vendors of such products must be flexible enough to adopt new signal processing technique to increase the coverage area, portability and combination of many more areas. An important application where a diverse range of frequency is required is a software defined radio (SDR). The first step towards the DSP area is usually initiated by sampling and then digitizing the analog signal to convert into a stream of numbers. Today FPGA technology is becoming an emerging technology to implement DSP algorithm among other traditional technologies like DSP processor, ASIC etc.

This paper is organized as follows. A brief discussion of DDC is presented in Sect. 2. Section 3 describes the proposed DDC structure. Co-hardware simulation of DDC model developed in Sect. 4 whereas results and analysis are shown in Sect. 5 and at the end conclusions are drawn in Sect. 6.

2 Digital Down Convertor

Down conversion of the signal involves the process of shifting a high rated signal to a require standard signal. In general the receivers receive a wide band of signal, but the receiving end user may only require a small portion of the entire band. So fulfilling the above requirement might involve prohibitively large filters. The deployment of DDC makes this process efficient and easier [1, 2].

A DDC consists of five basic blocks such as Direct Digital Synthesizer (DDS), Mixer, Cascaded integrate comb (CIC), Compensation FIR (CFIR) filter, Programmable FIR (PFIR) filter.

The area of digital signal processing plays a vital role in many fields of electronics, transmission and storage of digital date [3]. Apart from this, as far as complexity of the circuit is concerned the cost against performance is an important factor [4]. In wireless application, FIR filters, sampling technique is a primary concern and also in some multi rate DSP system where up sampling and down sampling is achieved by using interpolator and decimeters. In this variable sampling process, to attenuate the production of undesired signals associated with aliasing FIR filter is used [5]. As far as SDR based wireless base band receiver is concerned, DDC plays a vital role for the same. Basically a DDC shifts the high frequency carrier signal to an intermediate frequency and then converts to the baseband spectrum of interest. To achieve the maximum SNR and adjacent channel cancellation, DDC also performs decimation and filtering. Primarily for GSM application, DDC decimates the incoming spectrum [6] by filtering prior to decrease the sampling rate. With the advent of FPGA family and to explore the benefit, DDC is implemented in FPGA. A simplified architectural block diagram [7] is shown in Fig. 1.

In the first stage, a DDS along with a pair of multipliers (M1 and M2) [8] is used to translate a specific channel down to baseband spectrum by tuning the desired frequency of interest. Two cascaded integrated comb (CIC) filters are used in the second stage to reduce the sampling rate for matching the desired bandwidth at

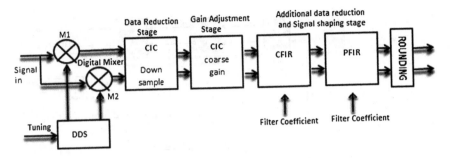

Fig. 1 Architectural block diagram of DDC

output and to provide a coarse gain adjustment. Finally a pair of polyphase filter one is compensation FIR (CFIR) and another is programmable FIR (PFIR) to provide the additional decimation and shaping the final output.

3 Proposed DDC Model Design

A model of Digital Down-Converter (DDC) is designed to meet the new generation application in communication like 3G or 4G in mobile communication, specifications using a multi-section CIC decimator and two equiripple based polyphase decimators [9] with the help of Xilinx block set in the Matlab®Simulink environment for a vertex family FPGA from Xilinx. The XSG is the environmental token to help the design to analyze and implement in a fixed float architecture, then convert to an integer based system architecture. The XSG help in developing Co-Hardware block with the help of XFLOW or generating the HDL code. These processes are very much helpful for DSP system analysis and implementation on FPGA. The Fig. 2 shows the design part in Simulink using Xilinx blockset, which is to be implemented on target FPGA. Equiripple window based technique is used to improve hardware complexity and speed with compare to other windows based process.

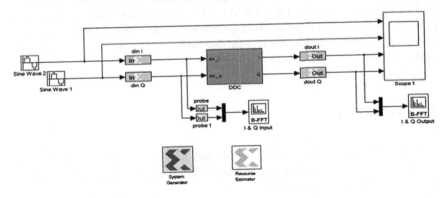

Fig. 2 Proposed DDC model

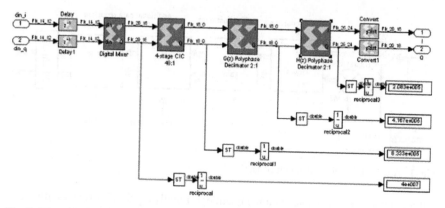

Fig. 3 Blocks in the DDC block

The functions performed by the developer model are waveform synthesis (DDS), complex multiplication and Multirate filtering. Figure 3 shows the three major blocks; mixer section, CIC section and decimator section are the three major parts of this design.

3.1 Mixer Section

This block consists of two multipliers M1 and M2, along with direct data synthesizer (DDS) as shown in Fig. 4. The multipliers M1 and M2 used in the proposed DDC model are implemented using the Virtex-II IP based embedded multipliers. To enhance the sample rate in excess of 200 MHz pipeline operation is used. In this design, with an input sample rate of 52 MHz, a single multiplier could be time-shared to implement the input heterodyne by nature. To store one quarter of a cycle of a sinusoid the FPGA block memory is used. Both the in-phase and quadrature components of the local oscillator to be generated simultaneously are enhanced by dual-port memory using a single BRAM. The single BRAM implementation can generate a 4 K sample full-wave 16-bit precision complex sinusoid. The synthesizer with phase dithering will generate a mixing signal with a spurious free dynamic range (SFDR).

3.2 CIC Section

As the baseband channel is highly oversampled, a simple cascade of boxcar filters, implemented as a cascaded integrator comb (CIC) is employed initially to achieve the require reduction factor of 48.

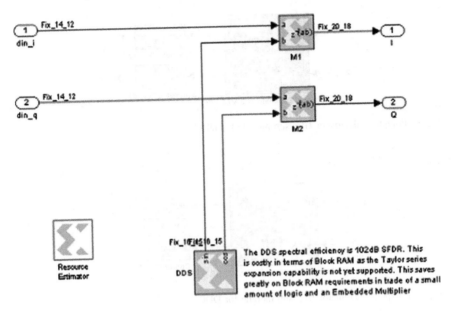

Fig. 4 Mixer sub module for DDC

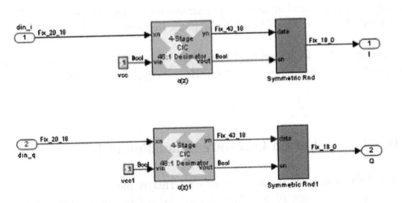

Fig. 5 CIC sub module for DDC

The CIC filter $C(z)$ is multiplier less consisting only of integrator and differentiator sections. For this application a cascade of 4 integrators followed by 4 differentiators, with an embedded 48:1 reduction in rate change is used as shown in Fig. 5.

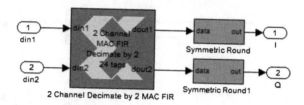

Fig. 6 Polyphase decimator $G(z)$ sub module for DDC

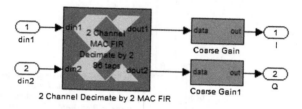

Fig. 7 Polyphase decimator $H(z)$ sub module for DDC

3.3 Polyphase Decimator Section

To produce the required input-to-output sample rate change of 192:1, the CIC filter is followed by a cascade of two 2:1 polyphase decimators shown in Figs. 6 and 7 respectively.

The polyphase decimator $G(z)$ is realized with a 24-tap MAC filter, while a 96-tap MAC filter is employed for $H(z)$. Here 96 cycles are required to implement this filter.

4 Development of Co-hardware Simulation of DDC Model

1. The DDC hardware library Generation—This is the process to generate the equivalent hardware of the above model for a targeting device and here we consider Vertex 2 (XC2V2000-4fg676). With the help of XSG by using XFLOW the equivalent hardware target device programmable bit file is generated.
2. Then the library model is integrated with the basic model as shown in Fig. 8.
3. The library file can be generated for a standalone JTAG method or an Ethernet JTAG mode where the standalone process is considered.
4. Then the device utility, timing parameter, and output are analyzed for JTAG Co-sim block as shown in Fig. 9.

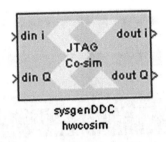

Fig. 8 Integration of library with basic model

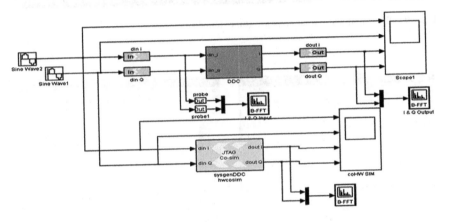

Fig. 9 DDC Co-hardware simulation model

5 Result and Analysis

The DDC model is designed with Xilinx blockset in Simulink. With the help of XSG, the model is analyzed through Co-hardware simulation process and HDL code for FPGA is generated by netlister process.

5.1 Result from Simulink Modeling

The developed DDC has been simulated and verified using Simulink whose output response is shown in Fig. 10. The first two waveforms show I & Q input signals. Third and fourth waveforms show the decimated I & Q output signals.

The deployed system estimator token suggest the number of devices with respected to total availed in an FPGA family, which is very much helpful to choose the FPGA. Figure 11 shows the input and the output response in the frequency domain, whereas Fig. 12 shows the resource estimation process both for pre and post estimator the parameters like slices, FFs, BRAMs, LUT are compared.

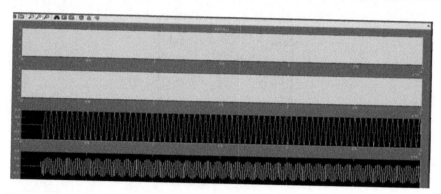

Fig. 10 DDC input and output response (time domain)

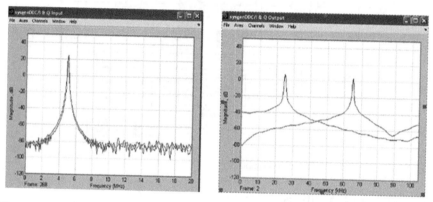

Fig. 11 DDC input and output response (frequency domain)

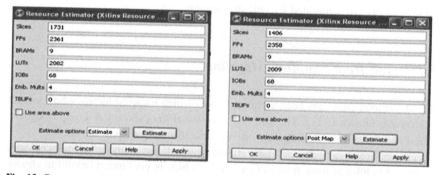

Fig. 12 Resource estimation process (pre synthesis and post synthesis estimation)

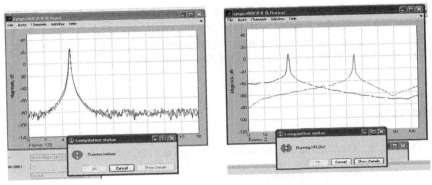

Fig. 13 Initialization of netlister process and output of co-hardware simulation of DDC using JTAG Sim-block

Fig. 14 Design utility in XFLOW process and timing parameters for DDC co-hardware simulation

5.2 Result for Co-hardware Simulation

Figure 13 shows the initialization of simulation in the netlister process, and also describes the output of Co-Hardware simulation using JTAG Sim block.

Figure 14 shows the design utility in Xflow process and timing parameters for DDC Co-hardware simulation respectively.

6 Conclusion

Xilinx System Generator based hardware co-simulation technique is presented to implement digital down convertor for software defined radios. The reconfiguration method of design for the design with XSG for DSP system design is very useful. An Equiripple window method based polyphase decomposition technique is used in DDC design to optimize the speed and area. The implementation of the proposed model on Virtex-II family (device: xc2v2000-4fg676) utilize less device resources to provide cost effective solutions for SDR. The Co-Hardware simulation is carried out and the results are verified with the Simulink environment result. Similarly the DDU unit should be designed with the same specification, which builds up DDC-DDU unit for an SDR application with GSM bandwidth.

References

1. Changrui, W., Chao, K., Shigen, X., Huizhi, C.: Design and FPGA implementation of flexible and efficiency digital down converter. In: International Conference on Signal Processing, pp. 438–441. IEEE, Beijing (2010)
2. Zhi-hai, Z., Shan-ge, L., Wen-guang, L., Ben-lu, L., Ze-chao, Z.: Implementation of high-performance multi-structure digital down converter based on FPGA. In: International Conference on Signal Processing, vol. 1, pp. 31–35. IEEE, Bejing (2012)
3. Haykin, S.: Adaptive Filter Theory, 3rd edn. Prentice-Hall Inc., Upper Saddle River (1996)
4. Fei-yu1, L., Wei-ming, Q., Yan-yu, W., Tai-lian, L., Jin, F., Jian-chuan, H.: Efficient WCDMA digital down converter design using system generator. In: Proceeding of International Conference on Space Science and Communication, pp. 89–92. IEEE (2009)
5. Beygi, A., Mohammadi, A., Abrishamifar, A.: An FPGA-based irrational decimator for digital receiver. In: International Symposium on Signal Processing and Its Applications, pp. 1–4. IEEE (2007)
6. Jack, H.: Hardware/Software Co-verification. EDA Café Weekly, 29 March 2004
7. Kuenzler, R., Sgandurra, R.: Digital-Down Converter implementation, FPGAs offer new possibilities, vol. 4, pp. 30–33. Military Embedded System (2008)
8. Kim, M., Lee, S.: Design of dual-mode digital down converter for WCDMA and cdma2000. ETRI J. 26, 555–559 (2004)
9. Chester, D.B.: Digital IF filter technology for 3G systems. Commun Mag, IEEE 37, 102–107 (1999)

Probabilistic Ranking of Documents Using Vectors in Information Retrieval

Balwinder Saini and Vikram Singh

Abstract On the web, electronic form of information is increasing exponentially with the passage of past few years. Also, this advancement creates its own uncertainties. The overload information result is progressive while finding the relevant data with a chance of HIT or Miss Exposure. For improving this, Information Retrieval Ranking, Tokenization and Clustering techniques are suggestive as probable solutions. In this paper, Probabilistic Ranking using Vectors (PRUV) algorithm is proposed, in which tokenization and Clustering of a given documents are used to create more precisely and efficient rank gratify user's information need to execute sharply reduced search, is believed to be a part of IR. Tokenization involves pre-processing of the given documents and generates its respective tokens and then based on probability score cluster are created. Performance of some of existing clustering techniques (K-Means and DB-Scan) is compared with proposed algorithm PRUV, using various parameters, e.g. Time, Accuracy and Number of Tokens Generated.

Keywords Information retrieval (IR) · Ranking/indexing · Tokenization · Clustering

1 Introduction

Information Retrieval (IR) is an essential part of Data mining. It mainly deals with the representation, storage, organization and access or retrieval of the information [1, 2]. Modern information retrieval (IR) system tries to provide better model which is responsible for finding the most relevant information with respect to the user's

B. Saini (✉) · V. Singh
Computer Engineering Department, NIT Kurukshetra, Chandigarh, Haryana, India
e-mail: me7saini@gmail.com

V. Singh
e-mail: viks@nitkkr.ac.in

613

© Springer India 2015
L.C. Jain et al. (eds.), *Computational Intelligence in Data Mining - Volume 1*,
Smart Innovation, Systems and Technologies 31, DOI 10.1007/978-81-322-2205-7_57

query when it is requested. Few years back, Information retrieval beginning was totally related to the knowledge stored in the textual form and information retrieval was processed manually. In libraries, information retrieval was in the form of book lists and in the books, it was like tables of contents and index pages etc. As it is difficult to maintain and build these indices so, the tables only have the index terms like Name, Writer, and Headings of given subject [3]. To differentiate between the relevant and irrelevant documents is the main uncertainty depends upon normal arrangement of the required documents done by a Ranking algorithm [4]. In the arrangement, the top most documents are taken to be more relevant. Therefore, ranking of given documents are the most important part of the process of information retrieval system. Thus, ranking algorithm deep reside in information retrieval systems. Working of ranking algorithm [1, 2] is according to the basic premises (regarding document relevance) yield distinct information retrieval models. In the modern IR systems there are various bases for the ranking. Probability score is one of the important factors, which is completely based on the identified tokens the document and documents vector's length.

2 Information Retrieval Model

Documents are considered as bunches of words under traditional retrieval modeling. This means that, the models acts like entity without structure in which relevance can be determined on the basis of no. of occurrences of terms only [5]. When the retrieval system input with a query, with respect to that query every document is scored. After the completion of this scoring process, user gets the final ranked list. These scores are produced by the IR model. Efficiency doesn't care under the general retrieval models; user's information need and the ranking process are the main focus of these models.

Modern Information retrieval system deals entire process of systems into two parts one deals with representation, organization and storage. Second part of information retrieval system deals with providing the mechanism for stored information items to the end users [6]. In the first part, Information retrieval system deals with extracting the useful data sources that is information and then it represent this information in a accurate and efficient way in the system memory, means organized in a proper manner like tabular form etc. Using suitable ranking or Indexing techniques. Finally, stored this managed information into a storages medium like hard disk, flash drives or on a cloud. In the second part, deals with the retrieval party of system with accessing the stored information from the storage medium when the user's query takes place in the information retrieval system. It also handles the errors which are generated when the information retrieval system is in an operational state. Therefore the IR models in information retrieval field are: Boolean Model, Vector-Space Model, and Probabilistic Model [1–3]. In the

Boolean model documents and queries are represented as set of index terms. The advantages of this is model is its simplicity. The Boolean model [7] suffers with too many documents retrieval during exact matching, this disadvantage decrease its popularity and make space for another model i.e. Vector-Space model [2]. Due to partial matches under the Vector-Space model, it gives better results from the Boolean model of information retrieval. For this partial match it uses a method of term weighting [5], in which frequency of each word in each document is found followed by the weight of words in the documents with respect to the other documents and the query as entered by the user is found. The vector-space model has one disadvantage that the unavailability of exactness in retrieval. The Probabilistic model [6] provides retrieval of matching documents in which nearly exact match is found. Composition of this paper is, Sect. 3 describes the probabilistic information retrieval model, ranking method and also discusses the tokenization process. Section 4 describes the related work and Sect. 5 discusses the proposed algorithm i.e. PRUV algorithm, and also describe its example. Section 6 gives the implementation results compared with K-Means and DB-Scan clustering algorithm.

3 Probabilistic Model and Ranking

Probabilistic information retrieval model [2, 6] uses the probability theory concepts for matching the given documents with user's query. It gives the matching results like either exact match is found or not exact. When this IR model is used in search engines or in other systems for the purpose of document retrieval gives better results as compared with other IR models. Although the results of retrieval are better when they are given with the use of the probabilistic model, they are retrieved with comparatively lesser speed because the retrieval is improved gradually. The Advantage of Probabilistic information retrieval model is, it uses simple probability theory for ranking the documents and the ranking order is in decreasing order of their probability of being relevant. Probabilistic information retrieval model has disadvantages also like: initially separate the relevant and irrelevant documents, it does not give any idea about the frequency of key-words/index terms which are in documents. Good information retrieval systems have some features which are as follows:

3.1 Ranking Procedure

Information Retrieval system or model should have a good ranking procedure because ultimately quality of a information retrieval model directly depend upon its ranking method by which it can index the given set of data items in the data repository so that when the user's query, for retrieving some information is

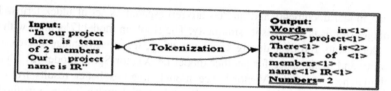

Fig. 1 Tokenization and input, output strings

requested then only relevant information should be retrieved to the user and decrease the irrelevant information from the retrieved result.

3.2 Representation and Organization of Data Items

The main part of all information retrieval is how they represent and organise its data items so that the final result of a user's query will be deliver in very short time period and it has good quality also means resultant document or information should be relevant to the user's need. Tokenization process plays an important role on representation of documents. Tokenization process [9] is of the most important aspect in information retrieval field in which simply all the words, numbers, and other characters etc. are formed these words, numbers, and other characters are called as tokens. This process also calculates the frequency value of all these tokens present in the given document. For example, there is a document in which the information likes "This is an Information retrieval model and it is widely used in the data mining application areas". The tokenization process, output of the process for a document, is identified words or numbers and words count etc., theses words and numbers are called tokens. This is shown in Fig. 1.

In output, values in angular braces indicate the frequency of a word in the given document, e.g. word "our" and "project" occurs two times in the document so their frequency is 2. Tokenization also facilitates by to separating the stop words and only gives the distinct words form the given document. Once the tokens are established by tokenization process, next step is to calculate the probability score, the value of probability score is with respect multiple topic's (vectors) and token's count on the document. Finally clustering process creates the cluster of tokens based on their probability score.

4 Related Work

Clustering of the given documents is a classical approach in data mining area. In modern era, many algorithms are there for document clustering process. Karthikeyan and Aruna [10] developed a method for document clustering on the basis of probability based semi-supervised document clustering. In this method, documents are clustered with respect to the user's need and this need is represented by multiple-attribute topic structure, it uses some topic annotations for each document for calculating similarity values between the documents. Cluster documents are based on maximum similarity value. Qiu and Tang [9] proposed a new topic oriented semi-supervised document clustering approach. In this approach, clusters are made-up on the basis of ontology and a dissimilarity creation function.

5 Proposed Algorithm (PRUV)

The probabilistic ranking using vectors (PRUV), addresses the issue of uncertainty in token identification process and exactness on numbers of tokens in documents. The Ranking of tokens is purely based on the occurrence of identified tokens from given documents. The probabilistic Ranking method is most suitable for the IR systems, where degree of uncertainty persist at data and user query level [6]. The probability score of tokens and defined threshold values for cluster are the basis for the clustering process. In case of K-means [11] and DB-Scan [12], Threshold the parameters are defined initially to the algorithm for implementations. In PRUV approach, along with threshold values for cluster the probabilistic rank (Occurrence of tokens) is also provided for implementation. The tokenization technique is used which tokenize all the documents and then applying the working principle of probabilistic information retrieval model on the output of this tokenization technique for finding their probability scores and it also uses the clustering process which makes the clusters of output (probability score) of probabilistic IR model i.e. it extends the overall ranking process for obtaining better results.

5.1 Motivation

Information Retrieval system or model should have a good ranking procedure, as quality of a information retrieval model directly depend upon its ranking method by which it can index the given set of data items in the data repository. On the user's query is requested to retrieve some information then only relevant information should be retrieved to the user and decrease the irrelevant information from the retrieved result. In following the basic details of algorithm are described below,

PRUV Algorithm (Input (Di, Dj)) = (Output (??))

Step 1:

   Identify Vector_Topics(**Vi**) for all **i**=1, 2, 3.....n // and add some keywords or reserve words to each vectors //

   Calculate Vector_Len(**VLi**)= (**Vi**) for **i**=1, 2, 3......n // The basis of number of keywords in it.//

Step 2:

   For each **Vi** & Training Document (**TDi**),

       Tokenization (**Ai**)= (**Vi**, **TDi**), // Apply Tokenization to extract words (**Wi**), numbers (**Ni**) & Special   symbols (**Si**) stored into some arrays(**Ai**). //

       Calculate Word_Count (**WCi**)= (**Ai**), for all **i**= 1, 2, 3...n, // for The total number of occurrences of each Wi //

       Calculate Similar_Words(**SWi**) = (**Wi**, **Vi**), for **i**=1, 2, 3.....n, // to store similar into vector's//

Step 3:

   For each **SWi**,

       Calculate Probability (**Pi**) = **SWi** / **SWj** , // Where **SWi** defines the occurrence of distinct words in a **TDi** & **SWj** defines the total number of distinct words in all **Vi**. //

Step 4:

   Clusters_Form (**c**), // Based on comparison between **Pi** & User's threshold values (**Uthres**) for each cluster. //

Step 5:

   Then finally execute for testing document (**TD**), **Repeat**, from **step 2** to **step 4**.

5.2 Example

The step-wise working example of PRUV algorithm is as follows:

Step 1: Initial part, Vectors (V) and words (Wv) will be created, as shown in Table 1. For given set of Input documents four vectors v1, v2, v3 and v4 are created, so value of i = 1, 2, 3, 4. In the Table 1 all the vector are shown (v1, v2, v3 and v4).

   In the Table 1, Column name represent the name of vectors e.g. v1 represents military background, v2 represents sports background, v3 represents science background and v4 represents engineering background. Next, calculation of each vector length (VL) that is total number of distinct words in a vector will be done. $VL_1 = 10$, $VL_2 = 10$, $VL_3 = 10$, $VL_4 = 10$ and $SW_j = 40$.

Step 2: Now, tokenize all these document (TD_1 to TD_4) text and find out all the words (W_d), all the numbers (N_d) and all special symbols (S_d). Calculate W_d

Table 1 Vector table

Military (v1)	Sports (v2)	Science (v3)	Engineering (v4)
Battle	Ball	Scientist	Computer
Soldier	Team	Atom	Engineer
Army	Player	Laboratory	Engine
Navy	Win	Research	Test lines
Airforce	Lose	Element	Construction
Defense	Play	Evolution	Professional
Weapons	Game	Experiment	Electrical
War	Competition	Particle	Mechanical
Battalion	World cup	Cell	Civil
Military	Toss	Energy	Diagrams

frequency also that is word count WC. So in TD1 the value of SW1 is equal to 4 because four distinct words are matched with Wv1 that are Military, Army, Navy and Air force. In TD2 the value of SW2 is equal to 3 because three distinct words are matched with Wv2 that are Player, Team and Ball. In TD3 the value of SW3 is equal to 2 because three distinct words are matched with Wv3 that are Scientist and Evolution. In TD4 the value of SW4 is equal to 4 because four distinct words are matched with Wv4 that are Engineer, Professional, Computer and Construction. Finally word count represents in angular braces (i.e. $\langle WC \rangle$).

Mathematically,

Vector (Vi)	Training documents (TDi)	SWi and WCi
Military (V1)	**Military** is a good option for a career builder for youngsters. **Military** is not covering only defense it also includes IT sector and its various forms are **Army, Navy,** and **Air force**	SW1 = 4 (Military $\langle 3 \rangle$, army $\langle 1 \rangle$, navy $\langle 1 \rangle$, air force $\langle 1 \rangle$)
Sports (V2)	In cricket a **player** uses a bat to hit the **ball** and scoring runs. It is played between two **teams**; the **team** scoring maximum runs will win the game	SW2 = 3 (Player $\langle 1 \rangle$, Team $\langle 2 \rangle$, Ball $\langle 1 \rangle$)
Science (V3)	Various **scientists** working on different topics help us to understand the science in our lives. Science is continuous **evolutionary** study	SW3 = 2 (Scientist $\langle 1 \rangle$, Evolution $\langle 1 \rangle$)
Engineering (v4)	**Engineers** are manufacturing beneficial things day by day, as the **professional engineers** of software develops programs which reduces man work, civil **engineers** gives their knowledge to **construction** to form buildings, hospitals etc. Everything can be controlled by **computer** systems	SW4 = 4 (Engineer $\langle 3 \rangle$, Professional $\langle 1 \rangle$, computer $\langle 1 \rangle$, Construction $\langle 1 \rangle$)

Table 2 Probability score table

TDi	Probability score (Pi) = $\sum SW_i/SW_j$
TD1	0.1
TD2	0.075
TD3	0.05
TD4	0.1

Table 3 Cluster table

Cluster no.	Probability values
C1{M, E, Sp}	{0.1, 0.075, 0.1}
C2{Sc}	{0.05}

Step 3: Calculate Probability Score Pi where i = 1, 2, 3, and 4 using formula as shown in Table 2.

Now, Clusters (C1, C2) are formed on the basis of these Probability values (P) and user defines threshold value i.e. 0.06 are shown in Table 3. In Table 3, C1 having Military, Engineering, Sports vectors and C2 has Science vector.

6 Implementation and Results

Java software programming tool i.e. Netbeans IDE 8.0 aims at building and manipulating the PRUV algorithm, K-Means algorithm, and DB-Scan algorithm on Windows system i.e. a dell Vostro laptop having windows 8 as operating system is selected as a system. The system has Intel Core 2 Duo processor and 4 GB RAM. However this is not the basic requirement for implementing the system. Any other system can also be used, but the computation time will be little bit higher.

6.1 Parameter Used for Comparison

6.1.1 Number of Token's

Document tokenization has its own importance. Tokenization helps to achieve better efficiency in IR system, as no of token are less but accurate. Initially all given documents are pre-processed and respective vectors are formed. Pre-processing of documents involves, removal of all the special symbols like (, . ? ! etc.), stop words like (is, am, are, we, as, it etc.) and stemming is applied in which stem is created. Stem is a portion of word after removing its affixes e.g. "Connect" is the stem of words

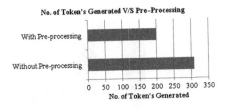

Fig. 2 Document tokenization graph

connections, connected, connecting for reducing the search time as well as search storage space. The tokens generated with processing or without processing are varies and has effect on overall efficiency. Comparative results are shown in Fig. 2.

6.1.2 Cluster Formation Time (CLUFo_T)

Time Value, defined as how much time is requires to create clusters of given documents (millisecond). In the implementation, it is observed that clustering of document is purely based on the token's probability score/values, in case of PRUV algorithm probability score are calculated based on accurately identified token's.

6.1.3 Cluster Accuracy with Vectors (CLUAcc_Vect)

Accuracy, defines as the measurement of correctness of a given process. In this paper, accuracy is defined how many documents are getting clustered accurately with respect to the vector categories, cluster accuracy with vectors (CLUAcc_Vect). Vectors are created by using training document and each of documents token's probability score are being calculated with using these vectors, this approach provide the two way of verification of accuracy of clustering.

6.2 Number of Token's Generated Versus Tokenization Strategy

Following of results are showing, the comparison on both cases with tokenization and without Tokenization on given documents. The Parameter used is "number of tokens generated" versus Pre-Processing, is shown in Fig. 2. In Fig. 2, Vertical axis represents the document with or without tokenization and horizontal axis represents the number of tokens generated so that with document tokenization number of tokens generated is equal to 200 and without document tokenization number of tokens generated is equal to 308.

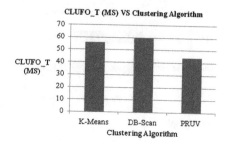

Fig. 3 CLUFO_Time versus clustering algorithm

6.3 CLUFO_T Versus Clustering Algorithm

In Fig. 3, graph shows the comparison between cluster formation time (CLUFO_T) and algorithm used for clustering e.g. K-Means, DB-Scan and PRUV algorithm. As results show PRUV algorithm perform significantly better than K-Means and DB-Scan, time elapsed by PRUV algorithm is 40 ms while K-Means 53 ms and DB-Scan 56 ms, to create same number of clusters of same type of input.

6.4 CLUAcc_Vect Versus Clustering Algorithm

In Fig. 4, graph shows the comparison between cluster accuracy with vectors formed (CLUAcc_Vect) and algorithm used for clustering e.g. K-Means, DB-Scan and PRUV algorithm. As shown in Fig. 4, when executing in the same environment, K-means gives less value of accuracy (83.33 %) whereas PRUV algorithm is giving 100 % accuracy, as PRUV algorithm uses document vectors for probability score.

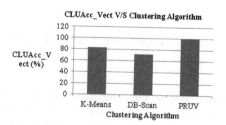

Fig. 4 CLUAcc_Vect versus clustering algorithm

Table 4 Result summary

S. No.	Parameter	PRUV	K-means	DB-scan
1	With document tokenization	200	200	200
2	Without document tokenization	308	308	308
3	Time (ms)	44	53	60
4	Accuracy (%)	100	83.33	72.33

6.5 Summary of Results

The final result analysis is summarized in Table 4: summarizing all the implementation results on the basis of these results, PRUV algorithm gives better results as compare with K-Means and DB-Scan clustering algorithm.

7 Conclusion

In information retrieval, Different model works for different applications. For a particular case, if there is a good weighting function is available so for this instance, vector space models are best fitted and if relevant and irrelevant documents are available, the probabilistic retrieval model will be a good option. The performance of different information retrieval models are governed by some conditions which are to be outlined. The PRUV algorithm, tokenization process is used to generate the tokens of given vectors or documents. Result shows that, PRUV algorithm gives 100 % accuracy for given document clustering as compare with K-means and DB-Scan clustering algorithm. As clustering is an extraordinarily great part so the incorporated approach of clustering should be used in probabilistic information retrieval model for information retrieval system. In the clustering approach model, it requires similarity score, so the identification of user's intent is done by information retrieval model. In PRUV algorithm approach, for the formation of clusters probability scores of different tokens and the user define threshold values are used.

References

1. Salton, G., McGill, M.J.: Introduction to Modern Information Retrieval. McGraw-Hill Book Co., New York (1983)
2. Yates, R.B., Neto, B.R.: Modern Information Retrieval. ACM Press, Harlow (1999)
3. Dong, H., Husain, F.K., Chang, E.: A survey in traditional information retrieval models. In: IEEE International Conference on Digital Ecosystems and Technologies, pp. 397–402 (2008)
4. Jarvelin, K., Kekalainen, J.: IR methods for retrieving highly relevant documents. In: Proceedings of SIGIR, pp. 41–48 (2000)
5. Robertson, S.E., Jones, K.S.: Relevance weighting of search terms. J. Am. Soc. Inf. Sci. **27**, 129–146 (1976)

6. Crestani, F., et al.: Is this document relevant? probably: a survey of probabilistic models in information retrieval. ACM Comput. Surv. **30**(4), 528–552 (1998)
7. Lashkari, A., Mahdavi, F., Ghomi, V.: A boolean model in information retrieval for search engines. In: IEEE International Conference on Information Management and Engineering, pp. 385–389 (2009)
8. Raman, S., Kumar, V., Venkatesan, S.: Performance comparison of various information retrieval models used in search engines. In: IEEE Conference on Communication, Information and Computing Technology, Mumbai (2012)
9. Qui, J., Tang, C.: Topic oriented semi-supervised document clustering. In: Proceedings of SIGMOD, Workshop on Innovative Database Research, pp. 57–62 (2007)
10. Karthikeyan, M., Aruna, P.: Probability based document clustering and image clustering using content-based image retrieval. J Appl Soft Comput **13**, 959–966 (2012)
11. Wang, J., Su, X.: An improved k-means clustering algorithm. In: IEEE Conference on Communication Software and Networks, Japan, pp. 44–46 (2011)
12. Senthesree, K., Daodaran, A., Appaji, S., Devi, D.N.: Web usage data clustering using DBSCAN algorithm and set similarities. In: IEEE Conference on Data Storage and Data Engineering, India, pp. 220–224 (2010)

An Approach for Computing Dynamic Slices of Structured Programs Using Dependence Relations

Madhusmita Sahu, Swatee Rekha Mohanty
and Durga Prasad Mohapatra

Abstract We propose a technique for computing dynamic slices of structured programs based on various types of dependence relations. First, we compute the static slice with respect to a variable given in the slicing criterion and store it in a slice file. Then, we execute the program for a given input and store various dependences in a table corresponding to the executed statements that are stored in the slice file. The slice is computed from the program path by taking dependences into account that are stored in a table. We have named our proposed algorithm as *Program Dependence* (PD) algorithm.

Keywords Dynamic slice · Slicing criterion · Program dependence relations

1 Introduction

Program Slicing is a decomposition technique used to extract statements from a program related to a particular computation. Weiser [1] first introduced the concept of program slice. A program slice is computed with respect to a slicing criterion. A slicing criterion is a tuple $\langle s, V \rangle$, s being a program point of interest and V being a subset of the program's variables used or defined at s. The concept of dynamic slicing was first introduced by Korel and Laski [2]. Jia et al. [3] proposed a dynamic slicing algorithm to simplify control and data dependences. Zhang et al. [4]

M. Sahu (✉) · D.P. Mohapatra
Department of CSE, National Institute of Technology,
Rourkela, Odisha 769008, India
e-mail: 513CS8041@nitrkl.ac.in

D.P. Mohapatra
e-mail: durga@nitrkl.ac.in

S.R. Mohanty
Department of CSA, Rourkela Institute of Management Studies,
Rourkela, Odisha 769015, India
e-mail: swatee_18@rediffmail.com

© Springer India 2015
L.C. Jain et al. (eds.), *Computational Intelligence in Data Mining - Volume 1*,
Smart Innovation, Systems and Technologies 31, DOI 10.1007/978-81-322-2205-7_58

implemented a dynamic forward slicer SPS, which can extract the part of the code influenced by the user. In this paper, we propose an algorithm for computing dynamic slices of structured programs using various dependence relations. The rest of the paper is organized as follows. Section 2 introduces some definitions. In Sect. 3, we describe our proposed algorithm to compute dynamic slices. We compare our proposed work with some other works in Sect. 4. Section 5 concludes the paper.

2 Background

In this Section, we briefly describe some definitions required to understand our work

Definition 1 *Control Dependence Relation (CDR)*: If there is a directed path p from statement m to n such that m may control all the statements in p except m, then, there exists a *control dependence relation* between m and n. For example, in Fig. 1, statement 24 controls the execution of statement 25. Thus, there exists a control dependence relation between statements 24 and 25.

Definition 2 *Data Dependence Relation (DDR)*: If there is a variable v defined at statement m and used at statement n and v is not defined again in between the directed path p from m to n, then, there exists a *data dependence relation* between m and n. For example, in Fig. 1, statement 18 uses the value of i, which is defined at statements 13 and 19. Thus, there exists a data dependence relation between statements 18 and 13. Similarly, there exists a data dependence relation between statements 18 and 19.

Definition 3 *Call Dependence Relation (ClDR)*: If a statement u calls a function p that is defined at v, then there exists a *call dependence relation* between statements u and v. For example, in Fig. 1, statement 9 calls the function *perm*, which is defined at statement 12. Thus, there is a call dependence relation between statements 9 and 12.

Definition 4 *Parameter Dependence Relation (PrDR)*: Suppose a statement u calls a function p by passing parameters to p and p is defined at statement v. Then, there exists a *parameter dependence relation* between actual parameters at statement u and formal parameters at statement v. For example, in Fig. 1, statement 9 calls the function *perm* by passing parameters n and k and *perm* is defined at statement 12. Thus, there is a parameter dependence relation between the actual parameters n and k at statement 9 and formal parameters x and y at statement 12 respectively.

Definition 5 *Return Dependence Relation (RDR)*: Suppose there is a statement u that calls a function p and there is a statement v inside function p that returns a

Fig. 1 A sample
interprocedural program

```
1    void main()
     {
          long int c;
          int n,k;
          printf("Enter n:");
2         scanf("%d",&n);
          printf("Enter k:");
3         scanf("%d",&k);
4         c=comb(n,k);
5         printf("%d",c);
     }
6    long int comb(int x,int y)
     {
          long int p,f;
7         if(x<0||y<0||y>x)
8              return 0;
          else
          {
9              p=perm(x,y);
10             f=fact(y);
11             return p/f;
          }
     }
12   long int perm(int x,int x)
     {
          long int p;
          int i;
13        i=x;
14        p=1;
15        if(x<0||y<0||y>x)
16             return 0;
          else
          {
17             while(i>x-y)
               {
18                  p=p*i;
19                  i=i-1;
               }
20             return p;
          }
     }
21   long int fact(int y)
     {
          long int f;
          int i;
22        f=1;
23        i=2;
24        if(y<2)
25             return 1;
          else
          {
26             while(i<=y)
               {
27                  f=f*i;
28                  i=i+1;
               }
29             return f;
          }
     }
```

value to u. Then, there exists a return dependence relation between statements u and v. For example, in Fig. 1, statement 9 calls the function *perm* and the statement 20 inside function *perm* returns the value of p to statement 9. Thus, there exists a return dependence relation between statements 9 and 20.

Definition 6 *Program Dependence Relation* (PDR): The *program dependence relation* is obtained by combining the *control, data, call, parameter,* and *return dependence relations* in case of *interprocedural programs.*

Definition 7 *closure*(X^p): The *closure*(X^p), where X^p represents a statement X running at pth step, is computed as $closure(X^p) = \{X^p\} \cup closure(\{Y^q | Y^q \in PDR(X^p)\})$. This is explained in Sect. 3.

3 Proposed Algorithm

In this section, we explain our proposed approach. In our approach, the static slice of the program is computed with respect to a slicing criterion and is stored in a slice file. Then, the program is executed for the given input and the executed statements corresponding to the statements in a slice file are stored in a table along with their dependencies. After execution, the dynamic slice is computed from the *closure* function.

Let the slicing criterion be $\langle q^k, v \rangle$, where q^k represents the statement q executed at the kth step and v is a program variable. The dynamic slice is computed from $closure(q^k)$ as follows:

$$closure\left(q^k\right) = \{q^k\} \cup closure\left(y^l\right), \text{ where } y^l \in PDR\left(q^k\right).$$

Let $dslice(q,v)$ denote the dynamic slice with respect to variable v for the execution of the statement q. Then, the dynamic slice $dslice(q,v)$ is given by

$$dslice(q, v) = closure\left(q^k\right).$$

Now, we propose our algorithm for dynamic slicing of interprocedural programs. We have named it *Program Dependence* (PD) algorithm.

PD Algorithm:

Input: The program P, the slicing criterion $<q^k,v>$ and a given input I.

Output: The dynamic slice corresponding to the slicing criterion $<q^k,v>$.

Begin

1. Find the static slice of program P w.r.t. the slicing criterion $<q^k,v>$.
2. Store the static slice in a slice file.
3. Execute the statements in P corresponding to the statements in the slice file for the given input I.
4. During execution, carry out the following after each statement x in the slice file is executed in the ith step.
   (a) If u is a variable used at a statement x^k and defined at y^j, where y is an executed statement in the jth step, then $DDR(x^k)=\{y^j\}$.
   (b) If a statement x^k is executed depending on the execution of a statement y^j, then $CDR(x^k)=\{y^j\}$.
   (c) If y^j, where y is an executed statement at the jth step, invokes a function defined at x^k, then $CIDR(x^k)=\{y^j\}$.
   (d) If y^j, where y is an executed statement at the jth step, invokes a function defined at a statement x^k with parameters, then $PrDR(x^k)=\{y^j\}$.
   (e) If a statement x^k receives a value from a function with a *return* statement defined at statement y^j, then $RDR(x^k)=y^j$.
   (f) Calculate $PDR(x^k)$ for a statement x executing in the kth step using the rule:
   $$PDR(x^k) = CDR(x^k) \cup DDR(x^k) \cup CIDR(x^k) \cup PrDR(x^k) \cup RDR(x^k)$$
5. Store *DDR, CDR, CIDR, PrDR and RDR* in a table.
6. For each statement q executed in the kth step, compute $closure(q^k)$ using the rule:
   $closure(q^k) = \{q^k\} \cup closure(y^l)$, where $y^l \in PDR(q^k)$.
7. If there exists two statements, p and q, such that $k<m$ and $p^m \in PDR(q^k)$, where q executed at the kth step is represented by q^k and p executed at the mth step is represented by p^m, then update
   $closure(q^k) = \{q^k\} \cup closure(y^l) \cup closure(p^m)$, where $y^l \in PDR(q^k)$.
8. For the slicing criterion $<q^k,v>$, look up $closure(q^k)$ for variable v.
9. Assign $dslice(q,v)= closure(q^k)$.

End

3.1 *Working of the PD Algorithm*

We illustrate the working of our algorithm with the help of an example. Consider the C program given in Fig. 1. This program finds binomial coefficients. For the slicing criterion $\langle 5^{31},c \rangle$, the static slice consists of the following statements: {1, 2, 3, 4, 5, 6, 7, 8, 9, 10, 11, 12, 13, 14, 15, 16, 17, 18, 19, 20, 21, 22, 23, 24, 25, 26, 27, 28, 29}. For our example program, the static slice contains all the statements which may not be always the case. For the input data $n = 4$, $k = 2$, the program will execute the following statements $1^1, 2^1, 3^3, 4^4, 6^5, 7^6, 9^7, 12^8, 13^9, 14^{10}, 15^{11}, 17^{12}, 18^{13}, 19^{14}, 17^{15}, 18^{16}, 19^{17}, 17^{18}, 20^{19}, 10^{20}, 21^{21}, 22^{22}, 23^{23}, 24^{24}, 26^{25}, 27^{26},$

28^{27}, 26^{28}, 29^{29}, 11^{30} and 5^{31}, in order. The corresponding statements and their dependences are stored in Table 1. For the slicing criterion $\langle 5^{31},c \rangle$, the dynamic slice obtained using PD algorithm is given by

$$closure(5^{31}) = \{5^{31}\} \cup closure(\{y^j | y^j \in PDR(5^{31})\}) \quad (1)$$

Equation (1) implies

$$closure(5^{31}) = \{5^{31}\} \cup closure(1^1) \cup closure(4^4)$$

Table 1 Dependency table for Fig. 1

Statements	DDR	CDR	ClDR	PrDR	RDR	PDR
1^1	ϕ	ϕ	ϕ	ϕ	ϕ	ϕ
2^2	ϕ	1^1	ϕ	ϕ	ϕ	1^1
3^3	ϕ	1^1	ϕ	ϕ	ϕ	1^1
4^4	$2^2, 3^3$	1^1	ϕ	ϕ	11^{30}	$1^1, 2^2, 3^3, 11^{30}$
6^5	ϕ	ϕ	4^4	4^4	ϕ	4^4
7^6	6^5	6^5	ϕ	ϕ	ϕ	6^5
9^7	6^5	7^6	ϕ	ϕ	20^{19}	$6^5, 7^6, 20^{19}$
12^8	ϕ	ϕ	9^7	9^7	ϕ	9^7
13^9	12^8	12^8	ϕ	ϕ	ϕ	12^8
14^{10}	ϕ	12^8	ϕ	ϕ	ϕ	12^8
15^{11}	12^8	12^8	ϕ	ϕ	ϕ	12^8
17^{12}	$12^8, 13^9$	15^{11}	ϕ	ϕ	ϕ	$12^8, 13^9, 15^{11}$
18^{13}	$13^9, 14^{10}$	17^{12}	ϕ	ϕ	ϕ	$13^9, 14^{10}, 17^{12}$
19^{14}	13^9	17^{12}	ϕ	ϕ	ϕ	$13^9, 17^{12}$
17^{15}	$12^8, 19^{14}$	15^{11}	ϕ	ϕ	ϕ	$12^8, 15^{11}, 19^{14}$
18^{16}	$18^{13}, 19^{14}$	17^{15}	ϕ	ϕ	ϕ	$18^{13}, 19^{14}, 17^{15}$
19^{17}	19^{14}	17^{15}	ϕ	ϕ	ϕ	$19^{14}, 17^{15}$
17^{18}	$12^8, 19^{17}$	15^{11}	ϕ	ϕ	ϕ	$12^8, 15^{11}, 19^{17}$
20^{19}	18^{16}	15^{11}	ϕ	ϕ	ϕ	$15^{11}, 18^{16}$
10^{20}	6^5	7^6	ϕ	ϕ	ϕ	$6^5, 7^6$
21^{21}	ϕ	ϕ	10^{20}	10^{20}	29^{29}	$10^{20}, 29^{29}$
22^{22}	ϕ	21^{21}	ϕ	ϕ	ϕ	21^{21}
23^{23}	ϕ	21^{21}	ϕ	ϕ	ϕ	21^{21}
24^{24}	21^{21}	21^{21}	ϕ	ϕ	ϕ	21^{21}
26^{25}	$21^{21}, 23^{23}$	24^{24}	ϕ	ϕ	ϕ	$21^{21}, 23^{23}, 24^{24}$
27^{26}	$22^{22}, 23^{23}$	26^{25}	ϕ	ϕ	ϕ	$22^{22}, 23^{23}, 26^{25}$
28^{27}	23^{23}	26^{25}	ϕ	ϕ	ϕ	$23^{23}, 26^{25}$
26^{28}	$28^{27}, 21^{21}$	24^{24}	ϕ	ϕ	ϕ	$21^{21}, 24^{24}, 28^{27}$
29^{29}	27^{26}	24^{24}	ϕ	ϕ	ϕ	$24^{24}, 27^{26}$
11^{30}	$9^7, 10^{20}$	7^6	ϕ	ϕ	ϕ	$7^6, 9^7, 10^{20}$
5^{31}	4^4	1^1	ϕ	ϕ	ϕ	$1^1, 4^4$

Table 2 closure(q^k) for all statements q executed at the kth step

q^k	Closure(q^k)		Final result
	Equation	Result of equation	
1^1	$1^1 \cup closure(\phi)$	$\{1^1\}$	$\{1\}$
2^2	$2^2 \cup closure(1^1)$	$\{1^1, 2^2\}$	$\{1, 2\}$
3^3	$3^3 \cup closure(1^1)$	$\{1^1, 3^3\}$	$\{1, 3\}$
4^4	$4^4 \cup closure(1^1)$ $\cup closure(2^2)$ $\cup closure(3^3)$ $\cup closure(11^{30})$	$\{1^1, 2^2, 3^3, 4^4, 6^5, 7^6, 9^7, 12^8, 13^9, 14^{10}, 15^{11}, 17^{12}, 18^{13}, 19^{14}, 17^{15},$ $18^{16}, 19^{17}, 17^{18}, 20^{19}, 10^{20}, 21^{21}, 22^{22}, 23^{23}, 24^{24}, 26^{25}, 27^{26}, 28^{27},$ $26^{28}, 29^{29}, 11^{30}\}$	$\{1, 2, 3, 4, 6, 7, 9, 10, 11, 12, 13, 14, 15, 17, 18,$ $19, 20, 21, 22, 23, 24, 26, 27, 28, 29\}$
6^5	$6^5 \cup closure(4^4)$	$\{1^1, 2^2, 3^3, 4^4, 6^5\}$	$\{1, 2, 3, 4, 6\}$
7^6	$7^6 \cup closure(6^5)$	$\{1^1, 2^2, 3^3, 4^4, 6^5, 7^6\}$	$\{1, 2, 3, 4, 6, 7\}$
9^7	$9^7 \cup closure(6^5)$ $\cup closure(7^6)$ $\cup closure(20^{19})$	$\{1^1, 2^2, 3^3, 4^4, 6^5, 7^6, 9^7, 12^8, 13^9, 14^{10}, 15^{11}, 17^{12}, 18^{13}, 19^{14}, 17^{15},$ $18^{16}, 19^{17}, 17^{18}, 20^{19}\}$	$\{1, 2, 3, 4, 6, 7, 9, 12, 13, 14, 15, 17, 18, 19, 20\}$
12^8	$12^8 \cup closure(9^7)$	$\{1^1, 2^2, 3^3, 4^4, 6^5, 7^6, 9^7, 12^8\}$	$\{1, 2, 3, 4, 6, 7, 9, 12\}$
13^9	$13^9 \cup closure(12^8)$	$\{1^1, 2^2, 3^3, 4^4, 6^5, 7^6, 9^7, 12^8, 13^9\}$	$\{1, 2, 3, 4, 6, 7, 9, 12, 13\}$
14^{10}	$14^{10} \cup closure(12^8)$	$\{1^1, 2^2, 3^3, 4^4, 6^5, 7^6, 9^7, 12^8, 14^{10}\}$	$\{1, 2, 3, 4, 6, 7, 9, 12, 14\}$
15^{11}	$15^{11} \cup closure(12^8)$	$\{1^1, 2^2, 3^3, 4^4, 6^5, 7^6, 9^7, 12^8, 15^{11}\}$	$\{1, 2, 3, 4, 6, 7, 9, 12, 15\}$
17^{12}	$17^{12} \cup closure(12^8)$ $\cup closure(13^9)$ $\cup closure(15^{11})$	$\{1^1, 2^2, 3^3, 4^4, 6^5, 7^6, 9^7, 12^8, 13^9, 15^{11}, 17^{12}\}$	$\{1, 2, 3, 4, 6, 7, 9, 12, 13, 15, 17, 19\}$
18^{13}	$18^{13} \cup closure(13^9)$ $\cup closure(14^{10})$ $\cup closure(17^{12})$	$\{1^1, 2^2, 3^3, 4^4, 6^5, 7^6, 9^7, 12^8, 13^9, 14^{10}, 15^{11}, 17^{12}, 18^{13}\}$	$\{1, 2, 3, 4, 6, 7, 9, 12, 13, 14, 15, 17, 18\}$

(continued)

Table 2 (continued)

q^k	Closure(q^k)		
	Equation	Result of equation	Final result
19^{14}	$19^{14} \cup closure(13^9)$ $\cup closure(17^{12})$	$\{1^1, 2^2, 3^3, 4^4, 6^5, 7^6, 9^7, 12^8, 13^9, 15^{11}, 17^{12}, 19^{14}\}$	$\{1, 2, 3, 4, 6, 7, 9, 12, 13, 15, 17, 19\}$
17^{15}	$17^{15} \cup closure(12^8)$ $\cup closure(15^{11})$ $\cup closure(19^{14})$	$\{1^1, 2^2, 3^3, 4^4, 6^5, 7^6, 9^7, 12^8, 13^9, 15^{11}, 17^{12}, 19^{14}, 17^{15}\}$	$\{1, 2, 3, 4, 6, 7, 9, 12, 13, 15, 17, 19\}$
18^{16}	$18^{16} \cup closure(18^{13})$ $\cup closure(19^{14})$ $\cup closure(17^{15})$	$\{1^1, 2^2, 3^3, 4^4, 6^5, 7^6, 9^7, 12^8, 13^9, 14^{10}, 15^{11}, 17^{12}, 18^{13},$ $19^{14}, 17^{15}, 18^{16}\}$	$\{1, 2, 3, 4, 6, 7, 9, 12, 13, 14, 15, 17, 18, 19\}$
19^{17}	$19^{17} \cup closure(19^{14})$ $\cup closure(17^{15})$	$\{1^1, 2^2, 3^3, 4^4, 6^5, 7^6, 9^7, 12^8, 13^9, 15^{11}, 17^{12}, 19^{14}, 17^{15}, 19^{17}\}$	$\{1, 2, 3, 4, 6, 7, 9, 12, 13, 15, 17, 19\}$
17^{18}	$17^{18} \cup closure(12^8)$ $\cup closure(15^{11})$ $\cup closure(19^{17})$	$\{1^1, 2^2, 3^3, 4^4, 6^5, 7^6, 9^7, 12^8, 13^9, 15^{11}, 17^{12}, 19^{14},$ $17^{15}, 19^{17}, 17^{18}\}$	$\{1, 2, 3, 4, 6, 7, 9, 12, 13, 15, 17, 19\}$
20^{19}	$20^{19} \cup closure(15^{11})$ $\cup closure(18^{16})$	$\{1^1, 2^2, 3^3, 4^4, 6^5, 7^6, 9^7, 12^8, 13^9, 14^{10}, 15^{11}, 17^{12}, 18^{13}, 19^{14},$ $17^{15}, 18^{16}, 19^{17}, 17^{18}, 20^{19}\}$	$\{1, 2, 3, 4, 6, 7, 9, 12, 13, 14, 15, 17, 18, 19, 20\}$
10^{20}	$10^{20} \cup closure(6^5)$ $\cup closure(7^6)$ $\cup closure(29^{29})$	$\{1^1, 2^2, 3^3, 4^4, 6^5, 7^6, 10^{20}, 21^{21}, 22^{22}, 23^{23},$ $24^{24}, 26^{25}, 27^{26}, 28^{27}, 26^{28}, 29^{29}\}$	$\{1, 2, 3, 4, 6, 7, 10, 21, 22, 23, 24, 26, 27, 28, 29\}$
21^{21}	$21^{21} \cup closure(10^{20})$	$\{1^1, 2^2, 3^3, 4^4, 6^5, 7^6, 10^{20}, 21^{21}\}$	$\{1, 2, 3, 4, 6, 7, 10, 21\}$
22^{22}	$22^{22} \cup closure(21^{21})$	$\{1^1, 2^2, 3^3, 4^4, 6^5, 7^6, 10^{20}, 21^{21}, 22^{22}\}$	$\{1, 2, 3, 4, 6, 7, 10, 21, 22\}$
23^{23}	$23^{23} \cup closure(21^{21})$	$\{1^1, 2^2, 3^3, 4^4, 6^5, 7^6, 10^{20}, 21^{21}, 23^{23}\}$	$\{1, 2, 3, 4, 6, 7, 10, 21, 23\}$
24^{24}	$24^{24} \cup closure(21^{21})$	$\{1^1, 2^2, 3^3, 4^4, 6^5, 7^6, 10^{20}, 21^{21}, 24^{24}\}$	$\{1, 2, 3, 4, 6, 7, 10, 21, 24\}$

(continued)

Table 2 (continued)

q^k	Closure(q^k)		Final result
	Equation	Result of equation	
26^{25}	$26^{25} \cup closure(21^{21})$ $\cup closure(23^{23})$ $\cup closure(24^{24})$	$\{1^1, 2^2, 3^3, 4^4, 6^5, 7^6, 10^{20}, 21^{21}, 23^{23}, 24^{24}, 26^{25}\}$	$\{1, 2, 3, 4, 6, 7, 10, 21, 23, 24, 26\}$
27^{26}	$27^{26} \cup closure(22^{22})$ $\cup closure(23^{23})$ $\cup closure(26^{25})$	$\{1^1, 2^2, 3^3, 4^4, 6^5, 7^6, 10^{20}, 21^{21}, 22^{22}, 23^{23}, 24^{24}, 26^{25}, 27^{26}\}$	$\{1, 2, 3, 4, 6, 7, 10, 21, 22, 23, 24, 26, 27\}$
28^{27}	$28^{27} \cup closure(23^{23})$ $\cup closure(26^{25})$	$\{1^1, 2^2, 3^3, 4^4, 6^5, 7^6, 10^{20}, 21^{21}, 23^{23}, 24^{24}, 26^{25}, 28^{27}\}$	$\{1, 2, 3, 4, 6, 7, 10, 21, 23, 24, 26, 28\}$
26^{28}	$26^{28} \cup closure(21^{21})$ $\cup closure(24^{24})$ $\cup closure(28^{27})$	$\{1^1, 2^2, 3^3, 4^4, 6^5, 7^6, 10^{20}, 21^{21}, 23^{23}, 24^{24}, 26^{25}, 28^{27}, 26^{28}\}$	$\{1, 2, 3, 4, 6, 7, 10, 21, 23, 24, 26, 28\}$
29^{29}	$29^{29} \cup closure(24^{24})$ $\cup closure(27^{26})$	$\{1^1, 2^2, 3^3, 4^4, 6^5, 7^6, 10^{20}, 21^{21}, 22^{22}, 23^{23}, 24^{24}, 26^{25}, 27^{26}, 28^{27}, 26^{28}, 29^{29}\}$	$\{1, 2, 3, 4, 6, 7, 10, 21, 22, 23, 24, 26, 27, 28, 29\}$
11^{30}	$11^{30} \cup closure(7^6)$ $\cup closure(9^7)$ $\cup closure(10^{20})$	$\{1^1, 2^2, 3^3, 4^4, 6^5, 7^6, 9^7, 12^8, 13^9, 14^{10}, 15^{11}, 17^{12}, 18^{13}, 19^{14}, 17^{15}, 18^{16}, 19^{17}, 17^{18}, 20^{19}, 10^{20}, 21^{21}, 22^{22}, 23^{23}, 24^{24}, 26^{25}, 27^{26}, 28^{27}, 26^{28}, 29^{29}, 11^{30}\}$	$\{1, 2, 3, 4, 6, 7, 9, 10, 11, 12, 13, 14, 15, 17, 18, 19, 20, 21, 22, 23, 24, 26, 27, 28, 29\}$
5^{31}	$17^{12} \cup closure(1^1)$ $\cup closure(4^4)$	$\{1^1, 4^2, 3^3, 6^5, 7^6, 9^7, 12^8, 13^9, 14^{10}, 15^{11}, 17^{12}, 18^{13}, 19^{14}, 17^{15}, 18^{16}, 19^{17}, 17^{18}, 20^{19}, 10^{20}, 21^{21}, 22^{22}, 23^{23}, 24^{24}, 26^{25}, 27^{26}, 28^{27}, 29^{29}, 11^{30}, 5^{31}\}$	$\{1, 2, 3, 4, 5, 6, 7, 9, 10, 11, 12, 13, 14, 15, 17, 18, 19, 20, 21, 22, 23, 24, 26, 27, 28, 29\}$

By evaluating the above expression recursively, we get $closure(5^{31}) = \{1^1, 4^4,$ $2^2, 3^3, 6^5, 7^6, 9^7, 12^8, 13^9, 14^{10}, 15^{11}, 17^{12}, 18^{13}, 19^{14}, 17^{15}, 18^{16}, 19^{17}, 17^{18}, 20^{19},$ $10^{20}, 21^{21}, 22^{22}, 23^{23}, 24^{24}, 26^{25}, 27^{26}, 28^{27}, 29^{29}, 11^{30}, 5^{31}\}$.

Table 2 shows the $closure(q^k)$ for all statements q executed at the kth step.

Then, $dslice(5,c) = closure(5^{31}) = \{1, 2, 3, 4, 5, 6, 7, 9, 10, 11, 12, 13, 14, 15, 17,$ $18, 19, 20, 21, 22, 23, 24, 26, 27, 28, 29\}$.

Thus, the dynamic slice with respect to the slicing criterion $\langle 5^{31}, c \rangle$ consists of the following set of statements: $\{1, 2, 3, 4, 5, 6, 7, 9, 10, 11, 12, 13, 14, 15, 17, 18, 19, 20, 21, 22, 23, 24, 26, 27, 28, 29\}$.

4 Comparison with Related Work

Jia et al. [5] developed an algorithm for computing dynamic slices for intraprocedural programs using dependence relations using a program dependence diagram algorithm. They have not developed any technique to compute dynamic slices in the presence of function calls. Jha et al. [6] developed a method to compute dynamic slices using static slicing of program dependence graph. Their approach takes more space and time due to the construction of program dependence graph (PDG) and modified PDG. Also, they have not addressed the issue of dynamic slicing in the presence of function calls. In our algorithm, we correctly compute dynamic slices for interprocedural programs where function calls are present.

5 Conclusion

In this paper, we presented an approach to compute dynamic slices of interprocedural programs. Our approach computes static slice with respect to a slicing criterion and store it in a slice file. Then, it executes the program for a given input and the executed statements corresponding to the statements in the slice file are stored in a table with their dependences. The dynamic slice corresponding to a slicing criterion is computed from the table. The advantage of our approach is that we don't need to construct the dependence graph for the program. We simply compute the static slice with respect to a slicing criterion and store it a slice file. The program is executed for a given input and the executed statements corresponding to the statements stored in the slice file are used to calculate dynamic slices. Now, we are working on slicing of Object-Oriented Programs (OOPs), Aspect-Oriented Programs (AOPs) and web applications.

References

1. Weiser, M.: Programmers use slices when debugging. Commun. ACM **25**(7), 446–452 (1982)
2. Korel, B., Laski, J.: Dynamic program slicing. Inf. Process. Lett. **29**(3), 155–163 (1988)
3. Jia, L., Jiao, H., Jie, L.: A dynamic program slice algorithm based on simplified dependence. In: Proceedings of 3rd International Conference on Advanced Computer Theory and Engineering (ICACTE '10), vol. 4, pp. 356–359. (2010)
4. Zhang, R., Zheng, Y., Huang, S., Qi, Z.: Structured dynamic program slicing. In: Proceedings of International Conference on Computer and Management (CAMAN '11), pp. 1–4. (2011)
5. Jia, L., Chen, Z., Jin, Y.: A dynamic slice algorithm based on program dependence diagram. In: Proceedings of 2nd International Conference on Consumer Electronics, Communications and Networks (CECNet), pp. 273–276. (2012)
6. Jha, L., Patnaik, K.S.: A new method to compute dynamic slicing using program dependence graph. Int. J. Comput. Appl. **75**(13), 30–36 (2013)

Sub-block Features Based Image Retrieval

Vijaylakshmi Sajwan and Puneet Goyal

Abstract In various domains like security, education, biomedicine etc., the volume of digital data is increasing rapidly, and this is becoming a challenge to retrieve the information from the storage media. Content-based image retrieval systems (CBIR) aim at retrieving from large image databases the images similar to the given query image based on the similarity between image features. This paper aim to discuss and solve the problem of designing sub-block features based image retrieval. Firstly, this paper outlines a description of the primitive features of an image. Then, the proposed methodology for partitioning the image and extracting its colour and texture is described. The algorithms used to calculate the similarity between extracted features, are then explained. Finally, we compared with some other existing CBIR methods, using the WANG database, which is widely used for CBIR performance evaluation, and the results demonstrate the proposed approach outperforms other existing methods considered.

Keywords Content-based image retrieval · CBIR · Histogram · Euclidian distance

1 Introduction

Imaging has played a major role in our life. Our ascendants used the walls of their caves to paint some pictures telling us different stories or other information. With the start of the twentieth century, imaging has grown in an unrivaled way in all our proceeds of life. Now, images play major act in many fields like security, medicine,

V. Sajwan (✉) · P. Goyal
Graphic Era University, Dehradun, Uttrakhand 248002, India
e-mail: vijaylakshmi@live.com

P. Goyal
e-mail: dr.puneet.goyal@iitdalumni.com

© Springer India 2015
L.C. Jain et al. (eds.), *Computational Intelligence in Data Mining - Volume 1*,
Smart Innovation, Systems and Technologies 31, DOI 10.1007/978-81-322-2205-7_59

journalism, education and entertainment. The evolution of the World-Wide Web facilitate users to access data from almost any region and provide the profiteering of digital images in any fields [1–4]. Naturally, when the size of data becomes larger and larger, it will be worthless unless there are powerful methods to access. Content based image retrieval (CBIR) refers to automatically retrieving some desired images from an image database on the basis of some features such as color, texture, and shape [1, 2]. The first theory of content based image retrieval was given by Kato to represent his experiments for accessing images from a database using color and shape features. There are various benefits of CBIR techniques over other simple image retrieval methods such as text-based retrieval techniques. CBIR provided key for many types of image information management systems such as medical imagery, criminology, and satellite imagery [4]. CBIR differs from traditional information retrieval as a digital image is represented using multi dimensional array of pixel values, with no inherent meaning.

Many CBIR systems use a single feature for image retrieval or rely on global features, and therefore these solutions do not perform well in terms of accuracy and efficiency [4–11]. For this reason, proposed approach uses combined features (such as colour, texture) of extended image sub-block with matching based on higher priority principle. And it provides better retrieving result than retrieval using some other existing methods.

2 Proposed System

Given a query image and a collection of images or say an image database, the aim of the proposed system is to find some images from the image database that are most similar to the query image. The proposed approach is based on combined features of image sub-blocks and gray-level co-occurrence matrix (GLCM) [11–16]. Similar to these methods, the image in proposed method is partitioned into sub-blocks which are of equal size and not coinciding with each other. In the proposed method, the images are partitioned into 24 equal sized blocks. For the sub-blocks matching, the histogram and GLCM metric values are computed for each sub-block of the image from image database and this is compared with similar computed values of the corresponding sub-block of the query image. Here, we explain in detail the steps which are used in the proposed approach (Fig. 1).

2.1 Convert the Image into Grayscale Image

First convert the image into grayscale image as shown in Fig. 2a and its histogram shown in Fig. 2b then equalize the image in order to increase contrast of values of an image as shown in Fig. 3a and its flat histogram shown in Fig. 3b. RGB and

Fig. 1 Partitioning the image considered in 24 different sub-blocks

(a)   **(b)**

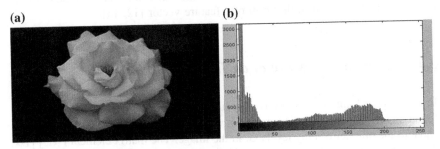

Fig. 2 **a** Grayscale image and **b** Histogram of grayscale image

(a)   **(b)**

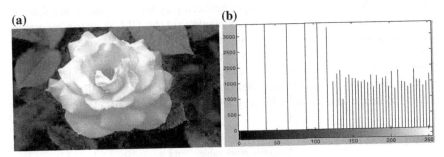

Fig. 3 **a** Equalized image and **b** Flat histogram of equalized image

indexed images bring high values that require more estimation time. Then, resulting image undergoes histogram equalization in order to raise contrast of values of an image by achieving its flat histogram [11].

2.2 Partitioning the Image into Sub-blocks

In second step, the image is divided into 24 (4 × 6) into sub-blocks which are of equal size and not coinciding with each other, as also shown in Fig. 1. If the image is of size 256 × 384, the size of sub-block is 64 × 64.

2.3 Extraction of Colour of an Image Sub-block

The histogram equalized image is split into four fixed bins in order to extract more specific information from it. The frequencies of 256 values of gray scale are split into sixteen (16) bins carrying 16 values each (0–15, 16–31, 32–47, …). Then we normalized the histogram which diminishes the number of bins by taking colors that are similar to each other and putting them in the same bin. The information from each sub-block is saved in the form of a feature vector [12, 13].

2.4 Extraction of Texture of an Image Sub-block

In the proposed approach, there is introduced a texture sample for image retrieval found on GLCM. A GLCM is a matrix in which the number of rows and columns is equal to the number of gray levels G, in the image. The matrix element P (i, j | Δx, Δy) is the corresponding frequency with which two pixels, distinct by a pixel distance (Δx, Δy), appear within a given area, one with intensity 'i' and the other with intensity 'j'. The matrix element P (i, j | d, θ) encloses the second order statistical probability values for changes between gray levels 'i' and 'j' at a specific displacement distance d and at a specific angle (θ) [16]. GLCM matrix generation can be explained with the example illustrated in Table 1 for four different gray levels. Here one pixel offset is used (a reference pixel and its immediate neighbor). If the window is large enough, using a larger offset is feasible. The top left cell will be filled with the number of times the combination 0,0 occurs, i.e. how many time within the image area a pixel with grey level 0 (neighbor pixel) falls to the right of

Table 1 GLCM calculation

Neighbor pixel value → ref pixel value:	0	1	2	3
0	0,0	0,1	0,2	0,3
1	1,0	1,1	1,2	1,3
2	2,0	2,1	2,2	2,3
3	3,0	3,1	3,2	3,3

another pixel with grey level 0 (reference). Elements in the matrix are measured by the equation shown below here.

$$p(i,j|d, \theta) = \frac{p(i,j|d, \theta)}{\sum_i \sum_j (i,j|d, \theta)} \tag{1}$$

Proposed approach used four important features; Entropy, Energy, Contrast, and Inverse Difference are selected for implementation.

2.4.1 Entropy

Entropy shows the amount of information of the image that is desired for the image compression. Entropy measures the image information.

$$Entropy = \sum_x \sum_y p(x, y) \log p(x, y) \tag{2}$$

2.4.2 Energy

It is the sum of squares of entries in the GLCM Angular Second Moment measures the image homogeneity. Angular Second Moment is high when pixels are very identical.

$$Energy = \sum_x \sum_y p(x, y)^2 \tag{3}$$

2.4.3 Contrast

It is a measure of the stain contrast between a pixel and its nearest pixel over the whole image. Contrast is 0 for a constant image.

$$Contrast = \sum_x \sum_y (x - y)^2 p(x, y) \tag{4}$$

2.4.4 Inverse Difference

Inverse difference moment (IDM) is the local homogeneity. $P(x, y)$ is the gray-level value of the align (x, y).

$$IDM = \sum_x \sum_y P(x,y) \frac{1}{1 + (x-y)^2} P(x,y) \qquad (5)$$

The texture characteristic are calculated for an image when d = 1 and ϴ = 0°, 45°, 90°, 135°. In each direction four texture features are computed. Then, combined feature vector of colour and texture is constructed.

3 Euclidean Distance Matching

With the decomposition of the image, the number of sub-blocks remains same for all the images. The sub-block approach proposed here is similar to one used in [15]. But proposed approach partitioned the image into 24 blocks (4 × 6) instead of 6 blocks. A sub-block from query image is allowed to be matched sub-blocks in the target image. However sub-blocks may play in the matching process only once. Then, sum all the distance between sub-block of query image and database image. If we have two feature vectors of query image and database image respectively, P and Q. The distance between two images is calculated by Euclidean distance is shown by following equation:

$$D(P,Q) = (\Sigma_{i=1}^{n} |(p_i - q_i)|^2)^{1/2} \qquad (6)$$

4 Experiment and Results

For comparison with some other existing CBIR methods, we use the images from the WANG database [17], which is widely used for CBIR performance evaluation. We use 50 images of each class, and we considered 10 classes; it means total 500 images are considered in our database.

4.1 Performance Evaluation Metrics for CBIR Systems

In CBIR, the most commonly used performance measure is Precision. Precision is defined as the ratio of the number of retrieved relevant images to the total number of retrieved images [11–16]. We denote precision by P and it is computed as follows:

$$P = \frac{No.\ of\ relevant\ images\ retrieved}{Total\ No.\ of\ images\ retrieved} \qquad (7)$$

A retrieved image is examined to be correct if and only if it is in the same category as the query. We compare proposed approach result with that of [11]. Precision (P) of retrieved results is given by Eq. (8).

$$P(k) = n_k/k \tag{8}$$

where, k is the number of retrieved images, n_k is the number of relevant images in the retrieved images. The average precision of the images belonging to the qth category A_q is given by

$$\bar{p}_q = \sum_{k \in A_q} P(I_k)/|A_q|, \, q = 1, 2, \ldots 10 \tag{9}$$

The final average precision is

$$p = \sum_{q=1}^{10} \bar{p}_q/10 \tag{10}$$

In order to check retrieval effectiveness of the proposed approach, we have to test it by selecting some images randomly and retrieve some images.

Table 2 shows the average precision of our proposed approach with other existing system. In experiment we found out that increasing the number of sub-blocks gives the better retrieval result. Proposed approach performs better than other approaches for a single object (like dinosaur, flower and bus) and we also observe that Average precision decreases with increasing the number of retrieved images.

Table 2 Average precision

Class	Average precision		Proposed approach
	Local colour histogram method	Local colour histogram + GLCM texture of image sub-block (using 6 blocks)	
Dinosaur	6.6	6.2	9.9
Flower	2.1	5.1	6.8
Bus	1.6	4.3	6.8
Horse	1.3	3.5	5.5
Mountain	1.8	2.4	4.0
Elephant	3.3	3.3	3.3
Building	2.1	2.6	3.9
Africa	2.4	2.2	3.4
Beach	1.8	2.6	3.3
Food	2.5	2.9	3.1
Average	25.5	35.1	59.5

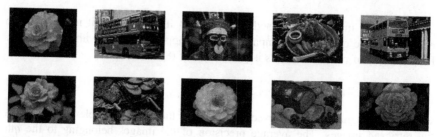

Fig. 4 Retrieval results based on color histogram method

Fig. 5 Retrieval results based on local histogram + GLCM texture of image sub-block (using 6 blocks)

Fig. 6 Retrieval results based on proposed approach

Figure 4 shows the result based on Local histogram method when the query image considered is an image from a Flower class (as shown as 1st picture in Fig. 4). We can observe that this method gives only four relevant and six irrelevant images from the database and the results obtained are shown in Fig. 4. Figure 5 shows the result based on the approach using on Local Histogram and GLCM Texture of image sub-block (6 sub-blocks). Here, the query image considered is from a Flower class and the results obtained are shown in Fig. 5. We can observe that it gives nine relevant and one irrelevant image from the database. Figure 6 shows the result based on proposed approach. In this also, the query image is from a Flower class and it gives ten relevant images from the database (as shown in Fig. 6).

5 Conclusions

With the role of images becoming increasingly important in different domains like social media, web advertisements, education, security, medicine etc., the research topic—CBIR is of great importance. In this work, an image retrieval method based on sub-block features is discussed. For comparison with some other existing CBIR methods, we use the images from the WANG database, which is widely used for CBIR performance evaluation. Experiments result show that the proposed approach outperforms other existing methods considered. New algorithms for further improving the performance of CBIR methods need to be explored.

References

1. Eakins, J., Graham, M.: Content-based image retrieval: Report 39, JISC technology applications programme (1999)
2. Dharani, T., Aroquiaraj, I. L.: A survey on content based image retrieval. In: International Conference on Pattern Recognition, Informatics and Mobile Engineering (PRIME), (Feb. 2013) pp. 485–490 (2013)
3. Ghanem, B., Resendiz, E., Ahuja, N.: Segmentation-based perceptual image quality assessment (SPIQA). In: 15th IEEE International Conference on Image Processing (ICIP), pp. 393–396. San Diego, (2008)
4. Datta, R., Joshi, D., Li, J., Wang, J.Z.: Image retrieval: ideas, influences, and trends of the new age. ACM Comput. Surv. 40(2), 1–49 (2008)
5. Mustaffa, M., Ahmad, F., Rahmat, R., Mahmod, R.: Content-based image retrieval based on color-spatial features. Malaysian J. Comput. Sci. 21, 1–12 (2008)
6. Hiremath, P.R., Pujari, J.: Content based image retrieval based on color texture and shape features using image and its complement. Int. J. Comput. Sci. Secur. 1, 340–345 (2010)
7. Nandagopalan, S., Adiga, B.S., Deepak, N.: A universal model for content-based image retrieval. World Acad. Sci. Eng. Technol. 22, 580–583 (2008)
8. Gopalakrishnan, S., Aruna, P.: Retrieval of images based on low level features using genetic algorithm. Int. J. Sci. Eng. Technol. Res. 3, 221–228 (2014)
9. Sreedhar, J., Raju, S.V., Babu, A.V.: Query processing for content based image retrieval. Int. J. Soft Comput. Eng. 1, 122–131 (2011)
10. Lin, T.X., Hung, C.S.: Quadrant motif approach for image retrieval. In: International Conference on Central Europe on Computer Graphics, pp. 209–215 (2006)
11. Kavitha, C.H., Rao, B.P., Goverdhan, A.: Image retrieval based on combined features of image sub-blocks. Int. J. Comput. Sci. Eng. 3, 1429–1438 (2011)
12. Vimina, E.R., Jacob, K.P.: CBIR using local and global properties of image sub-blocks. Int. J. Adv. Sci. Technol. 48, 11–22 (2012)
13. Kavitha, Ch., Rao, B.P., Rao, M.B., Goverdhan, A.: Image retrieval based on local histogram and texture features. Int. J. Comput. Sci. Inform. Technol. 2, 741–746 (2011)
14. Vimina, E.R., Jacob, K.P.: A sub-block based image retrieval using modified integrated region matching. Int. J. Comput. Sci. 10, 686–692 (2013)
15. Kavitha, Ch., Rao, B.P., Goverdhan, A.: Image retrieval based on color and texture features of the image sub-blocks. Int. J. Comput. Appl. 15, 33–37 (2011)

16. Mohanaiah, P., Sathyanarayana, P., GuruKumar, L.: Image texture feature extracting using GLCM approach. Int. J. Sci. Res. Publ. **3**, 2250–3153 (2013)
17. Jia, L., Wang, J.Z.: Automatic linguistic indexing of pictures by a statistical modeling approach. IEEE Trans. Pattern Anal. Mach. Intell. **25**(9), 1075–1088 (2003)

A CRO Based FLANN for Forecasting Foreign Exchange Rates Using FLANN

K.K. Sahu, G.R. Biswal, P.K. Sahu, S.R. Sahu and H.S. Behera

Abstract The trend in financial trading shows a significant growth in recent years due to globalization in financial market. Foreign Exchange Rate in these days plays a crucial role in financial marketing. The trend of foreign exchange rate follows nonlinear function which can be solved by artificial neural network. In this paper an adaptive CRO based FLANN forecasting model has been proposed for prediction of foreign exchange rate. This model predicts the dollar exchange rate of currencies in Rupees, Yen and Euro which varies over time. The experimental result shows that CRO based FLANN model trained with LMS performs better and efficient than FLANN model.

Keywords Forecasting · Foreign exchange rate · FLANN · Chemical reaction optimization · Back propagation

1 Introduction

Foreign exchange (FOREX) market has been the world's largest trading market. Triennial Central Bank Survey carried out by the Bank of International Settlements (BIS) showed that the global average trading volume is $5.3 trillion per day in April

K.K. Sahu (✉) · G.R. Biswal · P.K. Sahu · S.R. Sahu · H.S. Behera
Department of Computer Science Engineering and Information Technology,
Veer Surendra Sai University of Technology (VSSUT), Burla 768018, Odisha, India
e-mail: itkishore2000@gmail.com

G.R. Biswal
e-mail: grb2092@gmail.com

P.K. Sahu
e-mail: pprabin.sahu@gmail.com

S.R. Sahu
e-mail: mailsomuhere@gmail.com

H.S. Behera
e-mail: mailtohsbehera@gmail.com

647

© Springer India 2015
L.C. Jain et al. (eds.), *Computational Intelligence in Data Mining - Volume 1*,
Smart Innovation, Systems and Technologies 31, DOI 10.1007/978-81-322-2205-7_60

2013. And among all the currency traded, US Dollar remained at the top with 87 % of all traded currencies. EURO came the second highest with 33 %. And Japanese YEN came in third position. Recently India has become one of the central marketing hubs. In this work the FOREX rate of Indian Rupees (INR), EURO, and YEN to USD have been predicted.

FOREX rate depends upon various real world factors like economy, politics, investor psychology, etc. So, the task of FOREX rate prediction becomes cumbersome and challenging. Traditionally different statistical techniques are used to predict the FOREX rate [1]. In these statistical methods it was being assumed that the FOREX rate behaves in a linear way, which is practically the opposite. With the evolution of machine learning techniques the prediction of non-linear FOREX rate became possible.

This work makes use of artificial neural network (ANN) to scrutinize the historical data and predict the foreign exchange rate of future movements. Motivated by the challenging task of FOREX rate prediction a CRO based FLANN with an adaptive gradient descent method (Least Mean Square Method), has been proposed to predict FOREX rate. In FLANN model generally a single input can be expanded to more than one input. This paper shows the efficiency of FLANN-CRO model along with LMS technique over simple FLANN with LMS technique in predicting the foreign exchange rate. This paper is organized into 5 sections. The first section introduces the foreign exchange rate and motivation. Section 2 addresses the literature review. The proposed model with CRO and LMS used in this work are described in Sect. 3. Data collection and Experimental analysis, such as feature extraction and experimental result and their discussion are addressed in Sect. 4, with the conclusion in Sect. 5.

2 Related Work

FOREX market is very volatile in nature. So the development of a model for the problem of the FOREX rate prediction problem of real world finance is very crucial. The accuracy of the models highly essential as it predicts the most volatile assets of the world market. From the traditional statistical approach [1, 2] to modern artificial intelligence approach [3–5], many research works have been carried out to increase the accuracy of the prediction. So far the artificial neural network models outperformed the traditional approaches due to their ability to predict the nonlinearity of the FOREX market. A similar research work addresses how the nonlinear mathematical models of the MLP and RBFNN, and the Takagi–Sugeno (TS) fuzzy system are competent to provide a more accurate out-of-sample predict in comparison to the traditional auto regressive moving average (ARMA) and ARMA generalized auto regressive conditional heteroscedasticity (ARMA-GARCH) linear models [6].

Models other than MLP are also used to forecast the FOREX rate and their result shows that they perform better than MLP network. A study of Exchange rate forecast [7] has shown that a back propagation based neural network for RMB Exchange rate forecasting. Similarly Ritanjal et al. [8] used a single layer flat net ANN called FLANN, for FOREX prediction. Due to its single layered architecture, its complexity is very low. Further a study on nonlinear adaptive models [9] shows the comparison between FLANN and CFLANN on the basis of one, three, six and twelve months' ahead prediction of three different exchange rates. FLANN is also used in stock market prediction. Patra et al. [10] developed a computationally efficient FLANN-based intelligent prediction system for stock prices.

The use of optimization algorithm along with ANN to improve its performance became very popular. Sermpinis et al. [11] used an adaptive neural network with radial basis function and particle swarm optimization for FOREX rate prediction. The result of this work shows that the proposed model outperforms traditional neural networks model.

3 Proposed Work

In this paper, we have predicted foreign exchange rate using a CRO based FLANN model. The model and the optimization technique are described below.

3.1 FLANN Model

A FLANN (Function Linked Artificial Neural Network) is one single layered artificial neural network with a functional block whose job is to produce functional expansion of each input to the network, which is further fed to the network to produce the output. The functional block increases the number of inputs to the network which provides FLANN the ability to solve difficult non-linear problems.

A simplified FLANN model is shown in Fig. 1. Let Z is an input pattern vector with k element given by Z = [z(1), z(2) ... z(k)] when Z is applied to the functional Expansion block each element of the input vector is expanded to a number of inputs. An element z(k) is expended to $\emptyset(k) = [\emptyset_1(k), \emptyset_2(k), ..., \emptyset_{(2N+1)}(k)]$. Thus each element is split into 2N + 1 element which consists N number of cosine term and N number of sine term of the input element along with that element itself. The generic expansion formula is given below:

$$\phi(k) = \{z(k), \cos \pi z(k), \sin \pi z(k), \ldots, \cos N\pi z(k), \sin N\pi z(k)\} \tag{1}$$

Thus K number of elements in input vector Z passes FE block produce L = K (2N + 1) number of inputs, stores in Ø which will be fed to the network as shown in Fig. 1. These Ø terms are the final input to the FLANN model, which are then multiplied with corresponding weight to obtain the weighted input. These weighted inputs are then summed up to obtain the weighted sum S as given in Eq. 2.

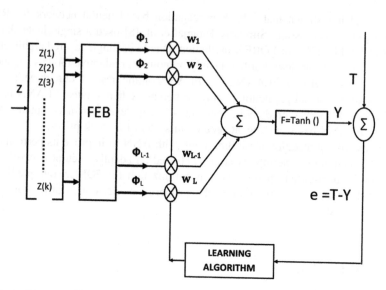

Fig. 1 Architecture of FLANN model

As shown in Fig. 1 the output of the model Y is obtained by passing 'S' through the activation function.

$$S = \sum_{j=1}^{p} \sum_{i=1}^{m} \phi(j,i) * W(j,i) \tag{2}$$

where m = (2N + 1) and 'p' represents the number of attribute in each pattern. Now the output of the model is compared with target vector T and the error is optimized by updating the weight using optimization techniques such as CRO, LMS, GA etc.

3.2 Least Mean Square Method (LMS)

Let k training patterns are applied to the model in a sequential manner and this process is repeated for 'l' experiments. After application of k patterns of input to the model, all the weights of the model are updated at the end of lth experiment by computing the average change mth weight as

$$W_m(l+1) = W_m(l) + \Delta w_m \tag{3}$$

The change of mth weight at lth experiments given by:

$$\Delta w_a(l) = \sum_{k=1}^{K} 2\mu\phi_a(k)\delta(k,l)/k \qquad (4)$$

The symbol $\emptyset_m$ (k) represented the mth expanded value at the application of kth pattern and the error term computed as

$$\delta(k,l) = [(1 - \dot{y}^2(k,l)/2)]e(k,l) \qquad (5)$$

$$e(k,l) = y(k,l) - \dot{y}(k,l) \qquad (6)$$

And $\dot{y}(k,l)$ stands for the estimated output of the model when the kth pattern of lth experiment is applied.

3.3 Chemical Reaction Optimization (CRO) Technique

CRO is a recent optimization technique based on the nature of chemical reactions. The reactants which are in an unstable state undergo a sequence of collision and produce a final state, which is chemically stable [12]. There are actually four elementary reactions involved in the chemical reaction. Basing upon the number of molecules taking part in the reaction the reactions are divided into two types. Such as uni-molecular collision (molecule hit on some external substance) and inter molecular collision (molecules collides with each other). For simplicity, we have considered one and two molecules for uni-molecular and inter-molecular reaction respectively. Both uni-molecular and inter-molecular collision can again be divided on the basis of the effectiveness of the collision. The resulting four elementary reactions are described as follows:

Decomposition: On wall effective collision, where one molecule produces two or more molecules.

On wall ineffective collision: In this case, the collision of the molecule on the wall is ineffective thereby leading to a small change in the molecular structure of the molecule.

Synthesis: More than one molecule combined with each other produces a single stable molecule.

Inter molecular ineffective collision: More than one molecule collides and results in same numbers of molecule with minor change.

3.4 Proposed Algorithm CRO Based FLANN with LMS

<u>**STEP 1**</u>

<u>*CRO ()*</u>

Randomly generate N number weight vectors and store it in population matrix p

For each P(i) calculate fitness value

While halting criteria not satisfied do

For I=1to size[p] do r1, r2   [0,1]

        If r1>0.5 **then**   // uni-molecular collision

        Choose P(i) randomly from P

                If r2>0.5

                [Px Py] = Decompose (P (i))

                    Evaluate fitness value for Px, Py and choose best among Px, Py, P (i) on basic of fitness value and assign   to P (i)

                Else

                [Px]=on_wall_ineff_col (P (i))

                    Evaluate fitness value for Px and choose best among Px, P (i) on basis of fitness value and assign to   P (i)

                End if

      Else

        Choose P (i), P (j) randomly from P

                If r2>0.5 **then**

                Px=synthesis [P (i), P (j)]

                    Evaluate fitness value for Px, Py and choose best two among Px, Py, P (i) ,P(j) on basis of fitness value and assign to P (i),P(j)

                Else

                  [Px, Py]=intmol_inef_col(P (i),   P (j))

                    Evaluate fitness value for Px, Py and choose best two among Px, Py, P (i) ,P(j) on basis of fitness value and assign to P (i), P (j)

                End if

      I=I+1

      End if

      End for

End while

<u>*Decompose (P (i))*</u>

For j = 1 to n

      Px (j) = rand ()*P (I,j)

      Py (j) = rand ()*P (I,j)

End for

Return Px, Py

<u>*on_wall_ineff_col (P (i))*</u>

Px = P (i) +rand ()

Return Px

<u>*Synthesis [P (i), P (j)]*</u>

for t= 1 to n

      Px (t) = P (i ,t) + P(j ,t) + rand()

Return Px

<u>*intmol_inef_col (P (i),   P (j))*</u>

Px = p (i) + rand ()

Py = p (j) + rand ()

Return Px, Py

STEP 2
Find the best weight vector from population matrix on basis of fitness value i.e. weight vector with highest fitness value.
w=bestmole (P)
bestmole (p): sort P in decreasing fitness value return p(1)

STEP 3
While halting criteria not satisfied do
       Calculate Y using FLANN model for M no of months
      e=T-Y
      d=(1-Y^2)*e/2
      If e ~=0
      For m=1: no of weights
      s=0;
      For j=1: no of months
                  s=s+2*learning rate*fe(m)*d;
    End
      dw=s/no of months
      w(m)=w(m)+dw
    End
    End
End while

4 Experimental Analysis

4.1 Feature Extraction from Raw Data

The raw data collected on a monthly basis and it contains the exchange rate of different currencies and the time in months. This information is not sufficient enough for predicting. So, to obtain better forecasting result, some statistical features are extracted from the raw data. The input is taken from 12th month onward so that features like moving average, standard deviation and variance can be calculated.

1. The N-month moving average is calculated as follows:

$$maN = \frac{1}{N}\sum_{i=1}^{N} X_i \qquad (7)$$

In our research N is set to 12 to obtain *ma12* and X_i represents the exchange rate for ith month.

2. The standard deviation for N-months is obtained using the following formula:

$$sdN = \sqrt{\frac{1}{N}\sum_{i=1}^{N} (X_i - \bar{X})^2} \qquad (8)$$

where $\bar{X}$ is the moving average for N-month period.

3. The variance for N-month period is calculated by squaring the standard deviation.

$$vN = sd^2 = \frac{1}{N}\sum_{i=1}^{N}(X_i - \bar{X})^2 \qquad (9)$$

After feature extraction the new dataset is normalized using min-max technique as follows:

$$N_r = (2*x - (\text{Max} - \text{Min}))/(\text{Max} - \text{Min}) \qquad (10)$$

where N_r is the normalized and x is the actual value and Max and Min are the maximum and minimum value of the dataset. Now the dataset is divided into two parts i.e. training set and testing set.

M_{tr} no of month for which data is available for training
M_{ts} no of months for which data is available for testing
$M = M_{tr} + M_{ts}$ no of month for which total data are available

For the (m + 11)th month mean and variance values are computed as.
The inputs to the model for any (m + 11)th month($1 <= m <= M_{tr-11}$) is thus X_{m+11}, $X_{m+11(ma12)}$, $X_{m+11(sd12)}$ and $X_{m+11(v12)}$ ($1 <= m <= M_{tr}-11$). the estimated output of the model is given by X_{m+12} as the model predicts for (m + 12)th month. The black box view of the proposed nonlinear model is shown in, Fig. 2. After the model is trained by CRO and LMS method, testing is carried out using data having M_{ts} month data.

4.2 Performance Metric

In this paper two performance metrics are used to evaluate the performance of the proposed model. Those are:

1. *Root Mean Square Error (RMSE)*: RMSE provides a clear view of the performance of the model as it provides movement of RMSE of the model. RMSE is calculated as follows:

$$RMSE = \sqrt{\frac{1}{N}\sum_{i=1}^{N}e(i,l)^2} \qquad (11)$$

where e(i,l) is the error of ith pattern for lth experiment.

Fig. 2 Black box view forecasting model

2. *Average Percentage Error (APE)*: In this work three different currencies are taken into consideration. The RMSE of FOREX prediction will be affected by the economic condition of the respective countries. Hence, to have a uniform conclusion, average percentage error is chosen as another performance metric. APE is calculated using the following formula:

$$APE = \frac{1}{N}\sum_{i=1}^{N}\frac{e(i,l)}{t(i)} * 100 \tag{12}$$

where e(i, l) is the error of the model at lth experiment t(i) is the target of ith patterns. Figure 3 shows the working principle of the proposed model. In this model CRO and the LMS method is hybridized to provide better results. Due to the presence of four different types of searching techniques CRO is able to overcome the local minima problem where as LMS method helps in refining the search result.

4.3 Result Discussion

The prediction is done taking three Different currencies as Rupees, Euro and Yen. For each currency the FOREX rate for 200 month is fed to both the model and the prediction of both the model were compared on the basis of actual-predicted value plot and RMS error plot. The comparison plot of one month ahead prediction of FOREX rate for rupees using FLANN and CRO based FLANN are shown in Figs. 4 and 5. The figures clearly indicates that CRO based FLANN model has better coinciding plot than its counterpart, FLANN model. The average percentage error comparison is shown in Table 1. The APE also shows the same result.

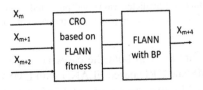

Fig. 3 Working of forecasting model

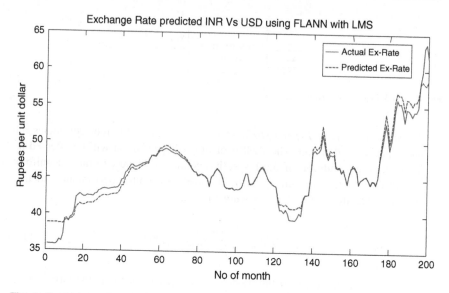

Fig. 4 Comparison of actual and predicted value and predicted value (equivalent Rupees for 1 USD) for one month ahead with test data set using back propagation with FLANN

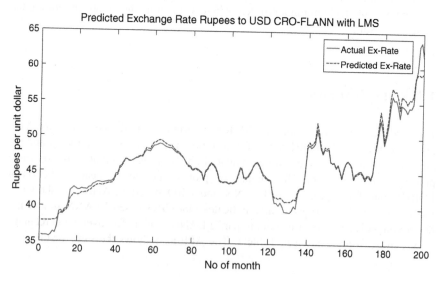

Fig. 5 Comparison of actual and predicted value and predicted value (equivalent Rupees for 1 USD) for one month ahead with test data set using back propagation in FLANN along with CRO

The Root Mean Square error in both the cases i.e. rupees exchange prediction using FLANN with LMS and CRO based FLANN with LMS shown in Figs. 6 and 7 respectively. In Fig. 6 RMS error decreases to 0.0255 in 100 iteration but in Fig. 8

Table 1 One month ahead FOREX rate prediction of Rupees to USD

Actual dollar price	FLANN with LMS		FLANN with LMS and CRO	
	Predicted dollar price	% Error	Predicted dollar price	% Error
55.5600	56.6567	1.9740	55.5510	0.0163
54.6060	56.7122	3.8572	55.6974	1.9987
53.0240	55.9669	5.5502	55.0679	3.8546
54.7760	54.4845	0.5323	53.9394	1.5273
54.6480	56.1229	2.6988	55.3556	1.2947
54.3170	56.0227	3.1403	55.3469	1.8960
53.7740	55.7431	3.6618	55.1817	2.6178
54.4050	55.2518	1.5564	54.8498	0.8175
54.3760	55.8330	2.6795	55.4104	1.9023
55.0110	55.8138	1.4594	55.4725	0.8390
58.3970	56.3525	3.5010	56.0398	4.0366
59.7750	58.2741	2.5109	58.6898	1.8155
63.2090	58.5880	7.3107	59.8369	5.3348
63.6830	58.2321	8.5594	62.2923	2.1838
		APE = 3.5018		APE = 2.1524

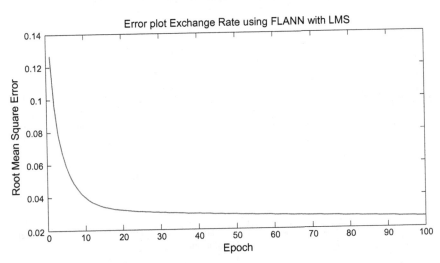

Fig. 6 Convergence characteristics of proposed forecasting model of Rupees with FLANN-LMS

the RMS error reaches down to 0.0226 in same100 iteration. In Figs. 8 and 9 one month advance actual-predicted plot of Euro for FLANN-LMS and CRO based FLANN-LMS taking 200 months former FOREX rate are shown. The percentage error is shown in Table 2.

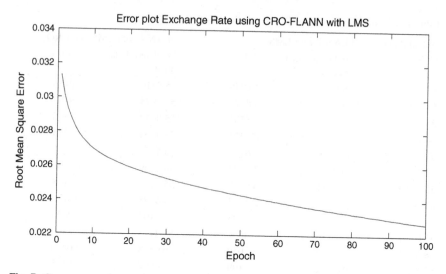

Fig. 7 Convergence characteristics of proposed forecasting model of Rupees using CRO based FLANN with LMS

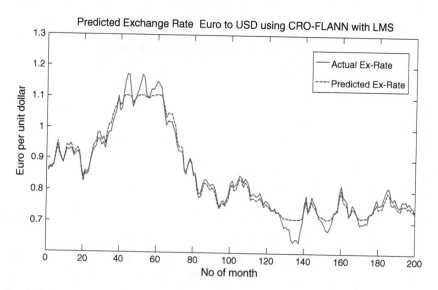

Fig. 8 Comparison of actual and predicted value (equivalent Euro for 1 USD) for one month ahead with test data set using back propagation with FLANN

The Root Mean Square error in both the cases of Euro exchange prediction i.e. FLANN with LMS and CRO based FLANN with LMS is shown in Figs. 10 and 11 respectively. In Fig. 10 RMS error decreases to 0.03 in 100 iterations but in Fig. 8 the RMS error converges to 0.0245 in 100 iterations.

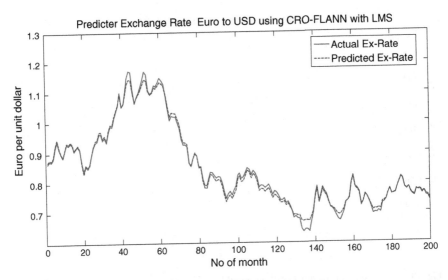

Fig. 9 Comparison of actual and predicted value (equivalent Euro for 1 USD) for one month ahead with test data set using CRO based FLANN with LMS

Table 2 One month ahead FOREX rate prediction of Euro to USD

Actual dollar price	FLANN with LMS		FLANN with LMS and CRO	
	Predicted dollar price	% Error	Predicted dollar price	% Error
0.8060	0.7842	2.7088	0.8177	1.4470
0.7780	0.7740	0.5135	0.8065	3.6618
0.7710	0.7385	4.2128	0.7675	0.4564
0.7800	0.7288	6.5672	0.7572	2.9260
0.7620	0.7391	3.0059	0.7689	0.9102
0.7530	0.7155	4.9794	0.7435	1.2679
0.7490	0.7032	6.1162	0.7305	2.4756
0.7710	0.6970	9.5987	0.7242	6.0674
0.7680	0.7232	5.8362	0.7534	1.8999
0.7700	0.7183	6.7108	0.7485	2.7898
0.7580	0.7198	5.0357	0.7506	0.9785
0.7650	0.7035	8.0385	0.7332	4.1591
0.7510	0.7112	5.2970	0.7421	1.1889
0.7480	0.6924	7.4359	0.7220	3.4801
		APE = 5.4326		APE = 2.3863

For Yen also the actual and predicted FOREX rate were plotted for FLANN-LMS and CRO based FLANN-LMS taking 200 months former FOREX rate values and the percentage error is shown in Table 3.

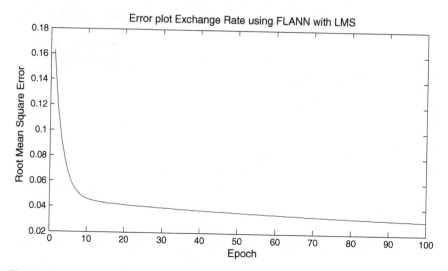

Fig. 10 Convergence characteristics of proposed forecasting model of Euro with LMS in FLANN

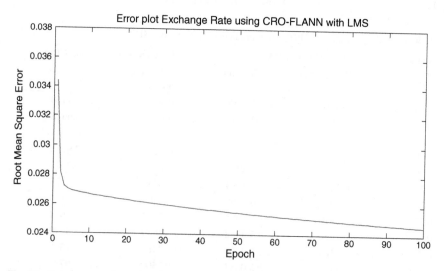

Fig. 11 Convergence characteristics of proposed forecasting model of Euro using CRO based FLANN with LMS

The Root Mean Square error in both the cases of Yen exchange rate prediction i.e. simple FLANN with LMS and with CRO based FLANN with LMS is shown in Figs. 12 and 13 respectively. In Fig. 14 RMS error converges to 0.033 in 100 iterations but in Fig. 15 the RMS error dropped to 0.02795 in 100 iterations.

For all the three currency, the actual-predicted value plots in CRO based FLANN with LMS (Fig. 13) maintain better convergence between the actual and

Table 3 One month ahead FOREX rate prediction of Yen to USD

Actual dollar price	FLANN with LMS		FLANN with LMS and CRO	
	Predicted dollar price	% Error	Predicted dollar price	% Error
78.7220	82.3250	4.5768	79.6628	1.1951
78.1400	82.2626	5.2759	79.4747	1.7081
79.0180	82.1726	3.9922	79.0256	0.0096
80.7900	82.2438	1.7995	79.7792	1.2511
83.5800	82.5475	1.2353	81.3630	2.6526
89.1600	83.3968	6.4639	84.0396	5.7429
93.0030	86.3945	7.1057	89.7954	3.4490
94.8040	89.3701	5.7317	93.8762	0.9786
97.7680	90.9831	6.9398	95.7916	2.0215
100.7770	93.9095	6.8145	98.8886	1.8739
100.7770	97.1473	3.6017	101.9756	1.1893
99.7020	97.1413	2.5684	101.9942	2.2991
97.5770	95.9504	1.6669	100.9238	3.4299
99.2180	93.6843	5.5774	98.7710	0.4506
		APE = 4.5249		APE = 2.3030

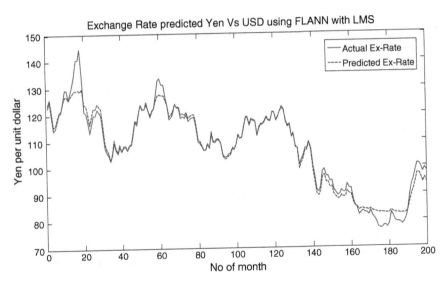

Fig. 12 Comparison of actual and predicted value (equivalent Yen for 1 USD) for one month ahead with test data set using LMS with FLANN

predicted curve as compare to simple FLANN with LMS (Fig. 12). This is because CRO helps in avoiding the local minima because of its different variant of reactions i.e. CRO search for solution in wide range as well as in short range, which make

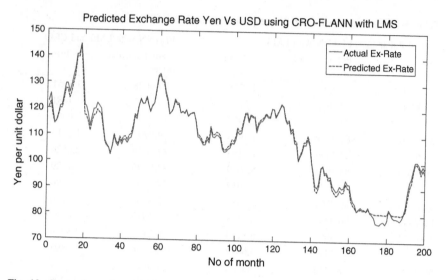

Fig. 13 Comparison of actual and predicted value (equivalent Yen for 1 USD) for one month ahead with test data set using CRO based FLANN with LMS

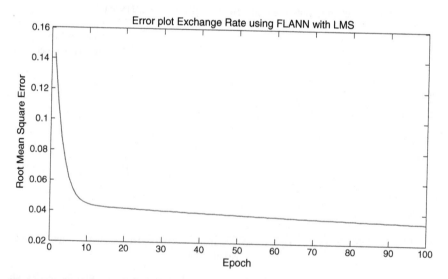

Fig. 14 Convergence characteristics of proposed forecasting model for Yen with back propagation in FLANN

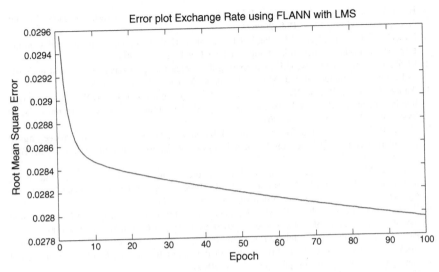

Fig. 15 Convergence characteristics of proposed forecasting model of Yen using CRO based FLANN with LMS

CRO faster technique. In the other hand LMS refines the search results by using the error as well as the input term to optimize error step by step. Therefore in all above cases the RMS error for CRO based FLANN decreases more in same number of epoch.

5　Conclusion

FOREX rate prediction has been a real challenging problem due to the volatile nature of the FOREX market. This paper compares the currency predication method based on single layer neural network using two different techniques i.e. FLANN with LMS and CRO based FLANN with LMS only. Through simulation study and analysis, it is observed that LMS with CRO produce better result. But still further research is also needed in this area to increase the accuracy of the prediction using different evolutionary network and optimization technique.

References

1. Harvey, A.C.: Forecasting structural time series models and the Kalman filter. Cambridge University Press, Cambridge (1989)
2. Box, G.E.P., Jenkins, G.M.: Time Series Analysis: Forecasting and Control. Holden-day, San Francisco (1976)

3. Boginski, V., Butenko, S., Pardalos, P.M.: Mining market data: a network approach. Comput. Oper. Res. **33**(11), 3171–3184 (2006)
4. Lubecke, T.H., Nam, K.D., Markland, R.E., Kwok, C.C.Y.: Combining foreign exchange rate forecasts using neural networks. Glob. Finance J. **9**(1), 5–27 (1998)
5. El Shady, M.R., El Shazly, H.E.: Comparing the forecasting performance of neural networks and forward exchange rates. J. Multinatl. Financ. Manage. **7**, 345–356 (1997)
6. Andre Santos, A.P., da Newton Costa Jr, C.A., Leandro Coelho, D.S.: Computational intelligence approaches and linear models in case studies of forecasting exchange rates. Expert Syst. Appl. **33**(4), 816–823 (2006)
7. Ye, S.: RMB exchange rate forecast approach based on BP neural network. In: Proceedings of the International Conference on Medical Physics and Biomedical Engineering (ICMPBE), pp. 287–293 (2012)
8. Ritanjali, M., Panda, G., Sahoo, G.: Efficient prediction of foreign exchange rate using nonlinear single layer artificial neural model. In: Proceedings of the International Conference on Cybernetics and Intelligent Systems (CIS), pp. 1–5 (2006)
9. Panda, G., Sahoo, G., Majhi, R.: Development and performance evaluation of FLANN based model for forecasting of stock markets. Expert Syst. Appl. **36**(3), 6800–6808 (2009)
10. Patra, J.C., Thanh, N.C., Meher, P.K.: Computationally efficient FLANN-based intelligent stock price prediction system.: Proceedings of the International Joint Conference on Neural Networks (IJCNN), Atlanta, pp. 2431–2438 (2009)
11. Sermpinis, G., Theofilatos, K., Karathanasopoulos, A., Georgopoulos, E.F., Dunis, C.: Forecasting foreign exchange rates with adaptive neural networks using radial-basis functions and particle swarm optimization. Eur. J. Oper. Res. **255**(3), 528–540 (2013)
12. Lam, A.Y.S., Li, V.O.K.: Chemical-reaction-inspired metaheuristicfor optimization. IEEE Trans. Evol. Comput. **14**(3), 381–399 (2010)

Performance Analysis of IEEE 802.11 in the Presence of Hidden Terminal for Wireless Networks

Anita, Rishipal Singh, Priyanka and Indu

Abstract IEEE 802.11 is the most important standard for wireless local area network (WLANs). In IEEE 802.11 the fundamental medium access control (MAC) scheme is the Distributed Coordination Function (DCF). DCF provides both Basic access and RTS/CTS (Request-to-send/ Clear-to-send) access. The previous work for DCF only provide analysis when all the nodes in WLAN lie in the carrier sense range of each other. In the real scenario of WLAN all the stations are in communication range of Access Point (AP) but not necessary with respect to each other, which results in hidden terminals (HT) and degrades the performance. This paper presents an accurate analytical model for DCF in presence of Hidden terminals through Markov Chain for throughput in unsaturated condition i.e. packet error, transmission error. This analytical model is suitable for both Basic access and RTS/ CTS access mechanism and is evaluated by extensive simulation results using MATLAB simulator.

Keywords WLAN · MAC · DCF · Basic access · RTS/CTS · Hidden terminals · Unsaturated conditions

Anita (✉) · Indu
Department of Computer Science and Engineering,
Banasthali Vidhyapith, Tonk, Rajasthan, India
e-mail: anitarana427@gmail.com

Indu
e-mail: indu.3027@gmail.com

R. Singh · Priyanka
Department of Computer Science and Engineering, Guru Jambheshwar
University of Science and Technology, Hisar, India
e-mail: pal_rishi@yahoo.com

Priyanka
e-mail: rathee.priyanka124@gmail.com

© Springer India 2015
L.C. Jain et al. (eds.), *Computational Intelligence in Data Mining - Volume 1*,
Smart Innovation, Systems and Technologies 31, DOI 10.1007/978-81-322-2205-7_61

665

1 Introduction

Wireless local area networks (WLANs) have been widely deployed in recent years. In WLANs, the most important standard is IEEE 802.11, where the fundamental medium access control (MAC) scheme is the distributed coordination function (DCF) which is a carrier-sense multiple access with collision avoidance (CSMA/CA) protocol. IEEE 802.11 support asynchronous data transfer in a basic service set (BSS). To understand the performance of WLANs, a critical challenge is how to analyze IEEE 802.11 DCF. There are two techniques used for transmission in DCF: basic access mode and RTS/CTS access mode.

The modeling of 802.11 has been a research focus since the standard has been proposed. Paper [1] gives the theoretical throughput limit of 802.11 based on a p-persistent variant. Later work [2] proposed a mechanism for Adaptive Contention Window, which dynamically determines the optimal backoff window according to the estimate of the number of contending stations and also investigated the CSMA/CA with the optional RTS/CTS technique, and show that these adaptive techniques reached better performance only when the packet size is short. They also concluded that the throughput performance is strongly dependent on the number of active stations, and on the total load offered to the system. The paper [2] presented a simple analytical model (Markov chain model) to calculate the saturation throughput performance in the presence of a finite number of terminals and with the assumption of ideal channel conditions. The model applied to both basic and RTS/CTS access mechanisms. Those works assume that each station lies in the carrier sense range of each other and ignore hidden terminals.

However, during access of WLAN all the stations are in the communication range of Access Point (AP), they may not be necessarily with respect to each other. The hidden terminal problem comes in focus when the stations out of the carrier sense range of each other transmit to Access Point. A lot of research [2, 3] has been done to show the performance degradation of DCF with hidden terminals. Compared to previous work, it has the following challenges: first the work model for DCF assume the shared wireless medium i.e. all stations observe the same channel status and freeze or decrease the backoff counter at the same time that is broken by hidden terminals. Second Carrier sense for DCF is performed both at physical layer and MAC layer. At physical layer it is referred as physical carrier sensing and at MAC layer referred as Virtual CS (VCS). The hidden terminals and the error present at packet challenge the accuracy of modelling. To address the challenges, we first propose a discrete time Markov system where each slot has a fixed small value taken by backoff counter instead of variable size slot. The fixed small slot represents that each station may observe different channel status in the same slot i.e. freezing, transmission or collision and also the presence of error. Here, we provide an equivalent transition state by using markov process.

The remaining paper is organized as follows. In Sect. 2, the hidden terminal problem is introduced briefly. In Sect. 3, introduced a Markov chain based model with equivalent transition state to analyze the average throughput performance of DCF with hidden terminals. The simulation result for the accuracy of model is shown in Sect. 3.3. The conclusion of paper is shown in Sect. 4.

2 Problem Statement

We first introduce the hidden terminal problem and then present the hotspot scenario that will be analyzed with our model.

2.1 Hidden Terminal

The hidden terminal problem is shown in Fig. 1, nodes that are hidden from one another may not be able to sense the carrier signal, so that their transmissions result in frequent collisions. This problem degrades network throughput because there will be numerous collisions and retransmissions.

2.2 Hotspot Scenario for DCF with Hidden Terminals

In this we assume that there is one AP in the center and v stations are scattering around AP. Some nodes lie in the carrier sense range of node n that is known as CS_n and the nodes that are hidden to node v_n are known as HS_n.

The standard is taken that each station can only transmit packets to AP, so the traffic flows only in two directions: from AP to stations and from stations to AP. The AP can always sense the traffic from stations and the traffic from AP to stations will be sensed by all the stations. We analyze the problem as: the given number of stations and the number of hidden stations to each node, to obtain the performance of the WLAN in terms of throughput.

Fig. 1 Hidden terminal scenario

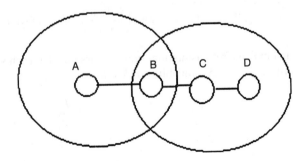

3 Proposed Analytical Model

The main purpose of this paper is to present analytical model for DCF to evaluate throughput in presence of hidden terminals. Here unsaturated condition is considered i.e. the station will always have packets waiting for transmission and there can be error in the packets. Imperfect channel conditions are assumed i.e. Wireless transmission error, fading, capture effect, etc. The analysis is divided into three parts: (1) A discrete time Markov chain model is proposed to study the behavior of each station with contention window and exponential backoff in presence of hidden terminals. By this we get a relation between transmission probability and collision probability; (2) The throughput of both basic and RTS/CTS access is calculated for the model in presence of error and the hidden terminals; (3) The analysis of collision probability under hidden terminals results another relation between collision probability and transmission probability.

For each station three dimensions are used to character its status at time slots t: (1) s(t) representing the number of backoff stages; (2) b(t) representing the backoff counter; (3) v(t) representing the residual time slot during either freezing, transmission or collision. The tri-dimensional process (s(t), b(t), v(t)) for the discrete-time markov chain is shown in Fig. 2. The s(t) = −1 denotes the post-backoff stage. The transition probability p_a and p_b are the corresponding mathematically equivalent state transition probabilities. The original states transit at every slot which is a fixed small value. The p_a and p_b model the freezing time between subsequent backoff counter between equivalent states. For example, for states (0, 0, 0), (0, 1, 0) and (0, 1, 1), p_a determines whether the state transits directly from (0, 1, 0) to (0, 0, 0) observing idle, or transits to freezing state (0, 1, 1) observing busy channel due to the four types of possible carrier sensing, at state (0,1,1), p_b determines the length of freezing time. Let CW_{min} and CW_{max} denote the minimal and maximal Contention Window (CW) in DCF, then the CW used in backoff stage b(t) = i can be calculated as

$$W_i = \begin{cases} 2^i W & 0 \le i \le m' \\ 2^{m'} W & m' \le i \le m'' \\ 2^{m''} W & m'' \le i \le m \end{cases} \tag{1}$$

where

$$W = (CW_{min} + 1), m' = \log_2\left(\frac{CW_{max}}{CW_{min}}\right), m'' = \log_2((CW_{max} + 1)/(CW_{min} + 1))$$

and m is the retransmission limit. We assume that each station transmits in any slot with a stationary probability τ. We have

$$L_s = \frac{T_s}{\sigma}$$

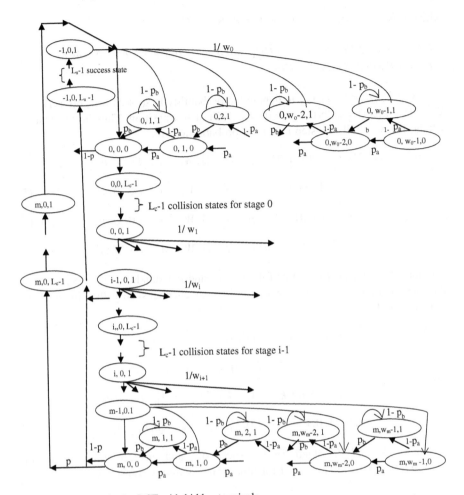

Fig. 2 Markov chain for DCF with hidden terminals

and

$$L_c = \frac{T_c}{\sigma}$$

in the Markov chain, where T_s is successful transmission time and T_c is the collision time. For Basic mode, T_s and T_c can be expressed as:

$$T_s^{bas} = DIFS + H + P + \delta + SIFS + ACK + \delta$$
$$T_c^{bas} = DIFS + H + P + SIFS + ACK$$

(2)

where δ is the propagation delay. For RTS and CTS

$$
\begin{aligned}
T_s^{rts} &= DIFS + RTS + SIFS + \delta + CTS + SIFS + \delta + H + P + SIFS + \delta + ACK + \delta \\
T_c^{rts} &= DIFS + RTS + SIFS + CTS
\end{aligned}
\tag{3}
$$

We use $P\{i_1, j_1, k_1 | i_0, j_0, k_0\}$ to denote the probability $P\{s(t+1) = i_1, b(t+1) = b_1, v(t+1) = k_1 | s(t) = i_0, b(t) = j_0, v(t) = k_0\}$ for state transition. Thus in this Markov Chain, the non-null one-step transition probabilities are:

$$
P\{i, j-1, 0 | i, j, 0\} = p_a \quad j \in [1, W_i - 1], \quad i \in [0, m]
\tag{4}
$$

$$
P\{i, j-1, 0 | i, j, 1\} = p_b \quad j \in [1, W_i - 1], \quad i \in [0, m]
\tag{5}
$$

$$
P\{i, j, 1 | i, j, 0\} = 1 - p_a \quad j \in [1, W_i - 1], \quad i \in [0, m]
\tag{6}
$$

$$
P\{i, j, 1 | i, j, 1\} = 1 - p_b \quad j \in [1, W_i - 1], \quad i \in [0, m]
\tag{7}
$$

The meaning of equation are as follows, (4) and (5) stands for the decrement of the backoff counter, (6) and (7) stand for the busy channel state as the virtual carrier sensing result.

$$
P\{i, 0, L_c - 1 | i.0, 0\} = P \quad i \in [0, m]
\tag{8}
$$

$$
P\{-1, 0, L_s - 1 | i.0, 0\} = 1 - P \quad i \in [0, m]
\tag{9}
$$

$$
P\{i, 0, k - 1 | i.0, k\} = 1, \quad k \in [2, L_c - 1], \quad i \in [0, m]
\tag{10}
$$

$$
P\{-1, 0, k - 1 | -1.0, k\} = 1, \quad k \in [2, L_s - 1]
\tag{11}
$$

Equations (8) and (9) stand for the successful and unsuccessful transmission respectively, (10) and (11) stand for time progress during transmission.

$$
P\{0, j, 0 | -1, 0, 1\} = \frac{1}{W_a} \quad j \in [1, W_0 - 1]
\tag{12}
$$

$$
P\{0, j, 0 | m, 0, 1\} = \frac{1}{W_a} \quad j \in [0, W_0 - 1]
\tag{13}
$$

$$
P\{i, j, 0 | i - 1, 0, 1\} = \frac{1}{W_i} \quad j \in [0, W_i], \quad i \in [1, m]
\tag{14}
$$

Equations (12)–(14) stand for the backoff and post-backoff stage. Let $b_{i,j,k}$ be the stationary distribution probability of the Markov chain. First we have,

$$
b_{i,0,0} = b_{0,0,0} P^i \quad 1 \le i \le
\tag{15}
$$

Since the chain is regular, so for each state, we have

$$
b_{i,j,k} = \begin{cases}
\frac{b_{i,0,0}(W_i-j)}{W_i} & k = 0 \\
\frac{(1-P_a)b_{i,j,0}}{P_b} & j \in [1, W_i - 1], \ k = 1 \\
b_{i,0,0} * P & i \in [0, m], \ j = 0 \\
b_{0,0,0}(1 - P^{m+1}) & i = -1
\end{cases}
\tag{16}
$$

Therefore by using normalization condition for stationary distribution, we have

$$
1 = \sum_{i=0}^{m} \sum_{j=1}^{W_i-1} \sum_{k=0}^{1} b_{i,j,k} + \sum_{i=0}^{m} \sum_{k=0}^{L_s-1} b_{i,0,k} + \sum_{k=1}^{L_s-1} b_{-1,0,k}
\tag{17}
$$

$$
1 = b_{0,0,0} \left[\frac{P_b - P_a + 1}{2P_b} * \left\{ W \left(\frac{1 - (2P)^m}{1 - 2P} \right) + \frac{1 + W(P^{m'} * 2^{m'} + P^{m'} * 2^{m'}) + P^{m'}}{1 - P} \right\} \right.
$$
$$
\left. + P * \frac{(L_c - 1)(1 - P^{m-1})}{1 - P} + (1 - P^{m-1})(L_s - 1) \right]
\tag{18}
$$

$$
b_{i,0,0} = \left[\frac{P_b - P_a + 1}{2P_b} * \left\{ W \left(\frac{1 - (2P)^m}{1 - 2P} \right) + \frac{1 + W(P^{m'} * 2^{m'} + P^{m'} * 2^{m'}) + P^{m'}}{1 - P} \right\} \right.
$$
$$
\left. + P * \frac{(L_c - 1)(1 - P^{m-1})}{1 - P} + (1 - P^{m-1})(L_s - 1) \right]^{-1}
\tag{19}
$$

Now the probability τ can be expressed as:

$$
\tau = \sum_{i=0}^{m} b_{i,0,0}
\tag{20}
$$

$$
\tau = b_{i,0,0} * \frac{(1 - P^{m+1})}{1 - P}
\tag{21}
$$

3.1 Collision Probability with Hidden Terminals

On the basis of the collisions only happen when two or more stations have backoff counters reduced to zero at the same time and have simultaneous transmission. In presence of hidden terminals, V_n station to make a successful transmission requires three steps: first V_n's hidden terminals are not transmitting. Second all the other stations in the network do not start transmission at the same time with station V_n. Third none of V_n's hidden terminals starts transmission before V_n transmission is finished. Let the success probability for these three steps is P_{s1}, P_{s2}, and P_{s3}. The probability of error present in the form of transmission error, fading and capture effect is P_s.

$$P = 1 - P_{s1}P_{s2}P_{s3}(1 - P_s) \tag{22}$$

For the first condition, none of hidden terminals started a successful transmission during the last L_s slots and an unsuccessful transmission during last L_c slots.

$$P_{s1} = (1 - L_s(1 - P)\tau - L_c P\tau)^h \tag{23}$$

where h is average number of stations in HS_n For the second condition we have

$$P_{s2} = (1 - \tau')^{s+h} \tag{24}$$

where s is average number of stations in CS_n and τ' denotes the conditional transmission probability i.e. a station starts transmission under the condition that previous slot is idle.

$$\tau' = \frac{\sum_{i=0}^{m} b_{i,0,0}}{\sum_{i=0}^{m} \sum_{j=0}^{W_i-1} b_{i,j,0}} \tag{25}$$

$$\tau' = 2(1 - P^{m+1})(1 - 2P)\left[2P(1 - P^{m+1})(1 - 2P) + W(1 - P)(1 - (2P)^m) \right.$$
$$\left. + (1 - 2P)\left\{(1 - P^{m'}) + P^{m'}(2^{m'}W + 1) + P^{m''}(2^{m''}W + 1)\right\}\right]^{-1} \tag{26}$$

$$\tau' = 2(1 - p^{m+1})(1 - 2p) \tag{27}$$

For the third condition, we define $L = T_d/\sigma$, where T_d is the transmission time for one data packet in basic mode and for one RTS packet in RTS/CTS mode.

$$T_d^{bas} = H + P \tag{28}$$

$$T_d^{bas} = RTS \tag{29}$$

we have

$$P_s = (1 - \tau'')^h \tag{30}$$

where τ'' is probability that a station in HS_n start transmission in continuous L slots with condition that station V_n has already transmitted.

3.2 Throughput Analysis

It may be defined as the expected number of successful transmission in time period t with average payload, so the throughput achieved is:

$$S = \tau(1 - P)(h + s + 1)L_p \tag{31}$$

where $L_p = \frac{T_p}{\sigma}$, T_p is the transmission time of payload.

3.3 Model Evaluation

To evaluate the proposed model, we use MATLAB as simulator. To obtain the unsaturated throughput performance, we vary both the total number of stations and hidden terminals. All the parameters are listed in Table 1. Note that our MAC Markov model equations are independent of the PHY parameters setup.

The Fig. 3 plots average throughput versus number of stations for both basic access and RTS/CTS access mechanism with hidden terminals. Throughput decreases as number of stations increases in basic access method but for RTS/CTS

Table 1 System parameters for simulation		
MAC header	224
PHY header	192
ACK	304
RTS	352
CTS	304
Propagation delay	1
Slot time	20
SIFS	10
DIFS	50
P	8,184

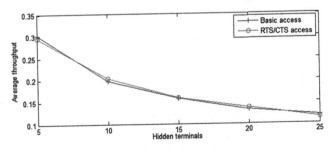

Fig. 3 Average throughput versus number of hidden terminals

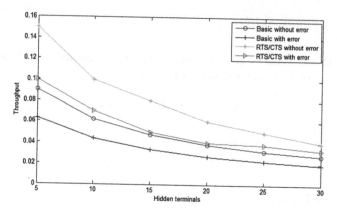

Fig. 4 Throughput versus number of hidden terminals

it decreases a little because there are more number of collisions in basic access method as compared to RTS/CTS. When the number of hidden terminals is 5 then basic access performance is better than RTS/CTS access. When the hidden terminals are 10, 15 and 20 the performance of RTS/CTS access is better than basic access but when terminals are 25 the RTS/CTS degrades.

The Fig. 4 presents throughput versus number of hidden stations for both basic access and RTS/CTS access mechanism in presence of error and absence of error. Throughput decreases as number of hidden stations increases. The Basic access mechanism throughput without error is better than basic access mechanism with error and the RTS/CTS access mechanism throughput without error is better than the RTS/CTS access mechanism with error.

The Fig. 5 shows throughput versus packet size for both basic access and RTS/CTS access mechanism. As the packet size increases throughput increases and the average throughput of the RTS/CTS access mechanism is better than basic access

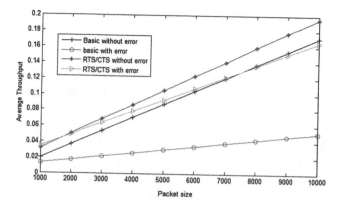

Fig. 5 Average throughput versus packet size

mechanism. Due to presence of error in the packet the performance degrades in basic access and RTS/CTS access mechanism. The performance of RTS/CTS access with error is better than basic access with error.

4 Conclusion

IEEE 802.11 is the standard used for Wireless LANs and DCF is one of the medium access control mechanism specified by this standard. DCF employs a random access mechanism based on CSMA/CA protocol with binary exponential backoff. There are many issues related to IEEE 802.11 WLAN and one of the issues is hidden terminal problem that exists in practice in any wireless network. In this work the performance of IEEE 802.11 DCF is evaluated in presence of hidden nodes using an analytical three dimensional markov chain model. Also the performance is analyzed with error and without error in RTS/CTS access mechanism and basic access mechanism. Due to presence of error performance degrades. For number of stations the performance of RTS/CTS access method is better than the basic access method and with hidden terminals the performance for both methods degrades. For packet size performance of both access methods is almost similar and it varies a little for basic access method with hidden terminals. The performance is also compared for transmission probability with both access methods by varying number of stations. For number of hidden terminals RTS/CTS access method has presented better performance than basic access method.

For future work performance can be evaluated in presence of exposed nodes also and thus by considering two issues together i.e. hidden as well as exposed nodes IEEE 802.11 DCF can be modelled.

References

1. Cali, F., Conti, M., Gregori, E.: Dynamic tuning of the IEEE 802.11 protocol to achieve a theoretical throughput limit. IEEE/ACM Trans. Netw. 8(6), 785–799 (2000)
2. Bianchi, G.: Performance analysis of the IEEE 802.11 distributed coordination function. IEEE J. Sel. Areas Commun. 18(3), 785–799 (2000)
3. Jiang, L.B., et al.: Improving throughput and fairness by reducing exposed and hidden nodes in 802.11 networks. IEEE Trans. Mob. Comput. 7(1), 34–49 (2008)
4. Wu, H., et. al.: Analysis of IEEE 802.11 DCF with hidden terminals. IEEE GLOBECOM. (2006)
5. IEEE 802.11, Part 11(Draft): wireless LAN medium access control (MAC) and physical layer (PHY) specifications. IEEE, New York (Aug 1999)
6. Chatzimisios, P., Vitsas, V., Boucouvalas, A.C.: Throughput and delay analysis of IEEE 802.11 protocol. In: Proceedings of IEEE International Workshop on Networked Appliances (IWNA), pp. 168–174. Liverpool, UK (Oct 2002)
7. Jeon, J., Kim, C., Kiseok Lee, L.: Fast retransmission scheme for overcoming hidden node problem in IEEE 802.11 Networks. J Comput. Sci. Eng. 5(4), 324–330 (2011)

8. Saeed, M.J., Merabti, M., Askwith, R.J.: Hidden and exposed nodes and medium access control in wireless ad-hoc networks. In: Proceedings of the PGNet. School of Computing and Mathematical Sciences, Liverpool John Moores University, UK (2006)
9. Ergen, M., Varaiya, P.: Throughput analysis and admission control for IEEE 802.11a. Mob. Netw. Appl. **10**, 705–716 (2005)
10. Jafarian, J., Hamdi, K.A.: Analysis of the exposed node problem in IEEE 802.11 wireless networks. In: Proceedings of the PGNet. School of Electrical and Electronic Engineering, University of Manchester, UK (2010)
11. Kapadia, V.V., Patel, S.N., Jhaveri, R.H.: Comparative study of hidden node problem and solution using different techniques and protocols. J. Comput. **2**(3), (2010)

Application of Firefly Algorithm for AGC Under Deregulated Power System

Tulasichandra Sekhar Gorripotu, Rabindra Kumar Sahu and Sidhartha Panda

Abstract In this paper, Proportional–Integral–Derivative controller with derivative Filter (PIDF) is proposed for Automatic Generation Control (AGC) problem of four area reheat thermal power systems under deregulated environment by considering the physical constraints such as Generation Rate Constraint (GRC) and Governor Dead Band (GDB) nonlinearity. The system is investigated in all possible scenarios under deregulated environment. The gains of the controllers are optimized using an Integral of Time multiplied by Absolute value of Error (ITAE) criterion employing of Firefly Algorithm (FA).The performance of some diverse classical controllers such as Integral (I), Proportional–Integral (PI) and PIDF controllers are compared under poolco based scenario. Simulation results reveal that the performance of the system is better with PIDF controller compared to others.

Keywords Automatic generation control (AGC) · Firefly algorithm (FA) · Generation rate constraint (GRC) · Governor dead band (GDB) · Deregulated

1 Introduction

Automatic Generation Control (AGC) plays an significant role in the large scale electric power systems with interconnected areas. The AGC is aimed to ensure the system frequency of each area and the inter-area tie line power within tolerable limits to deal with the fluctuation of load demands and system disturbances [1, 2].

T.S. Gorripotu (✉) · R.K. Sahu · S. Panda
Department of Electrical Engineering, Veer Surendra Sai University of Technology (VSSUT), Burla 768018, Odisha, India
e-mail: gtchsekhar@gmail.com

R.K. Sahu
e-mail: rksahu123@gmail.com

S. Panda
e-mail: panda_sidhartha@rediffmail.com

© Springer India 2015
L.C. Jain et al. (eds.), *Computational Intelligence in Data Mining - Volume 1*,
Smart Innovation, Systems and Technologies 31, DOI 10.1007/978-81-322-2205-7_62

These important functions are delegated to AGC due to the fact that a well-designed power system should keep voltage and frequency in scheduled range while supplying an acceptable level of power quality [3].

In restructured environment, GENCOs may or may not participate in the AGC task as they are independent power utilities. On the other side, DISCOs may contract with any GENCOs for power in different areas. Hence, in restructured environment, control is greatly decentralized and independent system operators (ISOs) are responsible for maintaining frequency oscillations and tie-line power flows [2, 3]. Several attempts have been made in recent past to study AGC issues in deregulated environment. Donde et al. [4] presented simulation and optimization in an AGC system after deregulation. They have demonstrated the concept of restructured power system and DISCO participation matrix (DPM). However, they have not dealt with reheat turbine, GRC in their work. Bhatt et al. [5] have presented the AGC problem in four area power system under deregulation. Hybrid particle swarm optimization is used to obtain optimal gains of PID controller. The authors have not considered the important physical constraints such as Generation Rate Constraint (GRC) and Governor Dead Band (GDB) in the system model which affect performance of the power system. Thus, to study the realistic power system, it is necessary to include the GRC and GDB nonlinearity [6, 7]. Therefore, this paper presents a comprehensive study on dynamic performance of reheat thermal deregulated power system by considering physical constraints such as GRC and time delay.

2 Material and Method

2.1 Power System Under Study

The study has been carried out on four control areas having different power capacities of 5,000, 2,000, 6,000 and 4,000 MW respectively. The schematic diagram of the system under study shown in Fig. 1. The system consists of four control areas in which each area has different combinations of GENCOs and DISCOs. Area 1 comprises of two reheat thermal power systems and two DISCOs, Area 2 consists of one reheat thermal power systems and two DISCOs, Area 3 comprises of three GENCOs with all reheat thermal power systems and two DISCOs, Area 4 comprises of two reheat thermal power systems and two DISCOs. The system is widely used in literature for the design and analysis of automatic load frequency control under deregulated power system [5]. The block diagram of a GENCO is shown in Fig. 2. In Fig. 2, Δu_i is control input of the power system of area i in p.u., ΔR_i is regulation parameter of area i in p.u. Hz, T_{Gi} is speed governor time constant of area i in sec, T_{Ti} is turbine time constant of area i in sec, K_{Ri} is steam turbine reheat constant of area i and T_{Ri} is steam turbine reheat time constant of area i in sec, where $i = 1,2,3,4$. In the present study, a GRC of 3 %/min and GDB of 0.036 Hz are considered. The relevant parameters are given in Appendix.

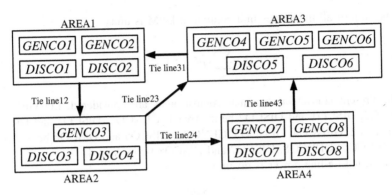

Fig. 1 Schematic diagram of a system under deregulation

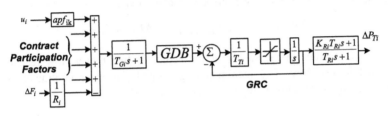

Fig. 2 Block diagram of a GENCO

2.2 *Power System Under Deregulation*

The electric power system has over the years been dominated by large utilities that had an overall activity in generation, transmission and distribution of power, its domain of operation is known as Vertically Integrated Utility (VIU). With emerge of deregulation environment the electric power industry is changing from a structure of regulated local vertically integrated organizations to one which competitive companies or Independent Power Producers (IPPs) generates electricity while the utilities maintain transmission and distribution networks. Thus, in deregulated environment generation, transmission and distribution is treated as individual sections. As there are several GENCOs and DISCOs in the deregulated environment, there can be various contracts between GENCOs and DISCOs. To know the contracts between GENCOs and DISCOs the concept of DISCO participation matrix (DPM) is introduced [4].

DPM is a matrix having no. of rows as no. of GENCOs and no. of columns as no. of DISCOs. The elements of DPM are indicated with cpf_{kl} which corresponds to fraction of total load contracted by a DISCO towards a GENCO.

The sum of all the entries in a column in DPM is unity

$$\sum_{k}^{n} cpf_{kl} = 1 \tag{1}$$

In the present study, four unequal control areas are considered. Let GENCO1, GENCO2, DISCO1 and DISCO2 be in Area 1, GENCO3, DISCO3 and DISCO4 be in Area 2, GENCO4, GENCO5, GENCO6, DISCO5 and DISCO6 be in Area 3 and GENCO7, GENCO8, DISCO7 and DISCO8 be in Area 4.

The actual tie-line power is given as

$$\Delta P_{Tie,ij,actual} = \frac{2\pi T_{ij}}{s} \left(\Delta F_i - \Delta F_j \right) \tag{2}$$

At any time, the tie-line power error between Area 1 and Area 2 is given by

$$\Delta P_{Tie12,error} = \Delta P_{Tie12,actual} - \Delta P_{Tie12,schedule} \tag{3}$$

$\Delta P_{Tie12,error}$ vanishes in the steady as the actual tie-line power flow reaches the scheduled power flow. The incremental generated power or contracted power supplied by the GENCOs is given as

$$\Delta P_{gk} = \sum_{k=1}^{8} cpf_{kl} P_{Ll} \tag{4}$$

This error signal is used to generate the respective Area Control Error (ACE) signals as in the traditional scenario. The $ACEs$ are defined as follows for the system considered for this particular study:

$$ACE_i = B_i \Delta F_i + \Delta P_{Tie,ij}^{error} \tag{5}$$

As there are two GENCOs in each area, ACE signal has to be distributed among them in proportion to their participation in the LFC. Coefficients that distribute ACE to GENCOs are termed as "ACE Participation Factors ($apfs$)".

2.3 Control Structure and Objective Function

The proportional integral derivative controller (PID) is the most popular feedback controller used in the process industries. It is a robust, easily understood controller that can provide excellent control performance despite the varied dynamic characteristics of process plant. PID controllers are used when stability and fast response are required. Derivative mode improves stability of the system and enables increase in proportional gain and decrease in integral gain which in turn increases speed of the controller response. However, when the input signal has sharp corners, the derivative term will produce unreasonable size control inputs to the plant. Also, any noise in the control input signal will result in large plant input signals. These reasons often lead to complications in practical applications. The practical solution to the these problems is to put a first filter on the derivative term and tune its pole so that the chattering due to the noise does not occur since it attenuates high frequency noise. In view of the above a filter is used for the derivative term in the present paper.

In the design of a modern heuristic optimization technique based controller, the objective function is first defined based on the desired specifications and constraints. Integral of Time multiplied Absolute Error (ITAE) criterion reduces the settling time which cannot be achieved with Integral of Absolute Error (IAE) or Integral of Squared Error (ISE) based tuning. ITAE criterion also reduces the peak overshoot. Integral of Time multiplied Squared Error (ITSE) based controller provides large controller output for a sudden change in set point which is not advantageous from controller design point of view. It has been reported that ITAE is a better objective function in LFC studies [8]. Therefore in this paper ITAE is used as objective function to optimize parameters of fuzzy PID controller. Expression for the ITAE objective function is depicted in Eq. (6).

$$J = ITAE = \int_{0}^{t_{sim}} (|\Delta F_1| + |\Delta F_2| + |\Delta F_3| + |\Delta F_4| + |\Delta P_{Tie12}| + |\Delta P_{Tie23}|$$

$$+ |\Delta P_{Tie43}| + |\Delta P_{Tie31}| + |\Delta P_{Tie24}|) \cdot t \cdot dt \qquad (6)$$

3 Simulation Results and Discussion

3.1 Implementation of Firefly Algorithm

FA is controlled by three parameters: the randomization parameter α, the attractiveness β, and the absorption coefficient γ. These parameters are generally chosen in the range 0–1. In the present study the controlled parameters taken as number of fireflies = 4; maximum generation = 100; $\beta = 0.9$; $\alpha = 0.2$ and $\gamma = 0.4$. The flow chart of proposed FA approach is clearly explained in [9, 10].

Initially, the performance of the system is studied under poolco based transaction by taking some diverse classical controllers. In the present work, the minimum and maximum values of PID controller parameters are chosen as −2.0 and 2.0 respectively. The range for filter coefficient N is selected as 10–300. The optimum values of the both the cases given in Table 1. The objective function (ITAE) value given by Eq. (6) is determined by simulating the developed model by applying a 1 % step increase in load in all unequal areas. The corresponding performance index in terms of ITAE value, and settling times (2 %) in frequency and tie line

Table 1 Tuned controller parameters under different scenarios

Controller parameters		Poolco based			Bilateral	Contract violation
		I	PI	PIDF	PIDF	PIDF
Area 1	K_{P1}	–	0.2800	−1.8749	−1.0060	−1.0379
	K_{I1}	−0.4010	−0.4262	−0.7531	−0.9148	−0.5695
	K_{D1}	–	–	−1.6023	−0.8145	−1.5969
	N_1	–	–	10.0005	10.1126	10.0066
Area 2	K_{P2}	–	0.2800	−0.7837	−1.1707	−0.8369
	K_{I2}	−0.4888	−0.2283	−1.4836	−0.3245	−0.9885
	K_{D2}	–	–	−1.4679	−0.9422	−0.6985
	N_2	–	–	10.0005	10.1126	10.0066
Area 3	K_{P3}	–	0.2800	−1.6027	−1.3853	−0.6787
	K_{I3}	−0.2290	−0.2001	−0.7072	−0.3579	−0.3260
	K_{D3}	–	–	−1.2633	−1.0745	−1.1265
	N_3	–	–	10.0005	10.1126	10.0066
Area 4	K_{P4}	–	0.2800	−0.8932	−0.7502	−0.8441
	K_{I4}	−0.3339	−0.5641	−0.3087	−0.9292	−0.2847
	K_{D4}	–	–	−1.8499	−0.9993	−1.5716
	N_4	–	–	10.0005	10.1126	10.0066

power deviations is shown in Table 2. From Table 2 it is observed that PIDF controller gives better performance indexes. So, for further system study only PIDF controller is considered.

3.2 Poolco Based Scenario

In this case the DISCOs having the contract with GENCOs of the same area. It is assumed that the load disturbance occurs in all areas. There is 0.005 (p.u. MW) load disturbance of each DISCOs and as result of the total load disturbance in areas is 0.01 (p.u. MW). In the case of Poolco based contracts between DISCOs and available GENCOs is simulated based on the following DPM:

$$DPM = \begin{bmatrix} 0.5 & 0.5 & 0 & 0 & 0 & 0 & 0 & 0 \\ 0.5 & 0.5 & 0 & 0 & 0 & 0 & 0 & 0 \\ 0 & 0 & 1.0 & 1.0 & 0 & 0 & 0 & 0 \\ 0 & 0 & 0 & 0 & 0.3 & 0.25 & 0 & 0 \\ 0 & 0 & 0 & 0 & 0.4 & 0.5 & 0 & 0 \\ 0 & 0 & 0 & 0 & 0.3 & 0.25 & 0 & 0 \\ 0 & 0 & 0 & 0 & 0 & 0 & 0.5 & 0.6 \\ 0 & 0 & 0 & 0 & 0 & 0 & 0.5 & 0.4 \end{bmatrix}$$

Table 2 Performance index values under different scenarios

Parameters		Poolco based			Bilateral	Contract violation
		I	PI	PIDF	PIDF	PIDF
ITAE		3.27	2.79	1.01	42.37	43.60
Settling time (s)	ΔF_1	29.26	21.51	9.18	7.80	7.47
	ΔF_2	30.42	25.09	14.08	7.16	10.74
	ΔF_3	26.68	23.34	9.8	5.51	5.99
	ΔF_4	25.31	20.14	15.38	6.68	8.28
	ΔP_{Tie12}	16.77	14.77	10.87	06.30	12.78
	ΔP_{Tie23}	16.97	17.29	9.5	08.80	21.38
	ΔP_{Tie43}	20.21	17.24	14.37	100.0	100.0
	ΔP_{Tie31}	11.34	11.6	3.54	9.06	6.94
	ΔP_{Tie24}	14.66	9.43	7.75	100.0	100.0
Peak over shoot ($\times 10^{-2}$)	ΔF_1	1.58	1.6	1.36	1.5	2.4
	ΔF_2	2.12	02.4	1.70	1.8	2.5
	ΔF_3	2.83	2.7	1.54	1.66	2.7
	ΔF_4	1.71	2.0	1.09	1.6	2.4
	ΔP_{Tie12}	0.18	0.1	0.13	0.1	0.2
	ΔP_{Tie23}	0.22	0.2	0.14	0.3	0.5
	ΔP_{Tie43}	0.51	0.5	0.33	0.4	1.0
	ΔP_{Tie31}	0.21	0.1	0.03	0.05	0.05
	ΔP_{Tie24}	0.22	0.22	0.17	0.20	0.29

Consider that the *ACE* participation of GENCOs as $apf_{11} = apf_{21} = 0.6$, $apf_{12} = apf_{22} = 0.3$, $apf_{13} = apf_{23} = 0.1$. The Fig. 3 shows the dynamic response of the system under poolco based transaction. From, Fig. 3 it is clear that PIDF controller gives better performance.

3.3 Bilateral Based Scenario

In this case, DISCOs have the freedom to contract with any of the GENCOs within own control area or with another control area. Now the DISCO participation matrix to be considered as

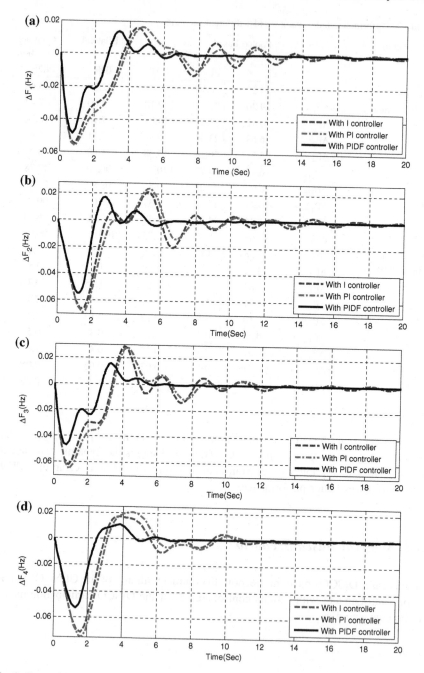

Fig. 3 Dynamic responses of the system under poolco based scenario **a** frequency deviation in area 1, **b** frequency deviation in area 2, **c** frequency deviation in area 3, **d** frequency deviation in area 4

$$DPM = \begin{bmatrix} 0.2 & 0.3 & 0.1 & 0.1 & 0.1 & 0.1 & 0 & 0 \\ 0.4 & 0.3 & 0.1 & 0.2 & 0.1 & 0.1 & 0 & 0 \\ 0.1 & 0.1 & 0.3 & 0.2 & 0 & 0.1 & 0.1 & 0.1 \\ 0.1 & 0.1 & 0.1 & 0.1 & 0.2 & 0.2 & 0.1 & 0.1 \\ 0.1 & 0.1 & 0.1 & 0.1 & 0.2 & 0.2 & 0.1 & 0.1 \\ 0.10 & 0.1 & 0.1 & 0.1 & 0.2 & 0.2 & 0.2 & 0.1 \\ 0 & 0 & 0.1 & 0.1 & 0.1 & 0 & 0.2 & 0.3 \\ 0 & 0 & 0.1 & 0.1 & 0.1 & 0.1 & 0.3 & 0.3 \end{bmatrix}$$

A 1 % of load disturbance is applied in all areas. The performance of the system is observed by using PIDF controller as it is found superior in poolco based scenario. The Fig. 4 shows the dynamic response of the system under this scenario.

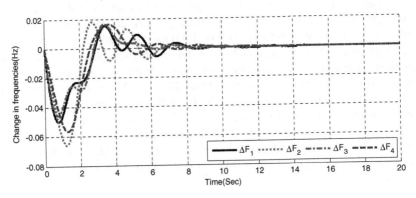

Fig. 4 Dynamic responses of the system under bilateral based scenario

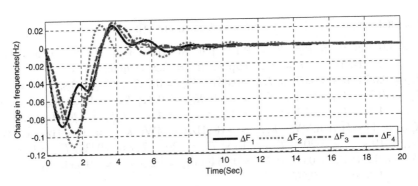

Fig. 5 Dynamic responses of the system under contract violation based scenario

3.4 Contract Violation Based Scenario

It may happen that a DISCOs may violate a contract by demanding more than that specified in the contract. This excess power is not contracted out to any GENCO. Consider, bilateral scenario again with a modifications that, DISCO1 and DISCO5 demands 0.005 p.u. MW excess of power. As a result the load disturbances of area 1 and area 3 are increases to 0.015 p.u. MW. The Fig. 5 shows the dynamic response of the system under this scenario.

4 Conclusion

In this paper, Firefly Algorithm (FA) is proposed to tune the different classical controllers such as Integral (I), Proportional-Integral (PI) and Proportional-Integral-Derivative with derivative filter coefficient (PID) for Automatic Generation Control (AGC) problem. A four area eight unit reheat thermal power system under deregulated environment is considered to demonstrate the proposed method and the physical constraints such as Governor Dead Band (GDB) and Generation Rate Constraint (GRC) are considered to show the ability of the proposed approach to handle nonlinearity in the system model. From the simulation results, it is observed that significant improvements of dynamic performance of the system in terms Integral of Time multiplied by Absolute value of Error (ITAE), settling time and peak overshoot are obtained under different transactions with PIDF controller.

Appendix

Nominal parameters of the system investigated are [5]:

$F = 60$ Hz; $B_1 = B_2 = B_4 = 0.125$ p.u. MW/Hz; $B_3 = 0.275$ p.u. MW/Hz; $R_1 = 2.4$ Hz/p.u.; $R_2 = 2.2$ Hz/p.u.; $R_3 = 2.5$ Hz/p.u.; $R_4 = 2.3$ Hz/p.u.; $T_{G1} = 0.08$ s; $T_{G2} = 0.078$ s; $T_{G3} = 0.081$ s; $T_{G2} = 0.082$ s; $T_{T1} = 0.3$ s; $T_{T2} = 0.5$ s; $T_{T3} = 0.7$ s; $T_{T1} = 0.4$ s; $K_{R1} = 0.34$; $K_{R2} = 0.31$; $K_{R3} = 0.32$; $K_{R4} = 0.33$; $T_{R1} = 4.2$ s; $T_{R2} = 4.1$ s; $T_{R3} = 4.0$ s; $T_{R4} = 4.3$ s; $K_{ps1} = 120$ Hz/p.u.MW; $K_{ps2} = 115$ Hz/p.u.MW; $K_{ps3} = 118$ Hz/p.u.MW; $K_{ps4} = 116$ Hz/p.u.MW; $T_{ps1} = 10$ s; $T_{ps2} = 20$ s; $T_{ps3} = 10$ s; $T_{ps4} = 20$ s; $T_{12} = 0.0231$; $T_{24} = 0.0231$; $T_{23} = 0.0231$; $T_{31} = 0.0231$; $T_{43} = 0.0549$; $a_{12} = -2.5$; $a_{23} = -0.333$; $a_{31} = -1.2$; $a_{24} = -0.5$; $a_{43} = -0.667$.

References

1. Elgerd, O.I.: Electric Energy Systems Theory—An Introduction. Tata McGraw Hill, New Delhi (2000)
2. Bevrani, H.: Robust Power System Frequency Control. Springer, Berlin (2009)
3. Bervani, H., Hiyama, T.: Intelligent Automatic Generation Control. CRC Press, Boca Raton (2011)
4. Donde, V., Pai, M.A., Hiskens, I.A.: Simulation and optimization in an AGC system after deregulation. IEEE Trans. Power Syst. **16**, 481–489 (2011)
5. Bhatt, P., Roy, R., Ghoshal, S.P.: Optimized multi area AGC simulation in restructured power systems. Int. J. Electr. Power Energy Syst. **32**(4), 311–332 (2010)
6. Sahu, R.K., Panda, S., Rout, U.K.: DE optimized parallel 2-DOF PID controller for load frequency control of power system with governor dead-band nonlinearity. Int. J. Electr. Power Energy Syst. **49**(1), 19–33 (2013)
7. Saikia, L.C., Mishra, S., Sinha, N., Nanda, J.: Automatic generation control of a multi area hydrothermal system using reinforced learning neural network controller. Int. J. Electr. Power Energy Syst. **33**(4), 1101–1108 (2011)
8. Shabani, H., Vahidi, B., Ebrahimpour, M.: A robust PID controller based on imperialist competitive algorithm for load-frequency control of power systems. ISA Trans. **52**(1), 88–95 (2013)
9. Yang, X.S.: Nature-Inspired Metaheuristic Algorithms. Luniver Press, UK (2008)
10. Chandrasekaran, K., Simon, S.P., Padhy, N.P.: Binary real coded firefly algorithm for solving unit commitment problem. Inf. Sci. **249**, 67–84 (2013)

Base Station Controlled Spectrum Allocation Technique to Detect the PUE Attack in CWSN

Pinaki Sankar Chatterjee and Monideepa Roy

Abstract Primary User Emulation (PUE) attack is a type of Denial-of-Service (DoS) attack which is commonly faced by Cognitive Wireless Sensor Networks (CWSNs). In CWSNs, malicious secondary users try to emulate primary users to maximize their own spectrum usage or to obstruct secondary users from accessing the spectrum. In this paper we have proposed a base station controlled spectrum allocation protocol for the secondary users to deal with PUE attacks in CWSNs. We have used the well-known lightweight hash function SHA-1 for the authentication process of a secondary user to the base station. We demonstrate that our authentication protocol can be implemented efficiently on CWSN nodes. Our experimental results show that the Base Station controlled Primary User Emulation Attack performs well in CWSN scenario.

Keywords Cognitive wireless sensor network · Hash function · Authentication · PUE · Dos

1 Introduction

Cognitive Wireless Sensor Network is a simple Wireless Sensor Network with Cognitive capabilities [1, 2]. Sensor nodes normally transmit data through the ISM band. But nowadays the ISM band is used by many other devices like Wi-fi, Bluetooth etc. [3]. That is why sometimes the ISM band becomes overcrowded. In such situation the CWSN nodes search for vacant channels in other spectrums. These are called white bands. A CWSN node places its own data in the white band. But a CWSN node has to vacate the channel when the original licensed owner of the

P.S. Chatterjee (✉) · M. Roy
Kalinga Institute of Industrial Technology, Bhubaneswar, India
e-mail: pinaki.sankar.chatterjee@gmail.com

M. Roy
e-mail: monideepa.roy@gmail.com

© Springer India 2015
L.C. Jain et al. (eds.), *Computational Intelligence in Data Mining - Volume 1*,
Smart Innovation, Systems and Technologies 31, DOI 10.1007/978-81-322-2205-7_63

spectrum wants to transmit through that channel. In such a scenario the original licensed owner of the spectrum is called a Primary User (PU) and the CWSN node is called the Secondary User (SU). This method of spectrum sharing is called Opportunistic Spectrum Sharing (OSS) [4].

The benefit of the OSS technique can be enjoyed by the user only when the network can be properly deployed and the security threats of such networks can be robustly handled. To ensure the trustworthiness of the network the spectrum sensing process is an important problem that needs to be addressed. The key to address this problem is being able to distinguish the PU signal from SU signals in a robust way. Otherwise it can led to a DoS attack in the network [5].

Suppose in a hostile environment a subset of SU forges the essential signal characteristics of the PU and generates enough power at the good SU location. This activity generally confuses other SUs to think that a PU transmission is under way. Such an attack is called a Primary User Emulation (PUE) attack [4, 6, 7]. Depending on the intention of the attacker this attack can be categorized as either Selfish PUE attack or Malicious PUE attack [4]. The task of distinguishing incumbent signals from secondary user signals becomes an even greater challenge when we consider the requirement described in FCC's NPRM 03–322 [8]. It says that there will be no modification required on the existing system to accommodate Opportunistic Spectrum Sensing. We have proposed a protocol which is based on the SU authentication to the base station. It doesn't need any positional information of the PU or the SU and does not interfere with the PU's communication process. The main objective here is to identify PUE attack without interfering with the Primary User's communication process to facilitate Opportunistic Spectrum Sensing.

2 Related Work

In the last few years some researches related to security on CRNs have appeared. But most of their studies are related to PUE detection technique.

PUE attacks in the CWSNs have been studied in [3–5, 9]. In [3] the authors have proposed a system to detect PUE using anomaly detection. Here Cognitive nodes sense the spectrum and create neighbor profiles to model their behavior. This information is used to inform other nodes about anomalous data. In [4] the authors proposed PUE identification using transmitter verification procedure. They need to know the position of the transmitter. They proposed two different location verification schemes namely distance ratio test (DRT) and distance difference test (DDT) which use the ratio and the difference, respectively, of the distances of the primary and malicious transmitters from the secondary user to detect a PUEA. In [9] the authors assume that the attacker is close to the victim and the real PU is much farther from the SUs than the attacker. Moreover, the position of each node, including that of the attacker, is fixed. It is assumed that the SUs can learn about the characteristics of the spectrum according to the received power. In [5] the authors proposed an analytical model of a lower bound on the probability of a successful PUEA on a secondary user in a cognitive radio network by a set of co-operating malicious users. They proposed that the probability of a successful PUEA increases with the distance between the primary transmitter and secondary users.

All the above approaches are based on the location of the Primary and Secondary User. In case the Primary and Secondary Users are mobile, these approaches are infeasible. In [10] the authors present a differential game approach to mitigate the PUE attack. Based on the assumption that a PUE attacker has lesser energy than the PUs, they look for the optimal sensing strategy of SU. The Nash equilibrium solution is obtained. Though this approach is suitable for mobile users the algorithm implementation requires extensive computational capabilities. This requirement is difficult to fulfil for sensor nodes.

In this paper we have proposed a lightweight hash function based authentication protocol for mobile Secondary User within a grid to identify Primary User Emulation Attack in Cognitive Wireless Sensor Network.

3 The Proposed System

CWSNs can be constructed in many different ways depending on the application for which it is being used. In our network architecture we have broadly two types of users, primary user (PU) and secondary user (SU). Additionally, a base station (BS) is a special type of node with computationally rich entities which control spectrum allocation process of SU. According to FCC rules [8] we cannot modify the PU's communication process. So in order to detect a PUEA we have proposed a base station controlled spectrum allocation process for SU. The base station maintains tokens for each spectrum band and a spectrum allocation table. The table maintain information about which SU is using which spectrum band and holding which token. The fields for the table are SU's ID, Spectrum band using, Token assign, Token returned.

Figure 1 shows the flowchart of the entire process of PUE attack detection. The base station first authenticates the SU by executing a lightweight hash function. When the SU is authenticated then the base station sends a token for that spectrum band to the SU as acknowledgement and updates its table. After getting the token, the SU can use the spectrum band. When a SU senses the PU signal on that channel then it leaves the channel and returns the token to the base station. The base station again executes the authentication process. It matches the returned token with the assigned SU's ID and updates the table. In this process by seeing the table the base station can keep track of the entire spectrum allocation among the SUs.

When the protocol marks a node as a possible PUE attacker, it sends a message through the Virtual Control Channel (VCC), a method for sharing information in cognitive networks [3]. The spectrum sensing process of the SU and the process of deciding potential PU signals at the base station is out of scope of this paper. Our focus is only to detect a PUEA in the network.

Fig. 1 The PUE attack
detection protocol

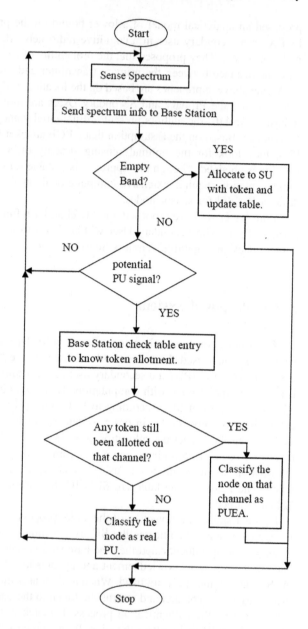

4 The Authentication Protocol

The Authentication Protocol is explained in detail in this section (Fig. 2).

4.1 Assumptions

Before proceeding further we present the system assumptions. These are:

- We assume that there will be a globally unique Identity available for each SU in the network. This unique ID will be allocated for each SU at setup time. This Identity of SU is known to the base station.
- We assume that each SU has it's own secret key (Sn). This key will be allocated to each SU at setup time. This secret key will be known only to the SU and the base station. We also assume that this secret key is stored in a tamper-resistant section of the SU as well as the base station.
- We assume that to avail of a spectrum band and to leave a spectrum band a SU will have to compulsorily authenticate itself to the Base Station. Without that no OSS facility is available for that sensor node.
- We assume that we can rely on the base station. This means the base station will never be compromised and its functionality will always be correct.
- We assume that the same hash function code is programmed for all the SUs and the base station.

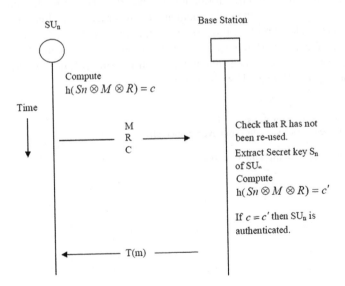

Fig. 2 Authentication of SU_n to base station

4.2 *The Steps of the Authentication Protocol*

When a Secondary User (SU_n) wants to gain access over the PU's spectrum band it has to prove its Identity to the base station as explained in Fig. 2.

- We assume that S_n, M and R are adequate in size to become inputs for our hash function h. where M is a message containing the ID of the SU and the channel it wants to use. R is the local time obtained from the LocalTime.get() command in TinyOS.
- Each SU_n XORs the secret Sn, the message M and the local time R to get x.
- SU_n computes $h(x) = C$.
- It transmits M, R and C to the Base Station.
- Base Station recomputes h(XOR of S_n, M and R) and checks whether it is C or not.
- If the value is true AND the R value has not been used for the authentication earlier then the Su_n is authenticated to the base station.
- Then base station will issue token (T_m) of spectrum band m to SU_n and update the table.

A simple hash function based authentication is been done here. For our experiment we have used SHA-1 as a lightweight hash function to run on the base station. Our base station is comparably richer in computational power than a sensor node.

The entire process is explained through three algorithms namely Spectrum_Sensing, PUE_Detector and Authenticate_node. The Spectrum_Sensing algorithm sends the sensed information to the PUE_Detector algorithm. If the spectrum band is free then the authentication process is done through the Authenticate_node algorithm. If the returned result is true then the spectrum band is assigned to the node. If the spectrum band is not free then the PUE_Detector algorithm checks whether the signal in the spectrum band is Emulated Primary User Signal or original Primary User Signal.

```
Algorithm Spectrum_Sensing( ) {
    for (Sense each spectrum after every t interval)
    {
      PUE_Detector (Spectrum Information)
    }
}
```

Algorithm PUE_Detector(Spectrum Information) {
If (Spectrum Band = = Empty) *then*
{

    Declare Empty band in the network.
    auth = Authenticate_node()
    If (auth = = TRUE) *then*

    {
    The SU is authenticated and the token for that band is assign to it.
    PUE_Detector then update the token allocation table.
    }
}

Else
{

    If (signal_detected = = Potential PU Signal) *then*

    {
    Declare the signal as PU signal in the network.
    If (still any token allotted for that channel) *then*
        {
        Classify the SU holding the token as PUEA
        }
    Else
        {
        Classify the signal as original PU signal
        }
    }
}

Algorithm Authenticate_node() {
SU execute the hash function with it's secret key and send the message digest (c) to the Base Station. Base Station retrieve the secret key of the SU and again execute the hash function to get a message digest (c') .
    If (c = = c') *then*
    {
    return TRUE
    }
    Else
    {
    return FALSE
    }
}

5 Result and Discussion

Our Base Station is a more computationally rich system than a CWSN node. At the SU side we have implemented the SHA-1 hash function [11], on TinyOS version 2.1.2. We take data input of 64 bit for SHA-1 to generate a 160 bit Message Authentication Code (MAC). This algorithm consumes 128 bytes of RAM, 4,000 bytes of ROM and approximately 7.2 ms of Execution time.

The scenarios have been executed in Castalia 3.2 to extract results

 (i) The scenario area is a 20×20 set of squares.
 (ii) The complete simulation time is 100 s.
(iii) The number of attacker nodes in the simulation is 20.
(iv) Sensor node sends 1 packet/s.
 (v) The attack starts at the beginning of the simulation.
(vi) The sensor nodes sense each channel for 50 ms.

In the simulation we have found that the execution time of the entire protocol takes about 4 s to identify PUEA. These 4 s include the HASH function execution by SU and the base station, Token allocation, PU signal emulation, PU signal sensing by other SUs, sending those information to the base station and PUEA detection by the base station. In Fig. 3 we found that two PUE attackers attack before authentication at time = 0 s. Till now we cannot handle them. At time = 2 to 4 s another 3 attacks happen. Those attacks have been detected at time = 6 to 8 s. From time = 10 s, another 4 attacks happen and so on.

The graph in Fig. 4 represents the average case scenario for 20 different attackers.

Figure 4 represents the best case scenario where all the attacks happen at time = 0 s and are solved at time = 4 s. Like WSN the attacks on CWSN are also broadly classified into passive attacks and active attacks. In a passive attack the attacker is able to interpret the information passing through the network. In an active attack the integrity and the availability of the network are disturbed. Our protocol is not designed to protect against passive attacks.

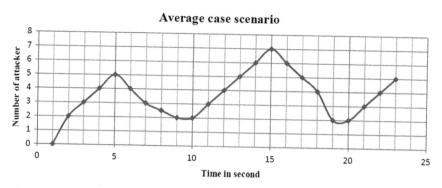

Fig. 3 Average case scenario

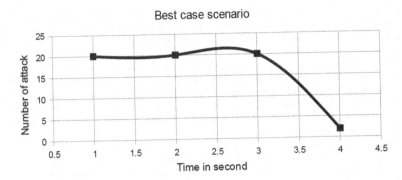

Fig. 4 Best case scenario

6 Conclusion

In this paper we have proposed a Base Station Controlled spectrum allocation technique to detect the Primary User Emulation attack in a Cognitive Wireless Sensor Network. The technique uses a lightweight hash function based authentication system of Secondary User to the Base Station. The process of handling the situation when the SU acts as PU before sending the authentication to the base station is kept as future work.

References

1. Haykin, S.: Cognitive radio: brain-empowered wireless communications. IEEE J. Sel. Areas Commun. **23**(2), 201–220 (2005)
2. Mitola, J.: Cognitive radio: an integrated agent architecture for software defined radio, Ph.D. Dissertation, Royal Institute of Technology (KTH), Stockholm, Sweden (June 2000)
3. Blesa, J., Romero, E., Rozas, A., Araujo, A.: PUE attack detection in CWSNs using anomaly detection techniques. Proc. EURASIP J. Wirel. Commun. Netw. **2013**(1), 1–13 (2013)
4. Chen, R., Park, J.-M.: Ensuring Trustworthy Spectrum Sensing in Cognitive Radio Networks. In: Proceedings of the 1st IEEE Workshop on Networking Technologies for Software Defined Radio Networks (SDR'06), pp. 110–119 (2006)
5. Anand, S., Jin, Z., Subbalakshmi, K.P.: An Analytical Model for Primary User Emulation Attacks in Cognitive Radio Networks. In: proceedings of 3rd IEEE Symposium on New Frontiers in Dynamic Spectrum Access Networks, pp. 1–6 (2008)
6. Chen, R., Park, J.M., Reed, J.H.: Defense against primary user emulation attacks in cognitive radio networks. IEEE J. Sel. Areas Commun. **26**(1), 25–37 (2008). (Special Issue on Cognitive Radio Theory and Applications)
7. Chen, R., Park, J.M., Bian, K.: Robust distributed spectrum sensing in cognitive radio networks. In: Proceedings of IEEE Conference on Computer Communications (INFOCOM) 2008 Mini-conference (2008)
8. Federal Communication Commission: Notice for proposed rulemaking (NPRM 03–22), facilitating opportunities for flexible, efficient, and reliable spectrum use employing cognitive radio technologies (2003)

9. Chen, Z., Cooklev, T., Chen, C., Pomalaza-Ráez, C.: Modeling primary user emulation attacks and defenses in cognitive radio networks. In: Proceedings of the IEEE 28th International Performance Computing and Communications Conference (IPCCC'09), pp. 208–2015 (Dec 2009)

10. Hao, D., Sakurai, K.: A differential game approach to mitigating primary user emulation attacks in cognitive radio networks. In: Proceedings of the IEEE 26th International Conference on Advanced Information Networking and Applications (AINA'12), pp. 495–502 (Mar 2012)

11. Li, B., Batten, L.M., Doss, R.: Lightweight Authentication for Recovery in Wireless Sensor Networks. In: Proceeding of 5th International Conference on Mobile Ad-hoc and Sensor Networks, pp. 465–471 (2009)

12. Sen, J.: Security and privacy challenges in cognitive wireless sensor networks. In: Cognitive Radio Technology Applications for Wireless and Mobile Ad hoc Networks (2013)

13. Shi, E., Perrig, A.: Designing Secure Sensor Networks. Proc. IEEE Wirel. Commun. 11(6), 38–43 (2004)

14. Araujo Pinto, A., Romero, E., Blesa, J., Nieto-Taladriz, O.: A framework for the design, development and evaluation of cognitive wireless sensor networks. Proc. Int. J. Adv. Telecommun. 5(3–4), 141–152 (2012)

Object Based Image Steganography
with Pixel Polygon Tracing

Ratnakirti Roy and Suvamoy Changder

Abstract The paper presents an object based image steganography technique which uses pixel polygonal area tracing in a cover image to select suitable pixels for embedding secret data. The polygon is generated using convex hull for specific selected image pixels and the distortion function is a hybrid between a high efficiency embedding scheme and LSB matching. The proposed technique is simple, yet effective. Experimental results show that the proposed method exhibits high fidelity of the stego-image and performs decently against well-known spatial domain steganalysis techniques for moderate payload capacity.

Keywords Image steganography · Region of interest · Convex hull · Polygon edge detection · High efficiency embedding

1 Introduction

The advent of the internet has brought upon a massive change in our lives. Communicating electronically is now easier and faster than ever before and as a result, huge amount of data transfer takes place over the internet. The nature of data communicated via various web applications range from user specific personal data to highly sensitive confidential data such as military information. Despite the convenience offered by the plethora of services over the web, information over the internet is susceptible to eavesdropping, theft, alteration and *anonimization* [1]. A possible solution to such a problem is to render the data inaccessible to any

R. Roy (✉) · S. Changder
Department of Computer Applications, National Institute of Technology, Durgapur, India
e-mail: rroy.nitdgp@gmail.com

S. Changder
e-mail: suvamoy.nitdgp@gmail.com

© Springer India 2015
L.C. Jain et al. (eds.), *Computational Intelligence in Data Mining - Volume 1*,
Smart Innovation, Systems and Technologies 31, DOI 10.1007/978-81-322-2205-7_64

unintended recipient or adversary. A common approach to implement such a solution is to *pack* the message in an envelope whose contents are revealed only to the intended recipient. Data security schemes such as cryptography and steganography are extensively used for the purpose of creating such envelopes.

Cryptography renders the secret data illegible by applying encryption functions and encryption keys. Encrypting information for security is useful except in cases where the transmitted data over a channel is constantly under vigil and there are *de jure* restrictions on public use of cryptography. Data hiding techniques such as steganography proves viable in such situations. Steganography refers to techniques that hide data within innocuous objects such that the very existence of the secret remains concealed to an adversary. A typical stego-system is commonly portrayed using *Prisoner's Problem* [2] where two inmates are plotting an escape plan but their communication is under the surveillance of a warden who would put them into solitary confinement if they communicate secretly. A general steganography system also follows *Kerckhoff's Principle* [3] of cryptography and is mathematically expressed as a quintuple $Q = \langle C, S, E_k, D_k, K \rangle$ where C and S are the sets of cover and stego objects respectively, E_k and D_k are the embedding distortion function and the extraction functions respectively with K being the decoding key [4].

Most of the contemporary research on steganography focuses on the development of better embedding distortion functions and aims to maximize undetectibility while minimizing embedding distortion. Any typical steganography system is concerned with two vital questions—*where to embed in the cover* and *how to embed?* In the case of image steganography, where an image is used as a cover medium, it has been observed that certain areas in the image are more efficient for data hiding than the others. Such areas in an image vary depending on the pixel selection method. It is evident that distortion functions which perform sequential embedding are more susceptible to steganalysis attacks. A common approach is to use a Pseudo Random Number Generator (PRNG) to populate a probable pixel set for embedding. This method ensures non-sequential embedding but cannot guarantee uniform distribution of the selected pixels as it depends on the PRNG function and the seed. Image steganography systems using the PRNG for pixel selection generally send the seed value of the PRNG as the decoding key at the receiver end. The seed value may not be long enough to serve as a key and thus has greater chance of not surviving *Brute-Force* attacks [5].

There are many strategies for locating non sequential pixels in an image. Most of these exploit properties of the pixels to mark areas in the image suitable for embedding. In [6–8] the authors have used the *skin regions* in an image as the *region of interest* for embedding the secret data. Authors argue that skin pixels have less visible distortions when embedded with data and can be used for embedding secret information with relative ease. Similarly, pixels in the edge areas in an image are also suitable for embedding data because distortion in the edges are rarely perceivable and also provide secure non-sequential pixel set for embedding. Many such image steganography schemes have been proposed and can be found in [9–11]. Edge adaptive methods despite being accurate involve extra overhead of creating the edge region map. Similarly, skin region detection based techniques lack accuracy because of the fact that it is extremely difficult to build a generalized skin classifier based on pixel characteristics given the large variation of human skin tones found worldwide.

Apart from the aforesaid techniques, non-sequential pixel positions suitable for data hiding may also be traced by carving out a polygonal area surrounding a small number of pixel positions spread throughout the image or in a specific region. These pixel positions may be randomly selected by the sender either through a Graphical User Interface (GUI) or any other numerical technique. Once these pixels are found, the surrounding area can be carved out by generating a polygon which encompasses all these points thereby generating the required region of interest for embedding. The procedure is simpler as compared to the previous methods and also allows segmentation of objects in the cover. This paper presents an image steganography technique where a convex hull of a set of points on an image is used to derive a polygonal area encompassing those points and acting as the *region of interest* for the actual distortion function. The method provides a possible alternative to conventional target pixel finding techniques for non-sequential data embedding through a simple, yet effective approach.

2 Computing the Region of Interest Polygon

2.1 Generating the Polygon Vertices

The first step in generating the region of interest polygon is to select a set of points that will act as the *bounding area descriptor*. These points (pixels in case of images) can be chosen arbitrarily as per requirement either by mouse click coordinate capture or by specifying pixel positions satisfying certain criteria. Once the set of points is chosen, an optimal polygon can be generated to find the bounding area using the *convex hull* algorithm [12]. The method works as in Algorithm 1 and it returns the set S comprising of the vertices of the polygon.

Algorithm 1: Compute $CHull(Q)$

Input: Initial Set of Points Q
Output: Convex Hull vertex set S

Steps:
1.   Let p_0 be the point in Q with the minimum y- coordinate or the leftmost point.
2.   Let $< p_1, p_2, ..., p_m >$ be the remaining point in Q, sorted in counter-clockwise order around p_0.
3. $Top[S] \leftarrow p_0$
4. $Push(p_0, S)$
5. $Push(p_1, S)$
6. $Push(p_2, S)$
7.   *for* $i \leftarrow 3$ *to* m
8.   *do while* the angle formed by points *Next-Top(S)*, *Top(S)* and p_i makes a non-left turn
9.   *Pop(S)*
10.   *Push(p_i, S)*
11.   *end do while*
12.   *end for*
13.   **Return** S

2.2 Computing the Polygon Edge Pixels

Once the vertices of the convex polygon are derived, they can be used to generate the edge pixels between the vertices. For finding the edge pixels, the method described in Algorithm 2 is followed. The algorithm accepts two pixel co-ordinates (x_1, y_1) and (x_2, y_2) returns the intermediate pixels. The procedure works similar to *Brassenham's Line drawing algorithm* [13].

Algorithm 2: Compute $GetLine(x_1, y_1, x_2, y_2)$

Input: End point vertices (x_1, y_1, x_2, y_2)
Output: Intermediate pixel set P, Q

Steps:
1.   Let $\epsilon' \leftarrow 0, y \leftarrow y_1$
2. $\Delta x = x_2 - x_1, \Delta y = y_2 - y_1$
3. $for\ x \leftarrow x_1\ to\ x_2$
4. $P(x) \leftarrow x$
5. $Q(x) \leftarrow y$
6. $if\ (2(\epsilon' + \Delta y) < \Delta x)$
7. $\epsilon' \leftarrow \epsilon' + \Delta y$
8. $else$
9. $y \leftarrow y + 1$
10. $\epsilon' \leftarrow \epsilon' + \Delta y - \Delta x$
11. $end\ if$
12. $end\ for$
13. **Return** P,Q

3 Proposed Method

The proposed steganography technique uses the polygon edges found in Algorithm 2 to embed secret data. To minimize the effect of embedding, a hybrid approach of combining a high efficiency embedding scheme [14] with a pixel bit matching method [15] has been adopted. The proposed method comprises of two phases, namely, Phase I and Phase II. Phase I deals with the embedding process and Key generation while Phase II deals with the extraction of hidden message from the stego image.

3.1 Phase I

3.1.1 Embedding

The embedding procedure is a hybrid embedding scheme having both high efficiency embedding and *no LSB replacement* characteristics. The scheme can be explained using an example as follows:

Let a_1, a_2, a_3 be three bit positions from consecutive pixels and x_1 and x_2 be two message bits to be embedded. There can be four possibilities denoted by P_1, P_2, P_3, P_4 as:

$$P_1 : x_1 = a_1 \oplus a_3, x_2 = a_2 \oplus a_3 \Rightarrow change\ nothing \tag{1}$$

$$P_2 : x_1 \neq a_1 \oplus a_3, x_2 = a_2 \oplus a_3 \Rightarrow change\ pixel\ value\ p_1\ to\ match\ P_1 \tag{2}$$

$$P_3 : x_1 = a_1 \oplus a_3, x_2 \neq a_2 \oplus a_3 \Rightarrow change\ pixel\ value\ p_2\ to\ match\ P_1 \tag{3}$$

$$P_4 : x_1 \neq a_1 \oplus a_3, x_2 \neq a_2 \oplus a_3 \Rightarrow change\ pixel\ value\ p_3\ to\ match\ P_1 \tag{4}$$

In all the above cases, it is not required to change more than one bit position to embed two bits. Generally, let there be a code word c with n modifiable bit places for m secret message bits x. If F be a hash function that extracts m secret bits from a code word, Matrix encoding can be used to find a modified code word c' for every c and x with $x = F(c')$ such that the Hamming distance between c and c' is at most d_{max}. Such a code can be denoted by an ordered triple (d_{max}, n, m) [16]. The example stated earlier is of a (1, 3, 2) encoding scheme. The pixel values are changed by LSB matching technique such that no *pair of values* forms between adjacent pixels. The distortion function *Embed()* takes P points (pixel positions) on a cover image C to generate the convex hull and embed data accordingly. The embedding function is elaborated in Algorithm 3.

Algorithm 3: $Embed(C, M, P)$

Input: Cover image C, Secret Message M, Polygon generating points P
Output: Stego Image S
Steps:
1. $S \leftarrow CHull(P)$
2. *for each pixel pair in S do*
3.   *[P,Q]=GetLine($s_k s_{k+1}$) //Each entry in S is a pixel (x,y)*
4. *end for*
5. $len \leftarrow size(bin(M))$ // *bin(X) converts X to its binary equivalent*
6. $B \leftarrow bin(M)$
7. *while $len \neq 0$ do*
8.   Extract k bits from B and embed into pixels of C specified by P,Q using process in *section 3.1.1*
9. $len \leftarrow len - k$
10. *end while*
11. $S \leftarrow C'$ //*C' is the modified cover image after embedding*
12.   *Return S //S is the stego image*

The vertices of the polygon obtained are also used to generate the decoding key. The key generation procedure is multi-stage and controlled by a pairing function which ensures delivery of a unique key which is long enough to withstand common *Brute Force* attacks and at the same time regenerates the original constituent integers when decoded successfully.

3.2 Phase II

3.2.1 Extraction

Extraction of the secret message follows a reverse procedure of the embedding mechanism. The extraction function (Algorithm 4) accepts a decoding key, the stego image and the length of the message and returns the secret. The key is derived from the original vertices and the polygon is redrawn to regenerate the ROI. The pixels containing the secret message are then used to extract the message depending on the original (d_{max}, n, m) distortion function adopted. It should be noted that here $d_{max} = 1$.

Algorithm 4: $Extract(S, K, L)$

Input: Stego-image S, Decoding key K, Message Length L
Output: Secret Message M

Steps:
1. Decode K to find polygon vertex set X
2. *for each pixel pair in X do*
3. $[P,Q]=GetLine(X_k, X_{k+1})$ *//Each entry in X is a pixel (x, y)*
4. *end for*
5. *len ← L*
6. *while len ≠ 0 do*
7.   Extract *m* bits from next *n* pixels of *S* specified by *P, Q* depending on the embedding scheme adotpted.
8. $T \leftarrow concat(T, [b_1, b_2, ..., b_m])$ *// $b_{1,2,...,m}$ are the bits extracted at each iteration*
9.   *len ← len − m*
10. *end while*
11. $M \leftarrow Dec(T, 8)$ *//Dec() converts binary string to equivalent 8 bit decimal values*
12. $M \leftarrow Reshape(M)$ *//Reshapes M to appropriate type*

4 Experimental Results and Analysis

The proposed method was coded with Matlab 7.0 on a 2.1 GHz Dual Core AMD Processor computer with 2 GB primary memory, 512 MB dedicated graphics memory and AMD Radeon HD 7290 graphics processor. *Lena* image of different sizes were used as the payload. A standard *Baboon* image and a general *Landscape* image (Fig. 1) (source: http://www.hasselblad.com/sample-file-downloads/landscape-samples.aspx) were used as the cover. The proposed technique was tested for *visual fidelity* and *sensitivity to statistical steganalysis*.

Fig. 1 Cover image and payload

Table 1 Fidelity metrics performance

Cover image	Payload size	PSNR (in dB)	SSIM
Baboon	16 × 16	83.07	0.997
	32 × 32	80.06	0.994
	64 × 64	78.67	0.979
Landscape	16 × 16	81.12	0.985
	32 × 32	80.01	0.982
	64 × 64	78.14	0.977

4.1 Visual Fidelity

Any distortion function used in image steganography is aimed at embedding secret data in a cover image with minimum visual artifacts produced. *Visual Fidelity* of the stego image is an important parameter that is taken into consideration by all image steganography systems. Measurement of the stego image quality is performed as a comparative measure of its similarity with the cover image. Image quality metrics like the Peak Signal to Noise Ratio (PSNR) and Structural Similarity Index Measure (SSIM) are some of the well-known techniques used to ascertain the quality of the stego image. Table 1 lists the performance of the proposed method with different sizes of the payload with respect to the Image Quality Metrics *PSNR* and *SSIM*.

It is evident from the results in Table 1 that the proposed technique exhibits high fidelity (both PSNR and SSIM are high) of the stego image for moderately sized payloads. To compensate for this shortfall, the proposed method may be applied to a video cover with each frame behaving as an individual cover image with similar data hiding capacity.

4.2 Sensitivity to Statistical Steganalysis

Steganalysis refers to techniques that detect the presence of hidden information in suspected cover objects. The proposed image steganography uses a threefold approach to reduce its sensitivity to statistical steganalysis namely, non-sequential

Table 2 Sensitivity to statistical steganalysis (in terms of detection statistic)

Cover image	SP	WS[a]	Triples	AUMP	HCF-COM
Baboon	−0.9303 (F)	0.0325 (P)	−0.0303 (F)	0.294 (P)	67.2978
Landscape	−0.0052 (F)	0.0046 (P)	−0.0013 (F)	0.1808 (P)	64.7364

F Negative, *P* Positive

[a] WS works for grayscale images. Results are of single channel

embedding, and high efficiency embedding paired with non-*PoV* (Pair of Values) forming LSB matching. In order to test the steganalytic sensitivity of the method, it was subjected to some of the most widely acclaimed structural detectors for LSB based embedding methods available in literature. These include the *Sample Pair Analysis (SP)* [17], *Weighted Stego-Image Analysis (WS)* [18], *Structural Steganalysis of LSB Replacement (Triples)* [19], *Asymptotically Uniformly Most Powerful (AUMP) detector* [20] and *HCF-COM* [21]. The results of steganalysis of the proposed method with the aforesaid techniques are presented in Table 2.

The results in Table 2 reveals that the proposed technique stands strong against steganalysis techniques like Sample Pair Analysis and Triples. However, Weighted Stego Analysis and AUMP show comparatively higher detection rates varying from 18 to 29 %. The discriminator values for HCF-COM steganalysis is supposed to be low in the presence of steganography [21] but results show *high* values for the proposed method with moderate capacity. The proposed method is hence secure against HCF-COM. It is noteworthy that HCF-COM is a targeted LSB matching attack which detects embedding in an image depending on the altered histogram characteristics of adjacent pixels. The pixel selection technique adopted combined with hybrid high efficiency embedding LSB matching technique helps to minimize the effect of the HCF-COM detectors. The overall result shows that the proposed technique is decently secure with respect to the common LSB detectors proposed in the recent times. Steganalysis with χ^2-test has not been performed as it is comparatively outdated than the methods taken into consideration here.

5 Conclusion

This paper presents an object based image steganography technique that uses the concept of a convex hull to generate a region of interest polygon for tracing out an area suitable for embedding secret data. The edges of the polygon are then chosen to hide data. The distortion function of the proposed method uses a hybrid data hiding scheme combining a high efficiency embedding scheme with LSB matching features.

Experiments with different cover images and moderate variable payload sizes show promising results for the visual quality of the stego image in terms of *Peak Signal to Noise Ratio (PSNR)* and *Structural Similarity Index Measure (SSIM)*. The sensitivity of the proposed method against some of the most widely acclaimed structural LSB detectors and targeted LSB matching detector is decently low. The low sensitivity imparts the proposed method with a high degree of transparency towards statistical steganalysis.

The proposed technique is simple, yet effective. The polygonal area traced out with pixel positions in the cover image can be scaled accordingly depending on the size of the payload thereby supporting variable data hiding capacity. Future research will focus on exploring the possibility of utilizing more geometric properties of an ROI polygon for developing steganography schemes with better key management and minimal side information sending requirement.

References

1. Kelly, G., McKenzie, B.: Security, privacy and confidentiality issues on the internet, http://www.ncbi.nlm.nih.gov/pmc/articles/PMC1761937/ (2014). Last accessed June 2014
2. Simmons, G.J.: The prisoners' problem and the subliminal channel. In: CRYPTO, pp. 51–67. Springer, Berlin (1983)
3. Kerckhoff, A.: La cryptographie militaire. J. des Sci. Mil. **9**, 161–191 (1883)
4. Katzenbeisser, S., Petitcolas, F.A. (eds.): Information Hiding Techniques for Steganography and Digital Watermarking, 1st edn. Artech House Inc, Norwood (2000)
5. Brute Force Attacks, http://codex.wordpress.org/Brute_Force_Attacks (2014). Last accessed July 2014
6. Shejul, A.A., Kulkarni, U.L.: A secure skin tone based steganography using wavelet transform. Int. J. Comput. Theor. Eng. **3**(1), 16–22 (2011)
7. Cheddad, A., Condell, J., Curran, K., McKevitt, P.: Biometric inspired digital image steganography. In: Proceedings of 15th Annual IEEE International Conference and Workshops on the Engineering of Computer-Based Systems (ECBS '08), pp. 159–168 (2008)
8. Roy, R., Changder, S., Sarkar, A.: SKINHIDE: A biometric inspired high delity steganography technique. In: 3rd International Conference on Advanced Computing and Communications, pp. 41–46 (2013)
9. Roy, R., Sarkar, A., Changder, S.: Chaos based edge adaptive image steganography. In: Procedia Technology, 1st International Conference on Computational Intelligence: Modeling Techniques and Applications (CIMTA), pp. 138–146 (2013)
10. Seivi, G.K., Mariadhasan, L., Shunmuganathan, K.L.: Steganography using edge adaptive image. In: Proceedings of the International Conference on Computing, Electronics and Electrical Technologies, pp. 1023–1027 (2012)
11. Luo, W., Huang, F., Huang, J.: Edge adaptive image steganography based on LSB matching revisited. IEEE Trans. Inf. Forensics Secur. **5**(2), 201–214 (2010)
12. Cormen, T.H., Stein, C., Rivest, R.L., Leiserson, C.E.: Introduction to Algorithms, 2nd edn. McGraw-Hill Higher Education, New York (2001)
13. Bresenham, J.E.: Algorithm for computer control of a digital plotter. IBM Syst. J. **4**(1), 25–30 (1965). doi:10.1147/sj.41.0025
14. Crandall, R.: Some notes on steganography. Source: http://www.dia.unisa.it/~ads/corso-security/www/CORSO-0203/steganografia/LINKS%20LOCALI/matrix-encoding.pdf (1998)
15. Li, X., Yang, B., Cheng, D., Zeng, T.: A generalization of LSB matching. IEEE Signal Process. Lett. **16**(2), 69–72 (2009)
16. Westfeld, A.: F5-a steganographic algorithm: High capacity despite better steganalysis. In: 4th International Workshop on Information Hiding, pp. 289–302. Springer, Berlin (2001)
17. Dumitrescu, S., Wu, X., Memon, N.D.: On steganalysis of random LSB embedding in continuous-tone images. In: Proceedings of ICIP, pp. 641–644 (2002)

18. Ker, A.D., Böhme, R.: Revisiting weighted stego-image steganalysis. In: Security, Forensics, Steganography, and Watermarking of Multimedia Contents X, Proceedings of SPIE Electronic Imaging, vol. 6819, pp. 0501–0517 (2008)
19. Ker, A.D.: A general framework for the structural steganalysis of LSB replacement. In: Proceedings 7th Information Hiding Workshop, vol. 3727, pp. 296–311 (2005)
20. Fillatre, L.: Adaptive steganalysis of least significant bit replacement in grayscale natural images. IEEE Trans. Signal Process. **60**(2), 556–569 (2012)
21. Ker, A.D.: Steganalysis of LSB matching in grayscale images. IEEE Signal Process. Lett. **12** (6), 441–444 (2005)

Author Index

© Springer India 2015
L.C. Jain et al. (eds.), *Computational Intelligence in Data Mining - Volume 1*,
Smart Innovation, Systems and Technologies 31, DOI 10.1007/978-81-322-2205-7

Author Index